[redacted]

CONTROL OF FLUID POWER:
Analysis and Design

2nd (Revised) Edition

[redacted]

ELLIS HORWOOD SERIES IN ENGINEERING SCIENCE

Editors:

Prof. John M. Alexander, *Head of Dept. of Mechanical Engineering University College, Swansea*

Dr. John Munro, *Reader in Civil Engineering Systems Imperial College of Science and Technology, University of London*

Prof. William Johnson, *Professor of Mechanical Engineering, Cambridge*

and **Prof. S. A. Tobias,** *Chance Professor of Mechanical Engineering University of Birmingham*

The Ellis Horwood Engineering Science Series has two objectives; of satisfying the requirements of post-graduate and mid-career education and of providing clear and modern texts for more basic undergraduate topics in the fields of civil and mechanical engineering. It is furthermore the editors' intention to include English translations of outstanding texts originally written in other languages, thereby introducing works of international merit to English language audiences.

STRENGTH OF MATERIALS
J. M. ALEXANDER, University College of Swansea.

TECHNOLOGY OF ENGINEERING MANUFACTURE
J. M. ALEXANDER, R. C. BREWER, Imperial College of Science and Technology, University of London, J. R. CROOKALL, Cranfield Institute of Technology.

VIBRATION ANALYSIS AND CONTROL SYSTEM DYNAMICS
CHRISTOPHER BEARDS, Imperial College of Science and Technology, University of London.

COMPUTER AIDED DESIGN AND MANUFACTURE
C. B. BESANT, Imperial College of Science and Technology, University of London.

THE NATURE OF STRUCTURAL DESIGN AND SAFETY
D. I. BLOCKLEY, University of Bristol.

BASIC LUBRICATION THEORY 2nd Edition
ALASTAIR CAMERON, Imperial College of Science and Technology, University of London.

ADVANCED MECHANICS OF MATERIALS 2nd Edition
Sir HUGH FORD, F.R.S., Imperial College of Science and Technology, University of London and J. M. ALEXANDER, University College of Swansea.

ELASTICITY AND PLASTICITY IN ENGINEERING
Sir HUGH FORD, F.R.S. and R. T. FENNER, Imperial College of Science and Technology, University of London.

TECHNIQUES OF FINITE ELEMENTS
BRUCE M. IRONS, University of Calgary, and S. AHMAD, Bangladesh University of Engineering and Technology, Dacca.

STRUCTURAL DESIGN OF CABLE-SUSPENDED ROOFS
L. KOLLAR, City Planning Office, Budapest and K. SZABO, Budapest Technical University.

CONTROL OF FLUID POWER: Analysis and Design 2nd (Revised) Edition
D. McCLOY, The Northern Ireland Polytechnic and H. R. MARTIN, University of Waterloo, Ontario, Canada.

DYNAMICS OF MECHANICAL SYSTEMS 2nd Edition
J. M. PRENTIS, University of Cambridge.

ENERGY METHODS IN VIBRATION ANALYSIS
T. H. RICHARDS, University of Aston, Birmingham.

ENERGY METHODS IN STRESS ANALYSIS: With an Introduction to Finite Element Techniques
T. H. RICHARDS, University of Aston, Birmingham.

STRESS ANALYSIS OF POLYMERS 2nd Edition
J. G. WILLIAMS, Imperial College of Science and Technology, University of London.

CONTROL OF FLUID POWER: Analysis and Design

2nd (Revised) Edition

D. McCLOY, B.Sc., D.I.C., M.Sc., Ph.D., D.Sc.

Dean of Technology
Ulster Polytechnic
Northern Ireland

and

H. R. MARTIN, B.Sc., M.Sc., Ph.D
Professor of Mechanical Engineering
University of Waterloo
Ontario, Canada

ELLIS HORWOOD LIMITED
Publishers Chichester

Halsted Press: a division of
JOHN WILEY & SONS
New York - Chichester - Brisbane - Toronto

First published in 1980 by

ELLIS HORWOOD LIMITED
Market Cross House, Cooper Street, Chichester, West Sussex, PO19 1EB, England

The publisher's colophon is reproduced from James Gillison's drawing of the ancient Market Cross, Chichester

Distributors:

Australia, New Zealand, South-east Asia:
Jacaranda-Wiley Ltd., Jacaranda Press
JOHN WILEY & SONS INC.
G.P.O. Box 859, Brisbane, Queensland 40001, Australia

Canada:
JOHN WILEY & SONS CANADA LIMITED
22 Worcester Road, Rexdale, Ontario, Canada.

Europe, Africa:
JOHN WILEY & SONS LIMITED
Baffins Lane, Chichester, West Sussex, England

North and South America and the rest of the world:
Halsted Press: a division of
JOHN WILEY & SONS
605 Third Avenue, New York, N.Y. 10016, U.S.A.

British Library in Cataloguing in Publication Data
McCloy, Donaldson
Control of fluid power: Analysis and Design 2nd(Revised) Edition. –
(Ellis Horwood series in mechanical engineering).
1. Fluid power technology
I. Title II. Martin, Herbert Robert
620.1'06 TJ840 80–40027
ISBN 0–85312–135–4 (Ellis Horwood Ltd., Publishers – Library Edition)
ISBN 0–470–27012–8 (Halsted Press)

Typeset in Press Roman by Ellis Horwood Ltd.
Printed in Great Britain by Fakenham Press.

Table of Contents

Chapter 1 INTRODUCTION

1.1 A Short History of Hydraulic Power 11
1.2 Pneumatics. 13
1.3 Fluidics 15
1.4 Control of Fluid Power. 18
1.5 Control by Means of Fluid Elements 19
1.6 Sequence Control 21
1.7 References 25

Chapter 2 POWER TRANSMITTING FLUIDS

2.1 Types of Fluid 27
2.2 Effective Bulk Modulus 31
2.3 Aeration 36
2.4 Cavitation 41
2.5 Filtration and Contamination Assessment 43
2.6 High Temperature Systems 45
2.7 Electrosensitive Fluids 46
2.8 Examples. 49
2.9 References. 50

Chapter 3 BASIC FLUID POWER SYSTEMS

3.1 Introduction. 53
3.2 The Basic Pneumatic System 53
3.3 The Basic Hydraulic System 55
3.4 Air Compressors 58
3.5 Hydraulic Pumps. 62
3.6 Motors and Actuators 68
3.7 Examples. 72
3.8 References. 72

Chapter 4 FLUID CONTROLLING ELEMENTS

4.1 Introduction 75
4.2 Valve Elements 75
4.3 The Variable Orifice 78
4.4 Characteristics of Jets 84
4.5 Theoretical Determination of Discharge Coefficients 86
4.6 Empirical Considerations 92
4.7 Reattached Flow in Spool Valves 96
4.8 Experiments with Hydraulic Poppet Valves 100
4.9 Experiments with Hydraulic Flapper-Nozzle Valves 101
4.10 Experiments with Hydraulic Short Tube Orifices 103
4.11 Experiments with Pneumatic Orifices 104
4.12 Examples 106
4.13 References 106

Chapter 5 MEASUREMENT AND SIGNAL PROCESSING

5.1 Introduction 109
5.2 Command Inputs 109
5.3 Proximity Detection 112
5.4 Position Measurement 115
5.5 Velocity Measurement 120
5.6 Force Measurement 122
5.7 Signal Processing 124
5.8 References 130

Chapter 6 VALVES AND SIMPLE CIRCUITS

6.1 Introduction 133
6.2 Directional Control Valves 133
6.3 Flow Control 144
6.4 Pressure Control 147
6.5 The Electric Field Valve 152
6.6 References 156

Chapter 7 STATIC CHARACTERISTICS OF VALVES

7.1 Introduction 159
7.2 The Variable Orifice 159
7.3 Application of Valve Characteristics 162
7.4 Linearisation of Valve Characteristics 164
7.5 Three-port Servovalves 166
7.6 Four and Five-port Valves 176
7.7 Examples 179
7.8 References 180

Chapter 8 FLOW FORCES IN VALVES

8.1 Introduction 181
8.2 Steady State Flow Forces in Spool Valves 182
8.3 Compensation of Flow Forces in Spool Valves 183
8.4 Transient Flow Forces in Spool Valves 189
8.5 Flow Forces in Poppet Valves 191
8.6 Forces on Flapper-nozzle Valves 194
8.7 Examples. 199
8.8 References. 200

Chapter 9 SOME ASPECTS OF CIRCUIT DESIGN

9.1 Introduction. 203
9.2 Pipe Line Design 203
9.3 Pressure Losses in Pipes 205
9.4 Losses Through Valves and Fittings. 207
9.5 Typical Circuit Design 209
9.6 Further Considerations of Efficiency. 217
9.7 Time Varying Loads. 221
9.8 Use of Accumulators to Reduce Energy Requirements 223
9.9 Examples. 226
9.10 References 227

Chapter 10 DYNAMIC CHARACTERISTICS OF CIRCUITS

10.1 Introduction. 229
10.2 Flow Networks 229
10.3 Flow Networks for Steady Conditions 231
10.4 Linear Resistance Networks. 233
10.5 Flow Networks for Dynamic Conditions 234
10.6 Fluid Impedance 236
10.7 Impedance Approach Applied to Systems 243
10.8 The Wave Equation and Distributed Parameter Models 245
10.9 Solution of the Wave Equations for Harmonic Excitation 250
10.10 Matched and Unmatched Transmission Lines. 253
10.11 Performance Characteristics of a Pump Exciting a Blocked Line 255
10.12 Examples 258
10.13 References 260

Chapter 11 HYDRAULIC SERVOMECHANISMS

11.1 Precision Control Systems in Hydraulics. 261
11.2 Simple Model of a Hydraulic Servo 262
11.3 Effects of Oil Compressibility. 265

11.4 Stability Considerations 270
11.5 Determination of Optimum Response 271
11.6 Response Characteristics 273
11.7 Effects of Load Disturbances 275
11.8 Designing a Hydraulic Servo 277
11.9 Performance Assessment 280
11.10 Choice of Control Valve 283
11.11 A Method for Evaluating Frequency Response 286
11.12 Cavitation in the Hydraulic Actuator 291
11.13 Effect of Underlap on Cavitation 294
11.14 Examples 296
11.15 References 297

Chapter 12 ELECTRO-HYDRAULIC SERVOMECHANISMS

12.1 Single Stage Hydraulic Valve 299
12.2 Two Stage Electro-hydraulic Valve 303
12.3 General Analysis of the Electro-hydraulic Servo 307
12.4 Simple Case 309
12.5 Results from Computer Study 312
12.6 Effects of Flow Forces 315
12.7 More Exact Simulation 319
12.8 Small Signal Responses 322
12.9 Methods of Improving Performance 325
12.10 Computer Study of Small Amplitude Response 329
12.11 Methods of Performance Improvement 332
12.12 Examples 335
12.13 References 335

Chapter 13 PNEUMATIC SERVOMECHANISMS

13.1 Introduction 337
13.2 Pneumatic Servos 338
13.3 Solution of the Model Equations 342
13.4 Linearised Analysis of the Pneumatic Servo 342
13.5 Load Sensitivity 346
13.6 Methods of Stabilisation 347
13.7 Stabilisation Using Auxiliary Tanks 350
13.8 Hot Gas Servomotors 353
13.9 Examples 355
13.10 References 356

Chapter 14 DYNAMICS OF PRESSURE AND SPEED CONTROL SYSTEMS

14.1 Introduction 357
14.2 Pressure Compensated Flow Control 357

14.3 Stability of Pressure Control Valves 359
14.4 Speed Control in Fluid Power Systems 365
14.5 Speed Control in Pneumatic Systems. 367
14.6 Pneumatic Meter-out Control of Mass Loads. 372
14.7 Hydraulic Meter-out Control of Mass Loads 376
14.8 Example 378
14.9 References 378

Chapter 15 BOOLEAN ALGEBRA AND LOGIC GATES

15.1 Introduction. 381
15.2 Boolean Algebra 381
15.3 Some Other Logic Gates 385
15.4 Manipulation of Logic Expressions 387
15.5 A Circuit Design 390
15.6 Logic with Conventional Valves 391
15.7 Counting with Valves 400
15.8 Examples 405
15.9 References 405

Chapter 16 CONTROL OF SEQUENCES

16.1 Introduction. 407
16.2 Fixed Programmes, Time Based 408
16.3 Fixed Programmes, Step Completion Based 414
16.4 A Simple Design Technique for Sequence Circuits 418
16.5 The Cascade Method 422
16.6 Other Means of Eliminating Locked-on Signals 425
16.7 The Logic of the Cascade Method. 426
16.8 Variable Programming 429
16.9 Examples. 432
16.10 References 432

Chapter 17 DISCONTINUOUSLY CONTROLLED AND DIGITAL SERVOS

17.1 Introduction. 435
17.2 Digital to Analogue Converters. 435
17.3 Pulse Controls. 438
17.4 Digital Actuators 440
17.5 On-Off Controlled Systems 442
17.6 Oscilating Control Systems – Pulse-Width Modulation 443
17.7 Response Improvements Using Digital Techniques 448
17.8 Some Phenomena Associated with On-Off Controlled
Hydraulic Servos 449
17.9 References 456

Chapter 18 NOISE IN FLUID POWER SYSTEMS
18.1 Introduction 459
18.2 Fundamentals of Acoustics 460
18.3 Signal Analysis 465
18.4 Energy Propagation Through Systems and its Detection 468
18.5 Pump Noise 472
18.6 Gear Pump Noise 478
18.7 Pipe Networks 481
18.8 Identification of Cavitation Noise 485
18.9 Pneumatic Circuits 487
18.10 Controlling Noise at the Design Stage 490
18.11 Noise Reduction in Existing Equipment 492
18.12 References 492

Appendix GRAPHICAL SYMBOLS FOR FLUID POWER COMPONENTS 495

Index 501

CHAPTER 1

Introduction

1.1 A SHORT HISTORY OF HYDRAULIC POWER

Fluid power has been in man's service for many centuries. It has propelled his boats, turned his windmills and water wheels, and nowadays hydro-electric power stations provide perhaps the most striking examples of the power of fluids [1, 2, 3]. But fluid power as it is known today does not include these examples, in all of which the dynamic fluid forces predominate. Fluid power is traditionally associated with the transmission of power by means of pressurised fluids in which the energy is predominantly stored as potential rather than kinetic energy as in the above examples.

The development of the subject owes a lot to the Frenchman Pascal who published a book on hydrostatics in 1648. This included his famous law which stated that pressure in a fluid is transmitted equally in all directions. This was to be the basis of power hydraulics, but Pascal was a theoretician and unfortunately made little attempt to apply the laws usefully. It was more than a century later before practical fluid power systems were devised by Bramah. He demonstrated that a relatively small force on a long stroke, small diameter piston could produce a very large force on a short stroke, large diameter piston. Although this force amplification effect was most impressive, Bramah did not lose sight of the fact that he also had a means for transmitting power as well as force. He appreciated that here was the basis for an efficient means of power transmission using the cycle: mechanical power – hydraulic power – mechanical power. In the earliest days the hydraulic power was developed by means of a hand pump and the power was reconverted to mechanical power by using an hydraulic ram.

So rapid was this growth of power hydraulics that by the late 1860s London and Manchester had central fluid power generating stations from which pressurised fluid was pumped to factories. But the growth rate was diminished somewhat at the end of the 19th century by the emergence of electrical power transmission.

However, the unique advantages of hydraulic power transmission soon reawakened an enthusiastic interest in the subject. For example the car braking system was at one time wholly mechanical. The transmission linkage was by necessity very complex to allow for relative movement caused by steering and suspension and to ensure that brake forces were balanced. The hydraulic system of Fig. 1.1 proved to be infinitely superior. It ensured that brake forces were balanced and that the forces were sufficiently large. The use of flexible pipes eliminated problems arising from relative motion of the various parts of the car.

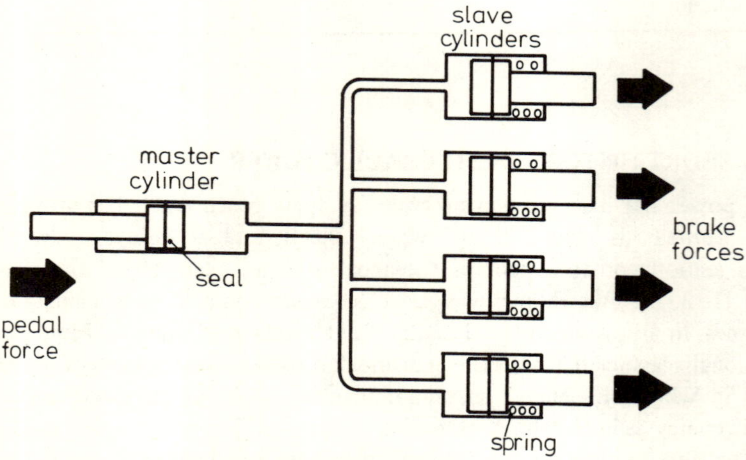

Fig. 1.1 – Car brake system using hydraulics.

The very large force capability still remains one of the big advantages of hydraulics. Brunel realised this when he used eighteen hydraulic rams, nine at the bow and nine at the stern of his ship, to give a total thrust of 45 MN when launching the *Great Eastern*. Nowadays some of the largest force applications are found in the metal processing industry where 500 Mg forges giving thrusts of 3 MN are used.

The high force-to-weight ratio is another important advantage of hydraulic equipment and makes its use attractive in situations where weight is at a premium – such as aircraft and missiles [4]. Hydraulic pumps, for example, are much smaller than electrical motors with the same power rating (Fig. 1.2). This difference in size and weight is attributable to the high working pressure of the hydraulic system – pressures of 280 bar being common. By comparison the magnet in an electric motor can only exert a tractive pressure of about 17 bar. Now, in a hydraulic system:

$$\begin{aligned} \text{Power} &= \text{force} \times \text{velocity} \\ &= \text{pressure} \times \text{area} \times \text{velocity} \\ &= \text{pressure} \times \text{flow rate}. \end{aligned}$$

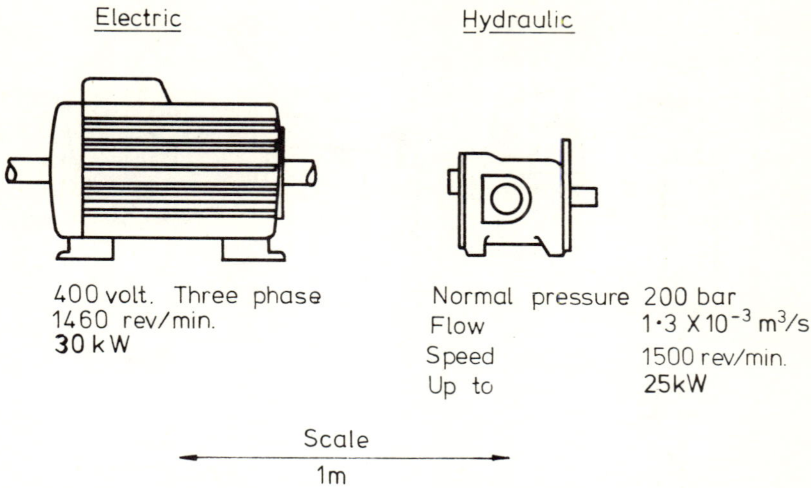

Fig. 1.2 – Comparison of electric and hydraulic motors.

Thus for a given power, a hydraulic pump working at a high pressure needs only to deliver a small flow rate and can thus be quite small. Indeed, nowadays hydraulic power is making an impact in the quest for miniaturisation. Pumps are available of 50 mm diameter and 75 mm length. A hydraulic motor capable of developing up to 0.6 kW has a 32 mm diameter and is only 50 mm long. The direction of rotation is reversible, the maximum speed is 30 000 rev/min and the mass is only 0.3 kg.

The conversion from fluid power to mechanical power can be achieved by either a rotary motor or a linear actuator. The linear output motion of the simple ram or cylinder of Fig. 1.3 can be used as the basis for a variety of motions. Fig. 1.3 shows (a) a simple clamping arrangement, (b) an indexing drive, (c) how three output positions can be determined by coupling two cylinders. Fig. 1.3(d) shows how the linear movement can be converted to a rotary motion: this is indeed the basic mechanism of many hydraulic pumps and motors. In addition, a simple valve and cylinder can produce output motions which in a wholly mechanical system would require a complicated linkage. For example, quick return motion is very easily achieved by ensuring that the flow of fluid to the cylinder is more restricted in one direction than the other.

The stiffness of the hydraulic drive is another important advantage. This arises because the oil is relatively incompressible. As a result, hydraulic drives are little affected by load disturbances; hence their popularity in the aerospace and machine tool industries.

1.2 PNEUMATICS

Pneumatic systems use pressurised gases for power transmission. The earlier systems used air as the working medium, but nowadays inert gases and hot gases

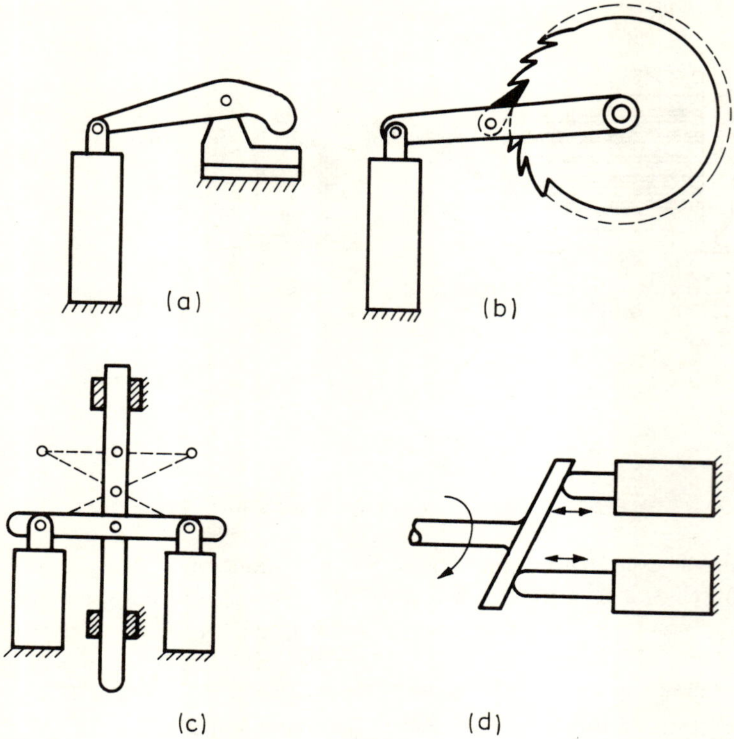

Fig. 1.3 – Some applications of cylinders.

are used in some applications. Working pressures are usually limited to 7.0 bar although at present some systems are being designed for 10 bar operation. Inefficiency of compression of gases and the dangers involved with storing high pressure gases necessitate these limits.

The use of compressed air as a source of power has grown rapidly over the past ten years and at present it is accepted by all branches of industry [5]. Its advantages, many of which are shared by hydraulics, are summarised below:

1. Most production plants have a compressed air supply on hand.
2. Pneumatic components are relatively cheap.
3. Components are readily available 'off the shelf'. The range of valves is extensive and there is a wide choice of cylinder sizes.
4. The biggest advantage of pneumatic equipment is its adaptability. Existing machines can be easily and quickly converted to pneumatic operation. Air cylinders can be mounted directly at the point where force is required – all that is needed is a few brackets and some means for linking the piston rod to the existing manually operated mechanism.

5. Components are reliable and are easily and cheaply maintained. Servicing can usually be carried out on site, since it only involves replacing valve seals or cylinder bearings.
6. Pneumatic tools, such as drills and grinders, do not burn out when stalled. Hydraulic tools also have this advantage.
7. There is no fire hazard when pneumatics are used. They can be used for instance in situations where risk of explosion would prohibit the use of electricity.
8. Compressed air is not inflammable and hence pneumatic equipment can be used in high temperature conditions where the use of hydraulics or electrics could be dangerous or costly. Most equipment can work up to 80-90°C, but higher temperatures can be tolerated if heat resistant seals are used.
9. Pneumatic systems are clean.

Of course, there are also associated disadvantages, and it is necessary to consider these carefully before installing pneumatic equipment:

(a) Compressed air is expensive to produce, and this must not be borne in mind if the plant does not already have a compressor. On a power basis it is considerably more expensive than electricity or hydraulics.
(b) Extreme accuracy of feed is difficult to achieve on account of the elastic nature of compressed air. It cannot compete with hydraulics in this respect.
(c) Transmission of air signals in pipes is much slower than the transmission of electrical signals in wires. Hence if signalling times are critical and lines are long (greater than 10 m) electrics should be used.
(d) Pneumatic cylinders are bulky and expensive if large forces are desired. Large forces are generated more conveniently by using hydraulics.

1.3 FLUIDICS

Fluidics is more concerned with the transmission and manipulation of information than with the transmission of power. It occupies the same place in the fluid field as does electronics in the field of electrical power. The success of electronics encouraged engineers to look at the possibility of using fluid streams instead of electron streams for amplification purposes. Such a fluid device would have considerable advantages over its electronic counterpart, such as immunity to radiation and perhaps tolerance of high temperatures.

The most significant advances in the field of fluidics did not come until 1959 when the results of reasearch at the Diamond Ordance Fuze Laboratories and at the Massachussetts Institute of Technology in the United States came to light. As in many other areas of invention, the bases of operation of most of the devices were so simple that one wonders why they had not been discovered earlier – perhaps because the need did not exist. A simple **proportional fluid**

amplifier, which behaves in an analogous fashion to the transistor, is sketched in Fig. 1.4. It consists of a supply tube, a receiver tube and an input or control tube. Variations in input signal deflect the main jet, hence changing the output signal (pressure, flow rate or power). The device is an amplifier and good design can, for example, give a change in output pressure which is 5-10 times the change in input or control pressure.

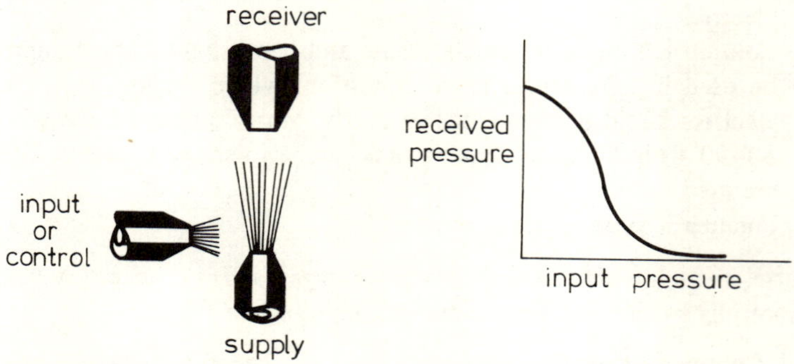

Fig. 1.4 – A proportional fluidic amplifier.

The other basic fluidic element is the **wall-attachment amplifier**. Its operation relies on the Coanda effect, which was first discovered by Coanda in 1932. It is found that if a jet is confined between top and bottom plates, Fig. 1.5, entrainment of the surrounding fluid makes the jet attach to an adjacent wall. This phenomenon is used in the bi-stable amplifier of Fig. 1.5. As shown the jet is attached initially to the right-hand wall and output B is on. If an input signal is then applied, the jet is deflected and attaches to the left-hand wall and output A becomes active. This output remains on after the input signal is removed – the jet is stable in two positions. Unlike the proportional amplifier it is thus ideally suited for digital controls and indeed this has been a most fruitful area of application.

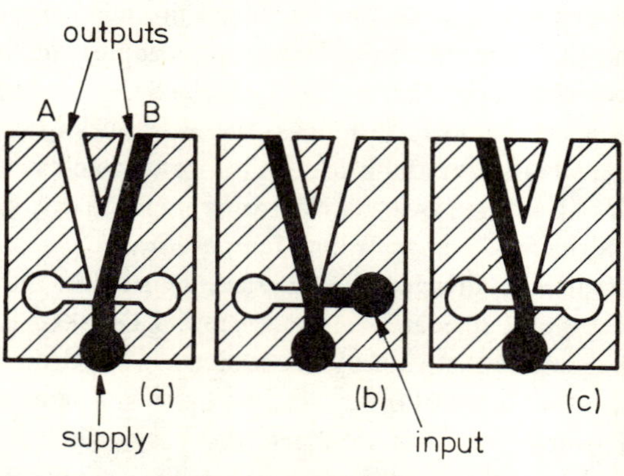

Fig. 1.5 – A bistable fluidic amplifier.

The word **FLUIDIC** is an elision of the words 'fluid' and 'logic', that is, logical operations using fluids. The language of symbolic logic was devised in 1854 by Boole and has been named Boolean algebra. This was originally an abstract mathematical form including a switching algebra, an algebra of sets and a propositional calculus. But in 1938, Shannon, an electrical engineer, revitalised Boole's work and showed how it could be used in the design of telephone exchanges and in computer switching circuits. The two state device with outputs true or false, ON or OFF, or (in binary) 1 or 0, is the basic element of any logic circuit. Thus when the wall-attachment devices were invented it was quickly evident that they could make a big impact on the logical control of machines and processes.

Initially in this rapidly expanding new technology most of the effort was devoted to investigating and developing non-moving part devices, or as they are often called, pure fluid devices. However, as a result of this new wave of enthusiasm for logic, engineers began to appreciate that the well established pneumatic and hydraulic valves were in fact logic elements. Unfortunately many of these were too large and expensive to allow their use in any but the simplest of logic circuits. However, miniaturisation is now an important fluid power development area and valves are presently available with a volume less than 16 cm^3. In addition many new moving part logic devices have been designed. Thus fluidics, although initially mainly concerned with pure fluid devices, now includes moving part devices. The pure fluid devices are often referred to as **FLUERIC** devices.

When the first flueric elements were introduced about 1958, they inspired a tremendous wave of interest, and within a short period many people began to think that here was a new technique that would solve all control problems at very low cost. Within a few years, however, it was evident that things were not going as well as expected. There were very few reported successful applications of fluerics, and doubts about the future of the subject began to grow. This was largely owing to the fact that there was a great temptation to use flueric components simply because they were new, and in many cases no attempt was made to compare the cost of the flueric system with a conventional system. Unproven components were used and matching of components presented a problem. As a result interest began to wane and fluidics has never quite achieved the successes that were anticipated in the early days. But fluidics is paying off in certain areas. One of the most fruitful areas is in non-contact sensing where fluerics offers advantages of cheapness and simplicity over other methods, such as the photoelectric cell. There is also advantage to be gained by using fluidics in fluid power systems. Interface devices are reduced to a minimum, both control and actuation using fluid power, and the high reliability of the fluidic element can result in a trouble free system. Applications of this type include sequence-controlled machine tools and automatic handling and inspection systems [10]. Other factors, such as environmental tolerance, have made fluidics an attraction.

Elements can be immune to extremes of temperature, illumination, radiation, corrosion, mechanical shock, electrical shock and vibration, and this has resulted in successful application in the aerospace field, in nuclear reactors, in steel mills and in the control of cryogenic pumps.

Another useful spin-off is that engineers now realise that fluid power valves are basically logic elements, and this has had a considerable impact on circuit design methods.

1.4 CONTROL OF FLUID POWER

The earliest applications of hydraulic power were concerned solely with the transmission of power from one station to another. However, it soon became evident that some means of controlling the power was essential. Since hydraulic power is the product of flow rate and pressure, control of either of these variables would in turn control power. Flow rate was easily controlled by using the variable restrictor in the form of the needle valve, Fig. 1.6. Pressure was controlled by the relief valve, Fig. 1.7. In this simple device the poppet lifts, relieving the system, when the pressure exceeds a value determined by the spring setting. Papin's pressure regulator [2] was the forerunner of the modern relief valve, but used gravity instead of a spring force for the controlling action. These methods of flow control and pressure control, although usually being wasteful of power, are both very popular in modern fluid power systems [6]. Where very large powers are used, so that low power efficiency cannot be tolerated, modern systems use a variable displacement pump whose power output can be varied to suit the demand.

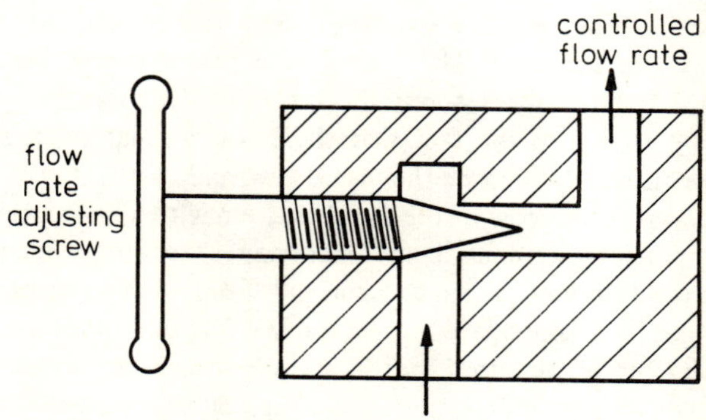

Fig. 1.6 – Needle valve for flow control.

The other main form of control within the fluid power circuit is directional control, by means of which the flow can be directed from one part of a circuit to another. The early directional control valves, some based on the old steam engine D valve, form the basis of many of today's moving part logic elements.

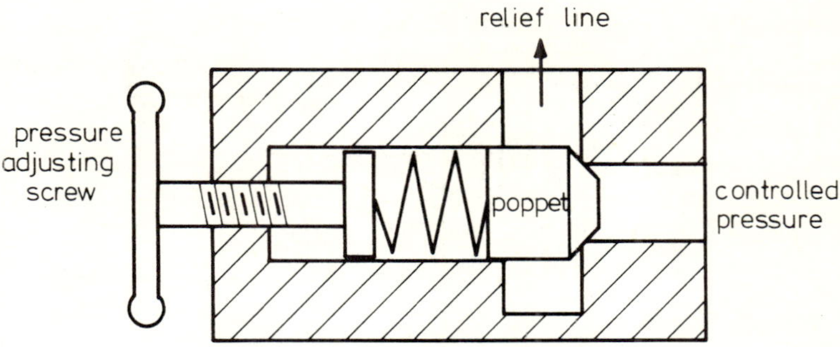

Fig. 1.7 – Relief valve for pressure control.

1.5 CONTROL BY MEANS OF FLUID ELEMENTS

The above mentioned devices were intended to control the internal behaviour of the fluid power circuit, but an equally important aspect of fluid power and fluidics in the control field is in the control of physical quantities, sequences, of events, etc.

At this stage it may be helpful to remind the reader of the difference between **open loop** and **closed loop** control [7]. An open loop speed control system is sketched in Fig. 1.8(a). This system could be initially calibrated, giving a relationship between control valve opening and output speed. Thereafter if a certain speed is demanded the operator has only to refer to the calibration chart to find the required valve opening. Such a controller would be adequate if there were no disturbances acting on the system. These could take the form of load changes, of variations in supply pressure or perhaps of valve wear, and their presence would result in the output speed being different from the desired value. Hence although open loop control is simple, it cannot cope with disturbances, basically because the output is not measured but is inferred. The system becomes a closed loop if the output affects the controller action. A closed loop control system employing a human operator is shown in Fig. 1.8(b). The output speed is now measured, is compared with the desired speed and corrective action is taken depending on the sign and magnitude of the error. The effect of disturbances is greatly reduced since the output is being continuously measured.

Closed loop control is often also referred to as **feedback** control, the output being fed back for comparison with the input, as in Fig. 1.9. The block diagram illustrates the three major components within a closed loop system; the measurement means, the error detector and the actuator. The speed control system of Fig. 1.8(b) is not an automatic control system since it involves a human operator. An automatic system would require some mechanical, electrical or fluidic means of error detection, measurement and actuation. The use of fluids for such purposes forms the basis of this book.

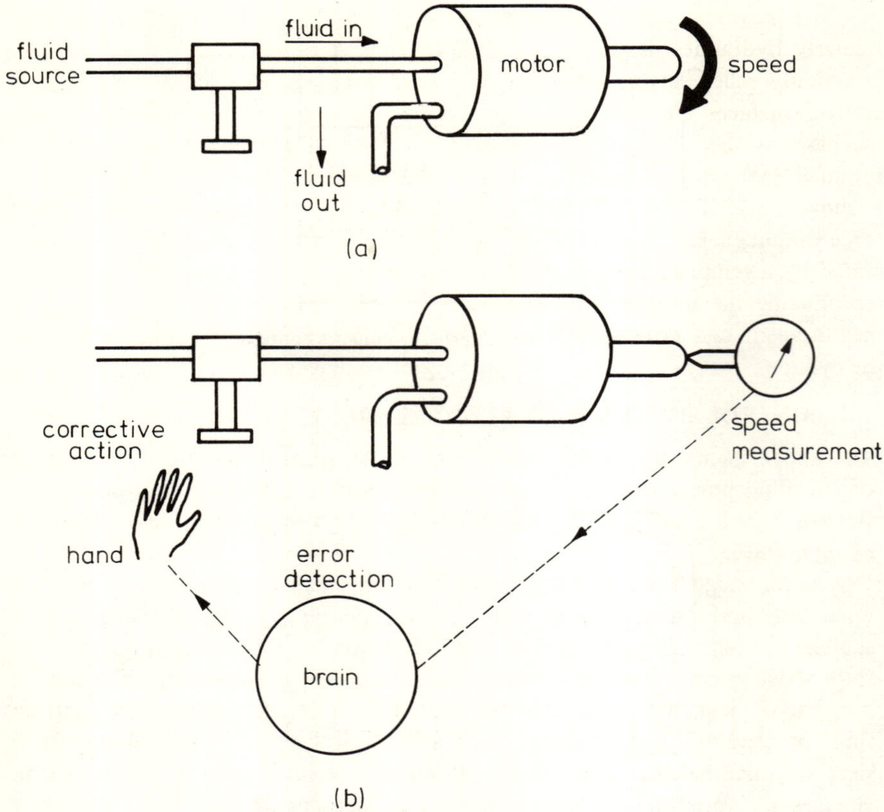

Fig. 1.8 – (a) open, and (b) closed loop systems.

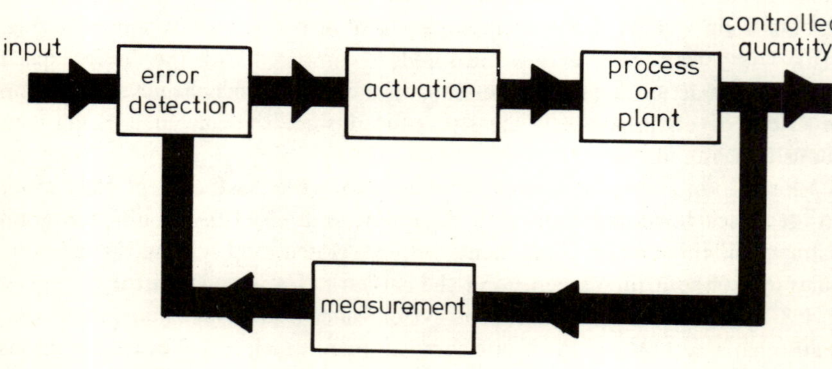

Fig. 1.9 – The elements of a closed loop system.

The previous example was concerned with speed control. Fluid power, particularly hydraulics, is also widely applied in position control devices – more commonly called servomechanisms. The control of machine tools and of aircraft control surfaces are two examples of position control using hydraulic servomechanisms [8]. These are accurate and rugged, have fast response and can produce large output forces. A sketch of a typical hydraulic servomechanism is shown in Fig. 1.10. The output y follows the input x, but with a very great force amplification. In a copy lathe, for example, the input x would be generated by a template follower, and the output y would be the tool position for reproducing the template shape. Note that the valve fulfils the roles of output measurement and error detection as well as supplying the corrective power to the cylinder.

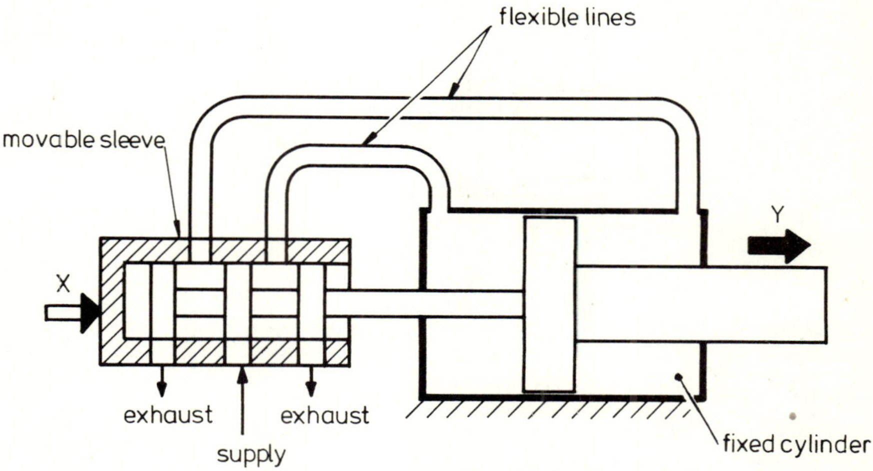

Fig. 1.10 – A closed loop hydraulic system.

Flueric devices have also found widespread application in control systems. A force controller is illustrated in Fig. 1.11. This uses a cantilever spring and a variable orifice to produce a pressure which is proportional to the error in force. If, for example, the force is too low, the variable orifice is reduced in size, the control pressure thus increases and this in turn deflects the supply jet into the bellows and increases the output force. Flueric elements are nowadays being used for measurement and control of temperature, pressure, flow rate, frequency and many other variables. They have made their greatest impact in those fields, such as reactor control, where environmental tolerance is essential.

1.6 SEQUENCE CONTROL

In many control systems it is not necessary to have proportional control over the controlled variable, as in the above examples. In simple clamping circuits, for example, it is only necessary either to operate the clamp or to release it. In many mechanical handling applications it is only necessary to extend or to retract a motor or cylinder. These are simple two state controls and fluidic and fluid power components are ideally suited for these applications. Two state devices are usually more easy to manufacture and hence are cheaper than proportional devices, mainly because they are not so sensitive to geometrical imperfections.

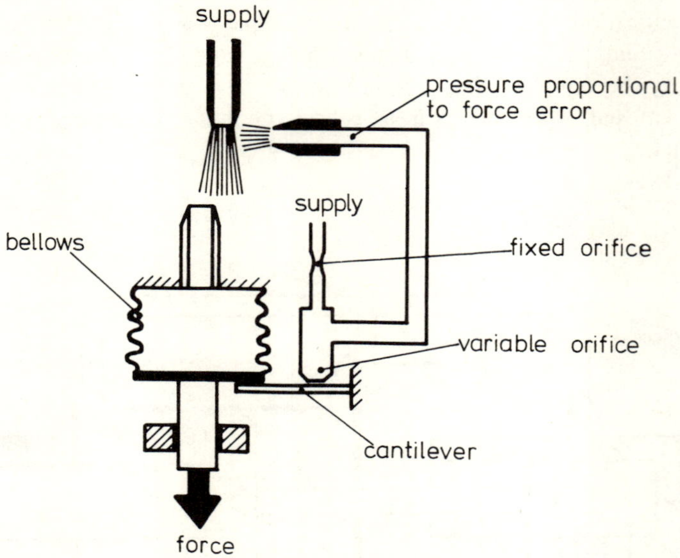

Fig. 1.11 – A force controller.

The use of a pneumatic valve and cylinder in a clamping circuit is illustrated in Fig. 1.12. Normally the cylinder rod is held in the retracted position by the spring. Manual operation of the valve makes the rod move to its other state – the extended position. The valve is designed for only two operating conditions, on or off, and is much simpler and cheaper than the proportional valve shown in Fig. 1.10. This simple system can be made partially automatic by including a limit valve, Fig. 1.13, which detects that the rod has reached the end of its stroke. Operation of this valve causes a pneumatic signal to be sent to the control valve, resetting it and making the rod retract. Thus the operator only needs to depress the control valve momentarily in order to initiate the complete cycle of extension and retraction. This simple concept can be extended to complicated systems with many motors and actuators and in which the sequence of events is the important thing. A two cylinder circuit is shown in Fig. 1.14. This is used for automatically lifting an object from one conveyor level to another. No human intervention is required. The sequence of operation is – (1) cylinder rod A extends then (2) cylinder rod B extends and then (3) rods A and B retract.

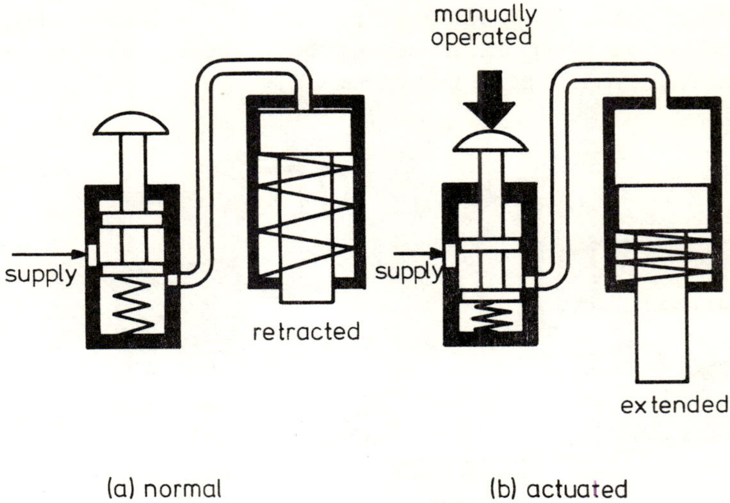

Fig. 1.12 – Simple two position control; (a) normal, (b) actuated.

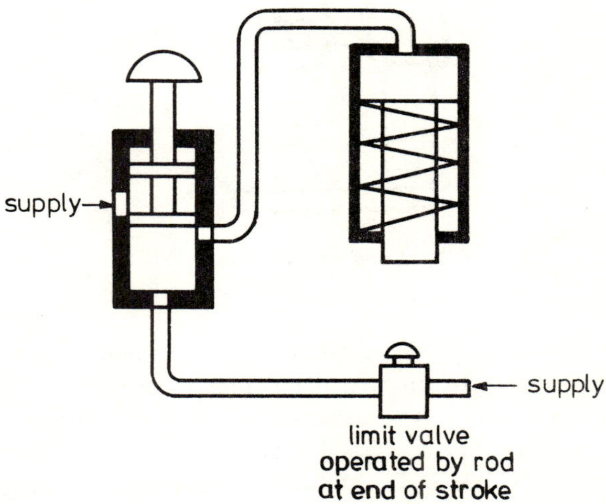

Fig. 1.13 – A partially automatic system.

Sequence controls can be either **event based** or **time based**. The examples quoted above are both event based sequences, one action taking place when the previous one is completed. This provides interlocking in the circuit so that if for some reason or other some action is not completed, the cycle will stop at that point. Flueric devices are replacing the conventional limit valves for the detection of event completion in many modern applications. One commonly used method involves air jets which are interrupted by the machine when a step is completed. Being non-contact sensors these devices require no operating force and are extremely rugged and reliable.

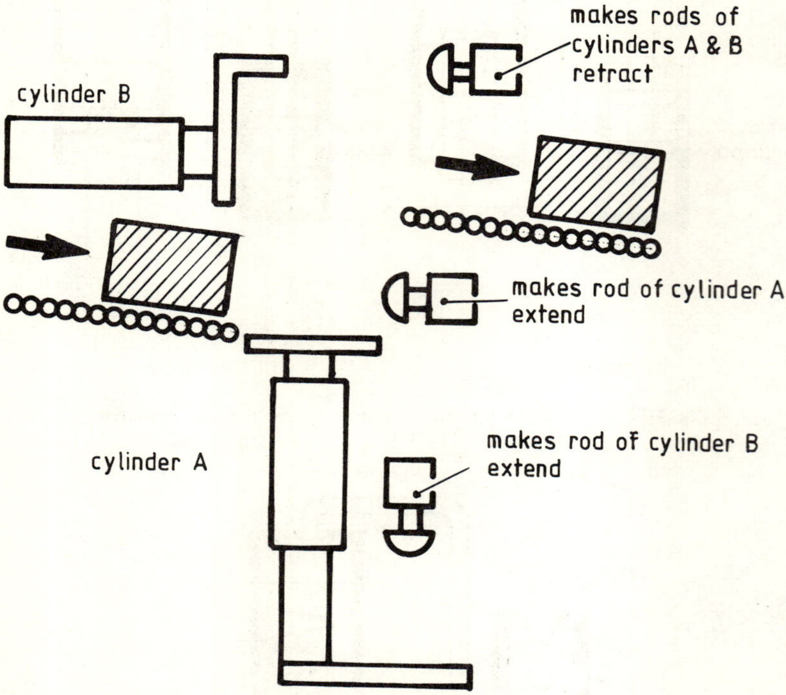

Fig. 1.14 – Event-based sequence control.

Fluid logic is also invaluable in circuits which require certain conditions to be met before the cycle of operations commences or continues. For example in a press operation it may be necessary to ensure that the component is in position, that the guard is down and that the start button is pressed before the press operates. A knowledge of Boolean algebra and fluidics enables the designer to solve such problems in a short time.

In the other form of sequence control, the time based sequence, the events are ordered on a time basis. The cam shaft programmer is one of this class of controllers. It consists of a motor driven bank of cams, each operating a switch, the order of operation being determined by the relative angular position of the cams on the cam-shaft. Although suitable for slow sequences the cam-shaft programmer has several shortcomings which preclude its use in fast acting systems. Electronic logic, driving electric relays, has in the past provided a solution to the control of fast systems, but flueric systems are now becoming popular in this area also. A block diagram of a time based sequence controller is shown in Fig. 1.15. The time base is derived from a flueric clock, which consists of an oscillator which sends constant frequency pulses to a binary counter. Digital flueric elements of the wall-attachment variety are suitable for this task. The output from the counter is an indication of elapsed time – as illustrated the counter has four outputs (counts to 15 before resetting) – and the combination of the values of the outputs is different for each time interval. The flueric decoder 'looks' for those combinations of counter outputs at which specific events should occur and, having found them, activates the appropriate output of the controller.

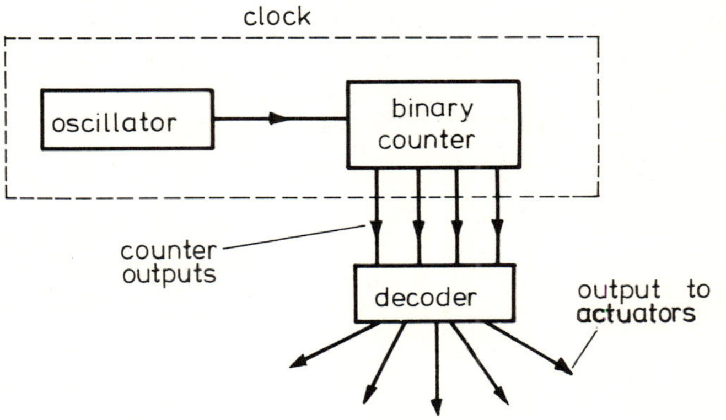

Fig. 1.15 – The elements of time based sequence control.

Fluid power, and fluidics to a lesser extent, play important parts in the field of control. In this book the emphasis will be on conventional fluid power components and systems. Fluidic components and logic techniques will be discussed occasionally, especially in relation to sensing devices and circuit design; the reader requiring more detailed information about fluidics is referred to the several existing text books [9, 10] on the subject.

1.7 REFERENCES

[1] *Proc. Conf. on Oil Hydraulic Power Transmission and Control* (1961), Inst. Mech. Engrs., London.

[2] McNeil, I. (1961), Hydraulic power transmission, *The Evolution of Engineering*, Inst. Mech. Engrs., London.

[3] Stowers, A. (1961), Pumps and water-raising machinery, ibid.

[4] Conway, H. G. (1949), *Fluid Pressure Mechanisms*, Pitman.

[5] Stewart, H. L. and Storer, J. M. (1968), *Fluid Power*, Howard W. Sams & Co.

[6] Blackburn, J. F., Reethof, G., Shearer, J. L. *(1960),*

[6] Blackburn, J. F., Reethof, G., Shearer, J. L. (1960), *Fluid Power Control*, M.I.T. Press.
[7] Martin, H. R. (1968), *Introduction to Feedback Systems*, McGraw-Hill.
[8] Guillon, M. (1969), *Hydraulic Servo Systems – Analysis and Design*, Butterworth.
[9] Foster, K. and Parker, G. A. (1970), *Fluidics – Components and Circuits*, Wiley-Interscience.
[10] Belsterling, C. A. (1971), *Fluidic Systems Design*, Wiley-Interscience.

CHAPTER 2

Power transmitting fluids

2.1 TYPES OF FLUID

One of the main differences between fluids and solids is that a fluid cannot sustain shear stress without motion, no matter how small a shearing force is applied. Fluids fall into two main groups, liquids and gases, which are identified by their relative magnitudes of compressibility. In addition, a liquid may have a bounding surface, while a gas will expand throughout its container.

Early types of hydraulic equipment and systems requiring large amounts of fluid favour the use of water. While this fluid is cheap and usually readily available, it does have the distinct disadvantage of promoting rusting, of freezing to a solid, and of offering poor lubrication properties.

In between mineral oil and water lie the different types of fire resistant fluids. There are two broad classes; those fluids where water provides the resistance to fire and those where fire is inhibited by the nature of the chemical structure. In the first group are water/glycol mixtures, water in oil emulsion (40-50% water), and oil in water emulsions (5-50% water). The second group includes synthetic materials in particular chlorinated hydrocarbons and phosphate esters.

The higher fluid pressure used in modern fluid power circuits have made fire hazards more serious when petroleum based oil is used, since a fractured component or pipe line could result in a fine readily flammable mist of oil travelling as far as 120 m. Fire resistant fluids are extensively used in the mining industry and in power generating stations [1].

Properties a designer would be interested in, when selecting a hydraulic fluid, are listed in Table 2.1.

Much of the success of mineral oils as hydraulic fluids has been due to the relative ease with which their properties can be altered by the use of additives, for example rust inhibitors.

Another major power transmitting fluid is compressed air. This fluid is very compressible compared with the so-called incompressible hydraulic fluid. A basic result of this property is that in a fixed volume of gas either the temperature or mass, or both, must change before pressure changes. Hence, when a pneumatic volume under a spring loaded piston has been vented, opening a valve does not move the piston immediately. Enough gas must flow into the volume to increase the pressure to such a level that it can overcome the spring pre-load first. In hydraulic systems a small opening will provide an almost instantaneous rise of pressure to the level of the supply pressure. The propagation delay in hydraulic systems due to fluid compressibility and container elasticity is small enough to be neglected, which is not the case in pneumatics. This propagation delay can cause switching problems in poorly designed logic circuits.

Gases have low viscosity, which makes sealing more difficult than in the case of systems using liquids. The use of low gas pressure is one way to overcome this problem, since gas flow is proportional to the upstream pressure.

Table 2.1.

Property	Comment
1. Viscosity	Variation with temperature and pressure. Low and high temperature performance.
2. Density	Variation with temperature and pressure. Used in flow calculations.
3. Bulk modulus	Assess system rigidity and natural frequency.
4. Lubricity	System accuracy and threshold.
5. Stability	In relation to heat, oxidation, hydrolysis and mechanical shear.
6. Compatibility with other materials	Ensure no problems with seal life. Paint adhesion.
7. Foaming	Can promote reduced system stiffness due to oil aeration.
8. (a) Specific heat (b) Thermal conductivity (c) Thermal expansion	Design calculations.
9. Vapour pressure	Cavitation prevention.
10. Inherent contamination level	Filter choice, tolerance selection.
11. Electrical properties	Design of electrical components associated with system.
12. Toxicity	Health and safety.

Low viscosity also leads to poor lubrication in pneumatic systems and hence it is often necessary to introduce a fine mist of oil into the air flow. This procedure must not however be introduced into fluidic circuits.

A further important characteristic of gases is their low density. Columns of gas are of relatively low mass and, as a consequence, forces developed by the sudden deceleration of fluid columns (water hammer) are not a serious matter in pneumatic systems. Naturally, not being liquids, gases do not suffer from cavitation. Gases can be used over a very wide range of temperatures well beyond the range of hydraulic fluids.

2.1.1 Density

The **density** ρ of a fluid is defined as the mass per unit volume:

$$\rho = \frac{M}{V}\ \mathrm{kg/m^3}. \tag{2.1}$$

For example, the density of water (at 4°C) is 1000 kg/m^3. Common hydraulic oils are lighter than water and their densities usually lie in the range 850–950 kg/m^3. While liquid densities do not vary significantly with pressure or temperature, this is not the case with gases;

For gases, pressure, density and temperature are related by the perfect gas law:

$$P = \rho R T \tag{2.2}$$

where P = pressure, N/m^2 absolute
ρ = density, kg/m^3
R = gas constant, J/kgK
T = temperature, K = 273 + °C.

For most pneumatic systems, equation (2.2) is sufficiently accurate up to about 20 MN/m^2. The gas constant for air is 287 J/kgK, while for any other gas it can be determined by dividing the universal gas constant (8.315 kJ/kgK) by the molecular weight; typical values are listed in Table 2.2.

Table 2.2.

Nitrogen	297 J/kgK
Oxygen	260 J/kgK
Hydrogen	4125 J/kgK

Air is a mixture of gases containing by volume approximately 78% nitrogen, 21% oxygen and about 1% of other gases. Water vapour is also present in amounts that can vary depending on the atmospheric conditions.

Since the characteristics of the atmosphere are variable, it is necessary to define a standard for pressure. The standard atmospheric pressure is 101.3 kN/m^2, so that at a temperature of 15°C, the density of air is 1.225 kg/m^3.

The bar, or 100 kN/m^2, is a convenient unit of pressure, and in most fluid power calculations it is sufficiently accurate to assume that atmospheric pressure is 1 bar absolute. The term **vacuum** is applied to any pressure level less than atmosphere, hence a perfect vacuum is 0 bar.

2.1.2 Viscosity

Viscosity is a measure of a fluid's resistance to flow. Low viscosity fluids such as air tend to promote high levels of leakage loss in equipment. On the other hand high viscosity fluids result in more sluggish operation and power wastage due to excessive pressure loss through pipe lines and components.

The **dynamic viscosity** (μ Ns/m^2) of a fluid is defined as:

$$\mu = \frac{\text{shear stress}}{\text{shear rate}} = \rho\nu \tag{2.3}$$

The **kinematic viscosity** (ν m^2/s) is the ratio of the dynamic viscosity to the fluid density (Table 2.3).

Table 2.3.

Property at 38°C†	F.R.F. (phosphate ester)	Hyd. oil	Water	Air at 15°C† and 1.0 bar
Density (kg/m^3)	1136	858.2	993.1	1.187
Viscosity (ν) (m^2/s)	4.6×10^{-5}	4.0×10^{-5}	0.9×10^{-6}	1.5×10^{-5}
Bulk modulus (N/m^2) at 35 MN/m^2	2.25×10^{9}	1.38×10^{9}	2.18×10^{9}	see section 2.2
Vapour pressure (kN/m^2) abs	6×10^{-5}	6×10^{-5}	6.89	–
Velocity of sound (m/s)	1407	1268	1481	343

† Average normal working temperature.

The rate of change of viscosity with temperature is expressed by a **viscosity index**. For wide changes in temperature a high viscosity index oil should be used if viscosity variations are to be minimised.

Another aspect of viscosity that is of importance to the designer is the fluid flow condition. Two distinct fluid motions are common:

(a) **Laminar** or **streamline** flow, where the flow pattern is smooth.
(b) **Turbulent** or **irregular** flow, where the fluid motion is random.

The flow condition which a fluid experiences is related to the **Reynolds number**:

$$R_e = \frac{v\rho d}{\mu} = \frac{vd}{\nu} \propto \frac{\text{inertia effects}}{\text{viscous effects}} \quad (2.4)$$

where v = mean flow velocity
d = a characteristic length (the diameter for pipes)
ρ = mass density
μ = dynamic viscosity
ν = kinematic viscosity.

The transition from laminar flow to turbulent flow in pipes normally takes place when the Reynolds number increases through the range 1200 to 2500, known as the transition region.

2.2 EFFECTIVE BULK MODULUS

The compressibility of power transmitting fluids is a major factor in determining systems resonant frequency. This in turn puts limitations on the speed of response of the output.

Consider a thick walled cylinder with movable piston as shown in Fig. 2.1.

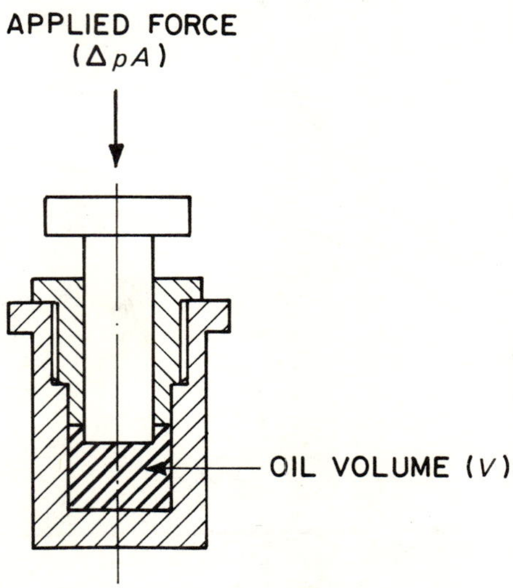

Fig. 2.1 – Thick walled cylinder with movable piston.

The volume within the container is filled with hydraulic oil. By applying a force F to the piston it is found that the oil volume compresses such that:

$$\frac{\Delta V}{V} = -\frac{\Delta p}{\beta} \tag{2.5}$$

where ΔV = volume change
V = initial volume
Δp = pressure increase due to applied force F
β = bulk modulus.

The term **compressibility**, which is the reciprocal of bulk modulus, is often used.

The actual value of the bulk modulus of oil is strongly dependent on the amount of air present in the form of bubbles. Data for some typical fluids are shown in Figs. 2.2 and 2.3.

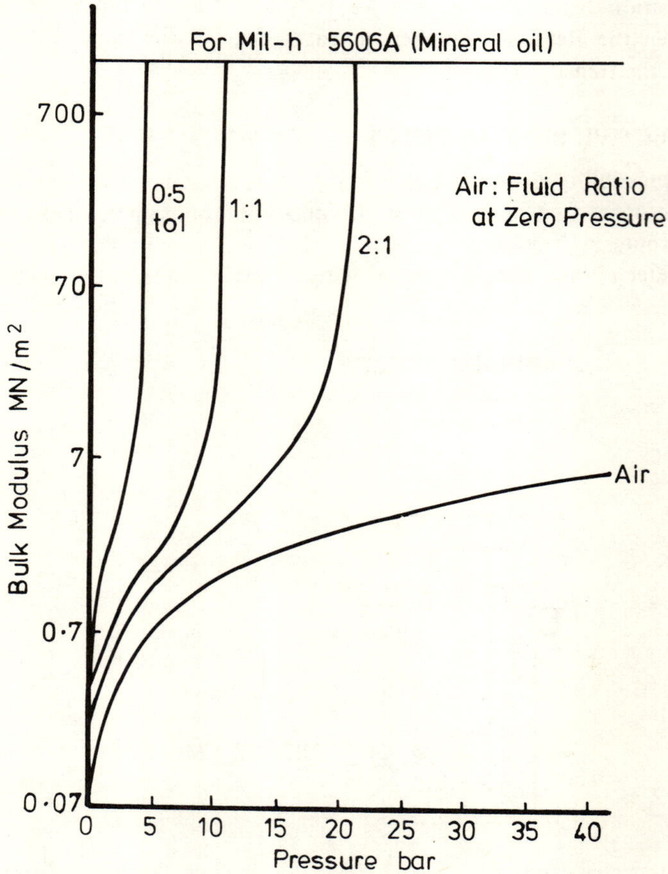

Fig. 2.2 – Effect of pressure on the bulk modulus of an air–oil mixture. The liquid is saturated with air at each pressure.

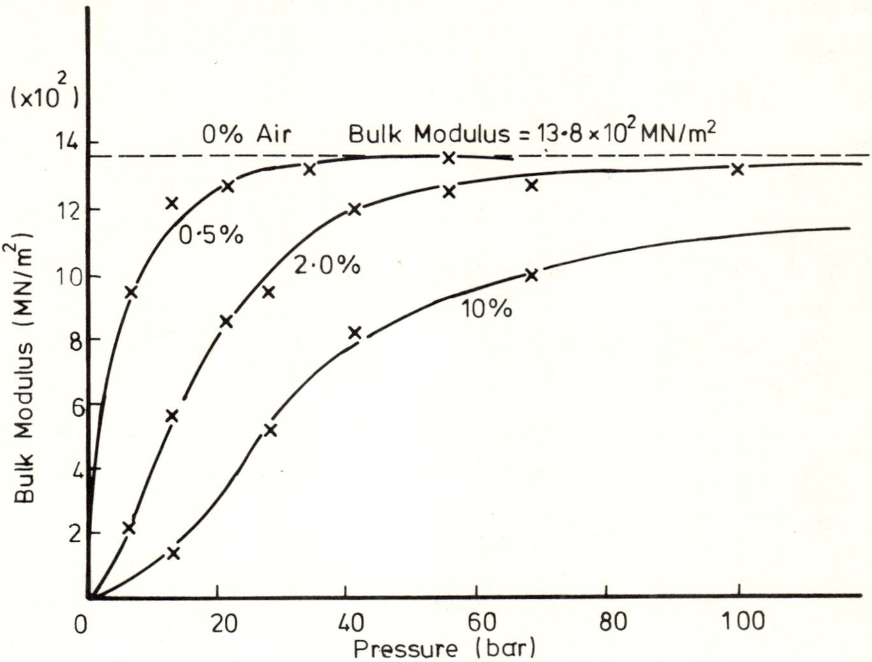

Fig. 2.3 – Bulk modulus for a typical hydraulic oil including the effect of free air.

Some level of entrained air is impossible to avoid even in the best circuit designs. From Fig. 2.3 it can be seen that the maximum value of bulk modulus, 1.38 GN/m^2, occurs above 70 bar. However, a practical design value would be more likely to be in the region of 0.7 GN/m^2.

An additional factor which will further reduce the effective bulk modulus will be the type of pipework used to contain the oil. Referring once more to Fig. 2.1, let the initial total container volume be:

$$V_t = V_l + V_g = V_c \tag{2.6}$$

where V_l = volume of liquid
V_g = volume of gas
V_c = volume of container.

As force is applied to the piston, the liquid and gas volumes will reduce, but the container, especially if it is a hydraulic hose, will expand. Hence equation (2.6) becomes:

$$\Delta V_t = -\Delta V_g - \Delta V_l + \Delta V_c. \tag{2.7}$$

According to equation (2.5), the total effective bulk modulus can be expressed as:

$$\beta_e = -V\frac{\Delta p}{\Delta V_t}\,.$$

But,

$$\beta_l = -\frac{V_l \Delta p}{\Delta V_l}\,; \qquad \beta_g = -\frac{V_g \Delta p}{\Delta V_g}\,; \qquad \beta_c = \frac{V_t \Delta p}{\Delta V_c}\,.$$

Combining these expressions with equation (2.7) gives:

$$\Delta V_t = \left(\frac{V_g \Delta p}{\beta_g} + \frac{V_l \Delta p}{\beta_l}\right) + \frac{V_t \Delta p}{\beta_c}$$

$$= \left[\frac{V_t}{\beta_c} + \left(\frac{V_g}{\beta_g} + \frac{V_l}{\beta_l}\right)\right]\Delta p\,. \tag{2.8}$$

Now $V_g \ll V_t$; $V_l \approx V_t$, so let the entrapped gas to liquid ratio be,

$$S = \frac{V_g}{V_l} \approx \frac{V_g}{V_t}\,.$$

Equation (2.8) can now be written:

$$\Delta V_t = \left[\frac{V_t}{\beta_c} + \left(\frac{S V_t}{\beta_g} + \frac{V_t}{\beta_l}\right)\right]\Delta p$$

or

$$\frac{\Delta V_t}{V_t \Delta p} = \frac{1}{\beta_c} + \frac{S}{\beta_g} + \frac{1}{\beta_l} = \frac{1}{\beta_e}\,. \tag{2.9}$$

When assessing a value of bulk modules for the entrained gas (β_g) it is necessary to define the conditions of compression. If the temperature is constant, by allowing heat to flow into or out of the volume of gas being compressed, the resulting quantity is termed **isothermal compressibility**. This is normally a slow process in which,

$$\beta_g = P$$

where P is the pressure absolute.

If, however, heat flow is restricted, **adiabatic compressibility** results, giving:

$$\beta_g = \frac{c_p}{c_v}P = \gamma P\,. \tag{2.10}$$

This is the more usual design condition. γ, the ratio of the specific heats, is 1·4 in the case of air.

More recent design practices, in relation to equipment noise reduction, have tended to encourage the use of hydraulic hose in circuit design. Most high pressure hoses are constructed of hard rubber or Teflon with an interwoven sheath of stainless steel. Such flexible hoses have a comparatively low bulk modulus.

Obtaining realistic design values for the bulk modulus of oil filled hydraulic hose is quite difficult. Values ranging from 70 MN/m^2 to 350 MN/m^2 can be found in the literature. Some experimental data are shown in Fig. 2.4 for 6.4 mm woven hose, where the effect of internal pressure can clearly be seen.

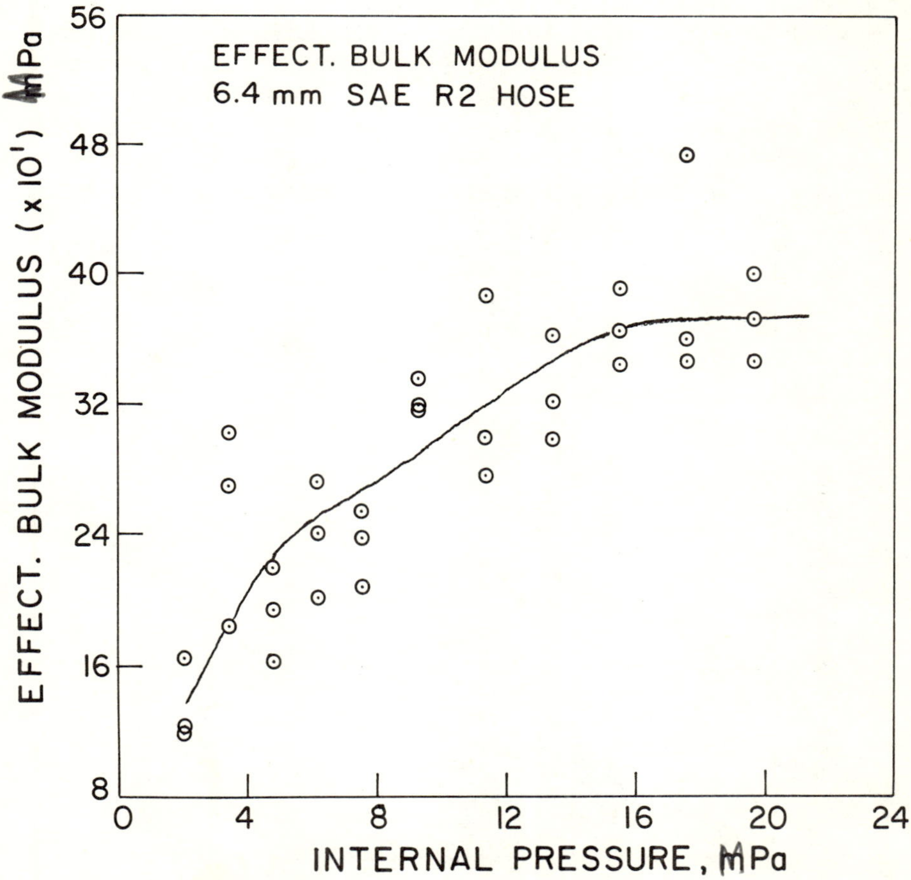

Fig. 2.4 – Effective bulk modulus of a typical hydraulic hose.

Hydraulic steel pipe can be treated as a thin walled cylinder so that:

$$\beta_c \approx \frac{tE}{D} \tag{2.11}$$

where t = wall thickness
D = pipe diameter
E = modulus of elasticity (195.3 GN/m^2).

2.3 AERATION

Air can be dissolved or entrained in hydraulic fluids, and it is possible for air to change from one of these forms to another depending on the conditions to which the fluid is subjected. Consider first the solubility of air in oil. Henry's law states that air solubility in any given liquid is directly proportional to absolute air pressure of the air above it. In a system reservoir where the pressure is one atmosphere, the air in solution exerts one atmosphere of pressure. The actual amount of air in solution must be determined experimentally, and Fig. 2.5 shows the results of some experiments conducted by Leroy [2]. From these

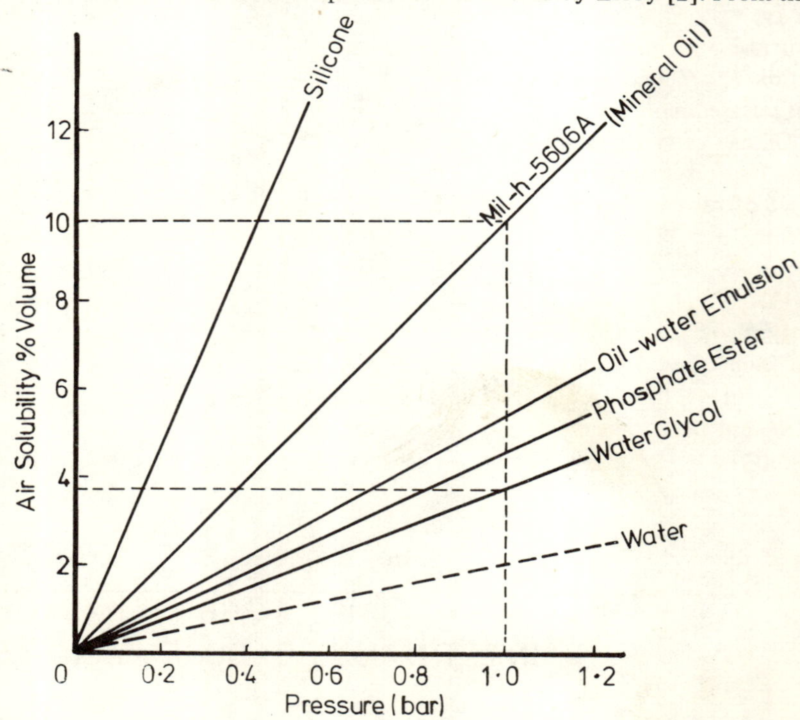

Fig. 2.5 – The solubility of air in various liquids.

data it can be seen that mineral oils can dissolve much more air than water glycol or water emulsions. The **solubility constant** k is defined at the precentage by volume of air dissolved per atmosphere, the air volume being reduced to standard temperature and pressure. Hence, from Fig. 2.5 the solubility constant for a conventional industrial hydraulic oil is approximately 10 and for a water glycol appoximately 4. Hayward [3] has produced similar results and has shown that the solubility constant for water is 2, between 6 and 12 for mineral oils, and between 15 and 25 for fuels and silicone fluids. The results shown in Fig. 2.5 relate to saturated fluids, that is fluids in equilibrium with the excess of undissolved air. Fluids may temporarily hold more than this equilibrium amount of air, and in this case the fluid is said to be **supersaturated**. On the other hand, if the liquid holds less than the equilibrium amount of dissolved air the solution is said to be **undersaturated**, and it will tend to dissolve more air until equilibrium is reached. In general, a liquid at high pressure can contain much more air than one at low pressure. If, for example, an industrial oil with a solubility constant k of 10 is saturated with air at atmospheric pressure, than a reduction of $\frac{1}{2}$ atmosphere in pressure will cause supersaturation and 5 per cent of air at standard temperature and pressure will come out of solution. Since the pressure is $\frac{1}{2}$ atmosphere, the actual volume of air will be about 10 per cent of the oil volume. This type of air release is easily seen in a bottle of carbonated beverage. If the top is removed, bubbles of carbon dioxide, which were originally forced into the solution under high pressure, are released and start to stream in the fluid. The gas is released more quickly if the liquid is agitated.

Oil can carry air in solution but in addition it can also carry air by entrainment. If a return line to the reservoir in a hydraulic system is not submerged, the jet of oil entering the liquid surface will carry air with it. This will cause the formation of air bubbles in the oil. However, if they rise very slowly they may be drawn into the pump intake, and this can cause damage (discussed later). Similarly, a small leak in a suction line can cause the entrainment of large quantities of air. Such leaks are difficult to spot as oil does not leak out, but air leaks in.

How does the presence of entrained and dissolved air affect fluid properties and system performance? Fortunately, air in solution has no measurable effect on fluid properties. Compressibility and viscosity are unaffected. Unfortunately the same cannot be said for entrained air. Hayward has measured the effect of bubble content on the viscosity of bubbly oils. He used a viscometer tube and showed that the effect of air bubbles upon viscosity is relatively small and can be expressed by the empirical equation:

$$\mu_\beta/\mu_o = 1 + 0.015B$$

where B is the percentage bubble content, μ_β is the viscosity of bubbly oil, and μ_o is the viscosity of the oil.

Air bubbles in oil can cause a great reduction in the bulk modulus. Smith [4] has investigated this phenomenon. Air, of course, is much more compressible than oil and it is to be expected that its presence in free form in the oil will reduce the overall bulk modulus. Fig. 2.2 shows the effect of air content on the adiabatic bulk modulus of a common hydraulic oil. Here at each pressure, equilibrium conditions are reached, that is the oil is saturated with air at each pressure. Clearly, as the pressure is increased, the oil dissolves more and more air.

For example, with a 2 to 1 air/oil volume ratio, all the air is dissolved at approximately 20 bar. This agrees with Hayward's results which show that mineral oils have a solubility constant of approximately 10, that is 10 per cent of air dissolved per atmosphere. Hence, to dissolve 200 per cent requires approximately 20 bar. If the air does not have time to dissolve, the bulk modulus can be greatly reduced even at very high pressures, as high as possible in order to:

(a) compress the air to give it a higher bulk modulus; see equation (2.10),
(b) cause the air to dissolve in the oil, in which case it does not affect the bulk modulus.

Air bubbles in the oil can have another indirect effect on its properties. If the bubbles pass through a pump they are compressed and heat is generated. There is still some doubt concerning the actual law of compression. Hayward [3] argues that it is nearer isothermal than adiabatic. For an adiabatic compression Schanzlin [5] has shown that the temperature of the air bubbles varies with pressure as shown in Fig. 2.6. For instance, at a pressure of 200 bar, the curve predicts a bubble temperature of 1100°C, although in practice the temperature will be less because of air going into solution and heat loss to the oil. However, the increase in temperature can cause oil oxidation and can overheat the system, and in some cases compressed air has raised the temperature of the reservoir oil by 25 per cent.

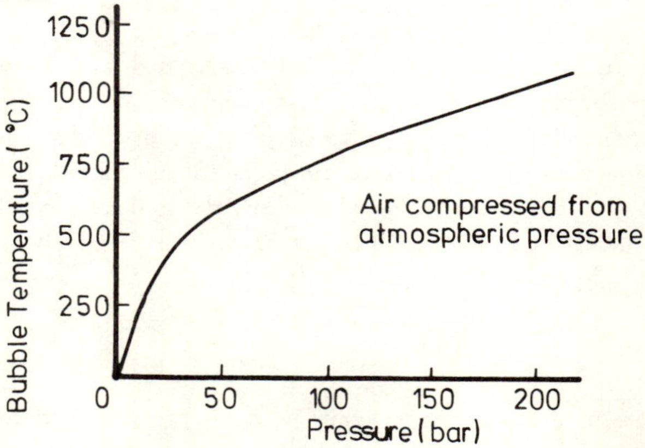

Fig. 2.6 – Temperature developed during adiabatic compression.

The presence of air can seriously impair pump performance. Pump delivery is decreased when the amount of dissolved air in the reservoir is increased. The reason for this is clear when it is realised that there is a pressure drop along the pump inlet line. Assume that the oil in the reservoir is saturated with air at atmospheric pressure and bad suction line design results in a pressure drop of 0.4 bar. The volume formed by air coming out of solution will be about 4 per cent of the oil volume, see Fig. 2.5. Clearly this amount of air entering the pump will reduce its volumetric efficiency. In addition to this, through poor reservoir design, there may be a substantial quantity of entrained air, and this too will contribute to the loss of pump efficiency. Losses in the suction line can be reduced by using a bell-mouthed entrance and by making the diameter of the pipe sufficiently large.

As these bubbles are compressed and passed through the pump they do not go into solution immediately. Hayward [3] has shown that the rate of solution depends on the pressure of the oil, and some of his results are shown in Fig. 2.7. These results were taken from a static apparatus. It consisted of a cylinder of bubbly oil on the surface of which was placed a mass. The position of the mass was related to the degree of saturation. Faster rates of solution can be achieved if the oil is agitated, as it is of course in a real system; and in a static apparatus, local saturation of the oil around each bubble occurs fairly quickly and tends to slow down the rate of solution. From this it is seen that a lower bulk modulus than expected will persist for some distance down the delivery pipe.

Both Pigott [6] and Smith [4] have shown that centrifuge action in pumps can cause the bubbles to coalesce to form larger bubbles. These do not go back

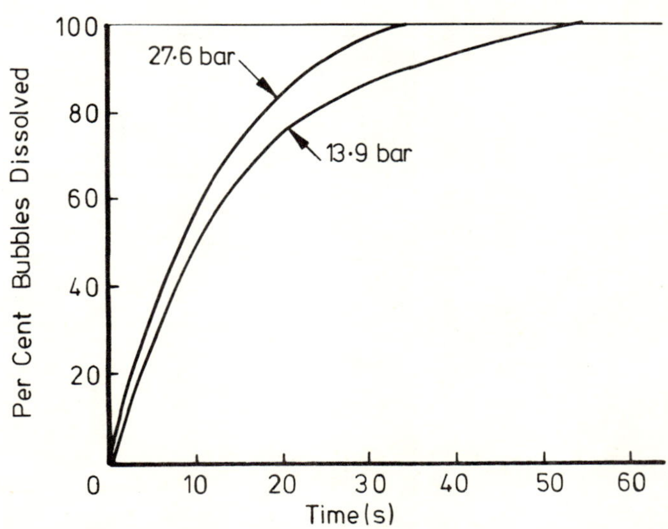

Fig. 2.7 – Rate of solution of air bubbles in hydraulic oil.

into solution readily on the discharge side. In fact under certain conditions, that is, very large rates of pressure change, these bubbles can be unstable and can collapse implosively on the high pressure side. This is related to the cavitation phenomenon and is sometimes called **gaseous cavitation** as opposed to **vapourous cavitation** which will be discussed later. The pressure developed at collapse can be large enough to cause mechanical damage, and Schanzlin [5] has shown this to occur on the high pressure outlet of a gear pump.

Attempts to reduce the level of dissolved air are not very effective, but some steps can be taken to reduce the amount of entrained air. Most of the air entrained in a hydraulic circuit is picked up in the sump or reservoir. It is important that the surface of the oil is not agitated by returning oil; therefore all oil return lines must be below the oil level. Return lines from relief valves must also be drowned. In addition to preventing bubble formation in the sump, the design must also prevent air bubbles in the returning fluid from entering the suction line. To this end a separator which encourages bubbles to return to the surface of the oil should be fitted. Tests [3] show that a simple gauze screen of 25 strands/cm fitted in the reservoir at an angle of 30 degrees stops all bubbles of significant size. The top of such a gauze should be below oil level since if there is any foam on the oil surface, it might well penetrate the gauze. Fig. 2.8, taken from [3], shows a good reservoir design.

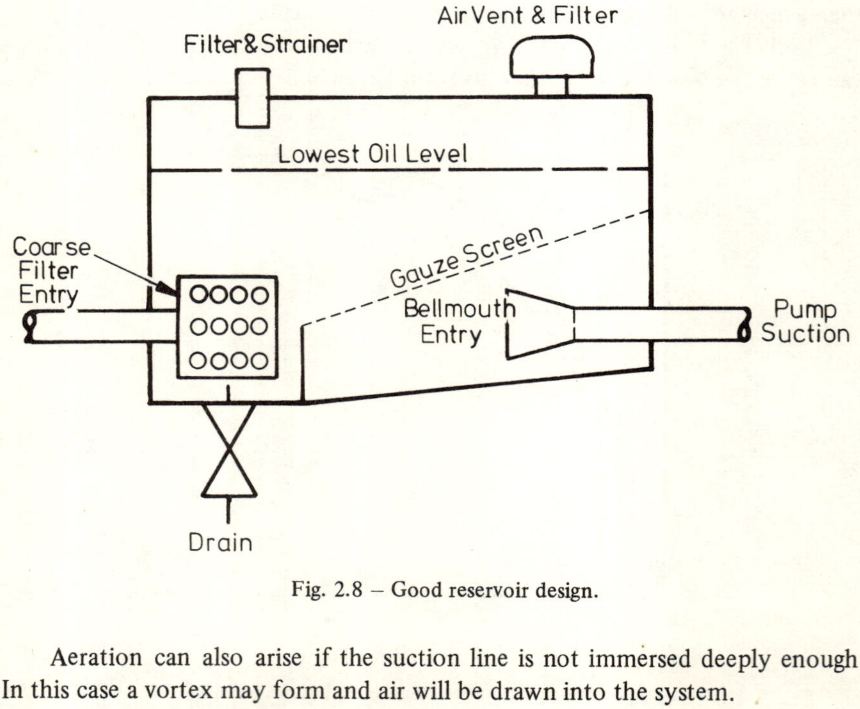

Fig. 2.8 – Good reservoir design.

Aeration can also arise if the suction line is not immersed deeply enough. In this case a vortex may form and air will be drawn into the system.

2.4 CAVITATION

In fluid systems the term **cavitation** usually refers to the formation and collapse of cavities in the liquid. The cavities may contain air or gas. If the pressure is reduced far enough hydraulic oil will vaporise and vapour cavities will form. The pressure at which vaporisation commences is called the vapour pressure of the liquid and is very dependent on the temperature of the oil. The phenomenon of boiling is related to cavitation. At atmospheric pressure water boils at 373 K, that is, the vapour pressure is 1.0128 bar. The vapour pressure is reduced to 0.025 bar absolute at 294.1 K. Hence, it is clear that the lower the temperature of a liquid the lower is the vapour pressure, and this means that the pressure must be reduced considerably before the liquid boils. The vapour pressures of mineral oils are much lower than that of water, typically 6×10^{-4} mbar at 40°C, and hence cavitation is expected to occur less readily with these fluids. In Fig. 2.9 taken from the paper by Leroy [2], the vapour pressures of common hydraulic fluids are displayed as a function of temperature This clearly shows the increase in vapour pressure as the temperature increases. Those fluids which contain water, such as emulsions and water glycol, have much higher vapour pressures than the other conventional fluids.

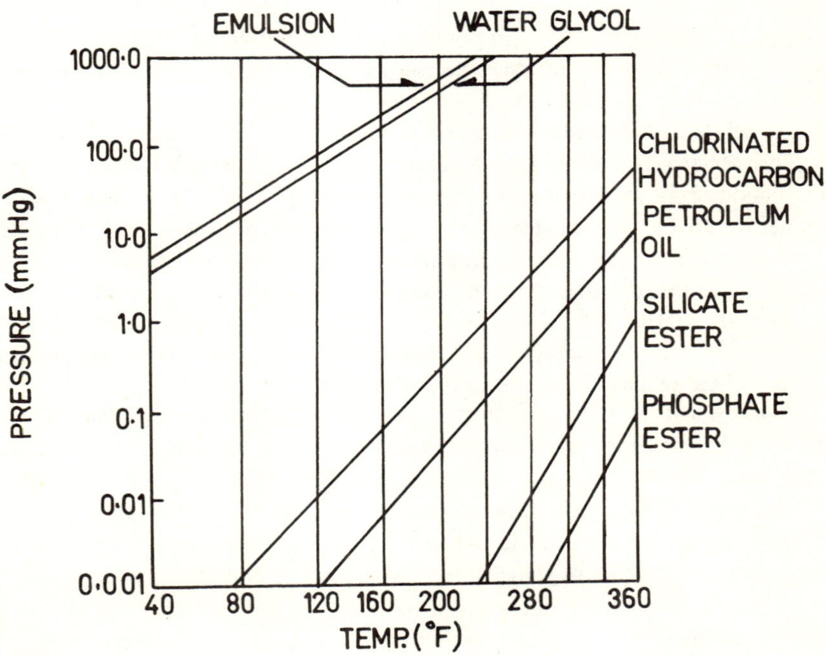

Fig. 2.9 – The vapour pressure of hydraulic fluids.

It is interesting to note that the formation of cavities in a liquid involves rupturing of the liquid. Since, for instance at 294.1 K, vapour cavities form in water at a pressure of 0.025 bar absolute, one might argue that the liquid has ruptured at this pressure and hence that its tensile strength is low. This is not so. It has been shown experimentally [7] that the tensile strength of water can be as high at 100 MN/m^2. The development of a cavity is now recognised as being associated with a nucleation centre such as a microscopic gas particle. When the liquid is subjected to a tensile stress, cavities do not form as a result of liquid rupture but are caused by the rapid growth of these nuclei [8]. As with aeration it is entrained air and not dissolved air that causes trouble. Experiments have demonstrated that the tensile strength of water saturated with dissolved air can have a very high value. However, conventional hydraulic fluids operating in a conventional system will contain sufficient nuclei to ensure that cavitation will occur when the pressure is reduced to vapour pressure.

The phenomenon of cavitation damage in hydraulic machinery, turbines, pumps and propellers is well known. It has been shown both analytically and experimentally that when cavities collapse as a result of increased hydraulic pressure, very large pressures, sometimes of the order of 7000 bar, can be developed. Unfortunately, there is still some controversy about the exact mechanism of the damaging process. The mechanical theory of damage suggests that prolonged exposure to cavity collapse results in brittle fracture and hence wear of the material. Thiruvengadam [9] proposes that the strain energy of the material is the controlling property of cavitation erosion. Materials such as aluminium of low strain energy are susceptible to severe cavitation erosion, while stainless steel, which has a high strain energy, withstands cavitation attack very well.

The rate of wear can be greatly accelerated if the cavitation erosion develops under conditions in which corrosion can occur [10, 11]. In such a case the hammering effect of imploding cavities removes the products, such as oxides, of the corrosion attack. At the same time the corrosive effect produces material which is more easily damaged by the hammering than the virgin material.

As with aeration the collapse of the bubbles results in increased temperature of the oil and this causes side effects such as oil oxidation. Resins and slags can form and these will cause deterioration of system performance.

It is clear that the fluid must experience a reduction in pressure before cavities will form. This reduction in pressure can come about in several different ways. Since the total energy in a fluid is a combination of dynamic and static pressure heads, a reduction in static pressure will arise if the dynamic pressure is increased. Such a situation arises when a liquid flows through a venturi section. At the point of maximum velocity the static pressure will be a minimum, and as the velocity of the fluid is increased a stage will be reached when the minimum pressure drops to the vapour pressure of the liquid. At this point cavitation will occur at the point of maximum velocity. As the cavities move downstream from the point of maximum velocity they will collapse and may well damage the surface of the container.

Thus aeration and cavitation are both undesirable phenomena. They will receive further attention throughout the book.

2.5 FILTRATION AND CONTAMINATION ASSESSMENT

Component failure or damage by contamination in the fluid is an area of major concern to the designer. The trend towards the use of higher system pressures generally means that clearances between moving parts must be reduced to control leakage. This in turn means that more emphasis must be attached to cleanliness of the fluid.

The care of a hydraulic fluid should start before it is introduced into a system. Containers should be stored indoors, and with the bung in a horizontal position so that it is below the oil level. This helps to prevent debris collecting around the seal.

The contamination level in a new batch of oil should be checked before pouring into a sytem reservoir.

Contamination is more likely to occur during filling of the system, than while the oil is in service. Dust can enter the system with the oil while filling, and the inlet should be protected with some form of paper filter. A new system will have a relatively high level of contaminants left from the machining and assembly procedures. It is good practice therefore to change the oil and filter fairly soon after initial start up. During service life contaminants can occur either as a result of atmospheric breathing or from wear particles. Experience has shown that as little as 0.02 per cent per weight of cement dust can cause accelerated wear [12]. Dirt in a system can result in increased friction and wear, silting of valves, and finally seizing of moving parts, often with catastrophic results. In valves, particles can round the corners of square edge orifices and so change the flow characteristics by altering the discharge coefficient. Fig. 2.10 shows this effect in a servo valve neutral leakage characteristic [13]. Valves employing very small orifices, such as nozzles in two stage valves, suffer from silting, as do capillary tubes, Bourdon tubes and dampers. The reduction of flow due to this form of contamination is shown in Fig. 2.11 [14].

Dirt in hydraulic systems consists of many different types of material ranging in size from less than 1 μm (micron) to greater than 100 μm. The smallest particle that can be seen by the naked eye is around 50 μm and therefore visual examination of the oil is not sufficient.

Most general industrial hydraulic systems operating below 140 bar are able to tolerate particles in the 25 to 50 μm range and therefore a 25 μm filter rating is satisfactory. Systems in the 140 to 210 bar range should have a 10 to 15 μm filter rating, while high pressure pumps and valves (above 210 bar), and in particular precision servo valves, should have a 5 μm rating. Typical clearances are summarised in Table 2.4.

The size distribution of particles is random, and generally speaking the smaller the size range the greater the number of particles per unit volume of fluid.

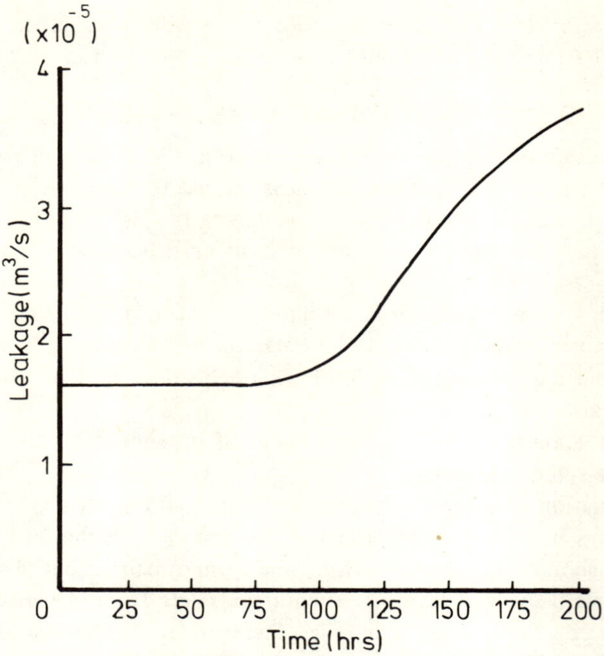

Fig. 2.10 – Variation of servo valve leakage with service life.

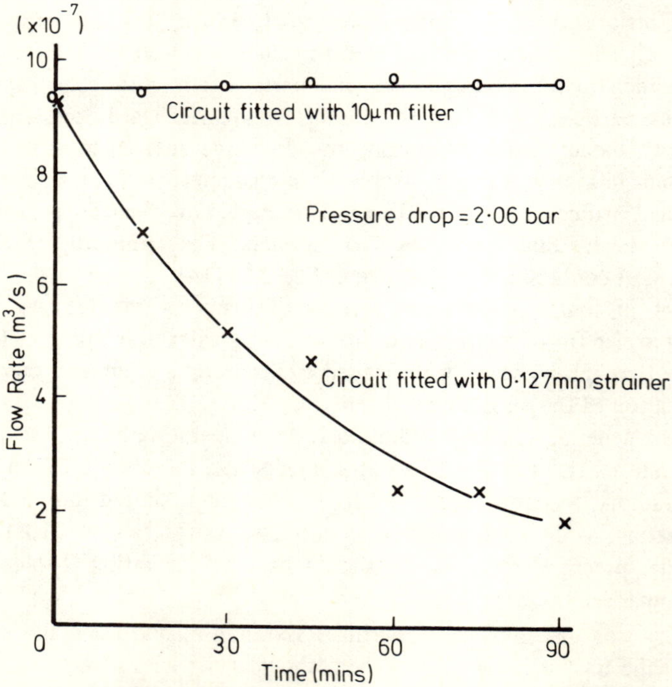

Fig. 2.11 – Variation of flow with time through a small orifice.

Filters are not capable of removing all the contaminants from the oil, but for example, a 10 μm filter is capable of removing 98 per cent of all particles exceeding 10 μm of a standard contaminant in a given concentration.

The test of contaminant levels in oils is a specialised field and the reader is directed to the references [13–16].

Table 2.4 – Typical clearances [15].

Component	Typical clearance μm
Gear pump	
gear to side plate	0.5-5
gear tip to housing	0.5-5
Vane pump	
tip of vane	0.5-1
sides of vane	5-13
Piston pump	
piston to bore	5-40
valve plate to cylinder	0.5-5
Servo valve	
orifice	130-450
flapper wall	18-63
spool-sleeve	1-4
Control valve	
orifice	130-10 000
spool-sleeve	1-23
poppet	13-40
Actuators	50-250
Hydrostatic bearing	1-25

2.6 HIGH TEMPERATURE SYSTEMS

Many modern fluid servomechanisms must operate under very high temperature conditions. Examples include nuclear reactors and re-entry vehicles in which controls must be capable of reliable operation at temperatures up to 1300 K. One method of approach is to use hot gases, discussed in Chapter 13. The problem of finding suitable liquid fluids is a difficult one, since they have not got the high temperature stability of hot gases. Polyphenyls and liquid metals have been used as heat transfer liquids in nuclear reactors, and their properties are fairly well known. In the investigation reported in [17] polyphenyls were on the whole found to be superior to the liquid metal alloy of sodium (23%) and potassium (77%). The organic polyphenyl is a fair lubricating fluid and has a bulk modulus of 647 MN/m^2 at 540 K and 270 bar pressure. The liquid metal alloy has a higher modulus, 2.756 GN/M^2 but poor lubricity. It is also hazardous, reacting violently with water.

2.7 ELECTROSENSITIVE FLUIDS

Electrosensitive liquids can be synthesised to exhibit shear resistances that can be controlled by applying an electric field across the fluid. This phenomenon occurs when oil is contaminated with small particles, for example, transformer oil Univolt N-36 and silica gel 7 of particle size around 40 μm. Detail of the composition of such oils is given in [18]. No such fluid is yet available commercially, and existing formulae still have many problems of settling, voltage breakdown, clogging, carbonisation and ageing. However, experiments show an interesting potential for the fluid for direct-acting valves, self-programming dampers [19], and high frequency vibration [20] equipment, to name a few.

In 1947 W. M. Winslow [21], the discoverer of the phenomenon, demonstrated that the suspension particles form fibres between the conducting plates carrying the electrical charge. Plate spacings are around 0.53 mm and the electrical circuit would supply typically 2500 V at 5 mA. Suspensions of colloidal silica are forced into a fibrous mass by electrical polarisation. The electrical sensitivity of the silica suspension is due to polarisation of a thin film of absorbed water on the surfaces of the particles. The particles then form fibres along the electrostatic lines of flux between the plates.

These fibres create a restriction to the flow of the fluid and therefore create an effect similar to viscosity but dependent on the magnitude of the magnetic field. The presence of a yield stress makes this apparent viscosity a function of local velocity gradient and severely complicates analytical studies. Experiments [18 and others] carried out on the fluid show that the fluid yield stress is proportional to the field strength squared, with a yield point dependent upon the local electrical field. Some investigators describe the fluid therefore, as electroplastic.

A theoretical model for the behaviour of the fluid has been developed in [22] and results in the relationship:

$$\Delta p = P_{\text{loss}} + \psi \epsilon_o \frac{L}{h} E_o^2 \qquad (2.12)$$

where
Δp = pressure drop across gap
P_{loss} = loss due to fluid flow only
ψ = electrosensitive factor
ϵ_o = a constant (8.82×10^{-12} N/V^2)

L, h = geometric dimensions ($L = 10.16$ cm, $h = 0.525$ mm)
E_o = magnitude of the applied field (V/m).

Hence equation (2.12) states that the total pressure drop due to flow through a gap is the sum of the normal losses together with an additional loss due to the electric field.

Fig. 2.12 shows the electroviscometer used in the experimental work described in [18]. The fluid mixture stored in chamber C_1 is forced into chamber C_2 and between cylinders A and B by air pressure (P_{air}). An electrical potential difference is created between cylinders A and B, which are isolated from each other by means of plexiglass spacers. The voltage is applied at V.

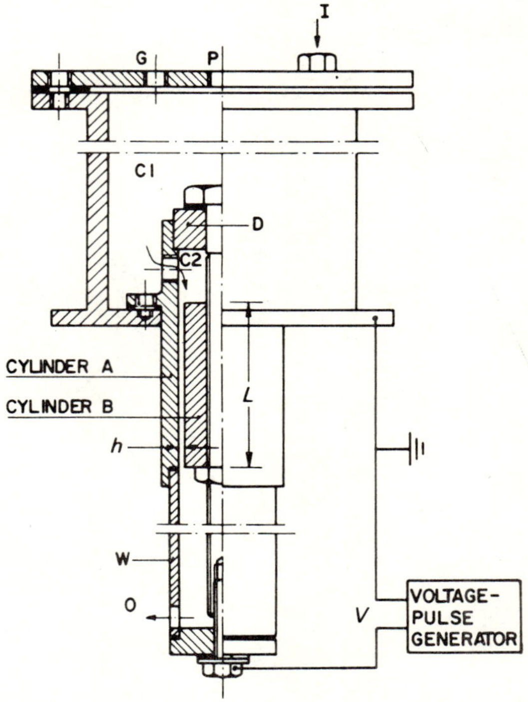

Fig. 2.12 – Electroviscometer design where $L = 0.101$ m, gap $h = 0.53$ mm, voltage V up to 2500 V, and input air pressure up to 4 bar.

For this arrangement equation (2.12) reduces to:

$$\Delta p = P_{air} = P_{loss} + \psi \epsilon_o \frac{L}{h} \left(\frac{V}{h}\right)^2 \tag{2.13}$$

where

$$\psi = 0.11 \times 10^{12} \frac{h^3}{L} (P_{air} - P_{loss}) \frac{1}{V^2} .$$

The results of experiments with different fluid compositions are shown in Fig. 2.13. In each case the experiment was continued until electrical breakdown occurred. In general, the fluids behaved as predicted by (2.12), following parabolic curves.

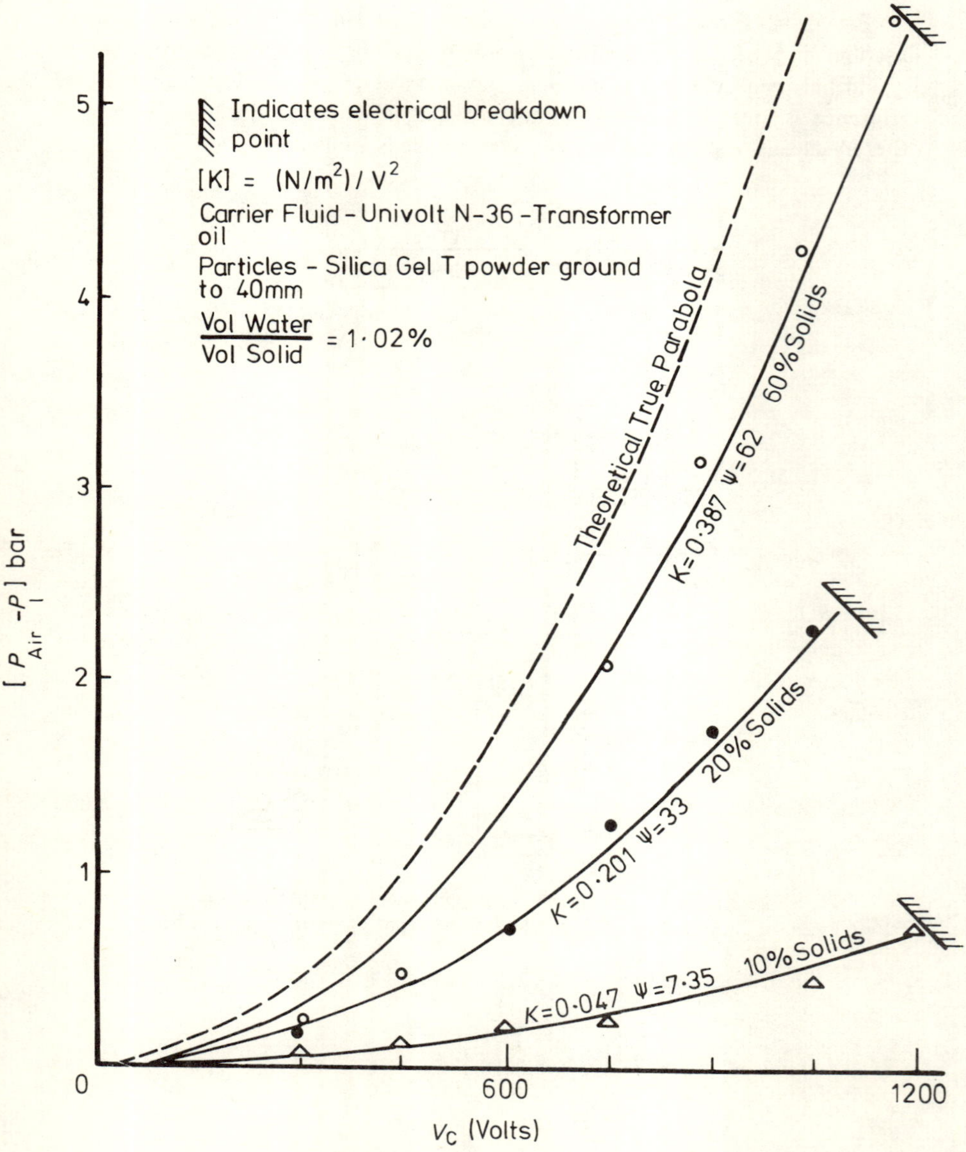

Fig. 2.13 – Characteristics of electrosensitive fluid mixtures.

The equation from the experimental results can be approximated to:

$$P_{air} - P_{loss} = KV^2 \tag{2.14}$$

giving

$$\psi = 161.6K. \tag{2.15}$$

where K is as defined in Fig. 2.13.

The results of this experiment show that pressures up to 5.16 bar can be controlled.

The general conclusions about these mixtures were:

1. For a given particle diameter, the maximum value of the field in the high field regions diminishes with increasing concentrations of solid particles. The total resistance of the particles to movement depends on the magnitude of the field and the volume concentration of particles in the fluid.
2. The properties of the mixture are affected by the amount of water absorbed by the suspension, and there appears to be an optimum concentration.
3. The most electrically sensitive mixtures have relatively high concentrations of solids (40–50%) by volume.
4. The particle size should be as small as possible.
5. The shear that a mixture can resist is limited by electrical breakdown.

Much effort is still needed to develop a realistic and commercially viable fluid. There is a major problem due to the particles settling out of the mixtures. Very small particles of the order of 0.5 μm, with a dielectric constant as high as possible and with a net electric charge of the same sign on each particle would help to overcome this problem. These properties would help to keep the particles separated and subject them to natural Brownian movement. If and when these problems are solved, there is great potential for the fluid since it allows direct interface with electrical signals. The recent development in low cost computing modules makes digital control an interesting and potential area for development. Some applications of electrosensitive fluids will be described in Chapter 6.

2.8 EXAMPLES

1. A hydraulic pipe 15 mm nominal bore, has an outside diameter of 21.3 mm and a wall thickness of 2.75 mm. Oil flow is delivered at a constant flow velocity of 7.59 m/s. The characteristics of the oil are:

Temp °C	37.8	15.6
Viscosity Ns/m^2	3.4×10^{-2}	12×10^{-2}

At which temperature will the pipe radiate the most noise? (see Chapter 18).

(*Answer*: 37.8°C because the flow is turbulent).

2. Part of a hydraulic circuit consists of a rigid pipe and a solenoid operated valve. Hydraulic oil, of density 878 kg/m³, is flowing in the pipe at 6.0 m/s. If the solenoid valve is suddenly closed, what is the value of the pressure increase in the pipe? (see Chapter 9).

   Suppose now the rigid pipe is replaced by a flexible hose. Calculate the new value of the pressure rise. Take $\beta_{oil} = 13.8 \times 10^8\,\text{N/m}^2$; $\beta_{hose} = 2.07 \times 10^8\,\text{N/m}^2$.

   (*Answer*: 66.4 bar; 23.8 bar).

3. In an oil hydraulic press installation the cylinder piston has a diameter of 20 cm and a stroke of 35 cm. It slides in a cylinder which is supplied by oil at a pressure of 200 bar. The total volume of oil in the connecting pipes is $4.0 \times 10^{-3}\,\text{m}^3$. Before the piston can begin its return stroke, an outlet valve must be opened, allowing the pressure in the cylinder to fall to atmospheric pressure. In addition, because of the oil compressibility, a certain volume will escape even before the piston moves. What will this volume be, and how much elastic energy will the oil carry away with it?

   (*Answer*: $2.159 \times 10^{-4}\,\text{m}^3$; $4.29 \times 10^4\,\text{N m}$).

2.9 REFERENCES

[1] (a) Stanley, C., Fire Resistant Hydraulic Fluids, paper F.2.
(b) Dalibert, A., Fire Resistant Hydraulic Fluids, paper F.3.
(c) Kelly, E. S., Erosive Wear of Hydraulic Valves Operating in Fire Resistant Fluids, paper F.4.
B.H.R.A. 2nd Fluid Power Conf., University of Surrey, Jan. 1971.

[2] Leroy, W. W. and Leslie, R. L. (1962), Hydraulic Fluid Properties and System Performance, *18th Nat. Conf. on Industrial Hydraulics*, U.S.A. Vol. 16.

[3] Hayward, A. T. J. (1961), Aeration in Hydraulic Systems, *Conf. Oil Hydraulic Power Transmission and Control*, Inst. Mech. Engrs., London, Nov.

[4] Smith, L. H., Peeler, R. L., and Bernd, L. H. (1960), Hydraulic Fluid Bulk Modulus Measurements. *16th Nat. Conf. on Industrial Hydraulics*, Vol. 14.

[5] Schanzlin, E. H. (1956), Higher Speeds and Pressures for the Hydraulic Pump, *12th Nat. Conf. on Industrial Hydraulics*, Vol. 10.

[6] Pigott, J. (1950), Cavitation in Reciprocating and Rotary Pumps, *6th Nat. Conf. on Industrial Hydraulics*, Vol. 4.

[7] Harvey, E., Newton, W. D., McElroy, A., and Whitely, A. (1947), On Cavity Formation in Water, *J. App. Physics*, **18**.

[8] Knapp, R. T. (1976), *Cavitation*, McGraw-Hill.

[9] Thiruvengadam, A. (1963), A Unified Theory of Cavitation Damage, *J. of Basic Eng., A. S. M. E.*, Sept.

[10] Plesset, M. S. and Ellis, A. T. (1955), on the Mechanism of Cavitation Damage, *Trans. A. S. M. E.*, 77.

[11] Hobbs, J. M. (1966), Practical Aspects of Cavitation, *Phil. Trans. Roy. Soc. (London)*, **A.260**.

[12] Jackson, T. L. (1962), Installing Industrial Hydraulic Systems, *Compress. Air and Hydraulics*, July.

[13] Davis, C. B. and Wenner, G. L. (1960), Hydraulic Fluid Contamination Sources; their Recognition and Control, *Proc. Nat. Conf. Industrial Hydraulics*, Vol. 14.

[14] Denney, D. F. (1961), Cleanliness in Hydraulic Systems. *Conf. Oil Hydraulic Power Transmission and Control.* Inst. Mech. Engrs., London, Nov.

[15] Kirnbauer, E. O. (1971), Contamination Control for Fluid Power Systems. *B. H. R. A. 2nd Fluid Power Conf.*, University of Surrey.

[16] Fitch, E. C. and Maroney, G. E. (1971), A Fundamental Method for Establishing Contaminant Tolerance Profiles for pumps, *B. H. R. A. 2nd Fluid Power Conf.*, University of Surrey.

[17] Damasco, F. and Spar, C. (1961), High Temperature Hydraulic Fluids, *Nat. Conf. Industrial Hydraulics*, Vol. 15.

[18] Arguelles, J., Martin, H. R., and Pick, R. (1973), Some Experiments with Electrosensitive Fluids. *3rd Fluid Power Conf.*, Turin.

[19] Eige, J. J. and Oleson, S. K. (1963), *Development of Automatically Hardening Shock Mounts*, Stanford Research Institute Project No. PHU-4109.

[20] Eige, J. J. and Peschon, J. (1960), *Vibration Shock System*, Stanford Research Institute, Project No. 3129.

[21] Winslow, W. M. (1949), Induced Vibration of Suspensions, *J. of App. Physics.*

[22] Arguelles, J., Martin, H. R., and Pick, R. (1974), A Theoretical Model for Steady Electroviscous Flow Between Parallel Plates, Inst. Mech. Engrs., *J. Eng. Sc.*, August.

CHAPTER 3

Basic fluid power systems

3.1 INTRODUCTION

A fluid power system consists basically of a fluid power source, a means of distributing the power, a means of controlling the power, and a means of reconverting the fluid power to mechanical power. This chapter discusses the fluid power source, both pneumatic and hydraulic, and indicates how the fluid, which is the distribution means, should be processed before use in a system. Motors and actuators are also briefly reviewed. Control elements are discussed in subsequent chapters.

3.2 THE BASIC PNEUMATIC SYSTEM

The compressed air source must provide clean air to a pneumatic system [1]. Failure to do so will result in choking, abrasion and erosion of valves and motors with their subsequent failure. Fig. 3.1, which uses standard symbols, shows a typical compressed air processing plant. It is general practice to use an inlet filter, especially if the compressor is housed in surroundings contaminated for example by dirt, dust, fumes or smoke. The filter may be of the dry type, in which the air passes through a corrugated or pleated element, perhaps made of paper or felt.

Oil and water are the main contaminants in the air after compression. The oil originates in the compressor itself, unless oil-free compressors such as the Roots blower are used. The water originates in the air and is a major nuisance in pneumatic systems. Water vapour content is measured in terms of humidity which is the ratio of the amount of vapour present to that needed to saturate the air. Warm air can hold more water than cold air (Fig. 3.2). A simple example illustrates the problem. Assume that a compressor draws in $60\,m^3/min$ free air and compresses it to 7 bar. At inlet the air is $20°C$ and the humidity is 70 per cent. Thus the water vapour drawn in per minute is $6 \times 0.17 \times 0.7 = 0.714\,kg$. After compression the volumetric flow rate is about $60/8 = 7.5\,m^3/min$ (assuming isothermal compression). If the temperature in

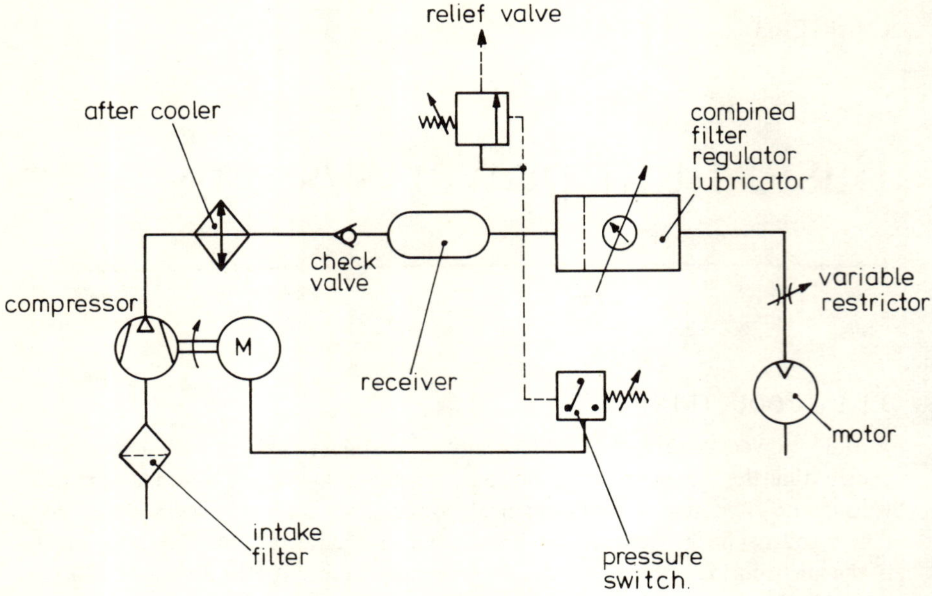

Fig. 3.1 – The basic pneumatic system.

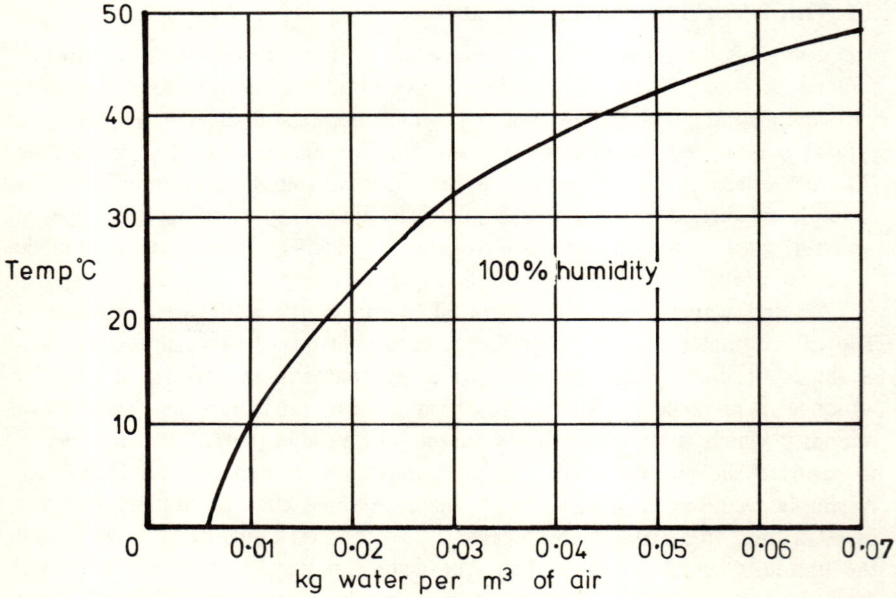

Fig. 3.2 – Effect of temperature on water vapour content.

the receiver is 25°C, the air can carry 0.022 kg water/m^3 or $0.022 \times 7.5 =$ 0.165 kg/min. The difference $0.714 - 0.165 = 0.549$ kg/min will separate out and must be extracted before the air goes to the working system.

Special separators can be used for this purpose, but quite often the receiver acts as an adequate separator. This usually takes the form of a large tank. It acts as a cooler, as an energy storing chamber, as an oil separator as well as water separator. A check valve is used to stop the receiver from blowing back through the compressor when the compressor is off-loaded.

The bulk of the contaminants are removed by the intake filter and the receiver but it is usually necessary to include additional safeguards in the form of air line filters. These are usually stationed at the actual machine. Many of these are so shaped that the incoming air is forced into a swirling motion. Contaminants, such as water, oil and dust are thrown outwards, impinge on the walls and fall to the bottom of the filter bowl from which they can be either manually or automatically drained at intervals. The air then passes through a filtering element, perhaps made of sintered bronze, before leaving the filter. Most industrial systems, hydraulic or pneumatic, can operate adequately with 25 μm filtration, but non-moving part fluidic systems require at least 5 μm filtration. Fluidic systems often use a silica gel purifier element to remove the last traces of water from the air.

It is also essential that all oil be removed from fluidic systems, but oil is necessary for lubrication purposes when conventional valves and motors are used. The filter, pressure regulator and lubricator are available as a standard package.

The remaining components in Fig. 3.1 are the pressure switch and the safety valve. The safety valve or relief valve protects the receiver from excessive pressures. The pressure switch is a pneumatic-electrical interface which turns off the compressor when a predetermined pressure has been reached. This avoids an unnecessary wastage of power when the load does not require it.

3.3 THE BASIC HYDRAULIC SYSTEM

Some of the necessary processing equipment is sketched in Fig. 3.3. The circuit is similar in many ways to that of Fig. 3.1, except that the hydraulic circuit is a closed one.

The reservoir or tank fulfils many functions besides that of containing sufficient oil to meet the system's demands. Fig. 3.4 shows a typical unit (c.f. Fig. 2.8). It acts as a cooler and also allows contaminants such as air or dirt to settle out. The tank's capacity is usually at least three times the pump's maximum delivery rate. In other words a pump delivering 20 l/minute would need a reservoir with a capacity of about 60 l. Cooling is improved if baffles are used to direct the oil along the walls of the tank. The inlet or suction strainer removes the larger debris (>100 μm). As discussed earlier, air bubbles can have

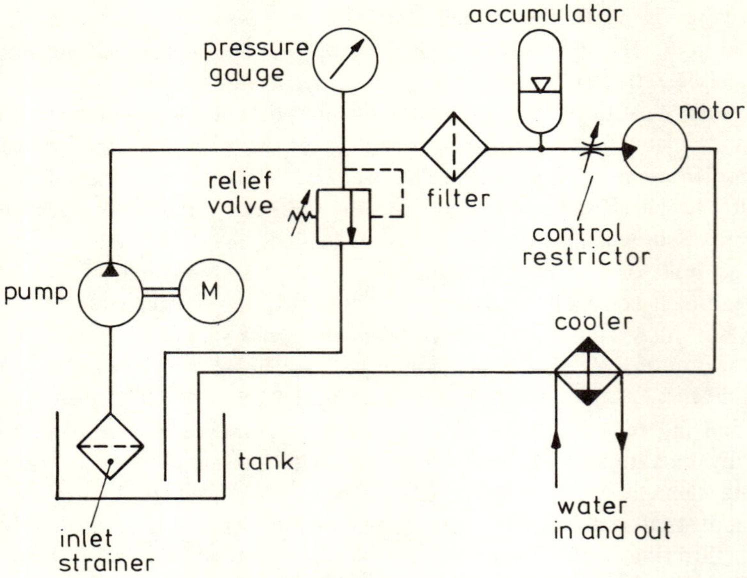

Fig. 3.3 – The basic hydraulic system.

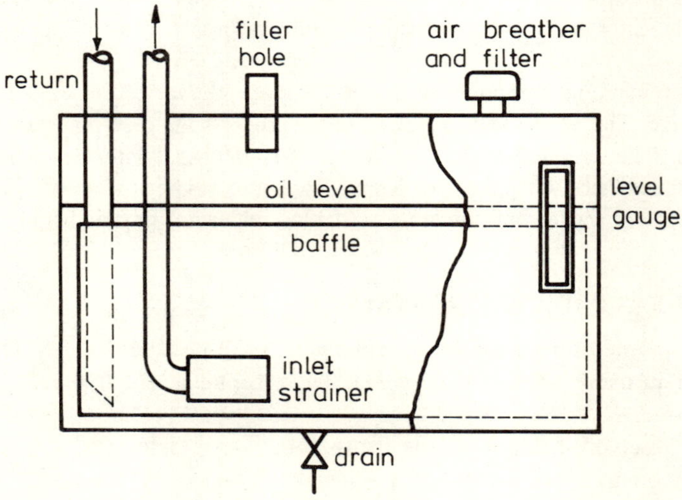

Fig. 3.4 – Layout of a typical hydraulic reservoir.

an adverse effect on the performance of a hydraulic system, and thus it is essential that all possible precautions be taken to avoid air ingress at the tank. There are several steps to be taken:

There are several steps to be taken:

(a) the return line is submerged so that aeration and foaming are minimised;
(b) the inlet is near the bottom so that an air vortex does not form;
(c) in some tanks the oil has to pass through an inclined wire gauze which collects air bubbles and allows them to rise to the surface of the oil.

As mentioned earlier further filtration is often required within the actual circuit, and the position of such filters within the circuit can be critical. They may be in the pump suction line, in the pressure line, in the return line or in a bypass. One must be careful to ensure that the pressure drop across a suction line filter does not cause cavitation at the pump inlet. The other filter positions have several advantages and disadvantages, but space does not allow a detailed discussion of them here.

Although the high bulk modulus of an oil is on the whole an advantage, occasions arise when a degree of compressibility is desirable. The hydraulic accumulator is an energy storage device, Fig. 3.5, which takes advantage of the high compressibility of a gas, usually air or nitrogen. The gas can be separated from the oil by a piston, by a diaphragm or by a rubber bag. The piston type has a large capacity but requires an excellent fit between piston and cylinder. The capacity of the bag type is limited, but it has the advantage of having no moveable mechanical parts. The use of an accumulator as a power reserve can allow a smaller pump to be used in systems which operate intermittently or irregularly. In addition it reduces water hammer effects, reduces or eliminates pump pulsation, and allows the system to function for a short time when the pump is stopped.

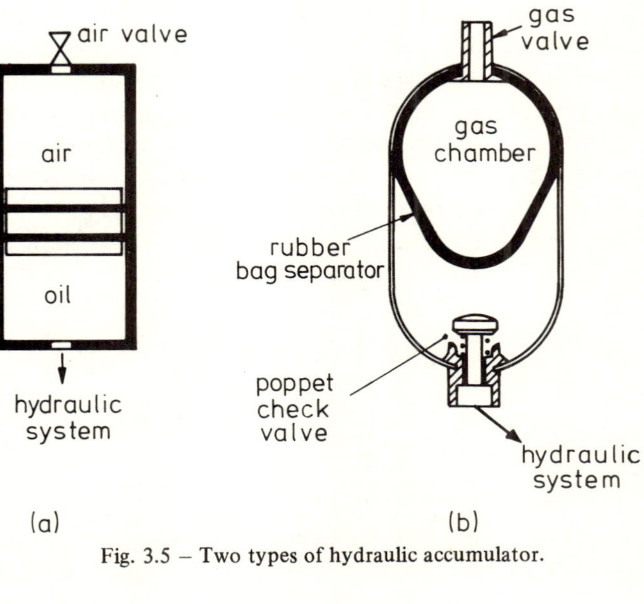

Fig. 3.5 – Two types of hydraulic accumulator.

Heat dissipation [2] is just as important in the hydraulic circuit as in the pneumatic one. Friction losses in components and, in particular, throttling in orifices are the major sources of heat generation. For example 0.73 kW are expended in forcing flow through a 1 mm orifice with a pressure drop of 70 bar. This energy is dissipated as heat, and if this is allowed to accumulate in the fluid, high temperatures can arise with subsequent damage to the fluid and the hydraulic seals. Best working temperatures are in the region 45–50°C. The oil is cooled to a certain extent in the tank and in the plumbing, but when this is not adequate a heat exchanger may be installed in the return line, Fig. 3.3.

3.4 AIR COMPRESSORS

The amount of air delivered by a compressor is usually quoted as a volume of free air, that is, air at normal atmospheric conditions. For example, if a load requires $0.2\,\text{m}^3/\text{min}$ of air at 7 bar gauge then, if isothermal compression is assumed, the required compressor capacity is $0.2 \times 8.013/1.013 = 1.582\,\text{m}^3/\text{min}$ of free air.

The ideal power required by a single stage compressor can be determined from the formula:

$$H = Q_f P_1 \left(\frac{n}{n-1}\right)\left(r^{(n-1)/n} - 1\right) \tag{3.1}$$

where Q_f = free air rating in m^3/s
P_1 = ambient pressure N/m^2
r = compression ratio (ratio of absolute pressures)
n = polytropic index or the index of compression.

In practice the actual power required would be up to 30 per cent more because of inefficiencies. A simple rule of thumb is that about 6 kW of power is needed to compress $1\,\text{m}^3/\text{min}$ of free air to 7 bar. The actual figure may be slightly higher for small compressors and lower for large compressors.

Wastage in pneumatic systems is often ignored, many engineers assuming that it is negligible, but reference to Fig. 3.6 shows that this can be an expensive assumption. The figure shows the required compressor capacity for an orifice load and illustrates the amount of air and power that can be lost through leaks in a system. For example at a pressure of 7 bar, $Q_f = 1.24\,\text{dm}^3/\text{s}$. Taking $n = 1.3$, $P_1 = 1$ bar and using equation (3.1) gives:

$$H = 1.24 \times 10^{-3} \times 10^5 \left(\frac{1.3}{1.3-1}\right)(8.013^{(1.3-1)/1.3} - 1)$$

$$= 0.340\,\text{kW}.$$

3.4.1 Non-positive displacement compressors

These are sometimes referred to as **aerodynamic compressors**. One such, a single stage centrifugal compressor, is shown in Fig. 3.7. The runner or impeller rotates

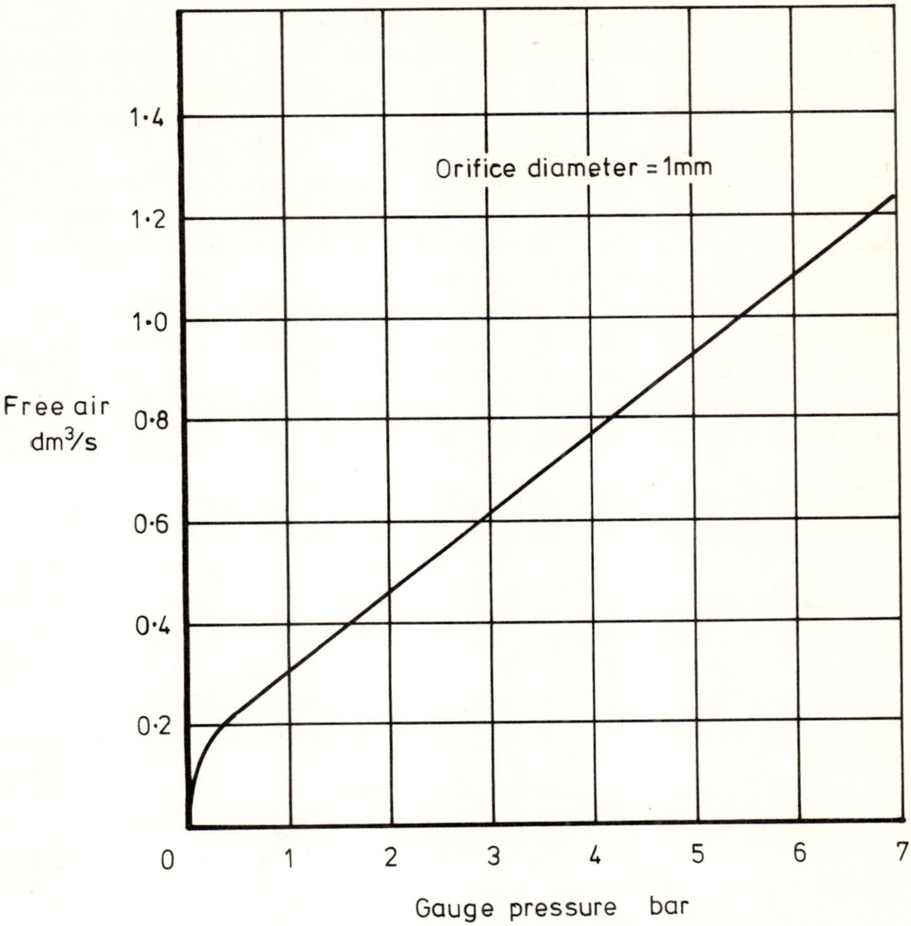

Fig. 3.6 – Air flow through a small hole.

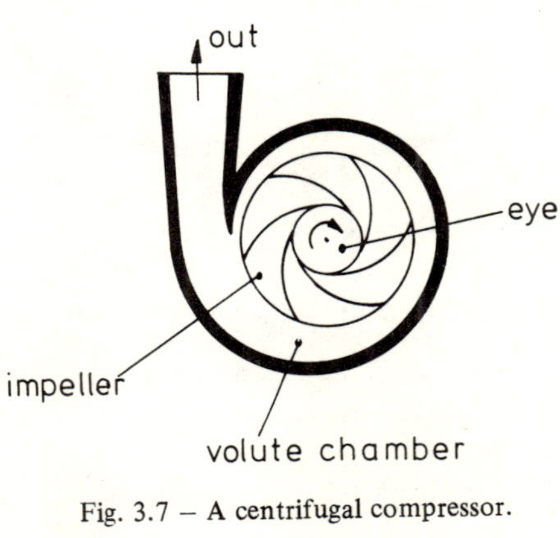

Fig. 3.7 – A centrifugal compressor.

inside a spiral casing. The inlet port is axial and the air enters the centre or eye of the compressor where it is accelerated and thown outwards by the blades. Its kinetic energy is gradually converted to potential energy by the diffuser action of the volute chamber so that the air arrives at the output with increased static pressure. A single stage compressor of this type can develop up to 0.5 bar, and depending on size can deliver up to 2500 m^3/min. They are quite adequate for the majority of systems using non-moving part fluidic elements and have the added advantage of providing oil-free air. Their delivery is load-sensitive, leakage causing the delivery to fall off with pressure.

The axial compressor also relies on the conversion of kinetic to potential energy, but in this case the air moves in a mean axial direction as it traverses a series of alternatively fixed and rotating blades. It has a very high delivery, up to 25 000 m^3/min and is usually found in jet engines, mines and quarries where such large flow rates are needed.

3.4.2 Positive displacement compressors

The Root's blower of Fig. 3.8 is a low pressure positive displacement compressor. It consists of two identical double lobed rotors which are driven in opposite directions by a pair of external gears. It finds use in diesel engine scavenging and supercharging. Pressures up to about 0.7 bar and deliveries up to 1200 m^3/min are possible.

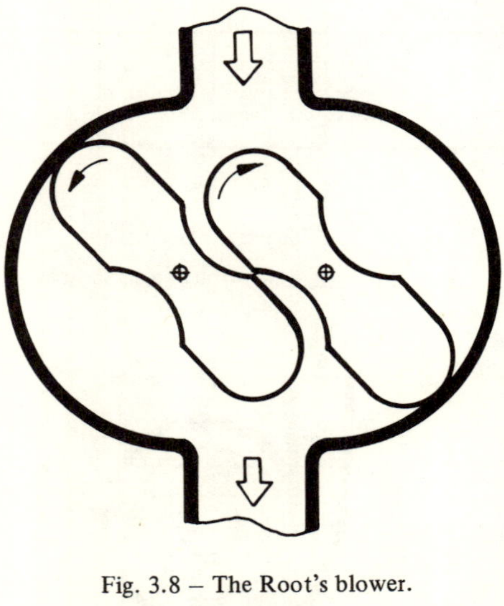

Fig. 3.8 – The Root's blower.

Higher pressures, up to about 7 bar, can be obtained from the rotary vane compressor, Fig. 3.9. The cylindrical slotted rotor contains spring loaded vanes which project radially. These are free to move in the slots and are held in contact with the casing by the combined effects of centrifugal action and spring force.

The rotor is placed eccentrically in the casing. As the vanes move from the region where rotor and casing are closest, the volume between a pair of vanes increases and air is drawn in through the suction port. Thereafter the volume decreases and the air is compressed and delivered to the outlet port. Leakage across the vanes and rotor can be reduced by mixing the incoming air with oil which acts as a sealant. The oil, especially in fluidic circuits, needs to be filtered out after compression.

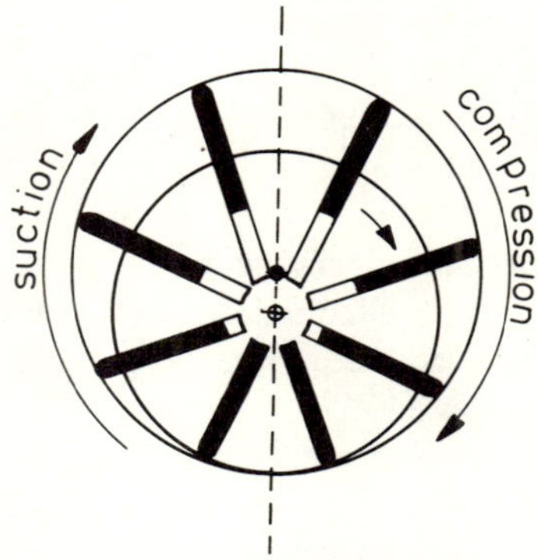

Fig. 3.9 – A vane compressor.

There are many types of reciprocating compressor [3]. In the single stage compressor maximum efficiency is achieved if the cylinder is kept as cool as possible, this approximating to isothermal compression which requires less work than adiabatic compression. Very high temperatures can be generated if the compression ratio is large, the temperature T after compression being related to the initial temperature T_1 by:

$$T = T_1 r^{(n-1)/n} .$$

Thus, if high compressor ratios are needed, multi-stage compressors are used with intercoolers between each stage.

Fig. 3.10 shows a two stage reciprocating compressor. The low pressure

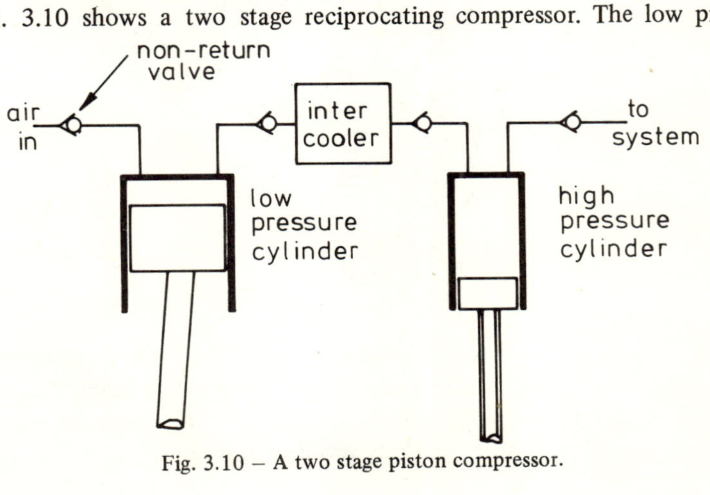

Fig. 3.10 – A two stage piston compressor.

cylinder usually raises the pressure to about 4 bar and the second stage raises it to 7 bar. Three stages are normally used for pressures in excess of 20 bar. The intercooler may consist of fixed tubes over which cold air is blown, or on larger compressors water cooling may be preferred.

3.5 HYDRAULIC PUMPS

Pumps are usually rated in terms of output flow rate at a given drive shaft speed. This can also be referred to as the pump's capacity or delivery. As with compressors, pumps fall into the two categories of positive and non-positive displacement, but since the latter type are rarely used in fluid power systems they will receive no further comment here. Unlike compressors, pumps can be very efficient in converting mechanical energy to fluid energy, and efficiencies in excess of 90 per cent are not uncommon. The power output of a pump is the product of pressure rise and flow rate. For example a pump delivery of 1000 cm^3/sec at 80 bar requires 8 kW.

The majority of positive displacement pumps use either gears, vanes or pistons. In its simplest form, Fig. 3.11, the gear pump consists of two meshing spur gears, usually with 8 or 10 teeth, enclosed in a closely fitting housing. Rotation of the gears from the inlet or suction port traps oil in the tooth spaces. As the teeth engage on the outlet side the volume between a pair of teeth decreases and the oil is forced out of the delivery port. As the teeth unmesh again, the volume increases and oil is drawn into the pump. Hence the actual pumping action takes place in the meshing area.

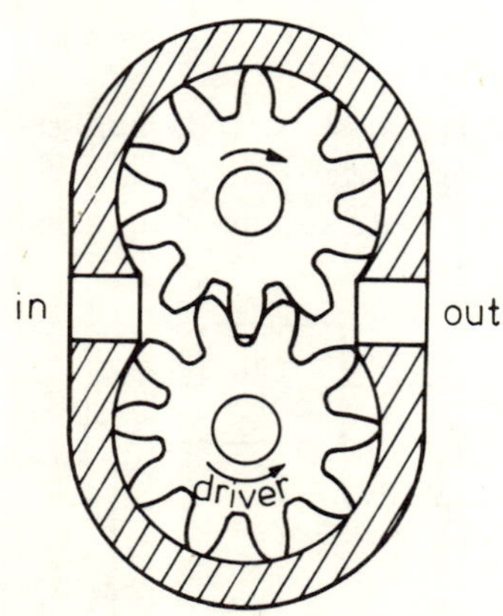

Fig. 3.11 – A gear pump.

The Gerotor pump, Fig. 3.12, is a form of internal gear pump in which the inner gear has one tooth less than the outer. The inner gear is placed eccentrically with respect to the outer and this gives rise to an alternative increase and decrease of the volumes of the pockets as the gears rotate.

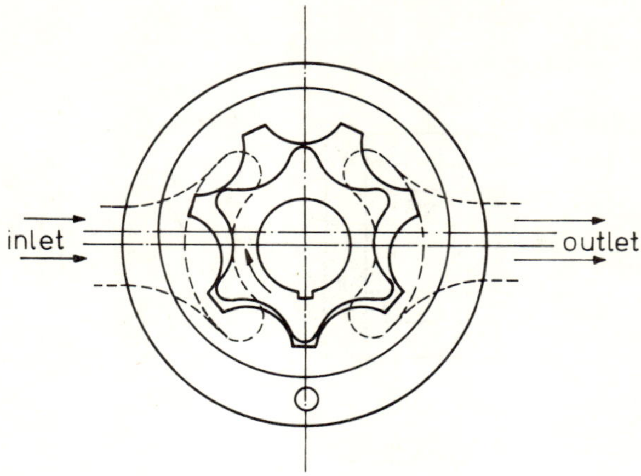

Fig. 3.12 – An internal gear pump.

The internal gear pump is particularly silent in operation. Gear pumps can deliver as much as 0.7 m^3/min and can develop up to 200 bar.

The operation of the hydraulic vane pump is similar to that of the vane compressor, Fig. 3.9. Pressures have generally been limited to 70 bar, but there are some two stage pumps that can develop 140 bar. Capacities are usually less than 0.3 m^3/min. The vane compressor of Fig. 3.9 is unbalanced in that the right-hand side is pressurised and the left is not. Thus the rotor experiences a side load which can cause excessive bearing wear in high pressure hydraulic systems. Balancing can be achieved by running the rotor concentrically in an elliptical chamber. This has two inlet and two delivery ports in diametrically opposite positions so that equal forces are generated on both sides of the rotor.

Piston pumps usually employ several small pistons reciprocating at high speeds. They are normally used for pressures in excess of 200 bar. There are many different types, including plunger pumps, radial pumps and axial pumps, but only the last will be discussed here. The axial piston pump is so called because the pistons move axially in the cylinder barrel. A swash plate pump is illustrated in Fig. 3.13. In such pumps piston reciprocation can be achieved by either:

(a) fixing the swash plate and rotating the cylinder block, or
(b) rotating the swash plate and fixing the cylinder barrel.

If the latter method is used then non-return valves can be used, as in Fig. 3.10, to control the flow of oil to and from the cylinders. If the former method is used a plate valve is needed to connect the cylinders to outlet and inlet at the correct times. This valve, Fig. 3.13, has two kidney shaped ports, one connected to outlet and the other to inlet. As the pistons revolve with the rotor they follow the path of the inlet kidney port and at the same time move further

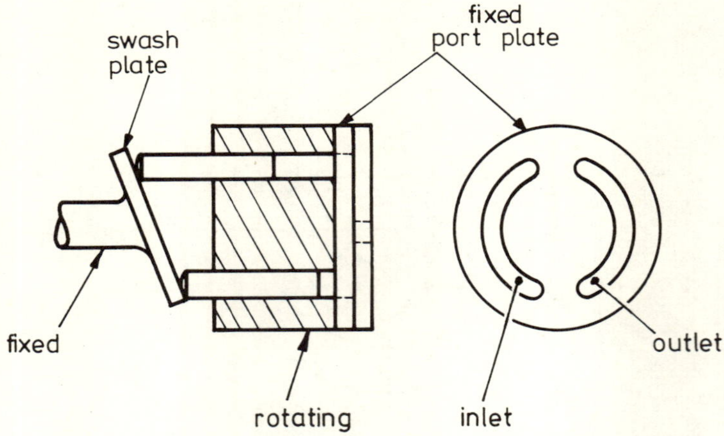

Fig. 3.13 – An axial piston pump.

out of the cylinders causing oil to flow from the inlet to the cylinders. Further rotation brings the pistons on to the path of the outlet kidney port and they now move into the cylinder, compresses the fluid and expelling it through the outlet kidney port.

The bent axis piston pump is shown in Fig. 3.14. In this case the cylinder is not in line with the drive shaft.

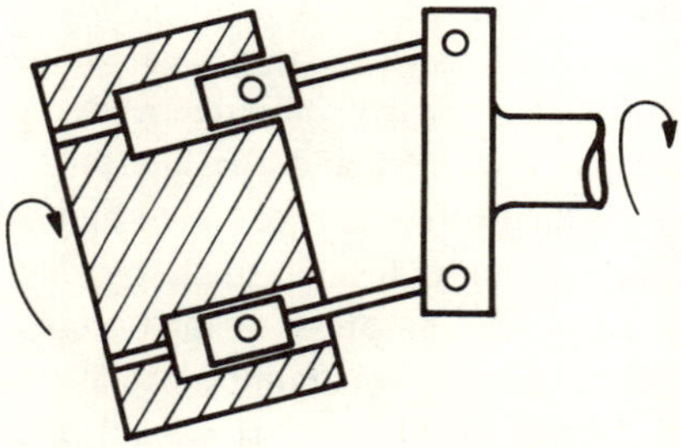

Fig. 3.14 – A bent axis pump.

The speed and the swept volume determine a pump's delivery. Delivery can be varied by changing the swept volume, and this can be done relatively easily with vane pumps and piston pumps. Unbalanced vane pumps can have a screw arrangement for varying the eccentricity between rotor and housing. When the eccentricity is zero the delivery is zero. Variable delivery swash plate pumps include a mechanism for varying the angle of the swash plate. This can be operated manually or by a pressure feedback loop within the pump as shown in Fig. 3.15. This is known as a pressure compensated pump. The precompression of the spring can be adjusted to give a sharp reduction in delivery when a certain pressure is exceeded.

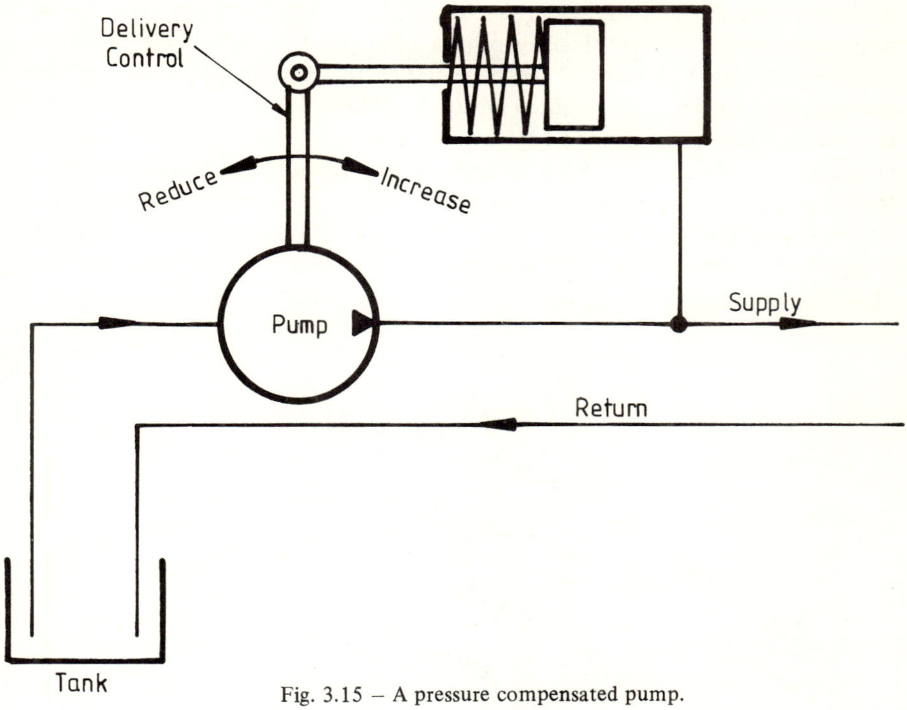

Fig. 3.15 – A pressure compensated pump.

Fig. 3.16 is a sectional view of a commercially available variable delivery pump capable of 6000 rev/min at 27 bar. The rotor, which is driven by the drive shaft, contains the pistons. The slipper heads of the pistons ride on the face of the adjustable cam plate and are retained in this position by the auxiliary cam plate. Variation of the angle of the adjustable cam plate changes the piston stroke and hence the flow rate output of the pump.

The stroke is controlled by a servo control mechanism. When, for example, the manual servo rod is moved inwards, the pilot piston uncovers a port allowing oil to pass from the small to the large area end of the servo piston. Thus the servo piston moves with the pilot piston and reduces the cam plate angle. Any fluid leaking past the pilot piston is returned to the pump body by means of a drain.

The relief valve sets an upper limit on system pressure but it is also a notorious heat generator, particularly when used in a circuit which uses a fixed displacement pump. It relieves the system to tank when a predetermined pressure is exceeded, but the relief flow passes to tank at maximum pressure drop and hence much heat is generated. As an extreme example, assume that the control restrictor of Fig. 3.3 is closed. The output from the fixed displacement pump must all pass back to tank through the relief valve, and in fact all of the hydraulic power developed by the pump is dissipated as heat in the relief valve. For this reason it is often necessary to direct the outflow from the relief valve to the heat exchanger. Greater efficiency can be achieved if a variable displacement pump is used since its output can be matched to the load. If in this case the control valve is closed, the pump delivery could be reduced to zero by changing the swash plate angle. Thus no heat would be generated in the relief valve. The pressure compensated pump, discussed earlier, unloads itself automatically. In spite of the possible inefficiency of the fixed displacement pump circuit, it is very popular for the lower power ranges (<12 kW). It can be much more compact than the variable delivery system. The variable delivery pump is more often found in high power systems, particularly in hydrostatic transmissions.

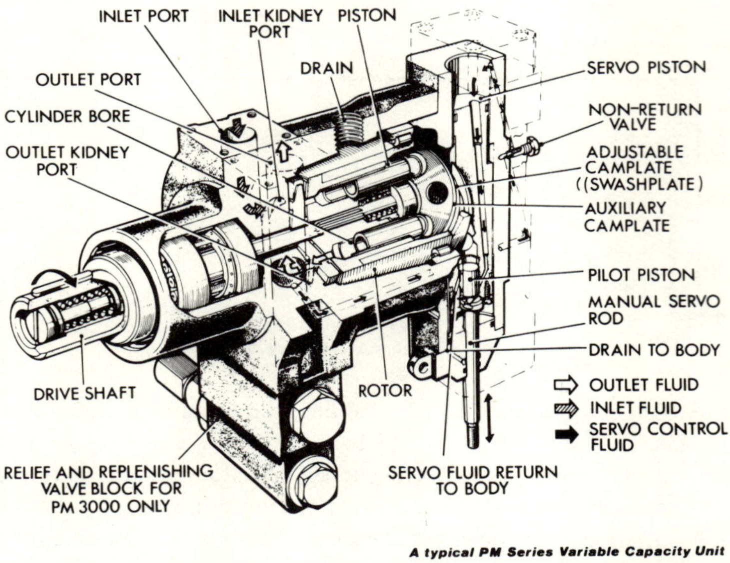

Fig. 3.16 – A variable delivery pump (Lucas Industrial Equipment Ltd.)

3.5.1 Pump efficiency [4]

The volume of fluid that the pump is theoretically capable of delivering can be determined from the geometry of the pump. For example in the axial piston pump of Fig. 3.13 the stroke of each piston is $D \tan \alpha$ where D is the pitch diameter of the cylinders and α is the angle of the swash plate to the vertical. Thus if N is the number of cylinders and d the diameter of each cylinder, the pump displacement D_p, defined as the delivery per radian of the input shaft, is given by:

$$D_p = \frac{N(\pi d^2/4)\, D \tan \alpha}{2\pi} \tag{3.2}$$

$$= 0.125\, d\, D\, N \tan \alpha\ .$$

Similar expressions can be derived for other pumps. For a vane pump, Fig. 3.9, the displacement is:

$$D_p = bDe \tag{3.3}$$

where b = width of a vane
D = internal diameter of the stator
e = eccentricity of the rotor within the stator.

Again, for the gear pump of Fig. 3.11:

$$D_p = 0.5\,b(r_a^2 - r_d^2) \tag{3.4}$$

where r_a = addendum radius of the gears
r_d = dedendum radius of the gears.

Other pump configurations are dealt with in references [2, 3].

If the pump displacement is known, the theoretical flow rate can be determined from:

$$Q_t = \omega D_p \tag{3.5}$$

where ω = angular velocity in rad/s.

Owing to leakage, the pump never quite delivers this theoretical flow rate, and the difference is expressed by the volumetric efficiency η_v:

$$\eta_v = \frac{\text{actual delivery rate}}{\text{theoretical delivery rate}} \tag{3.6}$$

Volumetric efficiencies usually lie in the range 80–90 per cent, being highest for piston pumps.

Turning now to a consideration of the torque T required to drive the pump we have:

$$\text{Mechanical power supplied to pump, } H_m = T\omega \tag{3.7}$$

$$\text{Hydraulic power delivered by pump, } H = PQ \tag{3.8}$$

where P = pressure rise across the pump ($\cong P_s$)
Q = delivery rate.

For an ideal, 100 per cent efficient pump:

$$T_t\omega = P_sQ_t = P_s\omega D_p$$

$$T_t = P_sD_p\,. \tag{3.9}$$

In practice the actual torque required to drive the pump will be greater than T_t and the difference is expressed by the mechanical efficiency η_m:

$$\eta_m = \frac{\text{ideal torque required}}{\text{actual torque required}} . \tag{3.10}$$

Mechanical efficiencies are usually in the range 80-95 per cent.

The overall efficiency η of the pump is given by:

$$\eta = \frac{\text{hydraulic power out}}{\text{mechanical power in}} = \eta_v \eta_m . \tag{3.11}$$

The following example will illustrate the use of these formulae.

Example

A hydraulic pump of displacement 5 cm^3 develops 12.5 kW hydraulic power when operating at 150 bar and 1800 rev/min. Under these conditions 15 kW are required to drive the pump. Determine: (i) the mechanical efficiency, (ii) the leakage flow rate.

(a) Theoretical torque needed $= PD_p = 150 \times 10^5 \times 5 \times 10^{-6} = 75$ N m.

Actual torque needed $= H_m/\omega = 15 \times 10^3/60\pi = 79.58$ N m.

Hence mechanical efficiency $\eta_m = 75 \times 100/79.58 = 94.2\%$.

(b) The overall efficiency $\eta = 12.5 \times 100/15 = 83.3\%$.

Hence volumetric efficiency $\eta_v = \eta/\eta_m = 83.3/94.2 = 0.885$.

$$\text{Now } \eta_v = \frac{\text{theoretical flow rate} - \text{leakage}}{\text{theoretical flow rate}}$$

$$\text{or } 0.885 = \frac{\omega D_p - \text{leakage}}{\omega D_p}$$

$$= \frac{60\pi \times 5 \times 10^{-6} - \text{leakage}}{60\pi \times 5 \times 10^{-6}}$$

therefore leakage $= 1.084 \times 10^{-4}$ m^3/s.

3.6 MOTORS AND ACTUATORS

Motors and actuators perform the opposite function of pumps and compressors in that they convert fluid energy back to mechanical energy so that useful work can be accomplished. The name **motor** is usually given to a device which produces a continuous rotary output, whilst **actuator** describes those devices which produce a limited linear or rotary output.

Many hydraulic pumps can be run as motors, and gear, vane and piston motors form the majority of the many types on the market today [3]. Motor speed is controlled by varying the flow rate of pressurised oil to the motor, or by varying the motor's swept volume.

The analysis of the efficiency of hydraulic motors is similar to that for pumps. The efficiencies are now:

$$\text{Mechanical efficiency } \eta_m = \frac{\text{actual torque developed } (PDp)}{\text{ideal torque developed}} \tag{3.12}$$

$$\text{Volumetric efficiency } \eta_v = \frac{\text{theoretical flow rate required } (\omega Dp)}{\text{actual flow rate required}}$$

$$\text{Overall efficiency } \eta = \eta_m \eta_v . \tag{3.13}$$

There is also a wide variety of pneumatic motors with output powers in the range 0–20 kW. Piston motors are usually preferred for heavier loads or where a good low speed torque is essential. Vane air motors are popular for medium power duties (<12 kW) and are used in portable tools, hoists, conveyor belt drives etc. Very high speeds can be achieved by the small turbine motors. For example pencil type grinders can be driven at 80 000 rev/min by these motors, and dental drills can go up to 700 000 rev/min.

The simplest linear acuator is the single acting cylinder or ram which is pressure energised in one direction only, Fig. 3.17. In some cases the return motion or retraction stroke may rely on the load or on gravity, Fig. 3.17(a), and in others a spring may be used, Fig. 3.17(b). Since the cylinder has to work against the spring load, spring return is usually only used in short stroke applications such as clamping. Sometimes in hydraulic servomechanisms the rod end is maintained at supply pressure thus providing a return force.

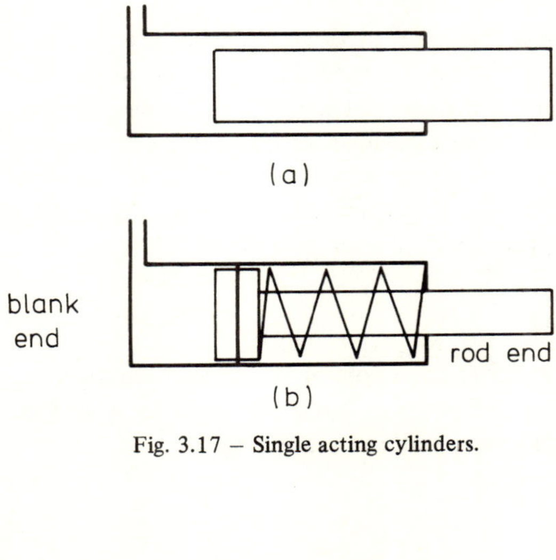

Fig. 3.17 – **Single acting cylinders.**

If control is required during both extension and retraction of the piston rod, then it is necessary to use a double acting cylinder, Fig. 3.18. This is usually driven by a valve which can raise the pressure on one side of the piston whilst simultaneously lowering it on the other side. The cylinder in Fig. 3.18 is a double rod cylinder and requires a space about three times its stroke, for installation. A single rod cylinder, like that of Fig. 3.17(b), can also be operated in a double acting mode. It requires less space and is probably the most commonly used type. However, because of the different piston areas on each side, the forces and speeds developed during extension and retraction will be different, and as a result the double rod cylinder tends to be the preferred arrangement for high performance servo-systems.

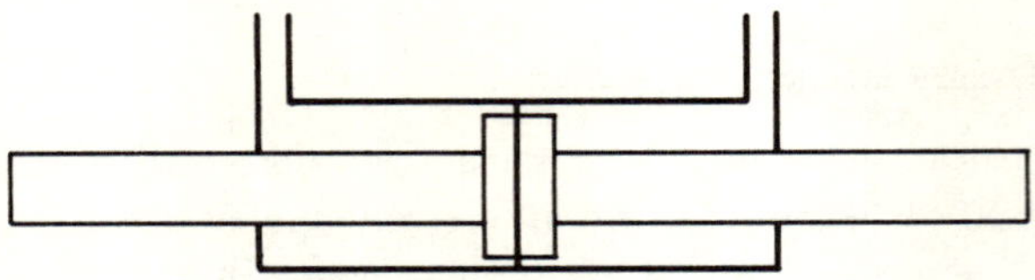

Fig. 3.18 – A double acting cylinder.

It is often necessary to include a deceleration cushion at the end of the stroke. This dissipates the kinetic energy of the load gently and reduces the possibility of mechanical damage to the cylinder, especially if fast acting cylinders are used. Fig. 3.19 illustrates the principle of operation. As the rod moves to the left the spigot enters the cushioning chamber and blocks the fluid's main escape path. The fluid must force its way past the needle valve whose setting determines the degree of deceleration. The load's kinetic energy is dissipated as heat. A non-return or check valve is also necessary in double acting cylinders so that, on reversal, the check valve unseats, allowing unrestricted flow into the cylinder when the spigot is in the cushioning chamber.

It was shown in Fig. 1.3 that there are many external means of converting the linear output motion of a cylinder to a rotary output. There are, however, some actuators on the market that are specifically designed to give a restricted rotary output. Two types are shown in Fig. 3.20. The vane type, Fig. 3.20(a), can be either single or double acting, but its output rotation is restricted to about 300°. The rack and pinion arrangement of Fig. 3.20(b) allows several rotations of the output shaft.

Example

A double acting cylinder has a bore of 50.8 mm, a rod diameter of 25.4 mm and a stroke of 254 mm. What air compressor capacity is needed to operate this cylinder at 20 cycles per minute at 4.12 bar (g)?

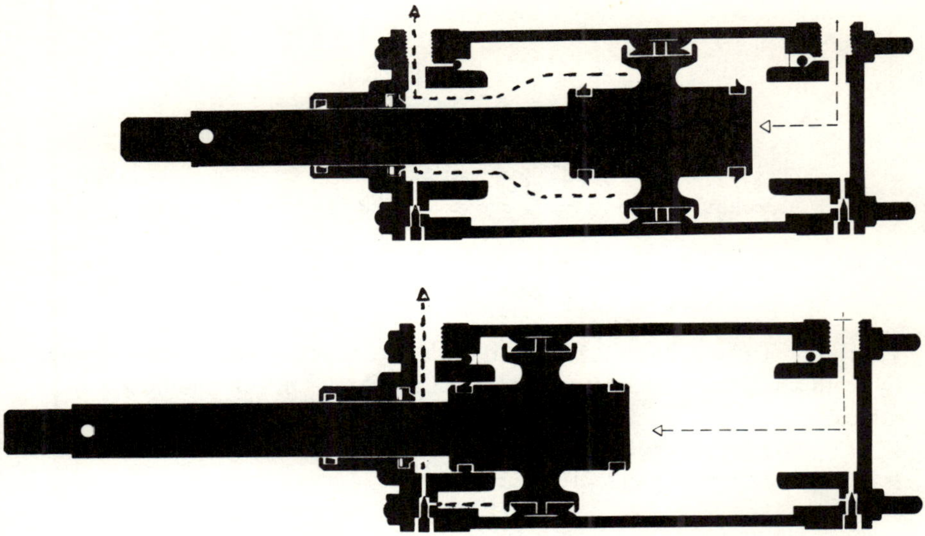

Fig. 3.19 – The cushioning mechanism (Martonair Ltd.)

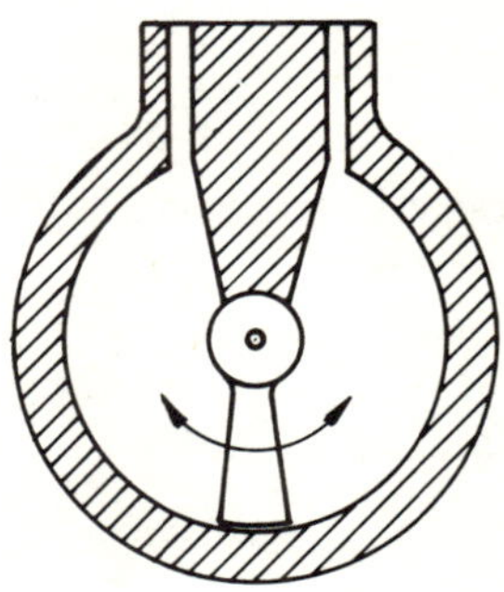

(a)

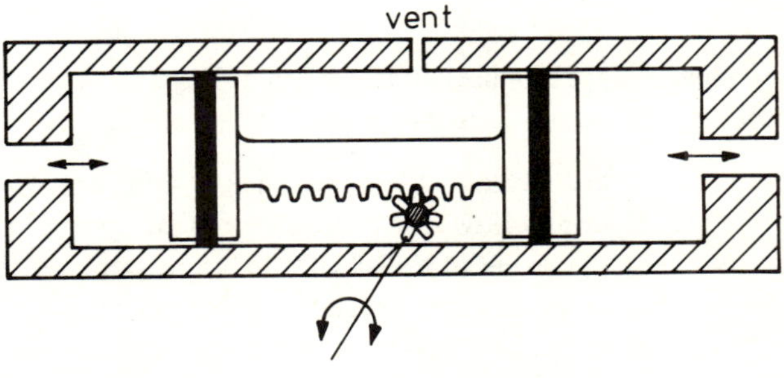

(b)

Fig. 3.20 – Means of achieving a restricted rotary motion.

$$\text{Volume of cylinder} = \frac{\pi \times (50.8)^2}{4} \times 254 \times 10^{-9}$$

$$= 5.14 \times 10^{-4}\,\text{m}^3.$$

$$\text{Volume of rod} = \frac{\pi \times (25.4)^2}{4} \times 254 \times 10^{-9}$$

$$= 1.286 \times 10^{-4}\,\text{m}^3.$$

Thus each stroke requires $3.854 \times 10^{-4}\,\text{m}^3$ of compressed air. A cycle requires twice this amount. Hence in one minute, $1.542 \times 10^{-2}\,\text{m}^3$ of compressed air is needed.

Compressors are rated in terms of free air, so it is necessary to apply the perfect gas law (equation (2.2)) in the form:

$$PV/T = \text{constant} .$$

Here $P_1 = 4.12 + 1.013 = 5.133$ bar
$P_2 = 1.013$ bar
$V_1 = 1.542 \times 10^{-2}\,\text{m}^3$.

It is assumed that $T_1 = T_2$, thus V_2 can easily be calculated:

$$V_2 = 7.814 \times 10^{-2}\,\text{m}^3 .$$

Thus the requirement compressor capacity is $7.814 \times 10^{-2}\,\text{m}^3/\text{min}$.

3.7 EXAMPLES

3.7.1

The capacity of a single stage compressor is 2 m^3/min. It draws in air at 25°C and 75 per cent humidity and compresses it to 7 bar.

(a) Determine the delivered flow rate if the index of compression n O
(a) Determine the delivered flow rate if the index of compression $n = 1.3$.
(b) Determine the temperature of the air after compression.
(c) Determine the amount of water vapour drawn in per minute.
(d) If the temperature of the receiver is 30°C determine the rate at which water separates from the air.

(*Answers*: (a) 0.404 m^3/min; (b) 208.53°C; (c) 0.033 kg; (d) 0.0217 m^3/min.)

3.7.2

(a) A gear pump has the following dimensions:

Addendum radius	= 50 mm
Dedendum radius	= 35 mm
Width	= 25 mm

What is its displacement?

(b) Tests show that the pump's volumetric efficiency and mechanical efficiency are respectively 80 and 90 per cent when turning at 1000 rev/min and delivery oil at 100 bar. Determines the delivered flow rate and the leakage flow rate under these conditions.

(c) Assuming that leakage is proportional to pressure and that mechanical losses are proportional to the angular velocity of the pump, determine the volumetric efficiency, mechanical efficiency and power required to drive the pump at 1500 rev/min while developing 140 bar.

(*Answers*: (a) $1.594 \times 10^{-5}\ m^3/rad$; (b) $1.3352 \times 10^{-3}\ m^3/s$, $0.3338 \times 10^{-3}\ m^3/s$, (c) 81.35%, 89.36%, 39.23 kW.)

3.7.3

A hydraulic pump drives a hydraulic motor supplying 3.7 kW at 500 rev/min to a machine. The pump delivery pressure is 52 bar and there is a pressure drop of 0.4 bar in the lines joining the pump to the motor. Assuming volumetric efficiencies of 90 per cent and mechanical efficiencies of 90 per cent for both pump and motor determine:

(a) the displacement of the motor;
(b) the delivery rate of the pump;
(c) the power required to drive the pump .

(*Answers*: (a) $1.521 \times 10^{-5}\ m^3/rad$; (b) $8.849 \times 10^{-4}\ m^3/s$; (c) 5.681 kW.)

3.8 REFERENCES

[1] Van Dijen, F. S. G. (1975), *Pneumatic Mechanisation*, Kemperman, Technical Publishers.

[2] Bradbury, F. (1972), *Hydraulic Systems and Maintenance*, Iliffe.

[3] Wolansky, W. D., Nagohosian, J., and Henke, R. W. (1975), *Fundamentals of Fluid Power*, Houghton Mifflin Co., Boston.

[4] Blackburn, J. F., Reethof, G., and Shearer, J. L. (1960), *Fluid Power*, MIT Press.

CHAPTER 4

Fluid controlling elements

4.1 INTRODUCTION

The variable orifice is the most popular device for controlling fluid flow. It is the fluid equivalent of the electrical resistor and like the resistor its use leads to energy dissipation. The resultant heat generation can require the use of special cooling circuits. Nevertheless the simplicity, reliability and ease of manufacture of the variable orifice still remain as overriding advantages.

The orifice can be used in analogue or discontinuous modes, depending on application. For example in hydraulic servomechanisms used for accurate template following in machine tools, the valve will include orifices which are opened and closed in a proportional manner and could then theoretically assume any one of an infinite number of positions. On the other hand many pneumatically powered systems only have to be capable of full cylinder extension or retraction, and hence the controlling orifices need only be fully open or fully closed.

Other controlling elements use the properties of jets. These include jet tube devices, jet pumps and jet pipe valves. Flueric devices also fall into this category, but their most distinguishing feature is their lack of moving parts which gives them the potential for long life and high reliability.

4.2 VALVE ELEMENTS

4.2.1 Shearing elements

Most common is the shearing action, which is the basis of operation of the spool valve, Fig. 4.1(a). Here the moving element is cylindrical, that is, the spool, and moves in a bore or sleeve containing holes or ports. Movement of the spool causes the 'land' to vary the area of port exposed to the flow, and if the pressure drop across the port is constant this will result in changes in flow rate through the valve. The same shearing action is present in the plate valve which is essentially an 'unwrapped' spool valve, Fig. 4.1(b). The rotary spool and rotary plate valves convert rotational inputs to changes in flow rate, Figs. 4.1(c) and 4.1(d).

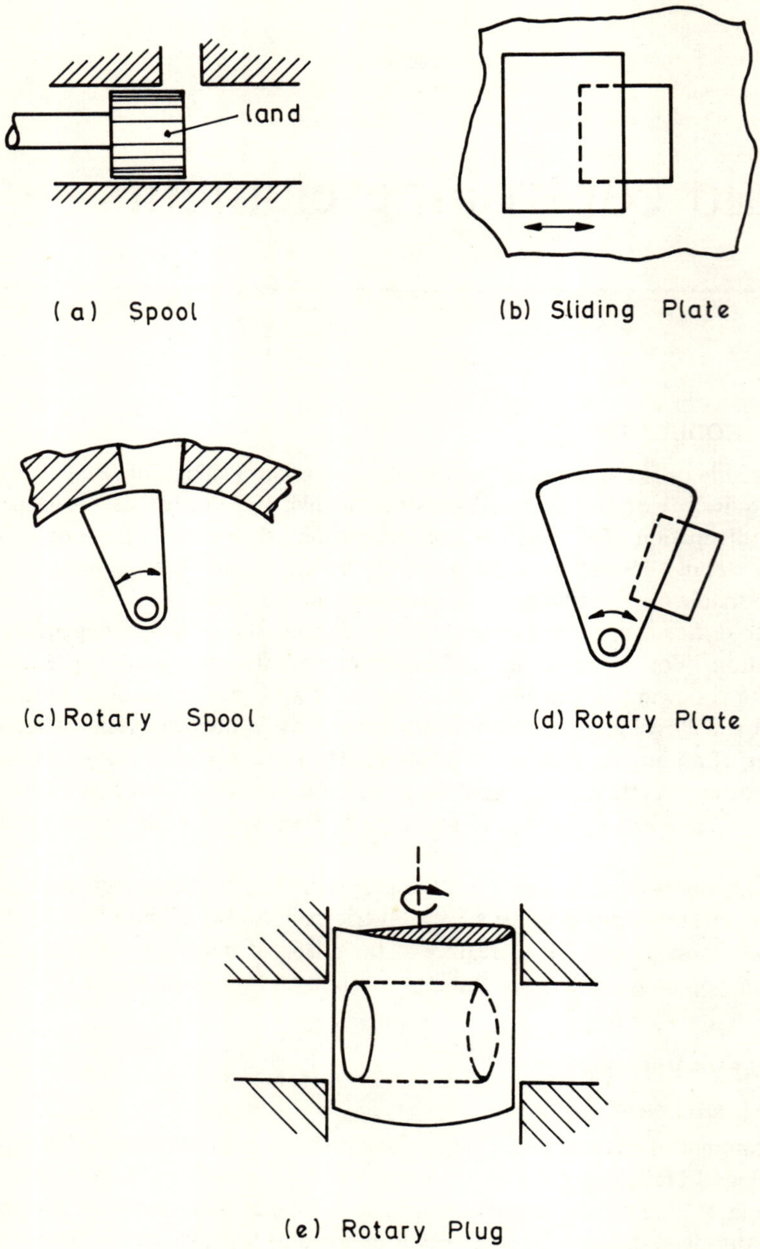

Fig. 4.1 – Examples of shearing elements.

The rotary plug valve, of which the glass stop cock is an example, is rarely used in power circuits. All shearing action valves are prone to damage by foreign bodies in the fluid. In addition, the relative movement caused by the shearing action necessitates a certain amount of clearance between the two surfaces, and this can in some cases lead to leakage problems. Hydrodynamic action in the clearance spaces of spool and rotary plug valves may lead to hydraulic lock, a condition in which it is very difficult to move the valve element.

4.2.2 Seating elements

Some valves using seating elements are shown in Fig. 4.2. These can provide a very good flow shut off and have the added advantage of being self cleaning. In addition a small movement of the seating element can expose a large orifice area to the flow, and this ability has made this type of valve element popular in relief

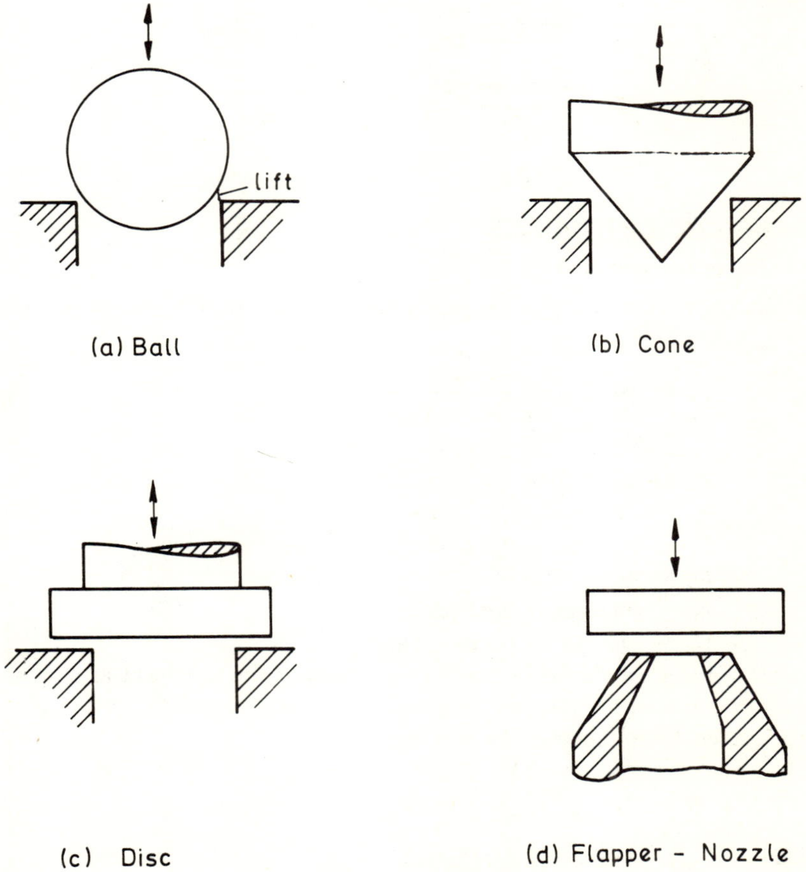

Fig. 4.2 – Examples of seating elements.

valves. Unfortunately seat wear can occur, causing changes in the flow rate at small lifts and precluding the use of such valves as accurate flow metering devices. Their main field of application is in simple on-off controls, but the flapper nozzle combination is often used in low power analogue systems.

4.2.3 Jet elements

The majority of flueric devices can be described as valves with jet elements. A device with a moving jet element is illustrated in Fig. 4.3. This is the jet pipe valve or Askania valve. Its action relies on the conversion of the dynamic pressure of the jet into static pressure at the load. Because of clearance between the receiver and jet pipe, a continual leakage flow exists, and this disadvantage restricts the use of this valve to low power applications.

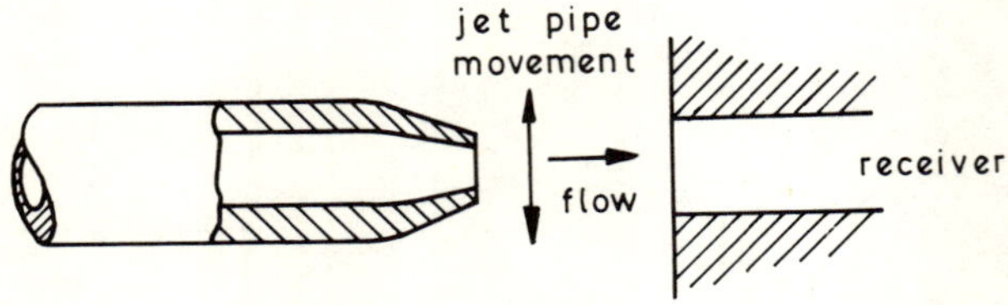

Fig. 4.3 – The basic jet pipe valve.

4.3 THE VARIABLE ORIFICE

4.3.1 The orifice equation

The volumetric flow rate Q of an incompressible fluid through an orifice is given by:

$$Q = \frac{C_c C_v a}{\sqrt{1-(C_c a/A)^2}} \sqrt{\frac{2(P_u - P_{vc})}{\rho}} \tag{4.1}$$

where a = orifice area
A = upstream area
P_u = upstream static pressure
P_{vc} = static pressure at the vena contracta
C_c = contraction coefficient
C_v = velocity coefficient, Fig. 4.4.

In most orifices employed in fluid power systems the upstream area is much greater than the orifice area and hence velocity of approach effects are negligible. In such cases the volumetric flow rate is given by:

$$Q = C_d a \sqrt{(2\Delta P/\rho)} \tag{4.2}$$

where C_d = discharge coefficient = $C_c C_v$
$\Delta P = P_u - P_{vc}$.

The orifice equation (4.1) was derived using the assumption that the fluid was incompressible. This is a reasonable assumption for the conventional hydraulic fluids. However, many modern control systems employ gases, particularly air, as the working fluid, and in these cases compressibility is an important factor. There are many excellent [1, 2] works dealing with the flow of gases through orifices and so it is unnesessary in the present book to go into compressible fluid dynamics. Suffice it to give the relevant formulae which have been shown

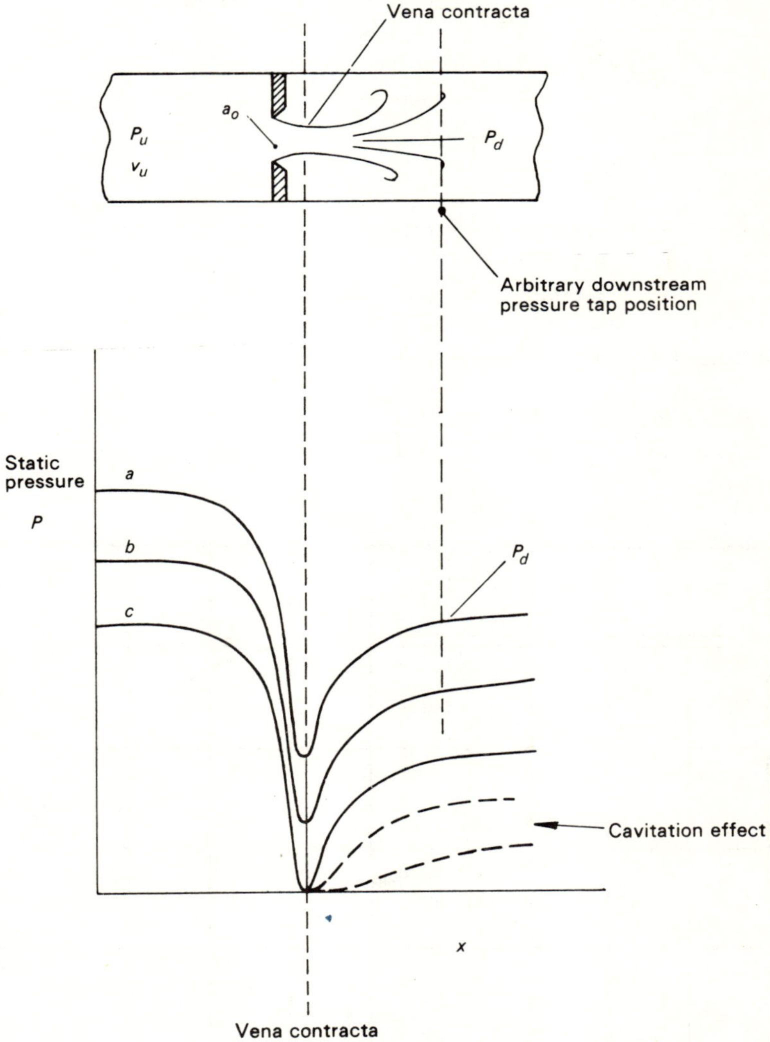

Fig. 4.4 – Static pressure distribution in a sharp edged orifice.

to predict mass flow rates fairly accurately. Mass rather than volume flow rate is generally used, since in the case of compressible flow, volume and pressure are interdependent. The mass rate of flow $\dot{M}$ can be written:

$$\dot{M} = C_d C_m a P_u^* / \sqrt{T_u} \tag{4.3}$$

where C_d = discharge coefficient
P_u^* = upstream stagnation pressure (absolute)
T_u = upstream stagnation temperature (absolute).

C_m, the mass flow parameter is given by:

$$C_m = \sqrt{\left[\frac{2\gamma}{R(\gamma-1)}\left\{\left(\frac{P_{vc}}{P_u^*}\right)^{2/\gamma} - \left(\frac{P_{vc}}{P_u^*}\right)^{(\gamma+1)/\gamma}\right\}\right]} \tag{4.4}$$

As before P_{vc} = static pressure at the vena contracta or throat
R = gas constant
γ = ratio of the specific heats (1.4 for air).

Equation (4.4) applies only when:

$$\frac{P_{vc}}{P_u^*} > \left(\frac{2}{\gamma+1}\right)^{\gamma/(\gamma-1)} = 0.528 \text{ for air.}$$

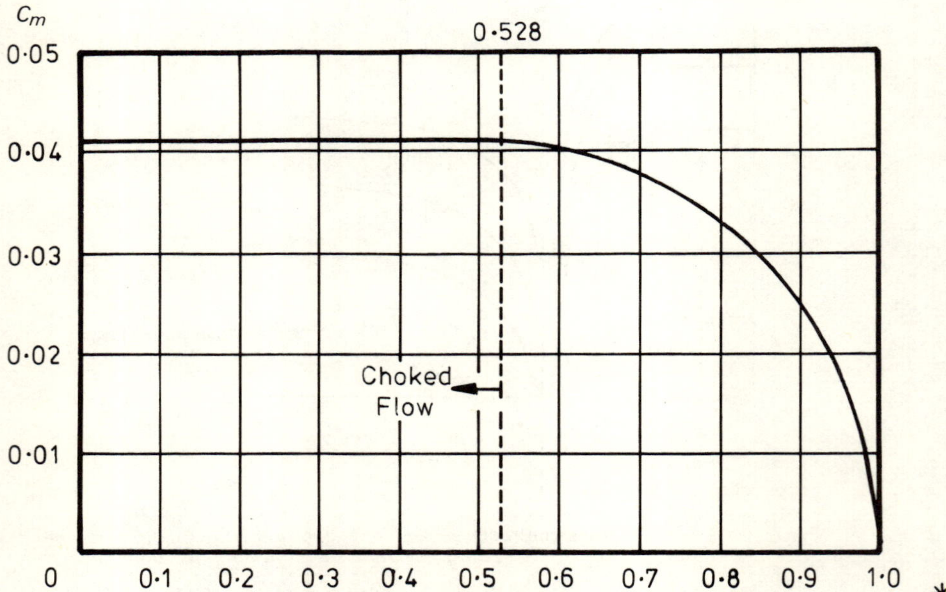

Fig. 4.5 – Variation of C_m with pressure ratio.

When P_{vc}/P_u^* is less than this value, the flow is said to be **choked** and C_m becomes:

$$C_m = \left(\frac{2}{\gamma+1}\right)^{1/(\gamma-1)} \sqrt{\frac{2\gamma}{R(\gamma+1)}} = 0.0405 \text{ for air.} \tag{4.5}$$

Under choked conditions, if P_u^* is fixed, decrease in P_{vc} will not result in an increase in $\dot{M}$.

The variation of C_m with pressure ratio is shown in Fig. 4.5.

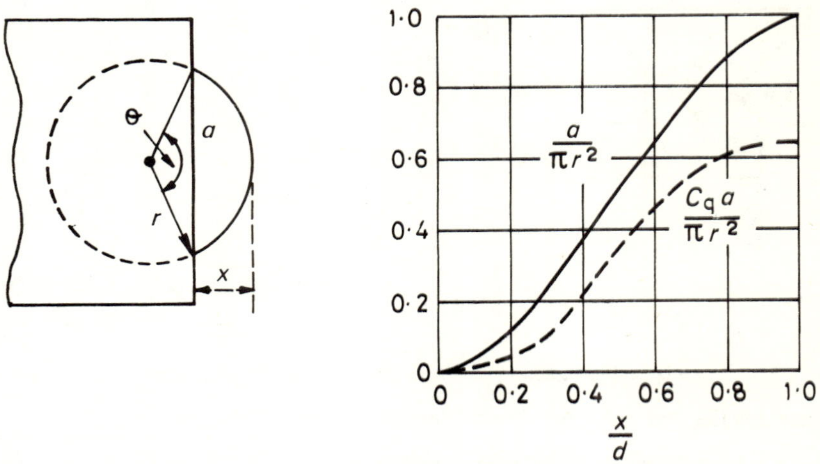

Fig. 4.6 – Variation of area and effective area with spool movement.

4.3.2 Area-displacement gain

The simplest shearing valve is a spool uncovering a circular port, Fig. 4.6. If the port is of radius r and the spool displacement from the closed position is x, the uncovered orifice area is:

$$a = (\theta - \sin\theta)\, r^2/2 \tag{4.6}$$

where $$\theta = 2\cos^{-1}[(r/x) - 1]\,. \tag{4.7}$$

The area-displacement characteristic is plotted in Fig. 4.6, and is clearly non-linear [1]. The slope of the curve, sometimes referred to as the **area-displacement gain**, varies with displacement and is a minimum at the closed and fully open positions. Such a variation can be undesirable in servo systems (they are gain sensitive) and hence rectangular ports with their linear characteristics are often used in spite of the increased cost of manufacture.

The conical poppet, Fig. 4.7, is an example of a seating valve. In this case the minimum flow area is the frustrum of a cone, the frustrum length being $s = x \sin\alpha$, where x is the poppet lift and α is the half poppet angle. The average circumference C of the frustrum is:

$$C = \pi(d - x \sin\alpha \cos\alpha) \tag{4.8}$$

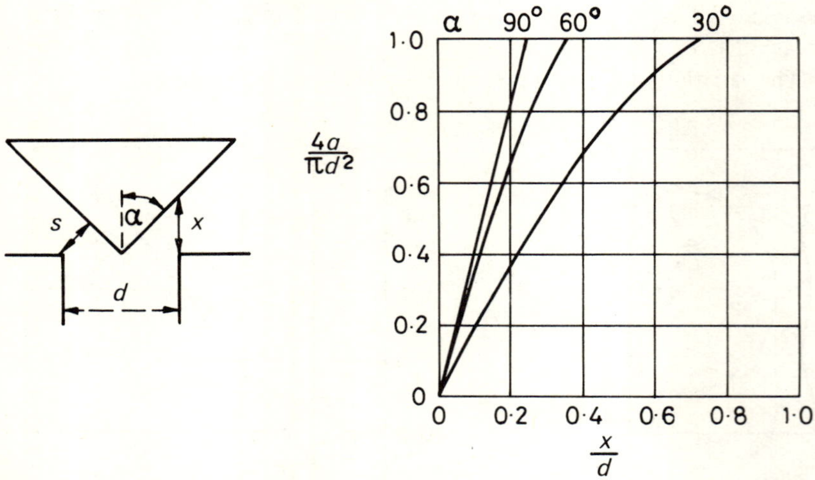

Fig. 4.7 – Variation of orifice area with poppet lift and angle.

where d is the seat diameter. The orifice area is then:

$$a = sC = \pi x \sin\alpha\,(d - x \sin\alpha \cos\alpha) \tag{4.9}$$

and normalising with respect to the area gives:

$$\frac{4a}{\pi d^2} = \left(4\,\frac{x}{d}\right)\sin\alpha\left(1 - \left(\frac{x}{d}\right)\sin\alpha\cos\alpha\right) \tag{4.10}$$

This is plotted in Fig. 4.7. A poppet valve is considered to be fully open when the orifice area equals the seat area, $4a/(\pi d^2) = 1$. The characteristics are non-linear except for the case $\alpha = 90^\circ$ (the flapper nozzle valve). The gain can be obtained by differentiating equation (4.6) with respect to x giving:

$$\frac{\mathrm{d}a}{\mathrm{d}x} = \pi d \sin\alpha\left(1 - \left(\frac{x}{d}\right)\sin\alpha\right). \tag{4.11}$$

For a given x/d this is a maximum when $\alpha = 90^\circ$. A comparison of Figs. 4.6 and 4.7 shows that the gain of the poppet valve (the slope of the curve) is greater than that of the spool if the fully open areas are the same. This ability to uncover a large orifice area for a small lift makes the poppet popular in digital elements and in safety devices such as relief valves.

The spherical poppet of Fig. 4.8 is a very simple device. The minimum flow area is again a frustrum of a cone, the average circumference being:

$$C = \pi(r + y) \tag{4.12}$$

and the area is:

$$a = \pi(r + y)s\,. \tag{4.13}$$

From the geometry:

$$(y/r) = R/(R + s) \tag{4.14}$$

where R is the ball radius.

Hence $$a = \pi r\left\{\frac{(R + s)^2 - R^2}{R + s}\right\}. \tag{4.15}$$

But $$(R + s)^2 = [x + (R^2 - r^2)^{\frac{1}{2}}]^2 + r^2\,. \tag{4.16}$$

Hence $$a = \frac{\alpha^2 + 1 - (R/r)^2}{\sqrt{\alpha^2 + 1}} \tag{4.17}$$

where $\alpha = (x/r) + \sqrt{(R/r)^2 - 1}.$

This is plotted in Fig. 4.8. The gain increases with ball size, the limiting case $R = \infty$ being equivalent to the flapper nozzle.

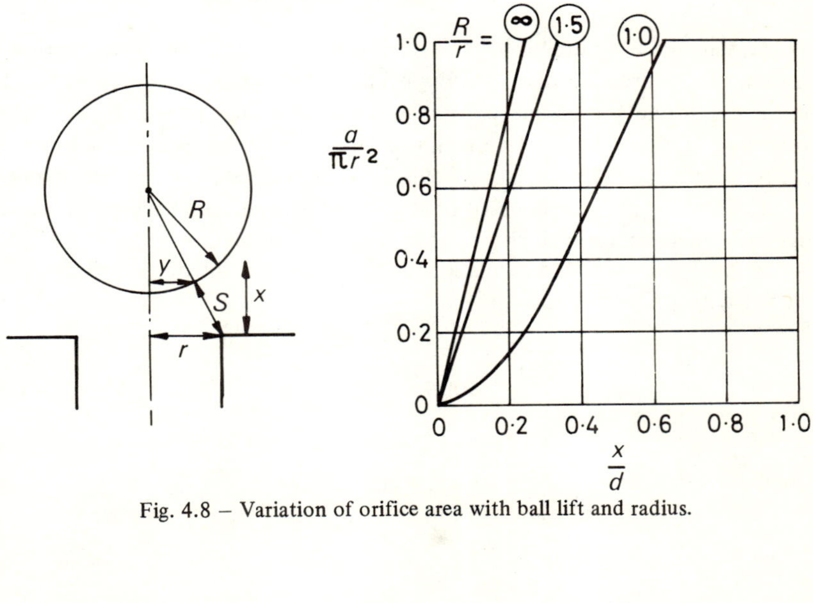

Fig. 4.8 – Variation of orifice area with ball lift and radius.

4.4 CHARACTERISTIC OF JETS

The jet tube device, Fig. 4.9, utilises the characteristics of the pressure distribution in the jet. The receiver is a narrow diameter tube placed with its axis parallel to the jet axis (other configurations are possible). This provides a means of determining the total pressure at any point in the jet, movement of the tube being achieved perhaps by air driven bellows. Because of the jet divergence, velocity vectors are not parallel to the jet axis except on the axis, but little error accrues if the tube pressure is taken as the pressure at a point. If the jet discharges into the atmosphere, the mean jet static pressure at any point can be assumed to be atmospheric pressure, and hence a gauge attached to the tube will register the dynamic pressure. These devices have no critical sealing surfaces, hence they can work over a wide temperature range, and manufacture is relatively easy. They are, however, open centre devices and can consume considerable quantities of air.

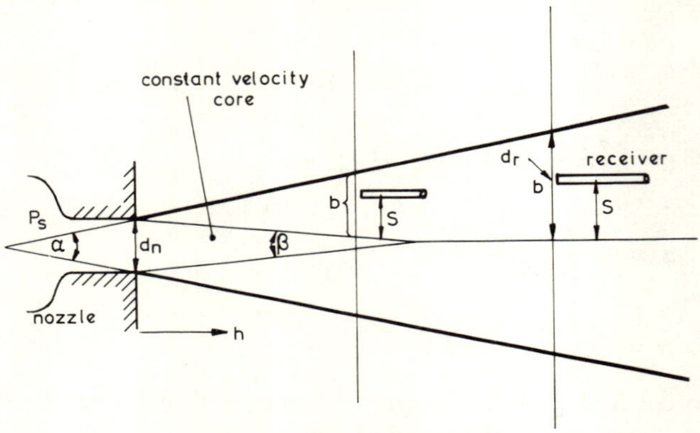

Fig. 4.9 – Geometry of a free jet.

Zalmanzon and Semikova [3] have produced an accurate mathematical description of the pressure distribution within a turbulent incompressible jet. They define the boundary of the jet as the locus of points at which the dynamic pressure is 5 per cent of the maximum dynamic pressure at a station h. This results in $\alpha = 14°$. The length of the constant velocity core is $5d_n$, giving $\beta = 11° 26'$. Thus, from the geometry of Fig. 4.9 the width of the zone between the edge of the constant velocity core and the jet boundary is $b = 0.252\,h$ for $h < 5d_n$. The pressure distribution in this region is adequately described by $P/P_s = 1$ within the core and

$$\frac{P}{P_s} = \left\{1 - \left(\frac{s}{b}\right)^{3/2}\right\}^4 \quad \text{for } h < 5d_n \tag{4.18}$$

where P = dynamic pressure at a point in the jet
P_s = pressure above ambient at the entry into the nozzle.

Beyond the core region $h > 5d_n$ and $b = (d_n/2) + 0.152h$ and

$$\frac{P}{P_s} = \left\{ \frac{\left[1 - \left(\frac{s}{b}\right)^{3/2}\right]^2}{0.3 + 0.14\dfrac{b}{d_n}} \right\}^2 \quad \text{for h} > 5d_n \tag{4.19}$$

These equations can be used to provide a variety of functional relationships between pressure output and input. For example, if the tube is constrained to move along the jet axis then the output is:

$$\frac{P}{P_s} = 1 \qquad \text{for } h < 5d_n$$

$$\frac{P}{P_s} = \left\{ \frac{1}{0.3 + 0.14\,h/d_n} \right\}^2 \qquad \text{for } h > 5d_n \tag{4.20}$$

If it is moved across the jet at fixed h, the characteristics shown in Fig. 4.10 will ensue. Other relationships can be achieved by combining these movements.

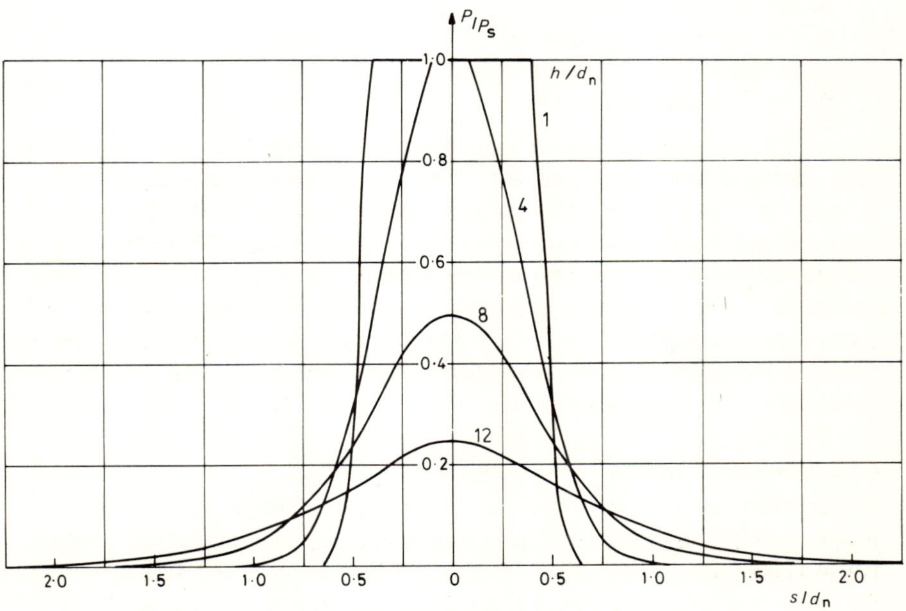

Fig. 4.10 – Pressure distribution in a free jet.

The above discussion assumes that the tube diameter d_r is small in relation to the jet diameter. As the tube diameter is increased it will not measure a point pressure but will indicate rather the mean pressure on the area of the tube face. This has been checked experimentally. Zalmanzon and Semikova [3] showed that theoretical calculation of the mean pressure agreed well with measured results when $d_n = 0.8$ mm and $d_r = 0.27$ mm over the range $6 < (h/d_n) < 10$. When the tube diameter is further increased a stage is reached when the tube interferes with the flow so that equations (4.18) to (4.20) do not apply. For example, with $d_n = d_r = 0.8$ mm, the recorded pressures were less than the theoretical, the error being about 16 per cent at $h/d_n = 6$.

4.5 THEORETICAL DETERMINATION OF DISCHARGE COEFFICIENTS

In general, the discharge coefficient of an orifice must be determined experimentally. It will be a function of Reynolds number and orifice geometry. However, in some special cases the coefficient can be predicted analytically, and these cases are of particular interest.

von Mises [4] analysed the flow through a general orifice configuration. He made the following assumptions:

1. the flow is steady
2. viscosity effects are negligible
3. the flow is irrotational
4. the flow is two-dimensional
5. gravity effects are negligible
6. the fluid is incompressible.

The use of assumption 2 implies that there is no friction loss from upstream to vena contracta and hence that the velocity coefficient C_v is unity. The flow in a control valve is generally far from being irrotational. However, vortices occurring upstream of the orifice, where the velocity is negligible compared with the velocity at the orifice, have a very small effect. Vortices occurring downstream of the vena contracta need not be considered. Hence assumption 3 is reasonable. Assumption 4 can be justified for most conventional fluid control valves. Assumption 6 restricts the results of the analysis to cases where compressibility effects are negligible.

4.5.1 Spool valve orifices

The contraction coefficients for spool valve orifices, poppet valve orifices, and flapper nozzles can be obtained from correct interpretation of von Mises' results. Fig. 4.11(a) shows a configuration described by von Mises as a side opening at the floor of a container. Fig. 4.11(b) shows how the results also apply to the flow through a spool valve orifice. His results, found by using potential functions and conformal transformation, are shown in Fig. 4.11(c), where the contraction coefficient and the jet efflux angle δ (which must be known if momentum forces are to be calculated) are plotted against the non-dimensional valve opening $\bar{x}$. C_c is 0.673 at very small openings and decreases to 0.20 at $\bar{x}_1 = 5.0$. The jet angle δ is 69° at small openings and decreases to 53°20′ at $\bar{x}_1 = 5.0$.

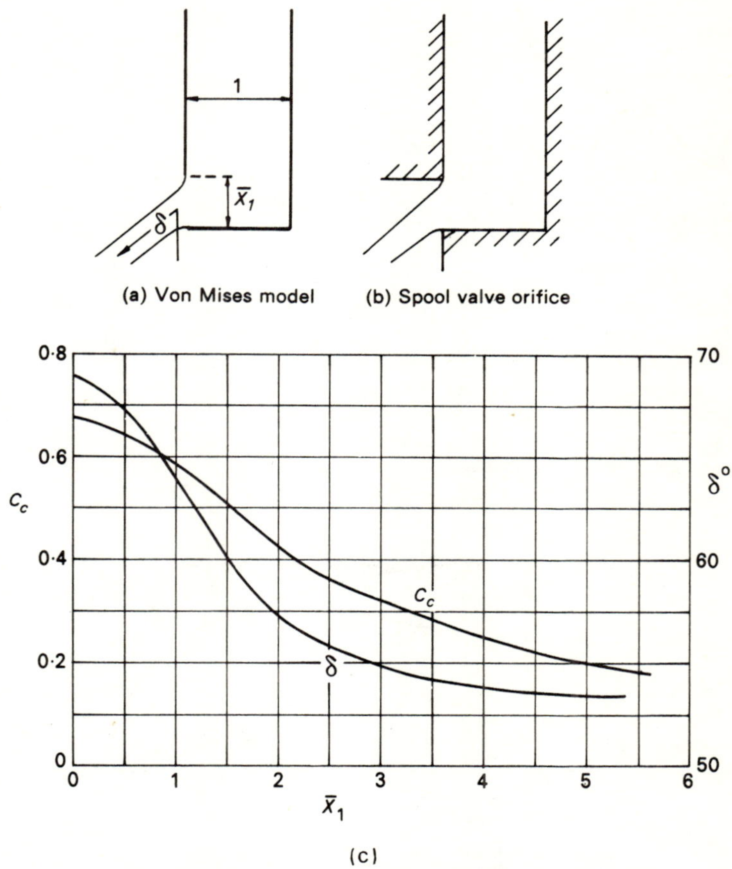

Fig. 4.11 – Contraction coefficient and discharge angle for a two-dimensional spool valve (after von Mises [4]).

Another configuration, the results of which can also be applied to spool valve orifices, is shown in Fig. 4.12. This is described by von Mises as a side lying opening in a horizontal floor. Fig. 4.12 shows the effect of non-dimensional orifice opening $\bar{x}_2$ on the contraction coefficient and jet angle. For small openings C_c is 0.673 and at the maximum opening, $\bar{x}_2 = 1.0$, C_c is 1.0. The jet angle is 21° at small openings and decreases to zero at maximum opening.

In conventional spool valves, the non-dimensional opening $\bar{x}$ is very small and hence the predicted contraction coefficient is 0.673 and the jet angle is 69°.

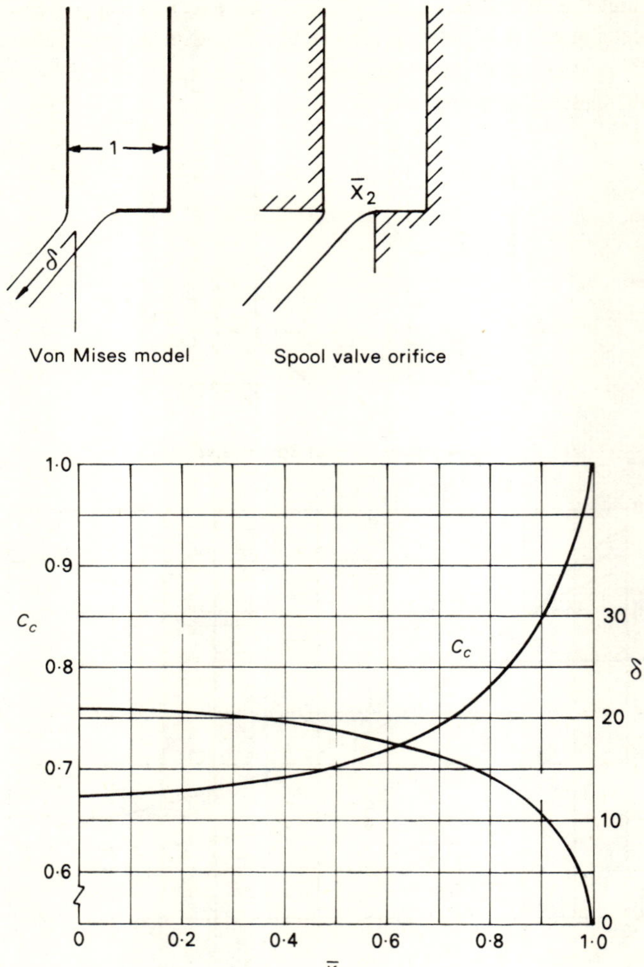

Fig. 4.12 – Contraction coefficient and discharge angle for a two-dimensional spool valve (after von Mises [4]).

The results quoted refer to valves in which there is no clearance between the fixed and moving parts. Of course in practice such clearance must exist and if the orifice opening is very small, clearance effects can be expected to be considerable. In this case we refer to von Mises' work on combined floor and side openings. Unfortunately the magnitude of orifice opening and clearance used by von Mises are unrealistic in this case and are not of great use. The present authors [5] have used the von Mises method to determine the jet angle and C_c for spool valve orifices with realistic openings and clearances, and these results are shown in Figs. 4.13 and 4.14. Here $\bar{x}_1$ is the non-dimensional clearance,

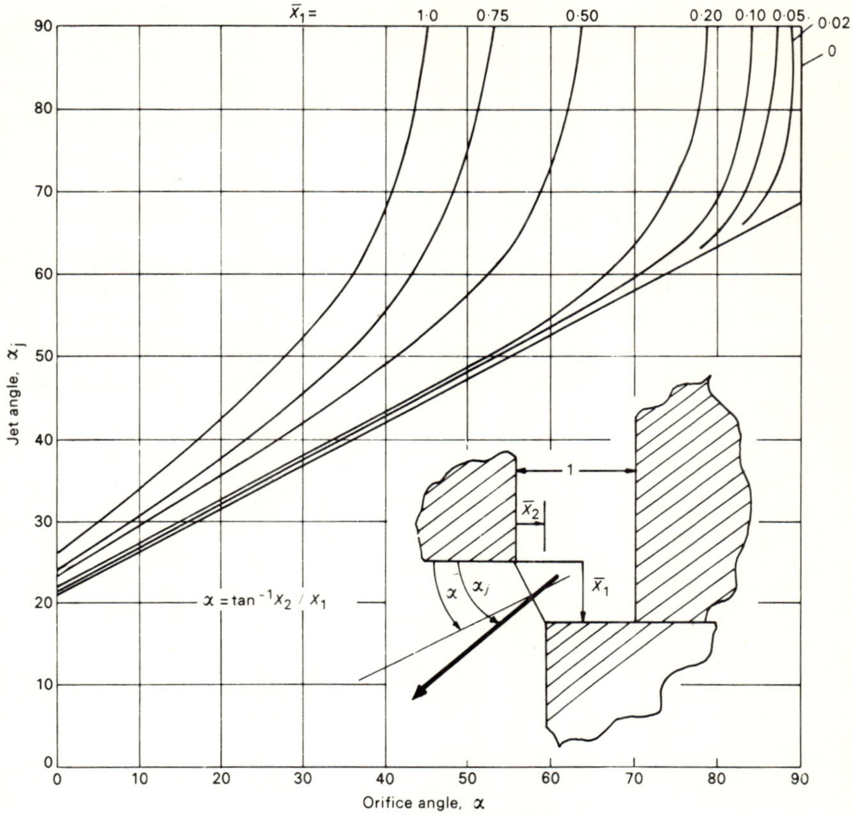

Fig. 4.13 – Jet angle at small clearance.

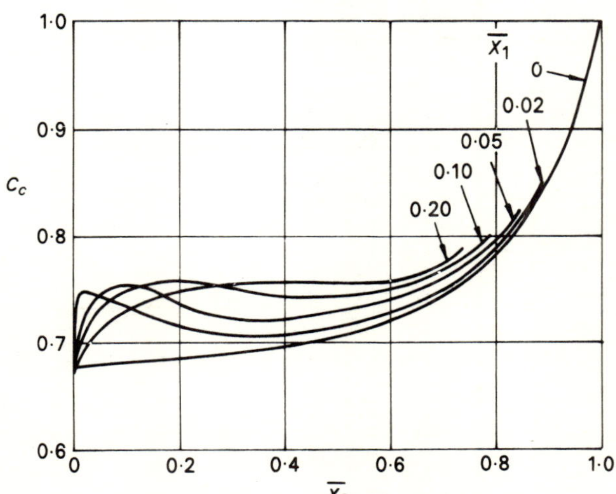

Fig. 4.14 – Contraction coefficients at small clearances.

and Fig. 4.14 shows that increasing clearance results in increasing C_c. The contraction coefficient is in this case referred to the total orifice area $\sqrt{(\bar{x}_1^2 + \bar{x}_2^2)}$. Fig. 4.13 shows the relationship between jet angle α_j and the orifice angle $\alpha = \tan^{-1}(\bar{x}_2/\bar{x}_1)$. These angles are only equal when the upstream chamber is infinite and $\bar{x}_1 = \bar{x}_2$, that is $\alpha = \alpha_j = 45°$.

4.5.2 Poppet valve orifices

von Mises' results for discharge from cones can be applied to poppet valves, Fig. 4.15. This is not immediately obvious when the cone configuration is examined, and Fig, 4.15(b) has been drawn in an attempt to make this clearer. When one appreciates that the face of the poppet is equivalent to the axis of symmetry of the cone, the rest is easy. It is important to note that this implies that the flow follows the poppet face, that is, the jet angle and poppet angle are equal.

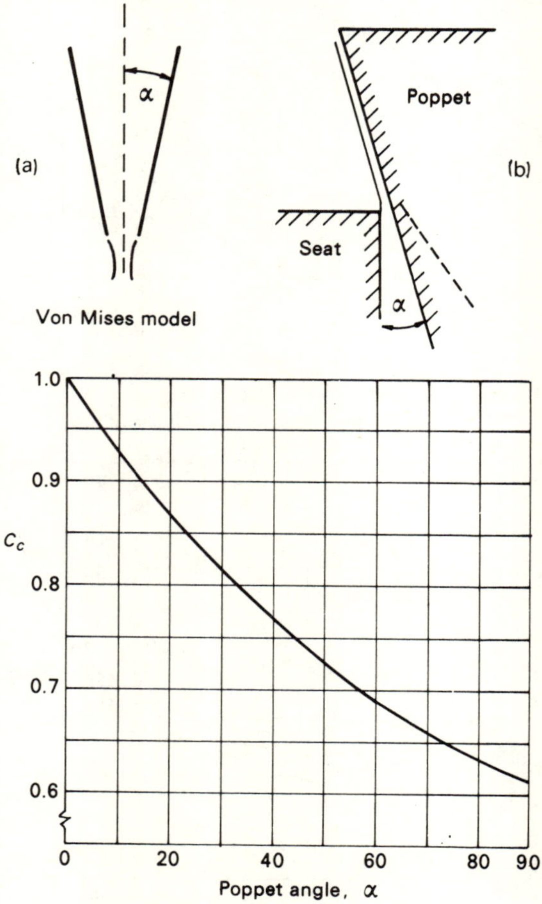

Fig. 4.15 – Contraction coefficient variation with poppet angle (after von Mises [4]).

4.5.3 Flapper nozzle valves

Flapper nozzle geometery is simulated if the opening $\bar{x}_2$ is negative, Fig. 4.16. The contraction coefficient for various geometries is shown in the figure. Some points worth noting are:

For a given flapper size (fixed $\bar{x}_2$), C_c reduces as the distance between flapper and nozzle increases.

For a given flapper nozzle separation (fixed $\bar{x}_1$), C_c increases as the flapper gets smaller.

When $\bar{x}_1 > 1$ the orifice area is greater than the supply area and control is lost. In the axisymmetric case the nozzle has unity radius, the supply area is π and the orifice area (curtain area) is $2\pi\bar{x}_1$. Control is lost in this case when $\bar{x}_1 > 0.5$.

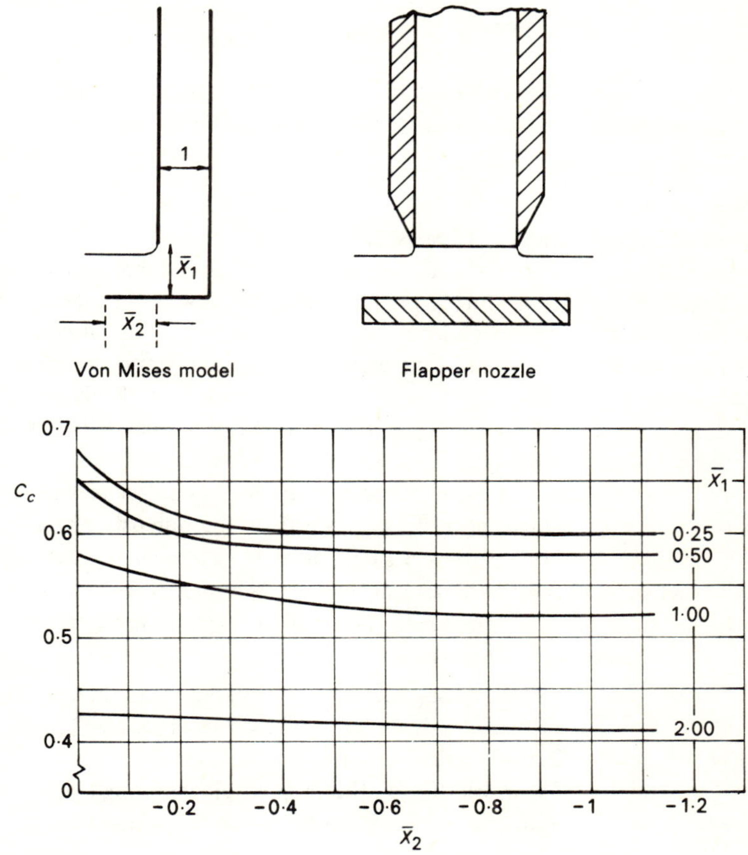

Fig. 4.16 – Contraction coefficient for flapper nozzle (after von Mises [4]).

4.6 EMPIRICAL CONSIDERATIONS

4.6.1 Experimental determination of discharge coefficients

The results of von Mises apply only to an inviscid fluid or in other words to a flow at infinite Reynolds number. The effect of Reynolds number must be determined experimentally. Reynolds number R for an orifice is defined as:

$$R = \frac{v_o D_h}{\nu} \tag{4.21}$$

where v_o = mean velocity through the orifice = Q/a

$$D_h = \text{hydraulic diameter} = \frac{4 \times \text{flow section area}}{\text{flow section perimeter}}$$

$$\nu = \text{kinetic vicosity} = \mu/\rho = \frac{\text{absolute viscosity}}{\text{mass density}}$$

For a circular orifice, the hydraulic diameter is equal to the actual diameter. For a rectangular slit of width w and length l the hydraulic diameter is:

$$D_h = \frac{4wl}{2(w+1)}.$$

This is nearly $2w$ for orifices of large aspect ratio l/w.

Reynolds number R is the ratio of inertia to viscous forces acting on the fluid. At high R the inertia forces predominate, the flow is turbulent and the analysis resulting in equation (4.1) is valid. However, when the viscous forces predominate, laminar flow conditions prevail and the analysis becomes invalid. Nevertheless it is general practice to apply this equation in the laminar flow regime also, and a plot of C_d for liquid flow through a circular orifice is shown in Fig. 4.17. Using flow visualisation techniques Johansen [6] showed that at low R (<10) the outermost streamlines closely followed the orifice profile. As R was increased the flow began to separate from the downstream side until at $R = 150$ a jet formed. This was surrounded by a region of reverse flow. Further increase in R caused the jet contraction to approach a maximum, and although the flow outside the jet consisted of a succession of vortex rings, the flow within the jet was not disturbed.

The figure shows that R must exceed approximately 1000 before the discharge coefficient settles to the value predicted by von Mises (0.611). A peak value is reached at $R = 200$ and below this value C_d falls rapidly.

An exact analysis by Wuest [7] of the laminar flow through sharp edged orifices shows that for a circular orifice (diameter d) in an infinite pipe the flow rate is:

$$Q = \frac{\pi d^3}{50.4\mu} \Delta P. \tag{4.22}$$

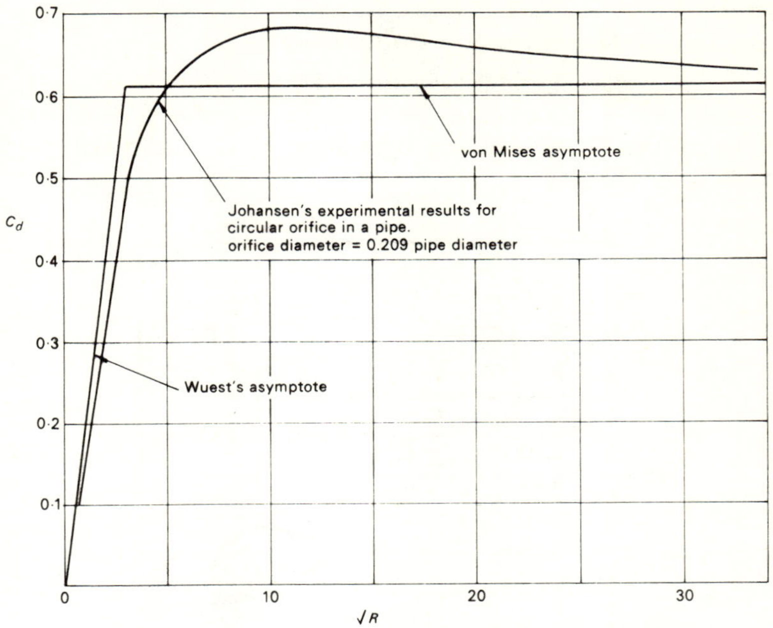

Fig. 4.17 – Variation of discharge coefficient with Reynolds number for a circular sharp edged orifice in a pipe.

If this is combined with equations (4.2) and (4.22) it can be seen that for laminar flow:

$$C_d = 0.2\sqrt{R} = \delta\sqrt{R} \tag{4.23}$$

where δ = laminar flow constant.

An approximate graph [8] of the variation of C_d with R can now be constructed using the low R asymptote defined by equation (4.23) and the high R value of 0.611 predicted by von Mises. This is shown in Fig. 4.17. The transition Reynolds number R_t is given by:

$$R_t = \left(\frac{0.611}{\delta}^2\right) = 9.3\,.$$

For a rectangular slit of height b and width w in an infinite plane Wuest showed that $\delta = 0.157$.

The majority of variable area orifices of spool servo valves are not circular but take the form of a segment of a circle, Fig. 4.6. The hydraulic diameter in this case is:

$$D_h = \frac{4(\theta - \sin\theta)r^2/2}{r(\theta + 2\sin\theta/2)} = \frac{2r(\theta - \sin\theta)}{\theta + 2\sin\theta/2}. \tag{4.24}$$

Results taken from [9] are shown in Fig. 4.18, and curve fitting indicates that δ is about 0.14 in this case.

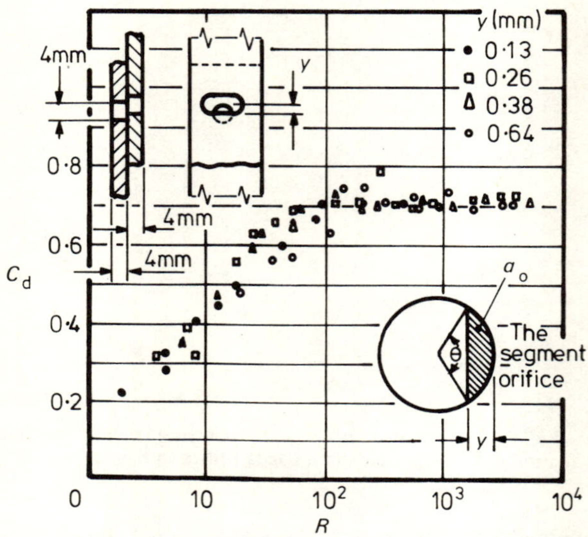

Fig. 4.18 – Experimental variation of discharge coefficient for segmental orifice (after Lewis and Stern [9]).

4.6.2 The flow coefficient

As far as fluid power control valves are concerned it can be difficult to measure the discharge coefficient C_d. In equations (4.2) and (4.3), C_d is effectively the area at the vena contracta, that is, $C_d \approx C_c$. In order to measure this it is necessary to measure the flow rate Q or $\dot{M}$, the orifice area and the static pressure at the vena contracta. In a real valve it is well nigh impossible to measure the pressure at the vena contracta, and it is usual practice to base a coefficient on the pressure drop from upstream to some convenient downstream point. Hereafter this coefficient will be referred to as the flow coefficient C_q.

Pressure recovery in the downstream chamber will result in P_d being greater than P_{vc} and hence in C_q being greater than C_d. Clearly the discharge and flow coefficients will be equal only when the downstream static pressure P_d is the same as the static pressure at the vena contracta. This occurs when the downstream chamber is very large compared to the orifice area. The degree of pressure recovery depends on Reynolds number, orifice geometry and downstream chamber size. It can also be affected by cavitation occurrence in the downstream chamber. Fig. 4.4 illustrates the differences that can exist between P_{vc} and P_d for oil flow through an orifice. Three static pressure distribution curves are shown, the flow rate being the same in each case. At the vena contracta the pressure falls to a minimum and recovery takes place downstream. The pattern is the same for upstream pressure (b), the flow rate being determined by pressure drop from upstream to vena contracta. However, for upstream pressure (c) the pressure at the vena contracta has fallen to the fluid vapour pressure, cavities form, pressure recovery is reduced and the flow coefficient falls. If cavitation is very severe so that the vapour cavities extend as far as the downstream tap, then $P_d = P_{vc}$ and $C_q = C_d$. Hence for such an orifice cavitation affects the flow coefficent. Note however that the discharge coefficient is unchanged.

4.6.3 The flow number

Reynolds number has been defined earlier (equation (4.21)). The mean velocity through the orifice must be known before this number can be evaluated and this is usually determined by dividing the measured flow rate by the orifice area. This is all very well if one is conducting a series of experiments on a particular orifice and measured flow rates are readily available. However, if one wishes to predict the flow coefficient from available information concerning the variation of C_q with Reynolds number, the Reynolds number must be evaluated without measurement. How then is the mean velocity estimated? It can clearly be obtained from the relationship:

$$v_o = C_c \sqrt{(2\Delta P/\rho)}$$

but unfortunately this cannot be evaluated unless the contraction coefficient is known, which it is not. The problem is not insoluble however, for if one eliminates the flow rate Q from the discharge equation (4.2) and from the Reynolds number equation (4.21), the following relationship evolves:

$$C_d = \left\{\frac{\nu}{D_h\sqrt{(2\Delta P/\rho)}}\right\} R = \beta R\,. \tag{4.25}$$

Thus lines of constant β can be superimposed on the C_d results. The designer then evaluates β and finds the discharge coefficient from the intersection of this β line and the C_d curve.

In an attempt to avoid this complication some researchers [10, 11] have presented the variation of C_q with a flow number λ where:

$$\lambda = (D_h/\nu)\sqrt{\{2(P_u - P_d)/\rho\}}\,. \tag{4.26}$$

This number can be evaluated without knowledge of C_q and the flow coefficient can then easily be found from graphs of C_q and λ.

4.7 REATTACHED FLOW IN SPOOL VALVES

Experiment has shown [10, 11, 12] that the flow pattern in a spool valve can assume several different configurations. This section examines these and discusses the effects of flow number, cavitation, geometry and dynamics.

4.7.1 Effects of chamber geometry

In 1956, Turnbull [12] showed photographs of flow patterns in a two-dimensional hydraulic valve, Fig. 4.19. His results indicated that, for inflow as shown, the flow reattached to the sleeve when the spool displacement x was less than three times the clearance w. When x was greater than $3w$ and also less than $0.1a$, the flow reattached to the vertical wall, that is, the face of the spool piston. Alpay *et al.* [13] continued this work on hydraulic valves and showed that as the valve opens the jet breaks away from the sleeve at $x/w = 2.32$ and exists as a free jet until $x/w = 3$ when it reattaches to the vertical wall. When the valve is opened even wider a stage is reached when the presence of the horizontal wall, the spool stem, interferes with the establishment of the reattachment bubble. This causes the jet to detach and form a free jet at an angle predicted by von Mises. A hysteresis zone was found to exist. When the valve was opening the reattached jet broke away when $x = 0.1a$, but when closing the reverse did not occur until $x = 0.075a$.

When the direction of flow was reversed, known as the outflow condition, the patterns were:

At small openings, $x < 3w$, the flow reattached to the horizontal land of the spool.

For $x > 3w$ and $x^2 < 0.11ac$ the flow reattached to the vertical wall of the cylinder port.

For $x > 3w$ and $x^2 > 0.11ac$ the flow existed as a free jet.

4.7.2 Effects of cavitation and flow number

There are two types of cavitation to be considered – (a) cavitation in the mixing zone of the turbulent jet and (b) cavitation in the reattachment bubble. The existence of the former was pointed out by Rouse [14]. The latter type can be of importance in flueric systems, and experiment has shown [15, 16, 17] that cavitation in the bubble tends to make it lengthen. The magnitude of the static pressure P_b within the bubble is related to the momentum of the jet and its radius of curvature r in the equation:

$$P_d - P_b = \rho u^2 w/r$$

where P_d = downstream pressure
u = jet mean velocity
w = width of the device.

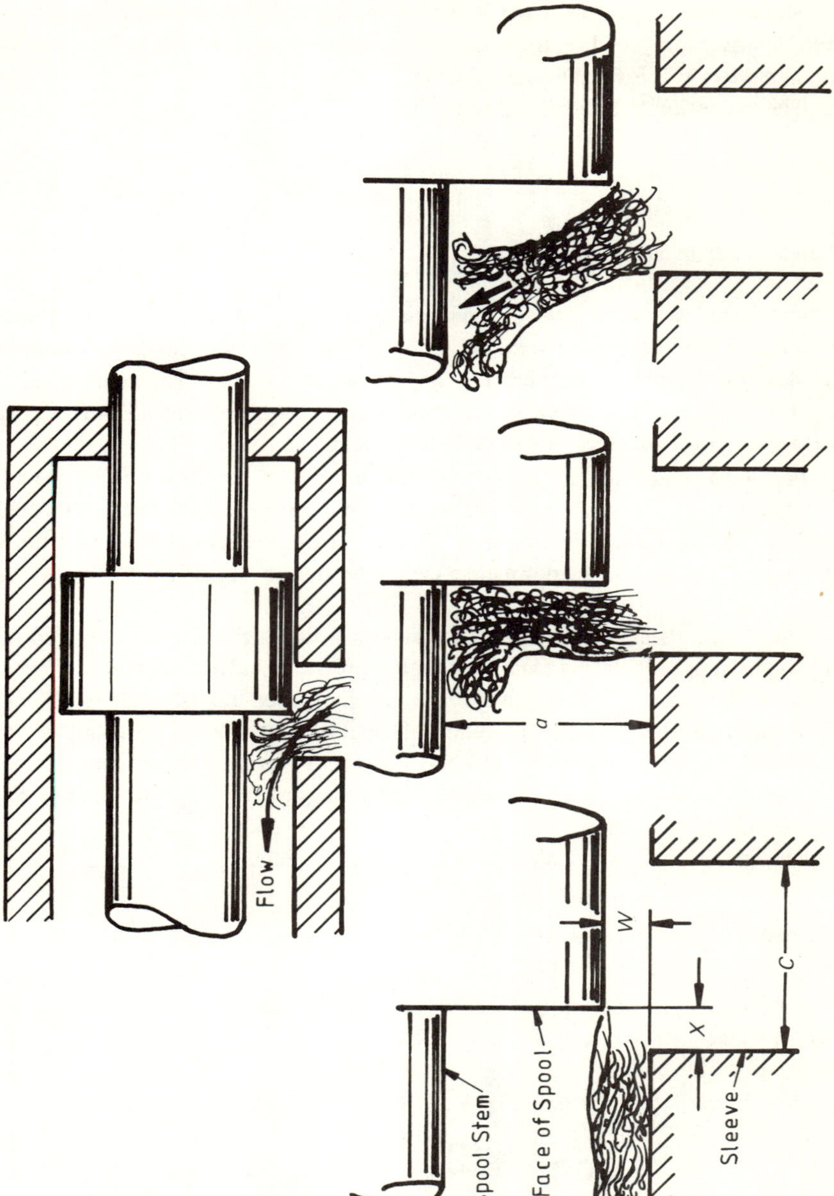

Fig. 4.19 – Different flow patterns in a spool valve.

Since vapour pressure is the lower bound on the fluid pressure, then for a given downstream pressure, the radius of curvature must increase and the bubble lengthen when this lower bound is reached.

The **cavitation number** K is defined as:

$$K = \frac{P_d - P_v}{P_u - P_d} \tag{4.27}$$

where P_v = vapour pressure.

If there is no cavitation and K is gradually reduced, cavitation will first arise when K reaches the critical value K_i, known as the **incipient cavitation number.** The **desinent cavitation number** K_d, on the other hand, is that value of K at which a cavitating flow ceases to cavitate. For practical purposes K_i and K_d can be assumed equal for fluid power control orifices.

Fig. 4.20 shows how K_i varies with flow number λ for three different types of orifice [18, 19]. For the spool valves the flow numbers have been calculated using $2x$ as the hydraulic diameter for the two-dimensional case and using equation (4.21) for the axisymmetric case. The poppet valve will be referred to later.

When cavitation occurs in a valve, violent noise is generated as the vapour cavities collapse on entering regions of higher pressure. At the incipient cavitation condition the noise is emitted in short randomly spaced bursts, while for $K < K_i$ the noise is continuous. The presence of cavitation can lead to erosion of the valve lands or ports, and one of the ways to avoid this is to ensure that $K > K_i$. This can be achieved by decreasing the upstream pressure or by increasing the back pressure.

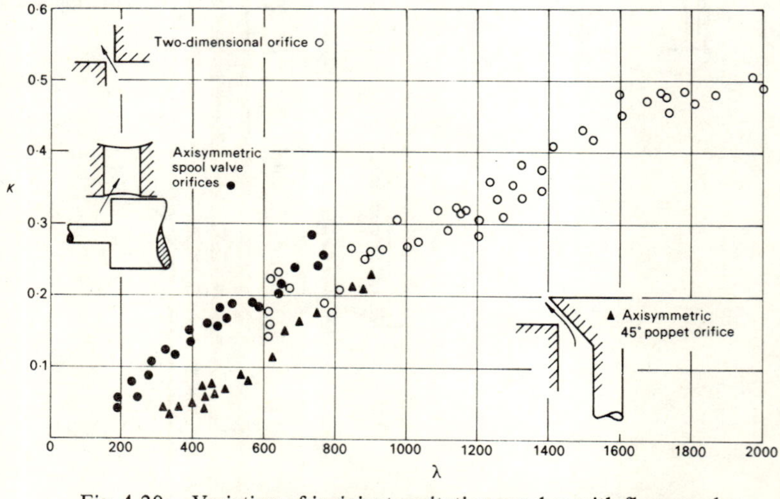

Fig. 4.20 – Variation of incipient cavitation number with flow number.

Fig. 4.21 shows how the flow coefficient C_q varies with K and λ for reattached flow. Note where the downstream pressure was measured. C_q settles out at 0.71 at large λ after reaching a maximum in the range $150 < \lambda < 400$. At large λ the flow coefficient is greater than the value predicted by von Mises for free jet flow (0.673). This is caused by the curved jet and has also been noted by Stenning in his work on pneumatic valves [20].

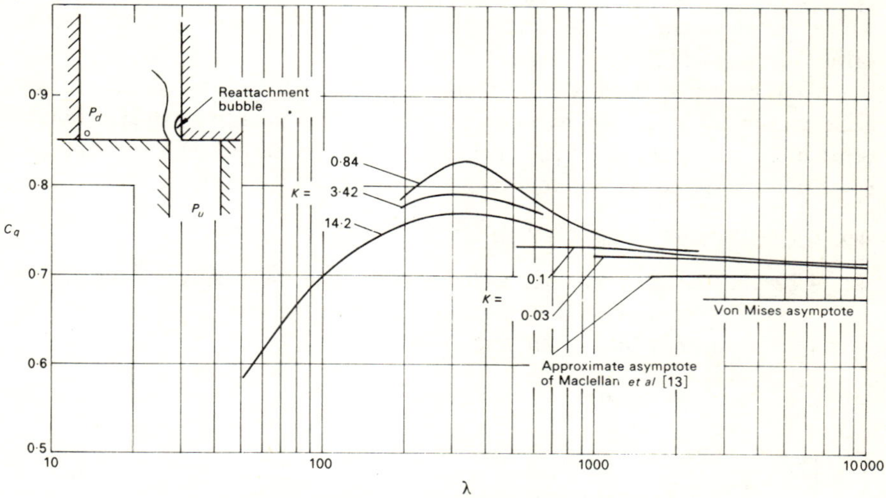

Fig. 4.21 – Flow coefficient of a two-dimensional spool valve orifice as a function of flow number and cavitation number.

4.7.3 Practical considerations

When applying these results to a particular valve it is important that the type of flow pattern be known. Is the jet free or does it reattach? Work reported to date on reattached flows has been considered with two-dimensional devices and models. The reasons for this is clear when one appreciates that low pressures can only be maintained in the reattachment bubble by the presence of the front and back plates of the device or model. However, a spool valve is not two-dimensional and although a two-dimensional model may exhibit reattached flow there is no guarantee that a real valve will behave similarly. For instance, if there are a number of circular ports in the sleeve, then for the inflow condition, a number of discrete jets will form and there is no possibility of reattachment bubbles forming under these conditions. However, for flow from spool to sleeve, reattachment flow may occur, but this is doubtful since the surface of the port is curved. A reattachment bubble could occur for the inflow condition if the ports were situated in a recess in the sleeve – effectively a rectangular orifice. In this case the fluid forms a continuous sheet after passing into the valve, and a pressure differential could exist across this.

Even when the static conditions allow the formation of a reattachment bubble dynamic conditions may alter this. Foster and Kulkarni [21] conducted tests on a two-dimensional model spool valve and they found that the flow did not reattach when the spool was oscillated quickly. The reattachment process requires time to establish, this being some 10–30 times the time for a particle to travel the length of the reattachment bubble, depending on the angle between jet and wall. Their results indicate that there is a limiting spool frequency above which reattachment will not occur, and that this frequency is a function of orifice opening and pressure drop.

Thus in many cases it is difficult to predict which type of pattern will exist, and under these circumstances the engineer will have to make an estimate, using the above data. Alpay [22] has shown that the statically determined flow coefficient has to be reduced to account for dynamic effects.

4.8 EXPERIMENTS WITH HYDRAULIC POPPET VALVES

Fig. 4.22 provides data on a poppet valve with a sharp seat and a 45° angle [19]. Flow visualisation showed that the flow followed the face of the poppet. The flow coefficient is defined by equation (4.22) with:

$$a = \pi x \sin\alpha (D - x \sin\alpha \cos\alpha) \tag{4.28}$$

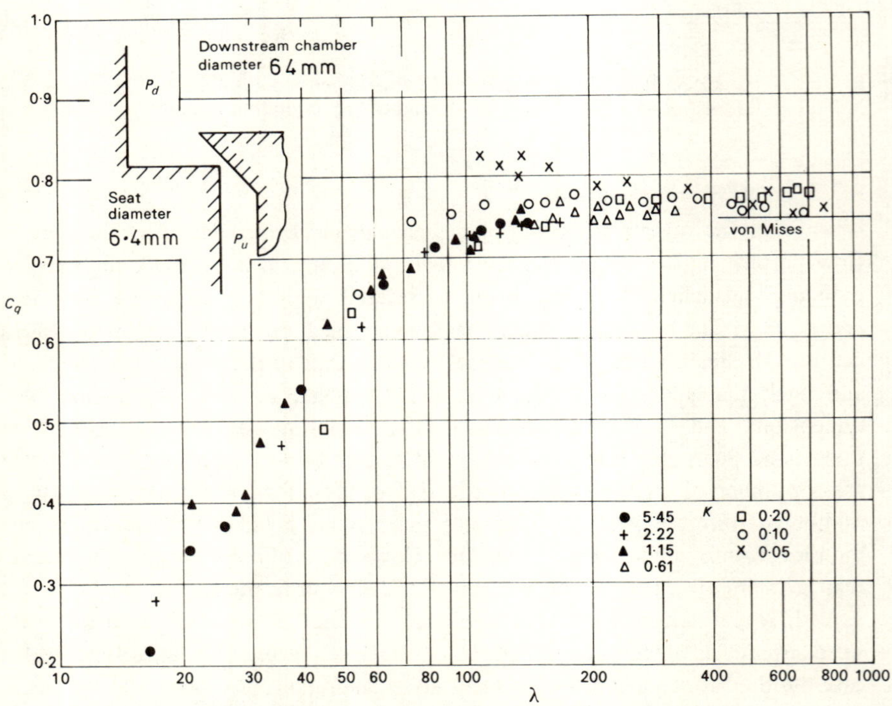

Fig. 4.22 – Measured flow coefficient of a 45° poppet valve.

and D and x are respectively the poppet seat diameter and the poppet lift. When calculating the flow number the mean hydraulic diameter is taken as $2x \sin \alpha$.

The variation of flow coefficient with flow number is shown in Fig. 4.22. These results show the same trend as the spool valve results. Again in this case, because of the large downstream chamber used in the experiments, C_q and C_d are approximately equal. Cavitation effects are slight. At the larger flow numbers $C_q = 0.77$ which shows reasonable agreement with von Mises' value of 0.746. This is not surprising since in the real valve the flow followed the face of the poppet and this was the flow model adopted by von Mises, Fig. 4.15. Work by Stone [23] produced considerable scatter and indicated values of C_q at large λ which were less than the von Mises value. This discrepancy is difficult to explain.

4.9 EXPERIMENTS WITH HYDRAULIC FLAPPER-NOZZLE VALVES

The analysis of the flow through such orifices is difficult, particularly when the edges of the nozzle are not sharp but have a flat or land [24, 25, 26, 27]. Unfortunately this is usually the case in practice since contact between flapper and nozzle tends to flatten the nozzle.

For values of the gap to land ratio $h/w > 1$ the geometry approaches that of the sharp edged orifice and the flow is separated at all times [24], Fig. 4.23. In this case the flow resembles that in a short orifice and the variation of C_q

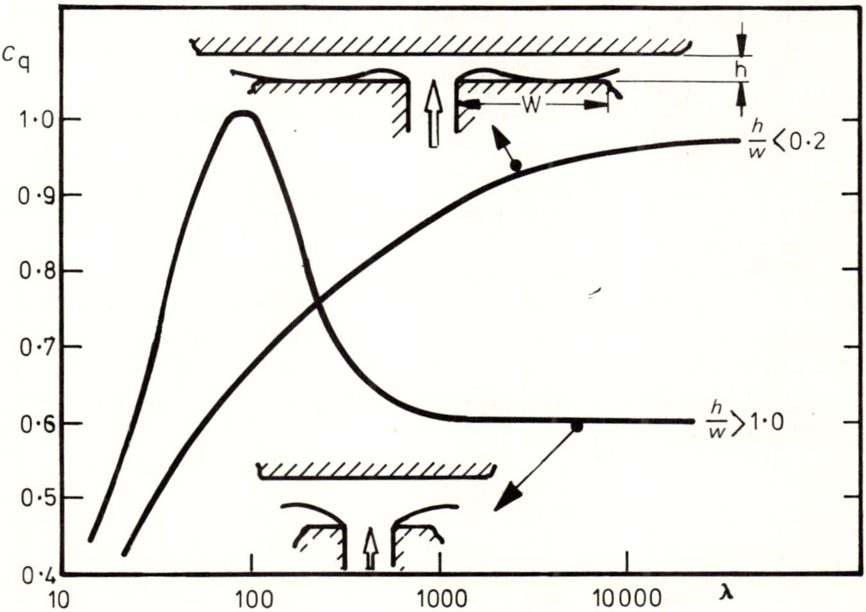

Fig. 4.23 – Effect of flapper-nozzle geometry on flow coefficient.

with flow number is typical of such orifices; there is an initial large variation in C_q for $\lambda < 400$, but above this value C_q settles out at the value of 0.6 predicted by von Mises.

At small gap to land ratios, $h/w < 0.2$, the flow is always reattached to the valve land, Fig. 4.23, and the flow coefficient increases gradually with λ to a limiting value 0.8 to 1.0. The high values arise because the flapper-nozzle gap acts like a diffuser, resulting in good pressure recovery from the vena contracta to the edge of the valve land. Fig. 4.24 shows the variation of C_q with h/w at large λ.

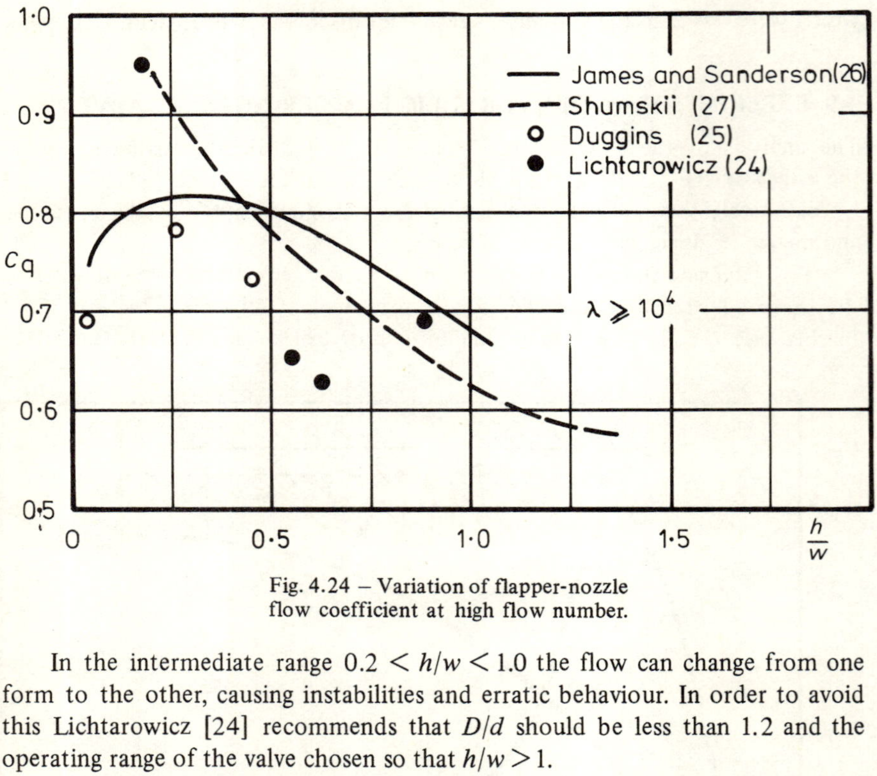

Fig. 4.24 – Variation of flapper-nozzle flow coefficient at high flow number.

In the intermediate range $0.2 < h/w < 1.0$ the flow can change from one form to the other, causing instabilities and erratic behaviour. In order to avoid this Lichtarowicz [24] recommends that D/d should be less than 1.2 and the operating range of the valve chosen so that $h/w > 1$.

At very low flow numbers laminar flow prevails and the flow rate between flapper and nozzle is:

$$Q = \frac{\pi h^3 \Delta P}{6\mu \ln (D/d)} . \tag{4.29}$$

Equating this to:

$$Q = C_q \pi d h \sqrt{\frac{2\Delta P}{\rho}} . \tag{4.30}$$

gives
$$C_q = \frac{h^3}{12 d\nu \ln (D/d)} \sqrt{\frac{2\Delta P}{\rho}} \tag{4.31}$$

and taking
$$\lambda = \frac{2h\sqrt{(2\Delta P/\rho)}}{\nu}$$

then
$$C_q = \frac{(h/d)\lambda}{24 \ln (D/d)}.$$

4.10 EXPERIMENTS WITH HYDRAULIC SHORT TUBE ORIFICES

The discharge of incompressible non-cavitating flow through this type of orifice is discussed in [28]. Shearer [29] points out that this type of orifice is usually employed in bridge type hydraulic circuits where the flow rates are very small and where it would be difficult to justify the expensive manufacture of small sharp edged orifices. Again with this type of orifice there is difficulty in deciding whether the measured coefficient is a discharge coefficient or a flow coefficient. In the longer orifices the vena contracta occurs within the tube and use of the pressure drop across the tube gives a flow coefficient. For the short orifices the vena contracta occurs outside the tube and the pressure drop across the tube (with the jet discharging to atmosphere) gives a discharge coefficient. For the sake of generality we will refer to the measured values as flow coefficients. The Reynolds number is defined in this case as:

$$R = Qd/a\nu \tag{4.32}$$

where Q/a = mean velocity through the orifice of diameter d.

Fig. 4.25 shows the effects of Reynolds number and of orifice length.

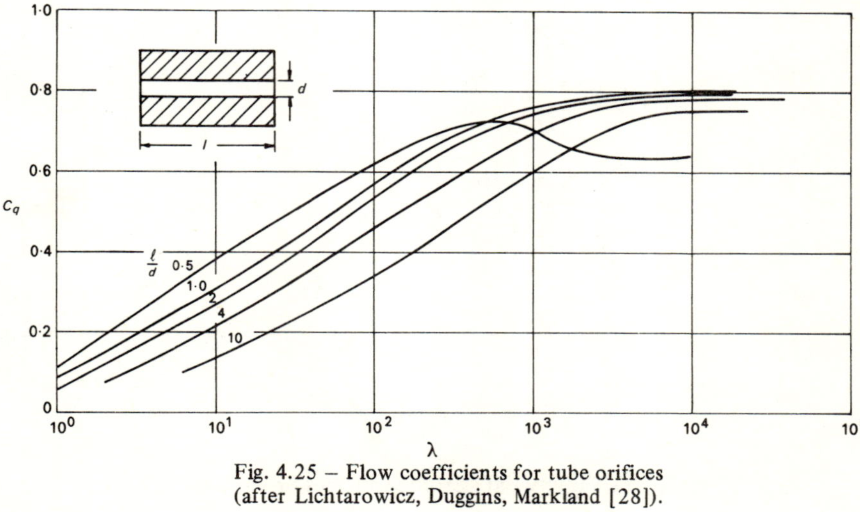

Fig. 4.25 – Flow coefficients for tube orifices (after Lichtarowicz, Duggins, Markland [28]).

The effects of cavitation on the performance of long tube orifices are reported in [30]. It was found that the flow coefficient remained constant as the cavitation number reduced to K_i. Thereafter the point of reattachment moved rapidly down the orifice and ultimately emerged downstream leaving a central jet of fluid surrounded by a bubbly shear layer. During this process pressure recovery was lost and C_q reduced.

4.11 EXPERIMENTS WITH PNEUMATIC ORIFICES

Usually when a gas is flowing through an orifice the Reynolds number is sufficiently large enough to justify the neglect of Reynolds number effects. This is mainly due to the low kinematic viscosity of air. As a result the main variable of concern is the pressure ratio P_d/P_u^* and several experimental programmes have been carried out to determine the variation of the flow coefficient with P_d/P_u^*.

Perry [31] has measured C_d for air through a sharp edged circular orifice, and his results are shown in Fig. 4.26. The coefficient is larger than that for incompressible flow. As the pressure drop reduces ($P_{vc}/P_u^* \rightarrow 1$) compressibility effects become less predominant and C_d tends to the value predicted by von Mises for the incompressible fluid. It should be noted, however, that at very small pressure drops ($P_{vc} \rightarrow P_u^*$) the flow may be laminar, in which case C_d will fall off quickly.

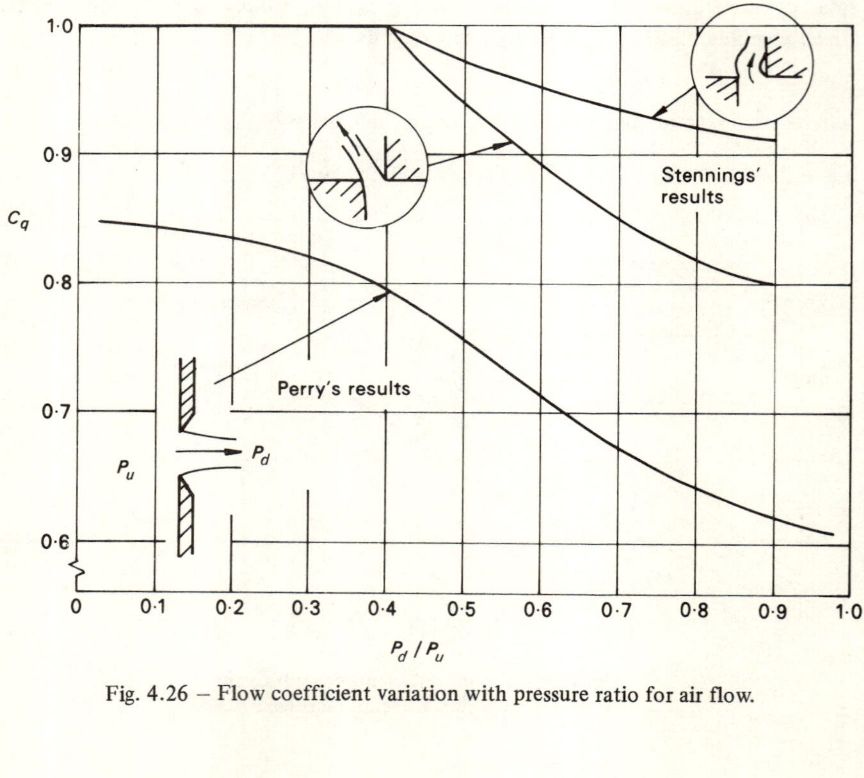

Fig. 4.26 – Flow coefficient variation with pressure ratio for air flow.

Stenning [20] has investigated 'two-dimensional' gas flow through spool valve orifices, and he found that the flow coefficient was greater for reattached flow than for jet flow. Both flow patterns were obtainable when $P_d/P_u > 0.3$–0.4, but the free jet become unstable for pressure ratios less than this value and the jet then reattached.

There are very little published data on gas flow through poppet valves. Tsai and Cassidy [32] have measured the characteristics of a 45° poppet. Unfortunately this did not have a sharp seat, so that at small lifts its behaviour is not strictly comparable with the incompressible results of Fig. 4.22. However, the results in Fig. 4.27 do again follow the same trend as those for sharp edged orifices and spools. As the pressure ratio P_d/P_u decreases, the flow coefficient increases, implying an increase in throat area. At large lift and pressure ratios near unity (incompressible) the flow coefficient agrees with that discussed earlier for oil flow. At small lifts, the finite seat area acts as a diffuser resulting in increased flow coefficients. For very small lifts the flow will be laminar and C_q will fall off.

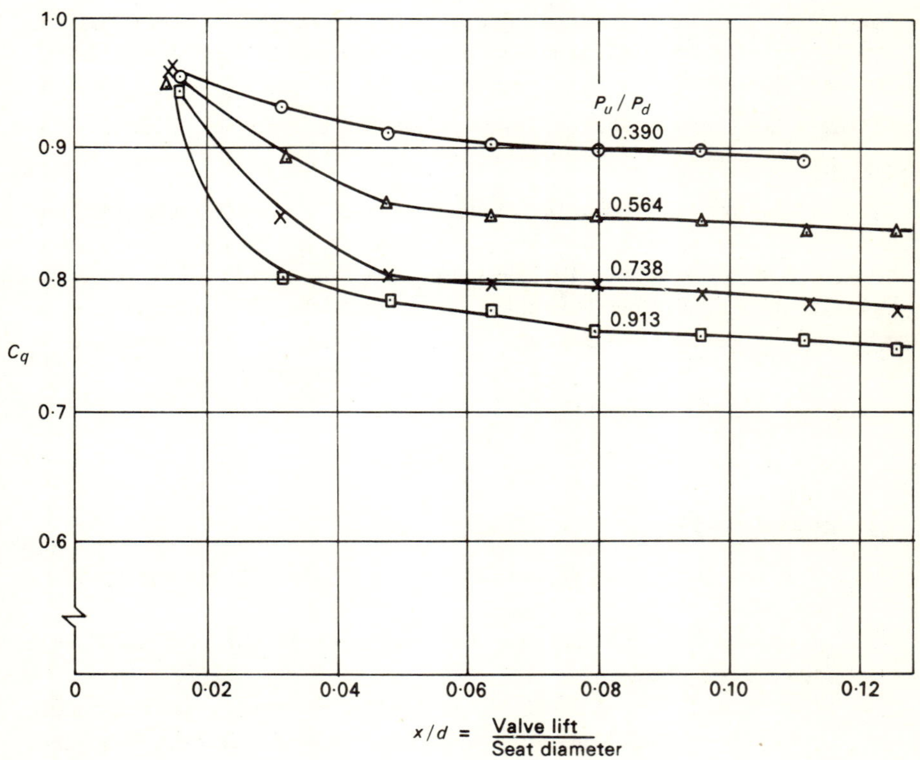

Fig. 4.27 – Flow coefficient for air flow through a poppet valve (after Tsai and Cassidy [32]).

Compressible flow in flapper-nozzle valves has not yet been adequately dealt with in the literature. However, the same trend can be expected as for the other orifices discussed above, that is, C_q should be larger than for oil flow, Fig. 4.16, and the difference should be largest at small pressure ratios. Flow coefficients of unity or more can then be expected at small pressure ratios and at small flapper lifts when the nozzle has a flat.

Further information on compressible flow can be found in [1].

4.12 EXAMPLES

4.12.1

A hydraulic poppet valve has a sharp seat of diameter 20 mm. The upstream and downstream static pressures are 60 bar and 10 bar respectively and the valve lift is 0.5 mm. If the kinematic viscosity of the oil is $50 \times 10^{-6}\,m^2/s$ and its density is $900\,kg/m^3$, determine (a) the hydraulic diameter, (b) the velocity at the vena contracta, (c) the Reynolds number – you may assume that there is negligible pressure recovery from the vena contracta to the position at which the downstream pressure is measured, (d) the cavitation number – does cavitation exist?, (e) the volumetric flow rate – you may assume that C_d is that predicted by von Mises.

(*Answers*: (a) 0.707 mm, (b) 105.4 m/s, (c) 1490, (d) 0.2, yes, (e) 1.711 *l*/*s*.)

4.12.2

Air flows through a circular orifice of diameter 1 mm. The upstream temperature is 30°C, the total upstream pressure 10 bar absolute, and the downstream static pressure is 8 bar absolute. The measured mass flow rate is $1.2 \times 10^{-3}\,kg/s$. Calculate the discharge coefficient, (a) assuming no pressure recovery from vena contracta to the downstream tap, (b) assuming that the downstream static pressure is 10 per cent greater than that at the vena contracta.

(*Answers*: (a) 0.802, (b) 0.723.)

4.13 REFERENCES

[1] Andersen, B. W. (1967), *The Analysis and Design of Pneumatic Systems*, John Wiley.

[2] Blackburn, J. F., Reethof, G., and Shearer, J. L. (1960), *Fluid Power Control*, MIT Press.

[3] Zalmanzon, L. A. and Semikova, A. I. (1968), Investigation of pneumatic jet-tube elements, *Pneumatic and Hydraulic Control Systems*, Pergamon Press.

[4] von Mises, R. (1917), The calculation of flow coefficients for nozzle and orifice, *1917 V.D.A.*, **61**, Nr. 21, 22, 23 May–July.

[5] McCloy, D. (1968), Discharge characteristics of servo-valve orifices, Paper 6, *Proc. 1968 Fluid Power International Conf.*, Olympia.

[6] Johansen, F. C. (1941), Flow through pipe orifices at low Reynolds number, *Proc. Roy. Soc.*, **126**, No. 801.

[7] Wuest, W., (1954), Stromung durch Schlitz-und Lochblenden bei kleinen Reynolds-Zahlen, *Ingenieur Archiv*, No. 22, pp. 357–367.

[8] Viersma, T. J. (1962), Designing load compensated fast response hydraulic servos, *Control Engng.*, May.

[9] Lewis, E. and Stern, H. (1962), *Design of Hydraulic Control Systems*, McGraw-Hill.

[10] MacLelland, G. D. S., Mitchell, A. E., and Turnbull, D. E. (1960), Flow characteristics of piston type control valves, *Proc. Symposium Recent Mech. Eng. Developments in Auto. Control, Inst. Mech. Eng.*, London, January.

[11] Beck, A. and McCloy, D. (1967–68), Some cavitation effects in spool valve orifices, *Proc. Inst. Mech. Engrs.*, London, **182**, Pt. I, No. 8.

[12] Turnbull, D. E. (1956), *Some Characteristics of Hydraulic Piston-type Control Valves in Relation to their Use as Components in a Feedback Position Control System*, Ph.D. thesis, Cambridge University.

[13] Alpay, S. A. and Fleming, G. K. (1967), Stabilisation of the flow in spool valves, *A.S.M.E. Conf. Philadelphia*.

[14] Rouse, H. (1953), Cavitation in the mixing zone of submerged jet, *La Houille Blanche*, **8**, **1**, pp. 9–18, January–February.

[15] Glaettli, H. H., Muller, H. R., and Zinag, R. H. (1964), Remarks on the limitations of pure fluid elements, *Proc. Fluid Amplification Symposium*, **1**, p. 217, Washington.

[16] McCloy, D. (1972), Oil hydraulic fluid logic, *Proc. 1972 Fluid Power International Conference.*

[17] Foster, K. and Parker, G. (1970), *Fluidics – Components and Circuits*, Wiley-Interscience.

[18] McCloy, D. and Beck. A. (1969), Flow hysteresis in spool valves, *First Fluid Power Symposium*, Cranfield, January.

[19] McCloy, D. and Beck, A. (1967), *Cavitation Effects in Poppet Valves*, A.S.M.E. paper No. 67-FE-27. Fluids Division meeting in Chicago, May.

[20] Stenning, A. H. (1955), *An Experimental Study of Two-dimensional Gas Flow Through Valve Type Orifices*, A.S.M.E. paper No. 54-4-45.

[21] Foster, K. and Kulkarni, M. M. (1966), *The Free Oscillations of an Inertia Loaded Hydraulic Spool Valve*, University of Birmingham, Mech. Eng. Dept. Research Rep. 50, May.

[22] Alpay, S. A. (1974), The flow coefficient of a hydraulic control valve

under dynamic conditions, *Flow, Its Measurement and Control in Science and Industry*, Instrument Soc. of America.

[23] Stone, J. A. (1960), Discharge coefficients and steady state flow forces for poppet valves, Trans.
poppet type valves, *Trans. A.S.M.E.*, **82**, Series D. No. 1.

[24] Lichtarowicz, A. (1973), Flow and force characteristics of flapper valves, *3rd International Fluid Power Symp.*, B.H.R.A.

[25] Duggins, R. K. (1973), Further studies of flow in a flapper valve, *ibid.*

[26] James and Sanderson, E. W. (1961/62), *Flow Through a Long Orifice*, Projects 1961, 1962, Dept. Mech. Eng. University of Nottingham.

[27] Shumskii, N. P. (1968), The results of experimental and theoretical investigations of control devices of the nozzle-baffle type, *Pneumatic and Hydraulic Control Systems*, Pergamon Press.

[28] Lichtarowicz, A., Duggins, R. K., and Markland, E. (1965), Discharge coefficients for incompressible non-cavitating flow through long orifices, *J. Mech. Engng. Sci.*, **7**, No. 2, London, January.

[29] Shearer, J. L. (1960), Resistance characteristics of control valve orifices, *Proc. Inst. Mech. Engrs.*, London, January.

[30] Pearce, I. D. and Lichtarowicz, A. (1971), Discharge performance of long orifices with cavitating flow, *Proc. 2nd International Fluid Power Symp.* B.H.R.A.

[31] Perry, J. A. (1949), Critical flow through sharp edged orifices, *Trans. A.S.M.E.*, October.

[32] Tsai, D. H. and Cassidy, E. C. (1961), Dynamic analysis of a simple pneumatic pressure reducer, *J. Basic Engng.*, **83**, 253, June.

CHAPTER 5

Measurement and signal processing

5.1 INTRODUCTION

Information flows around a system in the form of signals. The input signal, otherwise known as the **instruction** or **command signal**, is perhaps the most obvious. It must be analogue or digital and its function is to initiate action in the system. The form of the system is irrelevant; it may be open or closed loop.

The closed loop system, however, is characterised (Fig. 1.8) by the presence of another type of signal – the **measurement signal** that provides information about the output or controlled variable. Fluid power systems are often used to control position (for example, machine tools), to control velocity (for example, the speed of rolls) and to control force (for example, machines for testing material properties). Thus a knowledge of the means of measuring these particular variables is essential for the fluid power engineer.

In addition to measuring the above-mentioned outputs the fluid power engineer will have occasional need to measure variables within the system. Pressure and flow rate are examples of such intermediate variables that can be used to improve the stability and performance of a fluid power system.

The signals generated by the measuring device may require several forms of processing. They may need to be amplified. They may need to be changed from one form to another to make then compatible with signals elsewhere in the system – especially in the error detector. They may need to be shaped for stability improvement.

This chapter is a discussion of input signals, measuring devices, signal processing and error detection. It will emphasise the use of fluid devices for these functions, but brief reference will be made to electrical and mechanical methods. The interested reader will find further details on these aspects in [1, 2].

5.2 COMMAND INPUTS

The command input to a system may be generated manually, by a programmer or by another machine system. It informs the system of the desired state of the outputs. In many cases the command is simply to start or to stop, and conventional manually operated valves can fulfil these funtions adequately. Conventional valves are not suitable as inputs for low pressure fluidic systems, however, and the back pressure sensor of Fig. 5.1(a) has been developed for this purpose. This can be operated directly by the finger as shown, in which case there are no moving mechanical parts. However, if there is a danger of dirt being forced into the control orifice by the manual action, it may be wise to use a sensor in which the control port is covered and uncovered by a manually operated spring returned button.

Input data can also be presented in the form of punched cards or paper tape, and Fig. 5.1(b) shows how a simple reading head can be constructed from interruptible jets. The electrical equivalent could use photoelectric cells instead of air jets. An output signal is generated when a hole passes the reading head. A means for using a programme of this form will be discussed in Chapter 16.

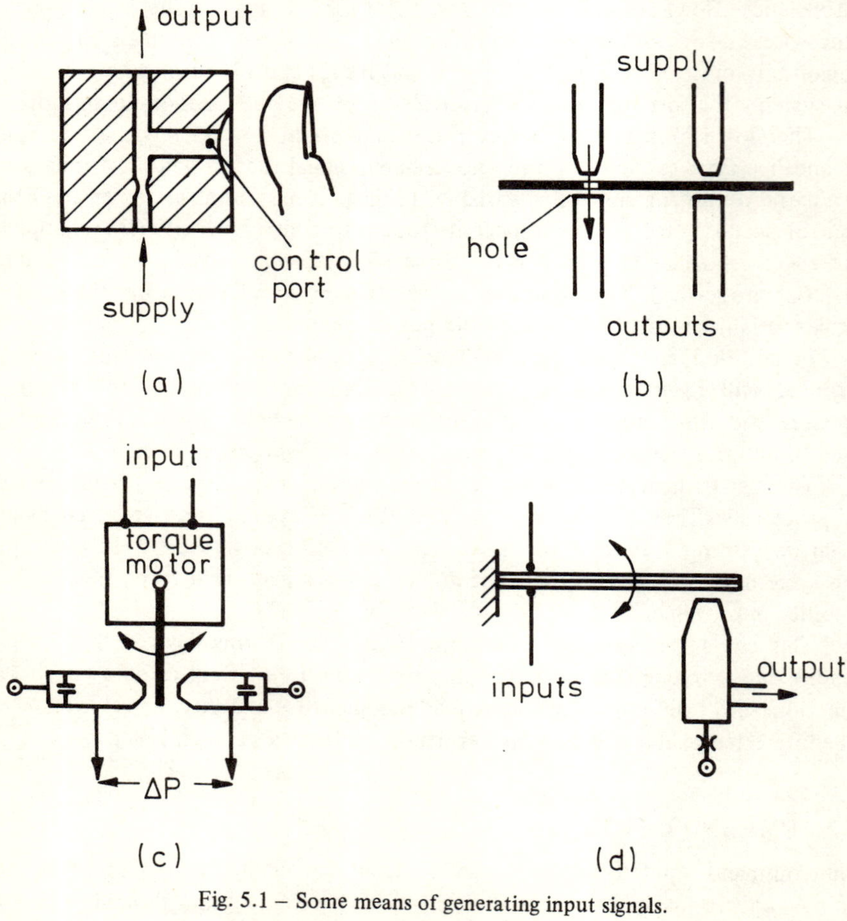

Fig. 5.1 – Some means of generating input signals.

Electrical-fluid interfaces are necessary when a fluid system is controlled from an electrical source, such as a computer or a radio link. The solenoid operated valve is a well established interface device and finds much use in on-off controlled systems. It has some drawbacks, however. For example it is difficult, without using expensive components, to achieve operating frequencies in excess of 10 Hz. Overheating due to prolonged switching, and sensitivity to wet or explosive environments, limit its areas of application. Better response can be achieved by an electrical relay operator in which a flapper-nozzle is used instead of the usual electrical contact. Such a device can switch at frequencies of 100 Hz. The electrically driven torque motor with flapper-nozzle pick-off, Fig. 5.1(c), can also respond adequately to signals in this frequency range [3, 4]. Its main use is in the first stage of electro-hydraulic servovalves where its proportional positioning accuracy make it attractive.

In some cases, Fig. 5.2, the torque motor is connected directly to the spool of the servovalve. This offers a relatively simple and compact electro-hydraulic interface but its effectiveness is limited by the magnitude of the fluid momentum forces in the valve (see Chapter 8). Such single stage valves are thus restricted to fairly low power applications.

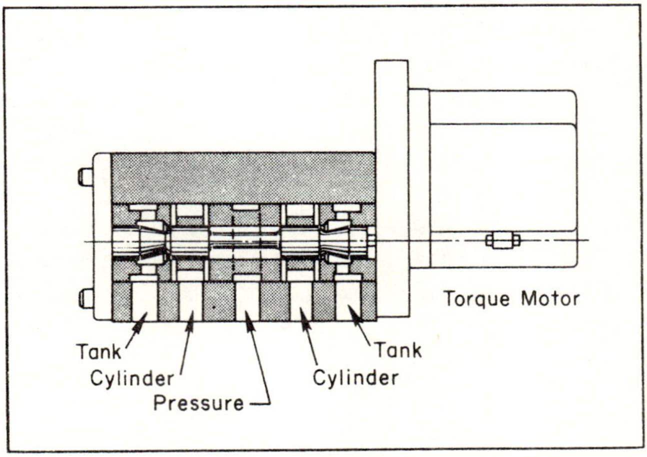

Fig. 5.2 – A single stage electrohydraulic servovalve.

The torque motor is basically a simple permanent magnet motor with limited rotational movement. The output torque T depends on the differential current i across the coils of the motor. It is also affected by the angle of rotation since the motor has an internal stiffness k_s. Thus the output torque can be written:

$$T = k_g i - k_s \theta$$

where k_g = torque motor gain
θ = rotation of the output shaft.

The piezoelectric bimorph bender element of Fig. 5.1(d) can transmit signals at even higher frequencies, in some cases up to 1 kHz. In this device a change in applied voltage causes the element to bend, the movement being converted to a fluid signal by means of a flapper-nozzle. At these high frequencies the dynamic behaviour of the fluid becomes important and signal attenuation will occur if, for example, the size of the flapper-nozzle chamber is too large.

5.3 PROXIMITY DETECTION

Conventional mechanically operated valves (see Chapter 6) and electrical microswitches are used extensively for detecting machine movements. There are problems, however, in using such components. For example the physical size of a pneumatic valve can preclude its use in some cases. The valve may fail owing to operation in a dirty or corrosive environment or because of faulty adjustment of the operating mechanism. Indeed limit valve breakdown is a major contribution to unreliability in pneumatic systems – although on the whole such systems have high reliability. Use of the whisker operated valve of Fig. 5.3(a) reduces the possibility of valve damage due to faulty adjustment. It can be operated by a very small force and is useful for object sensing. The electrical microswitch relies on mechanical action to make or break a contact. It finds much use in on-off systems but it is sensitive to its environment and has a rather flimsy operating mechanism which will not stand rough treatment. The electrical reed switch shows better environmental tolerance and is becoming increasingly popular in electro-fluid systems. They can be very cheap, consisting simply of two reeds inside a glass capsule. The reeds can be made to make or break contact by the presence of a magnetic field, generated either by a coil or by a permanent magnet. The overall length of the device can be as small as 25 mm and speed of switching is 1–2 ms. In fluid power systems it is usually a simple matter to attach a permanent magnet to the end of a cylinder rod, thereby providing the means of switching the reed.

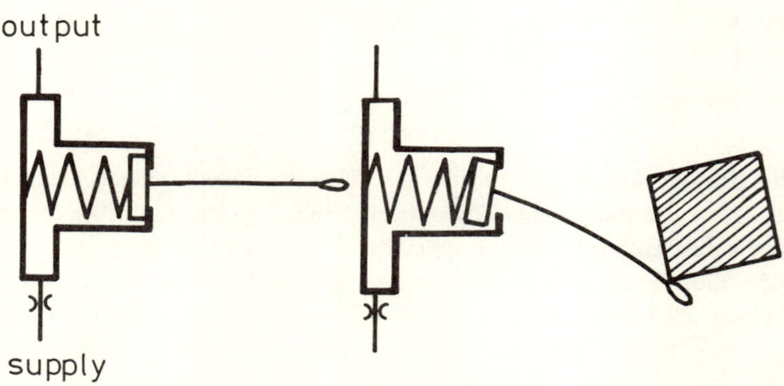

Fig. 5.3 – A 'Whisker' operated pneumatic valve.

Most of the hazards involved with the use of moving part or electrical sensors can be eliminated by means of the back pressure sensor [5, 6], Fig. 5.4. Here the moving object, which may be the end of a cylinder rod, blocks the control port of a flapper-nozzle valve, thus sending a signal to the control circuit. The hole can be very small, perhaps 1 mm, and the supply pressure is determined by the requirements of the rest of the circuit. This sensor is extremely reliable; it cannot be overloaded, is self cleaning, and has no moving parts.

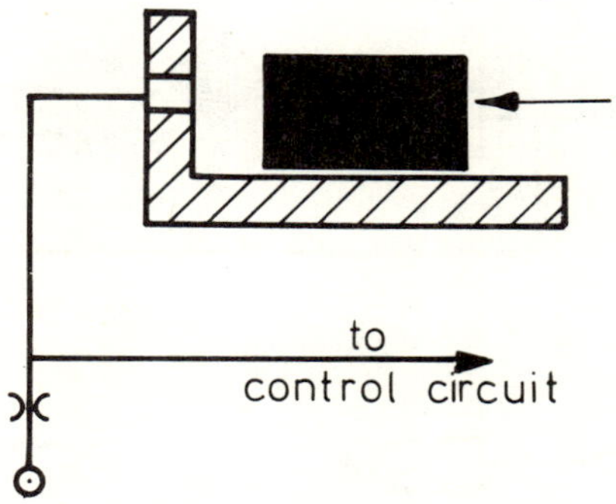

Fig. 5.4 – A back pressure sensor.

When the interruptible jet, Fig. 5.5, is used no physical contact with a component is necessary; hence it is ideally suited for the detection of light or fragile objects. It is the fluidic equivalent of the photocell and is indeed sometimes referred to as the pneumatic eye. It has some advantages over the photocell in that it can operate reliably in dusty, humid and explosive atmospheres. On the other hand it can be sensitive to stray draughts. Fig. 5.5(a) shows the principle. A low pressure air jet is transmitted from a transmitter orifice to a receiver orifice. Some pressure is recovered in the receiver. If the jet is interrupted the pressure in the receiver falls and this in turn can be used to actuate a step-up relay. Typical jets have transmitter diameters of 3 mm, receiver diameter 6 mm, gap 25–50 mm, and supply pressures 0.07–0.2 bar.

There are several possible configurations of jet and receiver. In Fig. 5.5(a) the jet entrains ambient air and the mixture enters the receiver. It is clear that if the device is operating in dusty conditions, such as flour mills or cement works, the receiver and the equipment it feeds can become clogged. This can be avoided by pressurising the receiver as shown in Fig. 5.5(b).

If greater separation of receiver and transmitter is required, the transmitter pressure must be increased so that sufficient signal reaches the receiver [7]. Pressure cannot be indefinitely increased since the jet is a continual source of power loss in the system. Pressures in excess of 0.07 bar are not recommended.

If large objects, 75–200 mm, are to be detected the configuration of Fig. 5.5(c) can be used.

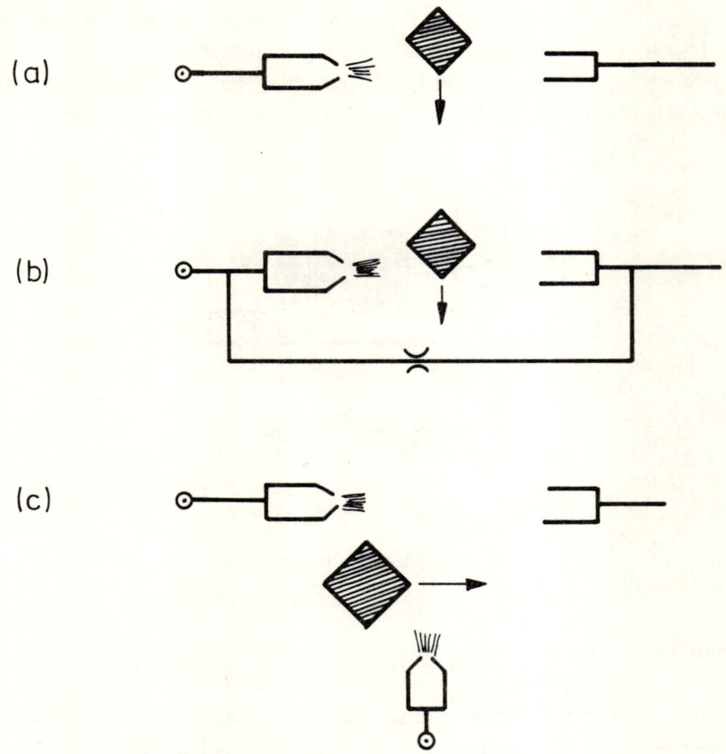

Fig. 5.5 – Interruptible jet sensor.

The cone sensor provides another means of object detection [8]. The basic converging cone sensor is shown in Fig. 5.6(a). It consists of an annular input jet which is concentric with the output passage. When there is no object interfering with the flow pattern, the jet converges to a conical shape, enclosing a bubble or region of reduced pressure. When an object approaches and disturbs this bubble, Fig. 5.6(a), some of the flow is directed into the output causing an increase in output pressure. The range of detection extends to approximately one annular diameter. This range can be increased to several diameters by varying the cone angle as shown in the diverging cone sensor of Fig. 5.6(b). In this case the interfering object destroys the symmetry of the flow pattern causing a change in output.

The range of jet sensors is limited to about 250 mm. Above this value, the power required to transmit a useful signal and vulnerability to stray draughts become serious considerations. When gaps are greater than this limit, the ultrasonic sensor can be useful. It can operate with distances up to 1.5 m. The principle of operation is fairly straightforward. An ultrasonic beam is directed at a turbulence amplifier. When the beam is interrupted, the flow in the amplifier is laminar with a resultant good pressure recovery. When the beam is allowed to reach the amplifier, the flow becomes turbulent and the amplifier turns off. The ultrasonic generator can itself be fluidic, consisting perhaps of an oscillator operating on the principle of the dog whistle.

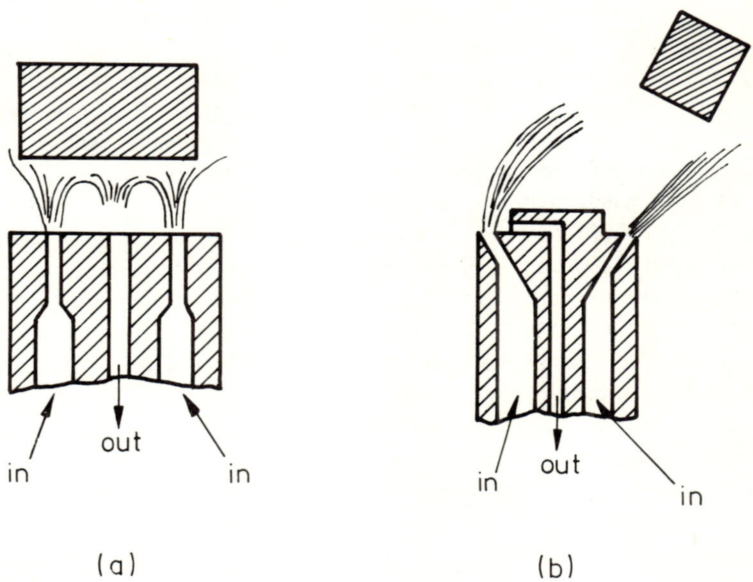

Fig. 5.6 – The cone sensor.

5.4 POSITION MEASUREMENT

Position can be measured by analogue or digital methods. The more common electrical transducers are analogue in operation and base their operation on variation in resistance, inductance, or capacitance, Fig. 5.7.

The potentiometer, Fig. 5.7(a), is the most familiar resistive element [1]. It can employ a wire-wound core, a carbon track or film deposit. Wire-wound potentiometers have the disadvantage that when used in a high gain loop the output signal can display a stepped effect. Linearity between output and input is reduced if the load resistance is high compared to the potentiometer's resistance.

If potentiometer resolution and wear are of concern the linear variable differential transformer (LVDT) may be used, Fig. 5.7(b). This is an inductive transducer and consists of a primary coil wound symmetrically on either side of the centre of the transducer. Alternating current is supplied to the primary and voltages v_a and v_b are induced in the secondaries, which are connected in series opposition giving an output $v_o = v_a - v_b$. The stroke of LVDTs normally varies from about ±100 μm to 25 cm and linearity is of the order ±0.25 per cent.

The third major form of electrical transducer relies on changes in capacitance, Fig. 5.7(c). The capacitance of a parallel plate capacitor is inversely proportional to the separation x and is proportional to the plate area A. The figure shows how change in position can be used to change x and A. It is possible to measure changes in capacitance as small as 10^{-18} F accurately, and so such transducers are often used for precise measurement of displacement.

There are several ways of measuring position by fluid means. In the hydraulic copying servo, Fig. 1.10, the follower which traverses the template profile is integral with the spool of the servovalve and this converts analogue position information to analogue fluid form. The flapper-nozzle can also give accurate analogue position information and has as a result found widespread use in the metrology field, where it is known as the air gauge. Fig. 5.8 shows how the blocked load pressure recovery varies with flapper displacement (or surface variation). When the nozzle is closed, the chamber pressure equals the supply pressure. When the flapper is approximately more than $\frac{1}{4}$ nozzle diameter away, the outlet curtain area exceeds the nozzle area and the device begins to saturate. Fig. 5.8 shows that the gain is approximately $P_s/(d/4) = 4P_s/d$. For example, $P_s = 7$ bar, $d = 1$ mm, gives a gain of 28 bar/mm. Thus a position change of 1 μm gives rise to an output change of about 0.028 bar. This is large enough to operate some fluidic elements and to be detected by manometers.

As an analogue device the flapper-nozzle is restricted to a range of $\frac{1}{4}$ nozzle diameter. For measurement of greater distances, such as the movement of a machine tool, digital methods are more practical. Fig. 5.9 shows how digital incremental encoders can be used for this purpose. The encoder subdivides the physical motion into incremental motions and converts it to digital form by counting the number of increments in the given motion. The incremental motion gives rise to fluid pulses generated either by back pressure sensors Fig. 5.9(a), or by interruptible jets, Fig. 5.9(b). The pulses are counted in a binary counter, the output being a digital measure of position.

This type of encoder needs to be checked against a reference or datum at intervals in order to ensure that a pulse has not been missed. The coded pattern encoder, Fig. 5.10, gives an absolute, rather than a relative measure of position, and hence does not suffer from this disadvantage [9]. In this case each position is defined by a particular arrangement or code of slots. In Fig. 5.10(a) a four bit binary code is used so that the movement can be divided into 16 increments, each being represented by a different four bit number. This system has another advantage over the incremental encoder in that the direction of movement is evident from the succession of the output numbers.

If the receivers in the reading head are not perfectly aligned, all outputs

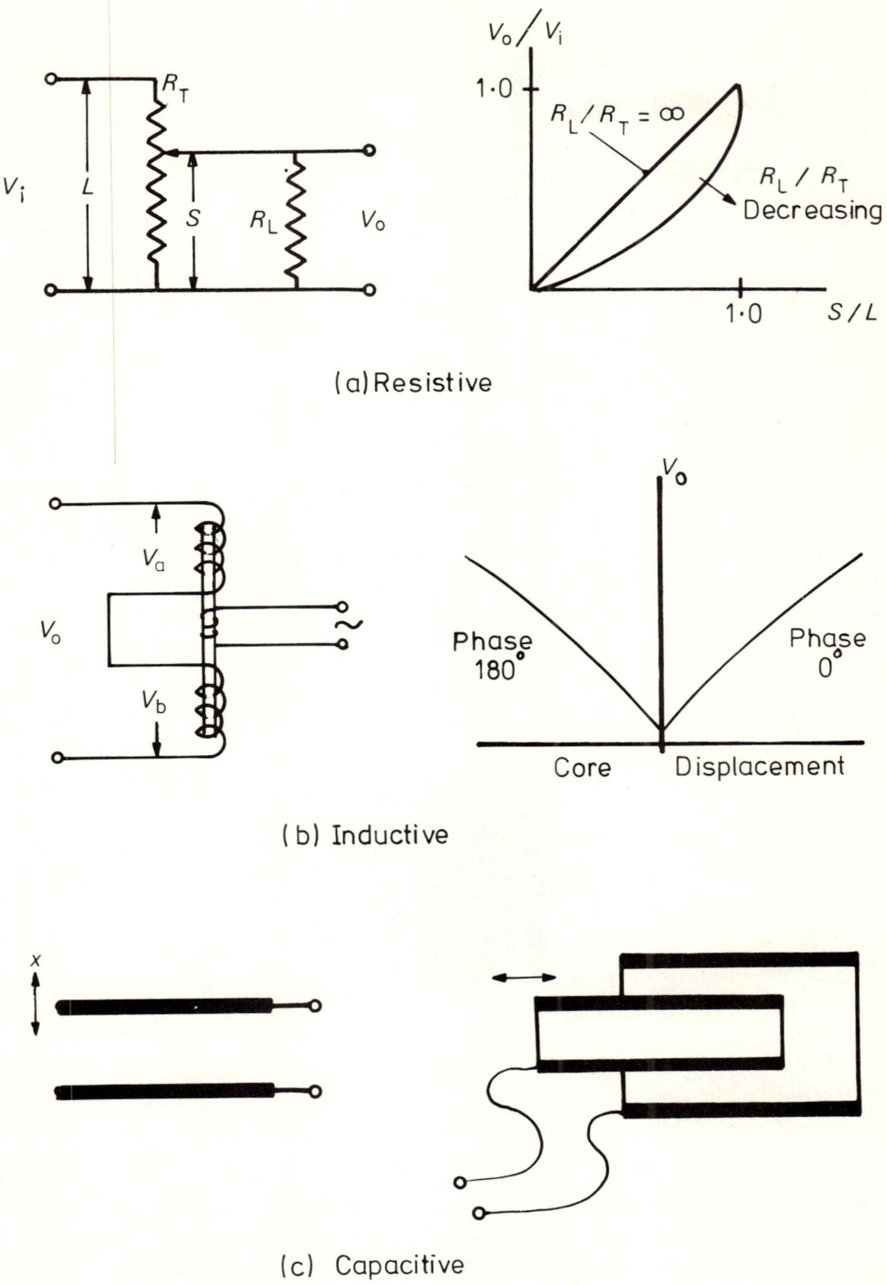

Fig. 5.7 – Some electrical methods of position measurement.

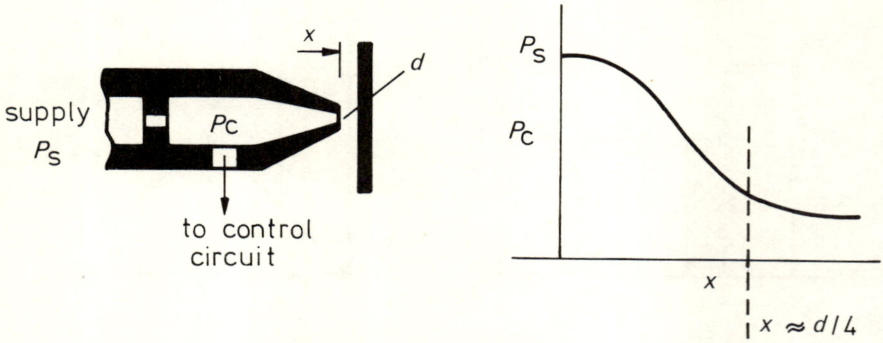

Fig. 5.8 – The back pressure sensor and its characteristic.

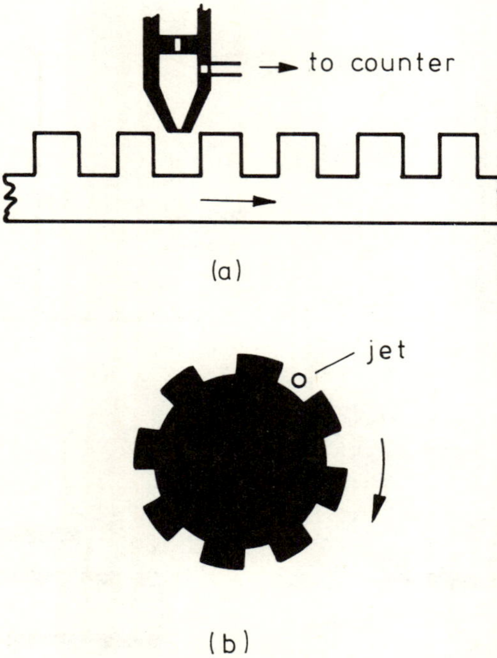

Fig. 5.9 – Examples of digital incremental encoders.

will not change simultaneously at a transition between increments. This results in a false indication of position. For example, station X in Fig. 5.10(a) is the transition point between the 7th and 8th increments. At this point the digital output should change from 0111 to 1000. Note that all the bits must change state simultaneously. Variations of alignment of jets and receivers can cause a random series of numbers to be generated during this small transition interval.

A false output 1111 or 0000 could well arise, and since these indicate movements of 15 or 0 rather that of 8 increments, it is evident that errors can be high.

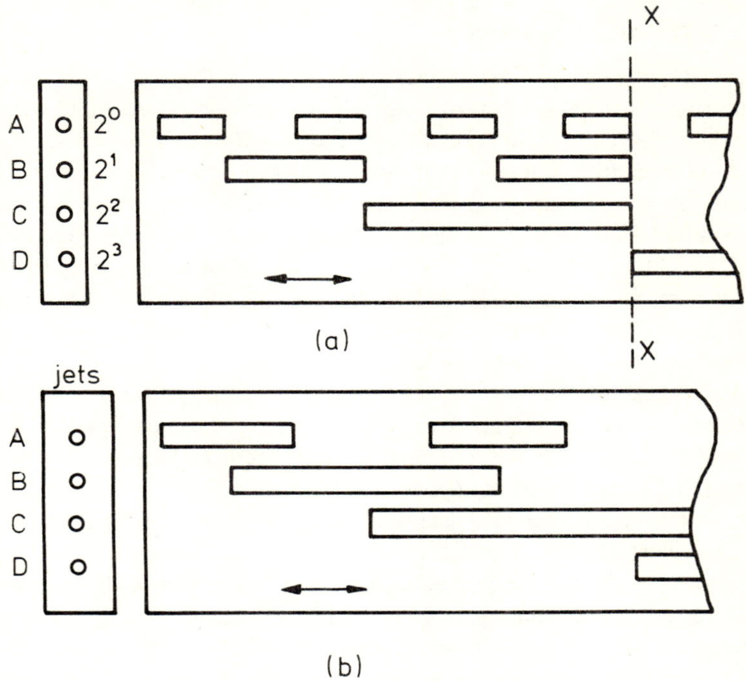

Fig. 5.10 – Examples of coded pattern encoders.
(a) binary code, (b) Gray code.

Use of the Gray code can eliminate these transition errors. This is one of the Unit Distance Codes, in all of which only one bit changes between increments, Fig. 5.10(b). In going from the 7th to the 8th interval in this case, the output changes from 0100 to 1100. Inspection of the figure shows that the ambiguity associated with the binary code cannot arise with the Gray code. Unfortunately, computing cannot be directly carried out with Gray code. It is necessary to convert the Gray code to binary code before computation is attempted, and this can require the use of many cascaded gates.

A high degree of resolution can be obtained by generating light pulses, rather than pressure pulses, using the Moiré interference patterns generated by optical gratings. Moiré fringes are an interference effect taking the form of alternating light and dark bands. They arise when a beam of parallel light is projected through two transparent diffraction gratings of the same pitch, mounted face to face but with their axes slightly misaligned. A translational movement of one grating relative to the other results in a movement of the fringes, thus generating periodic fluctuations in light intensity at some fixed point. This phenomenon is used for position measurement in machine tools, particularly the large ones, and can give accuracies of ±0.001 per cent over a range of 1 cm–3 m.

5.5 VELOCITY MEASUREMENT

Measurement and control of the angular velocity of a shaft is a common problem in engineering. A signal proportional to velocity is often required to stabilise position control systems. The d.c. tachogenerator is probably the most popular device for this purpose and its use is well established. Basically it relies [1] on the generation of a voltage in a coil rotating in a fixed field. A typical tachogenerator could have 11 coils, a size of about 3 cm diameter by 7 cm long, a range of ±6000 rev/min, a gain of 5 V per 1000 rev/min, and a linearity of ±0.01 per cent. Since the voltage generated in each coil varies with the sine of the angle of rotation there is a resultant ripple on the output, but in most cases this is less than 5 per cent of the d.c. output.

There are translational as well as rotational electromagnetic transducers for velocity measurement. A popular one is the moving coil velocity pickup. It can have a gain of 10 mV per mm/s, linearity ±1 per cent and a capability of responding to frequencies in the range 10–1000 Hz.

The vortex rate sensor [10] is one of the best known fluidic elements for measuring angular velocity, Fig. 5.11(a). It provides a differential output which is proportional to the rate of turn about its axis of symmetry. Without rotation, the fluid passes through the porous ring and flows radially to the outlet port. If the sensor is rotated about its axis, a tangential velocity is imparted to the fluid as it passes through the porous ring. As the fluid flows towards the central outlet, tangential velocity increases and the resultant vortex motion in the outlet can be used to give a measure of angular rate. Fig. 5.11(b) shows one type of pick-off in which the total head differential is proportional to angular rate. Clearly the differential is zero when the flow if purely axial.

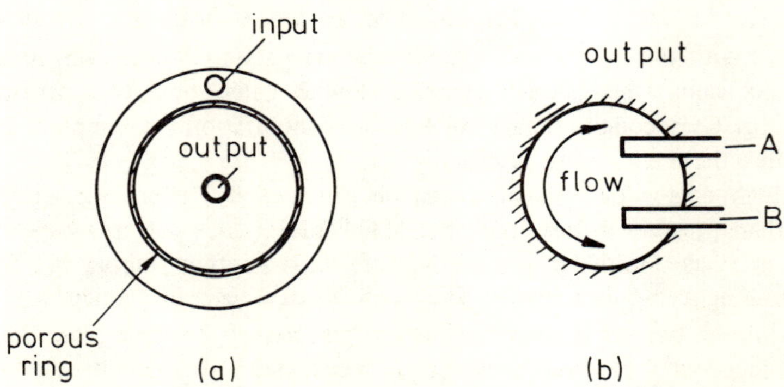

Fig. 5.11 – The vortex rate sensor and a typical pick-off.

Digital speed sensors require some means of producing pulses at a rate proportional to speed. The interruptible jet and slotted plate or disc can be used for this purpose, Fig. 5.12(a). The variation of receiver signal with speed change is shown in Fig. 5.12(b). How can a measure of speed be obtained from these outputs? One method could involve counting pulses over a given time interval, say one second. If there were only one slot in the disc, the counter output would give the speed in rev/sec. This method would require the use of an accurate timing device with a fluidic output in order that the counter could be reset every second. Resolution could be improved by increasing the number of slots.

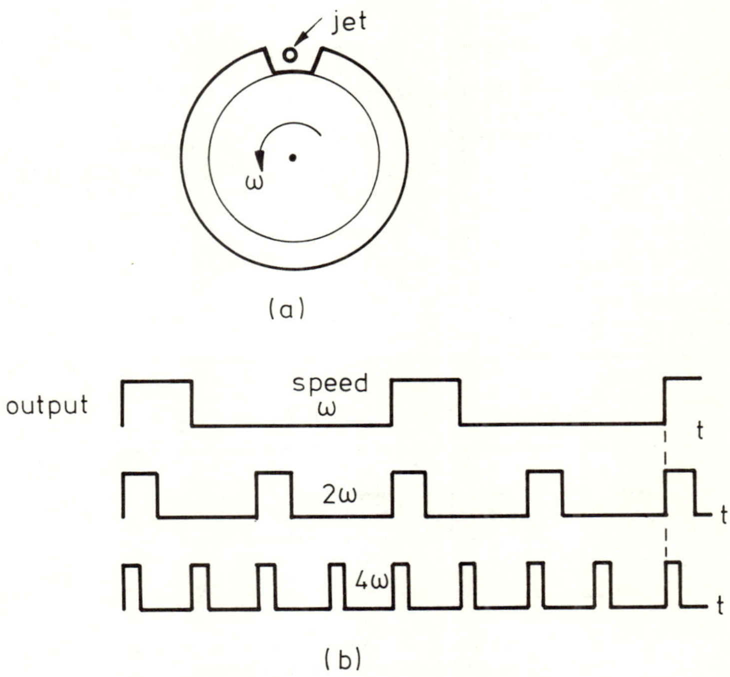

Fig. 5.12 – A method for producing pneumatic pulses at a rate proportional to angular velocity.

Experiments has shown that it is easier to present the output in analogue form. This requires the use of a pulse shaper [5] Fig. 5.13(a). If the output pulses of the interruptible jet of Fig. 5.12(b) are fed to the pulse shaper, trains of constant width pulses are produced.

The incoming pulses initially switch the bistable output to the right, but the build up in pressure at the right-hand control port ultimately switches the output to the left. The time required for this to happen (the duration of the output pulse) is determined by the time constant R_1C. The pulses are still at the shaft speed frequency, Fig. 5.13(b). These are then fed to a fluid volume or capacitance C_2 and orifice or resistance R_2, which in combination act as an approximate integrator (see section 5.8.2). The integrator output is proportional to shaft speed. The combination, pulse shaper plus integrator, is often known as a frequency converter. One might wonder why the pulse shaper is necessary; why not feed the interruptible jet outputs directly to the integrator? Close examination of Fig. 5.12(b) shows that the average of the output pulses is the same at all speeds, and hence if these were fed to the *RC* circuit, its output would be practically independent of speed.

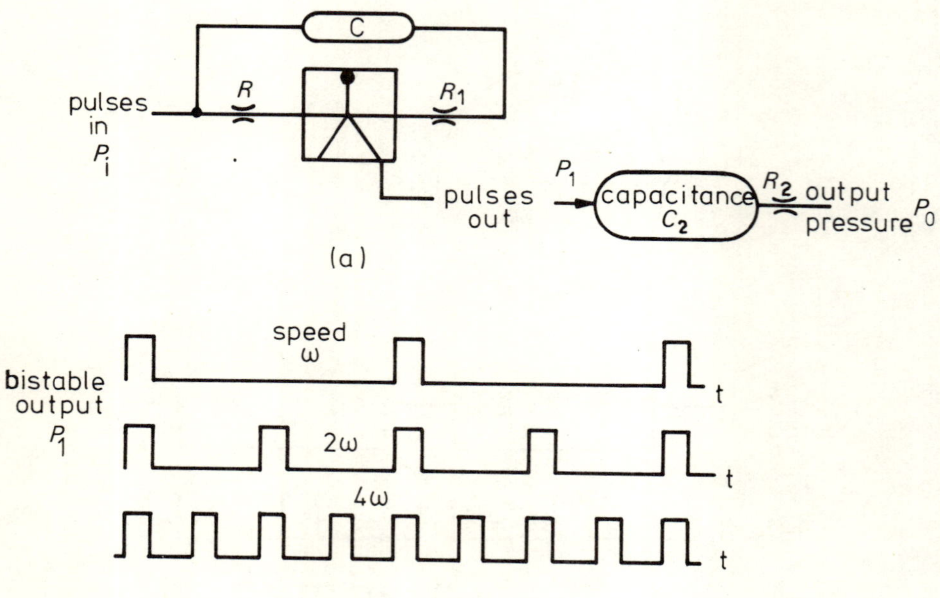

Fig. 5.13 – A frequency convertor giving an analogue output P_0 proportional to the input pulse rate P_i.

The back pressure sensor can also be used to generate pulses at a rate proportional to shaft speed. The wobble plate in Fig. 5.14, is attached to the shaft and as it rotates the outputs from the back pressure sensors are 180° out of phase. Pulse shaping and integration can again be used to give an analogue speed signal.

5.6 FORCE MEASUREMENT

Force sensors use the resulting mechanical deflections to generate signals. In the strain gauge this deflection causes a change in electrical resistance which in turn

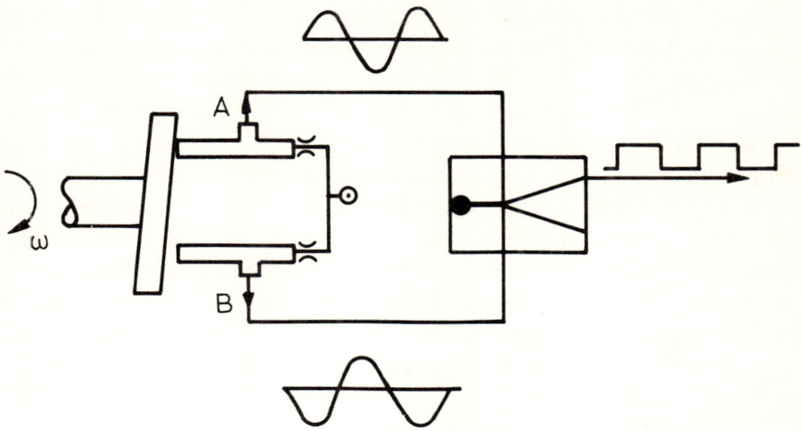

Fig. 5.14 – Use of back pressure sensors to generate pulses at a rate proportional to shaft speed.

changes an output voltage. In force and pressure measuring systems it is usual to employ unbonded gauges rather than bonded gauges. The resistance wires are stretched between several points and are strained by the flexure of a spring element.

Piezoelectric crystals are also used as force transducers [1]. In this case deformation of the crystal generates a potential difference across its faces. Lead zirconate titanate, quartz and rochelle crystals are commonly used piezoelectric materials. These transducers provide no output for steady loads; they are normally used for dynamic situations with frequencies in the range 2 Hz–1 MHz. Forces up to 200 kN can be measured and gain can be as high as 20–100 V/kN.

Fig. 5.15 shows two methods of measurement using fluids. In Fig. 5.15(a) a swash plate and back pressure sensor are used to provide a fluid output proportional to applied torque. Force is measured by the system of Fig. 5.15(b) which uses a cantilever as the flapper in a flapper nozzle valve. This method has been used for controlling thread tension and is also used for providing force

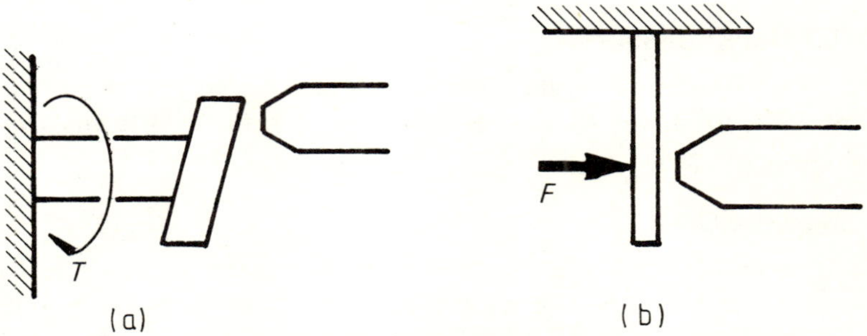

Fig. 5.15 – Use of back pressure sensors for measuring force and torque.

feedback in two stage electrohydraulic servovalves. For example, the valve shown in Fig. 5.16 is so designed that the spool movement is proportional to the current input to the torque motor. It is basically a position control system, the position feedback being achieved by using a spring, the feedback wire, to convert the spool position to a force signal which is fed back to the torque motor. The action can be explained by reference to Fig. 5.16. Assume that the input current generates a torque that drives the flapper to the left. This closes off the left-hand nozzle and opens the right-hand one, and the resultant pressure differential across the spool drives it to the right. The spool will continue to move until the torque on the flapper caused by the feedback wire balances the torque generated by the input current. At this point the flapper will be at mid-position and the main spool will have taken up a position proportional to the input current.

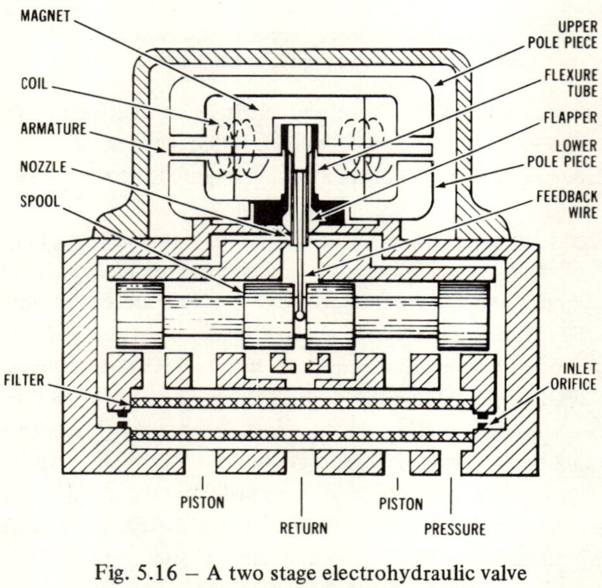

Fig. 5.16 – A two stage electrohydraulic valve with force feedback (Dowty Series 4551).

5.7 SIGNAL PROCESSING

This section will briefly deal with four main aspects of signal processing – amplification, comparison for the purposes of error detection, static and dynamic manipulation of signals.

5.7.1 Amplification

There is room here only for the briefest of discussions of the wide field of electronic amplifiers. Power amplifiers are used to boost power but in the instrumentation amplifier the objective may be to (a) increase the amplitude of the signal, (b) buffer the signal, that is, eliminate interference from the subsequent part of the circuit, or (c) convert a current to a voltage. The d.c. operational amplifier is popular as an instrumentation amplifier. When d.c. amplifiers are used great care must be taken to minimise the effects of drift due to variations in power supplies and changes of component characteristics with time and temperature.

The a.c. amplifer is much less prone to drift. In general, a.c. systems are essential when variable inductance and capacitance transducers are used. They are also recommended when there are considerable pickup effects due to leads from transducers to amplifiers being over long (>3 m).

In fluid power systems the conventional pilot operated valves are used as amplifiers. Depending on the application, they can be viewed as power amplifiers, pressure amplifers, or flow rate amplifiers. Although many fluidic devices are called amplifiers it is usually necessary to employ some additional form of amplification when they are used, and Fig. 5.17 shows one way of doing this. Use of the diaphragm has several advantages; it allows different fluids to be used at input and output; it allows a large amplification of the input; it does not suffer from the stiction forces found in spool valves. Assuming that, for small diaphragm displacements from the seated position, the force on the right-hand side of the diaphragm is $P_o a$ it is easy to show that $P_o/P_i = A/a$. An element with a gain of 100 has operated successfully, inputs being in the range 0-100 mm H_2O. Linearity was excellent over the range 20-100 mm H_2O.

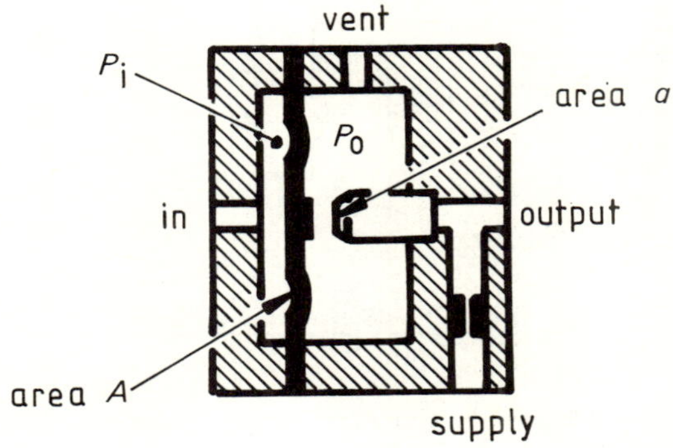

Fig. 5.17 – A pneumatic amplifier.

A disadvantage of this amplifier is the fact that the capacity of the diaphragm chamber makes the response rather slow. For example there are several diaphragm operated three and five-port valves on the market at present and these, although being able to detect inputs as low as 1 mbar, have response times as long as 50 ms. If faster responses are required it is necessary to use the so called step-up relays, made by most manufacturers of fluidic elements. These are rugged, have a three-port valve action, and can switch in about 10 ms.

5.7.2 Signal shaping

This topic covers the application of arithmetic operations to fluid signals and the conditioning of fluid signals for the purposes of improving the dynamic behaviour of the system.

In process controllers, particularly extremal controllers that attempt to maintain an optimum plant output, it is necessary to carry out arithmetic operations using pneumatic signals. The multiplying/dividing device [11] of Fig. 5.18, relies on the properties of air jets for its operation. There are three inputs P_1, P_2 and P_3 and one output P_4. Force balance ensures that the pressure in chamber B equals that in chamber A (P_1). Hence pressure P_1 is picked up by the tube positioned on the axis of the jet supplied at pressure P_2. Now, if the two sets of supply and receiver tubes are identical and are equidistant from their jet nozzles, then:

$$\frac{P_4}{P_3} = \frac{P_B}{P_2} = \frac{P_1}{P_2}$$

or

$$P_4 = P_1 P_3 / P_2 \; .$$

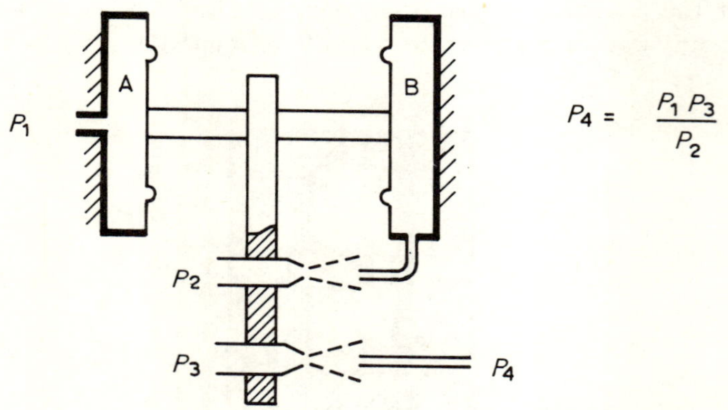

Fig. 5.18 – A pneumatic device for multiplying and dividing, after [11].

For multiplication purposes, P_2 is held constant whilst for division either P_1 or P_3 can be fixed.

The device [13] shown in Fig. 5.19 uses flapper nozzles. There are again three inputs; P_1 applied to the top of the diaphragm, P_3 applied through resistor Z_1 to the bottom of the diaphragm and also to chamber C, P_2 applied through resistor Z_2 to chamber D and also to the output P_4. Z_3 and Z_4 are respectively the impedances of the flapper nozzle connected to chambers C and D. In the steady state, continuity gives:

$$(P_3 - P_c)/Z_1 = P_c/Z_3$$

and $(P_2 - P_4)/Z_2 = P_4/Z_4$

assuming that the load takes no flow.

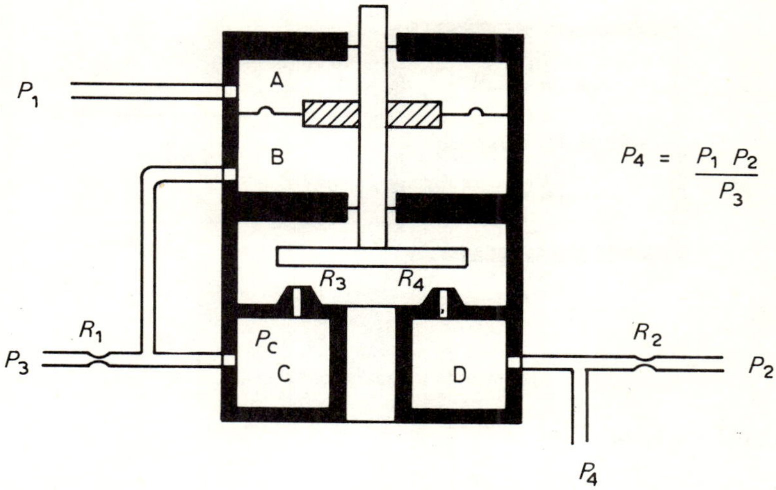

Fig. 5.19 – A pneumatic multiplying and dividing device, after [13].

Force balance ensures that:

$$P_c = P_1 .$$

If the nozzle areas are equal, then since the flapper is common:

$$Z_3 = Z_4 .$$

Combination of these equations yields:

$$P_4 = P_1 P_2 / P_3 .$$

Another example of pneumatic computing devices is the square root calculator [12,13] shown in Fig. 5.20. This uses a combination of laminar (L) and turbulent (T) restrictors. The ouput is pressure P_2 and the input is P_1. If the output impedance is infinite, that is it takes no flow, then:

$$a_T \sqrt{(P_a - P_2)} = a_L P_2$$

where a_T and a_L = appropriate constants for the turbulent and laminar restrictors.

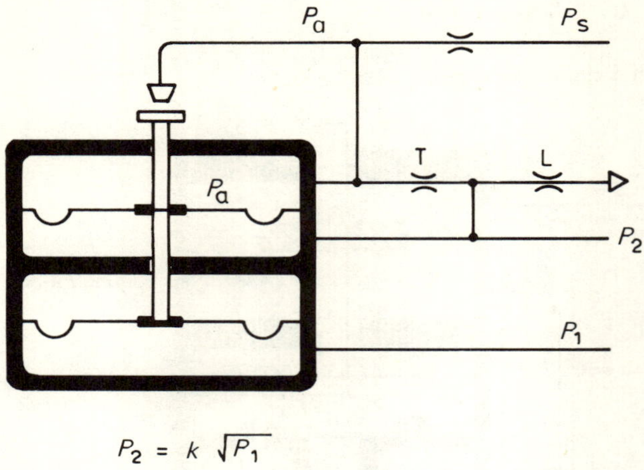

Fig. 5.20 – A pneumatic square root calculator, after [12].

Force balance gives:

$$P_1 + P_2 = P_a\,.$$

Combination of these equations gives:

$$P_2 = k\sqrt{P_1}$$

where $$k = a_T/a_L\,.$$

In addition to these quasi-steady arithmetic manipulations of fluid signals it is often necessary to alter their dynamic characteristics in order to improve the stability of the overall system. This whole field of signal shaping was first developed in relation to electrical circuits, but the vast amount of knowledge gathered there can be applied to fluid systems using the well-established analogies between the two types of system. This will be discussed in greater detail in Chapter 10, so at this stage it is sufficient to state that a fluid system can be reduced to resistive, capacitive and inductive elements; these are related respectively to energy dissipation, compressibility and fluid inertia properties. These elements can be interconnected to produce the required shaping of a fluid signal. For example, in Fig. 5.13 the capacitance and resistance together provide a first order lag between the pressures P_o and P_1, and this approximates to an integration over a limited frequency range.

5.7.3 Error detection

Control systems are error acuated, the error tending to produce a control action to reduce the error. In analogue fluid systems, errors can be measured by a proportional amplifier acting in an analogous manner to an electronic differential amplifier. A signal proportional to the desired value of the variable is maintained on one input, whilst the other input is supplied from a sensor. The net input is then proportional to error. In on-off systems, biased bistable amplifiers or Schmitt triggers can indicate when the variable has exceeded the desired value [5].

The subtraction of two signals can be accomplished by using the two-diaphragm device of Fig. 5.21 [12]. The diaphragms are connected to common stem, to which the flapper of a flapper-nozzle valve is connected. Force balance gives the steady state relationship:

$$P_o = P_1 - P_2 + P_3 .$$

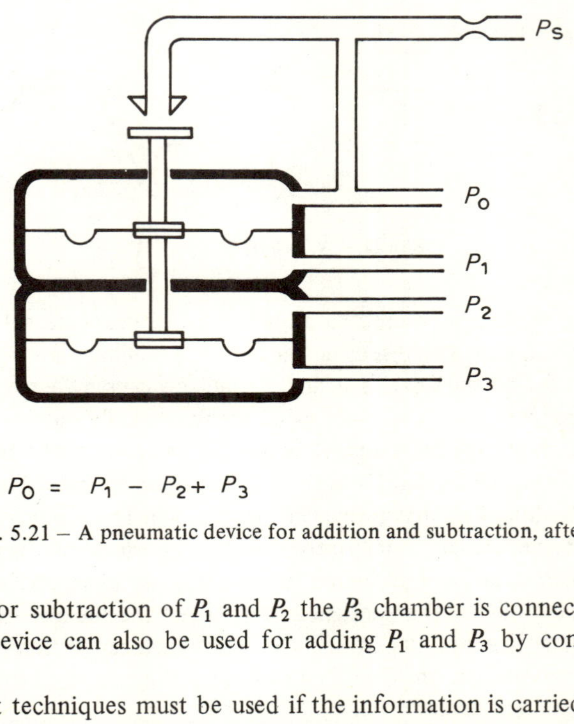

Fig. 5.21 – A pneumatic device for addition and subtraction, after [12].

When used for subtraction of P_1 and P_2 the P_3 chamber is connected to atmosphere. The device can also be used for adding P_1 and P_3 by connecting P_2 to atmosphere.

Different techniques must be used if the information is carried in a.c. form, such as the output of the wobble plate speed sensor of Fig. 5.14. A beat frequency detector, Fig. 5.22, can be used in such cases. This is basically a proportional amplifier which compares the sensor output with a reference or desired output signal. The reference may be generated by a tuning fork heterodyne or by another output sensor operating at the desired conditions. The amplifier is used to subtract the signals, a typical output being shown in Fig. 5.22(b). A beat frequency is developed, being the difference of the two input frequencies.

This is in fact the error signal in a.c. form. It is an easy matter to filter out the input frequencies, provided that they are arranged to be much higher than the beat frequency. The transmission line from the amplifier to the next component may itself be an adequate filter, resulting in a filtered signal like that of Fig. 5.22(c). The filtered signal can now be fed to a frequency convertor whose analogue output will be proportional to error.

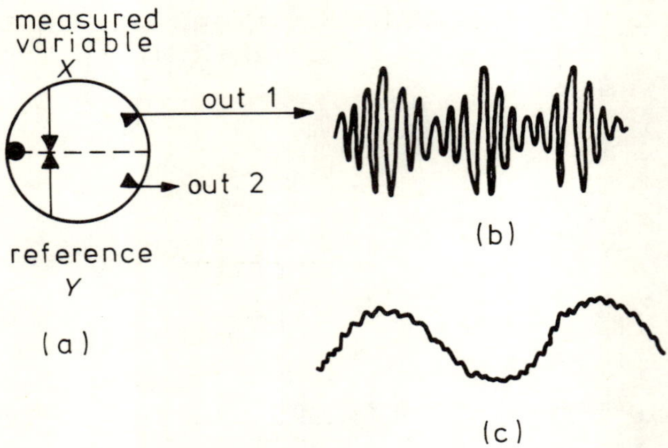

Fig. 5.22 – A beat frequency detector.

Oscillating reeds can be used to provide a signal proportional to frequency error, Fig. 5.23. In shaft speed sensing, air from the wobble plate signal generator of Fig. 5.14 is fed to the nozzles A and B, and this causes the reed to vibrate at shaft speed frequency. As the shaft speed approaches the natural frequency of the reed, reed amplitude increases, allowing the jet from nozzle S to be picked up by receiver O_1. The output O_1 will be a maximum at the natural frequency, the response looking like that shown in Fig. 5.23(a). The receiver is so designed that its recovered signal is relatively steady, the vibration frequency being filtered out. Error can be determined if two reeds are used, one with a natural frequency slightly higher than designed shaft speed and the other slightly lower, Fig. 5.23(b). The two outputs O_1 and O_2 of the reed sensor can be used as the inputs of a proportional amplifier; its output in turn is proportional to speed error over a restricted range.

5.8 REFERENCES

[1] Jones, B. E. (1967), *Instrumentation, Measurement and Feedback*, McGraw-Hill.

[2] Martin, H. R. (1969), *Instrumentation Relating to Mechanical Engineering*, DATA Tech. Publishers.

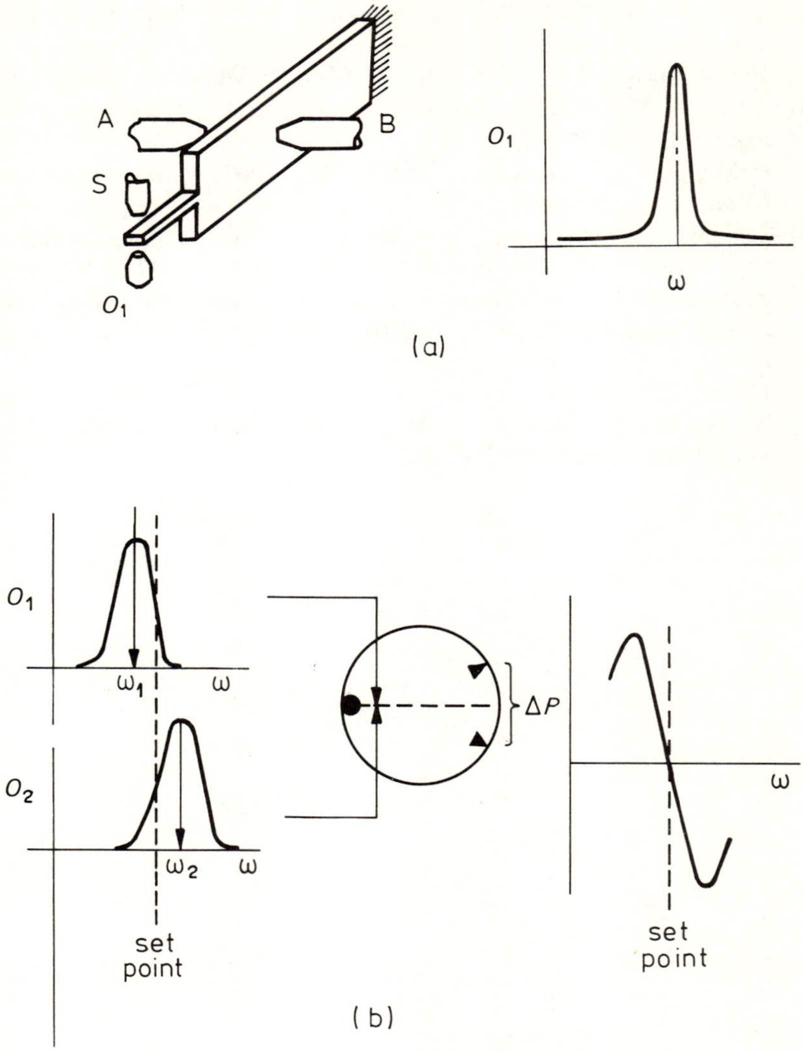

Fig. 5.23 – Use of an oscillating reed to provide a signal proportional to an error in frequency.

[3] Guillon, M. (1969), *Hydraulic Servo Systems – Analysis & Design*, Butterworths.

[4] Blackburn, J. F., Reethof, G., and Shearer, J. L. (1960), *Fluid Power Control*, MIT Press.

[5] Foster, K. and Parker, G. A. (1970), *Fluidics – Components and Circuits*, Wiley-Interscience.

[6] Tanney, J. W. (1972), Fluidics, *Progress in Aero Sciences*, **10**, Pergamon Press.

[7] Fluidic sensors, (1968), *Instrument and Control Eng.*, September.

[8] Auger, R. N. (1967), A fluidic proximity detector, *Proc. 3rd Cranfield Fluidics Conf.*, Paper E7.

[9] Parker, G. A. (1965), Digital fluid position encoders, *Proc. 1st Cranfield Fluidics Conf.*, Paper E2.

[10] Sarpkaya, T., Goto, J. M., and Kirschner, J. M. (1967), A theoretical and experimental study of a vortex rate gyro, *Proc. A.S.M.E. Fluidics Symp.*

[11] Zalamanzon, L. A. and Semikova, A. I. (1968), Investigation of pneumatic jet-tube elements, *Pneumatic and Hydraulic Control Systems*, Pergamon Press.

[12] Ferner, V., Pneumatic controller elements, ibid.

[13] Ivlichev, Y. I. and Nadzhafov, E. M., Universal pneumatic multiplying device and device for square rooting, ibid.

CHAPTER 6

Valves and simple circuits

6.1 INTRODUCTION

Valves fall into three main categories – directional control, flow control and pressure control. This chapter introduces the commoner types and shows their application in simple circuits [1, 2]. More complicated circuitry will be met in subsequent chapters.

6.2 DIRECTIONAL CONTROL VALVES

These consist of two or more ports so arranged and connected that a controlling or actuating signal can direct the fluid to various places in a system. Such valves are classified by the number of ports they are designed to interconnect.

6.2.1 Two-port valves

Fig. 6.1(a) shows a spool valve in which static pressure balance is achieved by the use of two lands or pistons. There are two external ports in the body of the valve and hence it is referred to as a two-port valve. Two-port poppet valves are also popular. The valve shown is held in the closed position by a spring and is operated mechanically. It is closed in its normal or unactuated position. The number of operating positions of a valve is important and must be specified. Most fluid power valves are used in sequential on-off circuits, and in these cases the valve need only have two or sometimes three operating positions. At the other extreme are proportional action valves, in which the flow rate varies continuously with the input signal. These valves have theoretically an infinite number of operating positions. The valve of Fig. 6.1 is designed to operate in only two positions, as shown, and its complete description is two-port, two position (2/2), mechanically operated, spring return, normally closed spool valve. Fortunately there is a simple symbol (Appendix A) for the valve, Fig. 6.1(b), in which a basic envelope is divided into two squares, representing the two positions of the valve [3]. In circuit diagrams valves are usually drawn in the unactuated position. When actuated, the envelope is envisaged to move to the right relative to the two external ports.

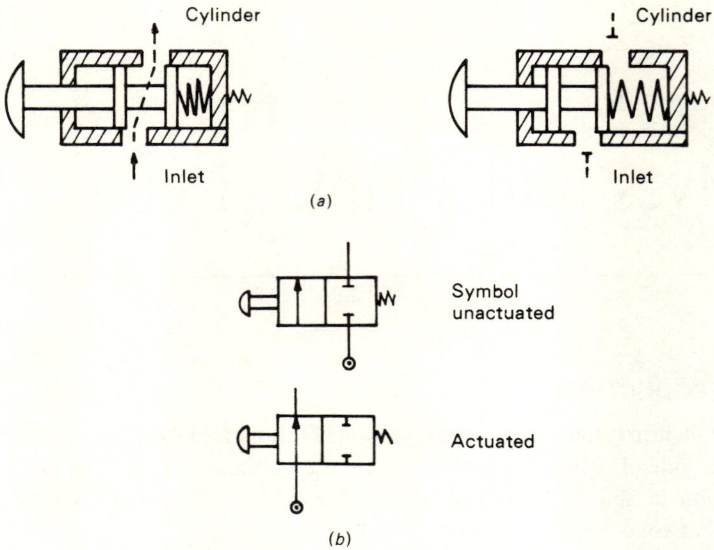

Fig. 6.1 – A two-port valve and its symbol.

Two two-port valves can control a single acting cylinder, Fig. 6.2. One is used to pressurise and the other to exhaust. In sequential control the directional control valve should offer as little resitance to flow as possible; it is used to change the direction of flow, not to throttle it. There are, however, cases where on-off or two position control is not satisfactory and proportional control must be used. Throttling is achieved by making the valve element capable of assuming any position between fully open and closed. Such valves are symbolised by the addition of horizontal bars at the top and bottom of the envelope in Fig. 6.1(b) (as in Fig. 6.4).

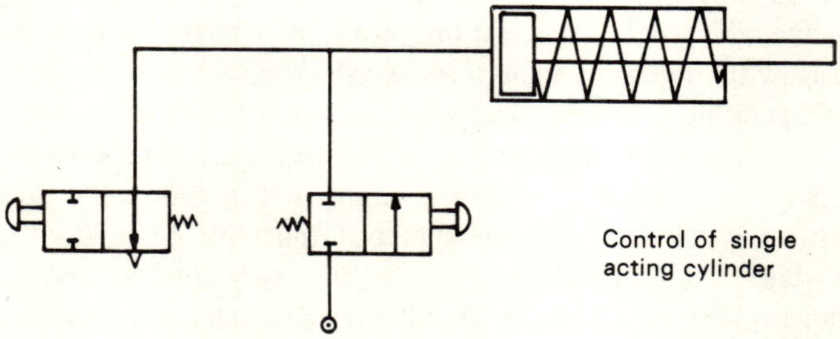

Fig. 6.2 – Two two-port valves controlling a single acting cylinder.

6.2.2 Three-port valves

The three-port valve has three external ports. It is used alternatively to pressurise and exhaust the working port or cylinder port. Fig. 6.3(a) shows a three-port, two position (3/2) mechanically operated, spring return, spool valve. Fig. 6.3(b)

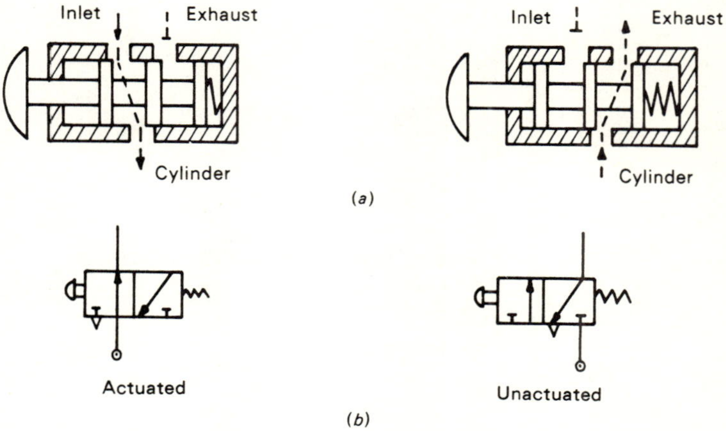

Fig. 6.3 – A three-port valve and its symbol.

shows the standard symbol. Static pressure balance in both positions requires the use of three lands in this case, Fig. 6.4 shows how a proportional three-port valve can be used to control the speed of a single acting pneumatic cylinder. A two position valve can be connected in this manner if only two cylinder positions are required. If the load needs to be stopped at intermediate positions, then a three position, three-port valve could be used. In the third position the fluid is trapped in the cylinder when the load has reached the desired position.

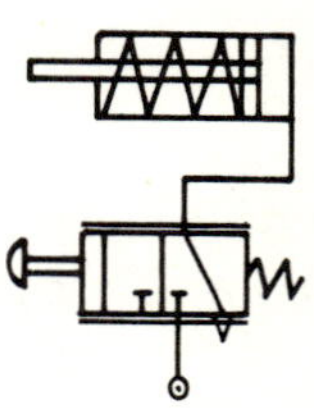

Fig. 6.4 – A proportional three-port valve controlling a single acting cylinder.

A three-port, three position valve and its symbol are shown in Fig. 6.5. There are other types of mid-positions but these will be discussed later. Double acting cylinders can be controlled with two three-port valves so arranged that when one valve pressurises one end of the cylinder, the other valve exhausts the other end and vice-versa, Fig. 6.6.

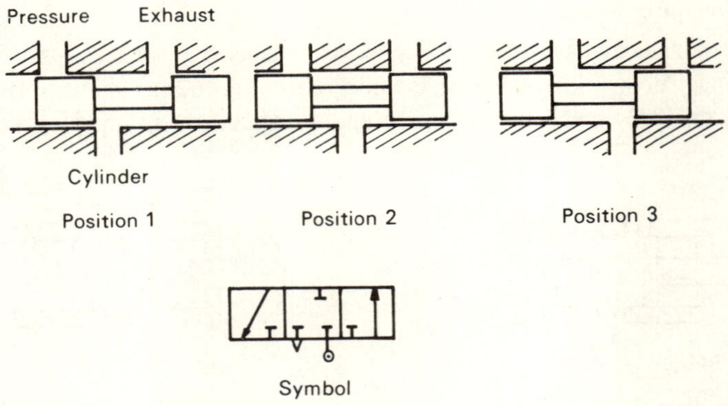

Fig. 6.5 – A three-port valve with closed mid-position.

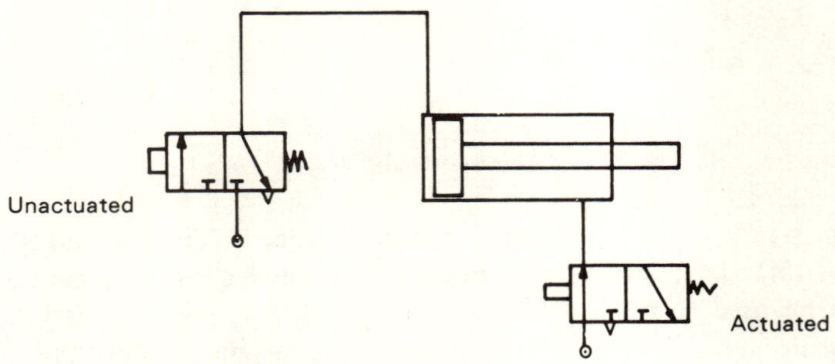

Fig. 6.6 – Two three-port valves controlling a double acting cylinder.

6.2.3 Four and five-port valves

The five-port valve, Fig. 6.7(a), combines the functions of two three-port valves. It alternatively pressurises or exhausts two working or cylinder ports. Note that four lands are needed to ensure static pressure balance in both operating positions. Some of these valves, although having five internal ports, only have four external ports; for example in the valve in Fig. 6.7(a) the two exhaust lines could have been connected internally and brought out as one external port in the valve block. Fig. 6.8(a), the well known D valve, and Fig. 6.8(b), the rotary spool, illustrate other types of four-port valves. Only one four or five-port valve is needed to control a double acting cylinder, Fig. 6.9.

As a four or five-port valve shifts between its extreme positions it passes through the very important neutral position. If the spool can be held in this position for any length of time, the valve is a three position valve. As with the

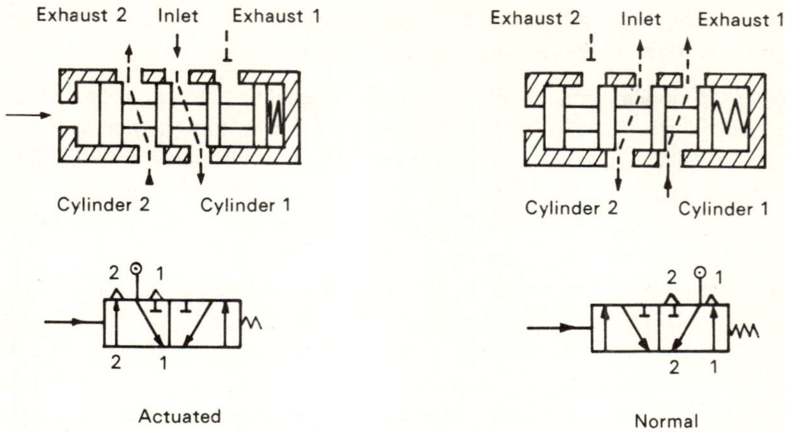

Fig. 6.7 – A five-port valve and its symbol.

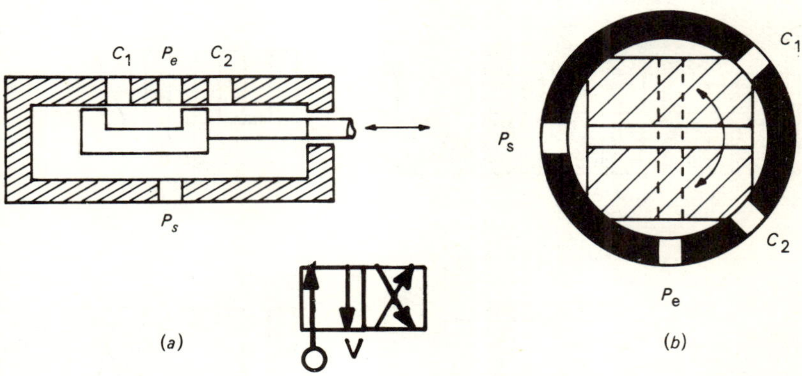

Fig. 6.8 – Four-port valves.

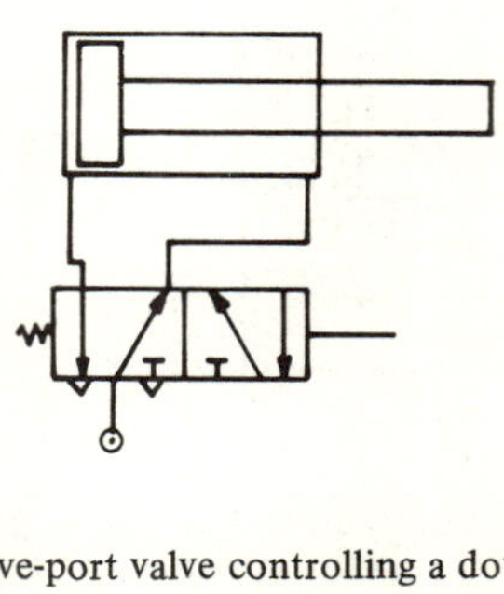

Fig. 6.9 – A five-port valve controlling a double acting cylinder.

three-port valve, there are many different possible connections between ports in the mid-position. Three such will be discussed here:

(a) In the open centre valve, Fig. 6.10(a), all lines are interconnected in mid-position. The width of a spool land is less than the port width and the valve is said to be **underlapped**. Hence in hydraulic systems flow is permitted back to tank at relatively low pressure, eliminating the heat generated in relief valves when closed centre valves and constant delivery pumps are used.

(b) In the closed centre valve, all lines are closed off in the mid position, Fig. 6.10(b). Here the valve is **overlapped**, the land width being greater than the port width. This enables a cylinder to be held at an intermediate position, but there are other reasons for using this type of valve. For example, if a hydraulic circuit employs an accumulator, then a closed centre valve will prevent loss of accumulator fluid while the spool moves from one extreme to the other.

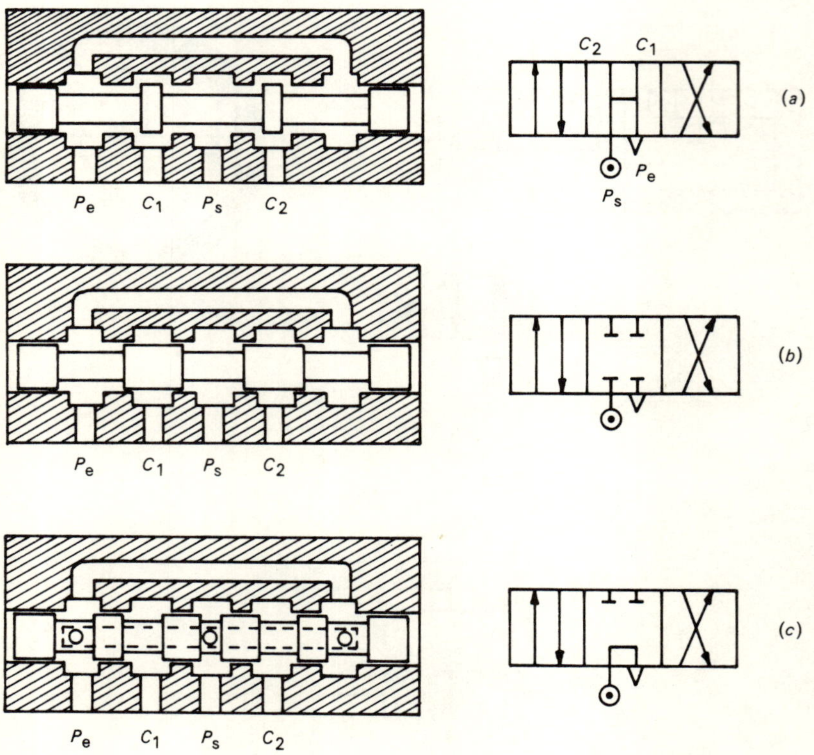

Fig. 6.10 – Three-position, four-port valves (a) open centre, (b) closed centre, (c) tandem centre.

(c) The tandem centre valve, Fig. 6.10(c), combines the advantages of closed and open centre valves. In this case, although the pump is connected to tank in mid-position, minimising heat generation, the cylinder ports are blocked, holding the load rigidly in position.

6.2.4 Servovalves

Where accurate control of position is demanded, a servovalve is essential. The two and three position valves discussed above find their application in what are essentially open loop systems. In a closed loop system the valve must respond not only to input changes but also to output changes. This is the principle of feedback. Most automatic control systems are proportional, in the sense that the controller output is proportional to controller input, and experience has shown that such systems can be very accurate. If the controller uses fluid power then the valve must be proportional in its action and infinite position valves are used. Because of their association with servo systems, these valves are called servovalves [4, 5, 6].

6.2.5 The three-port servovalve

Fig. 6.11 shows three different types of three-port valve, driving a single acting cylinder. Each valve has three ports or connections: supply, exhaust and cylinder.

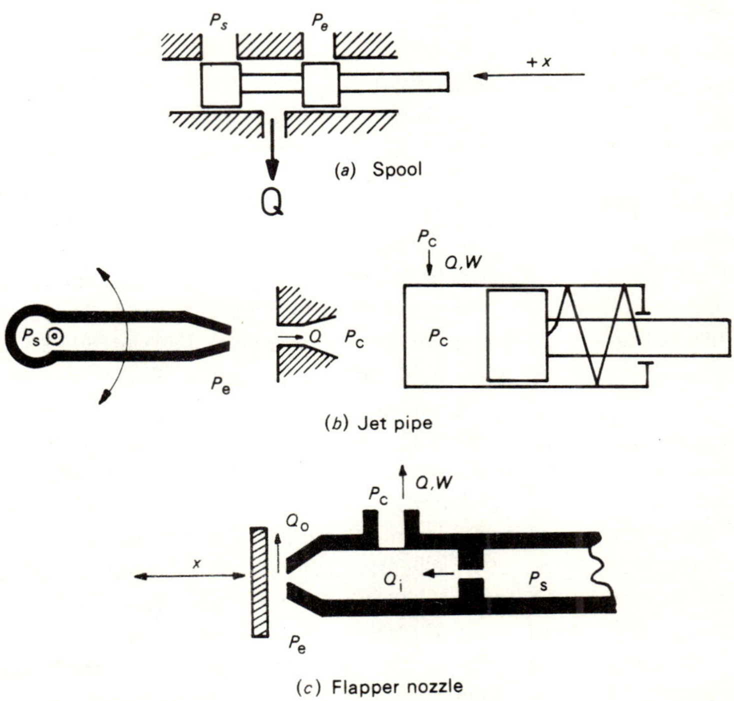

Fig. 6.11 – Three-port servovalves driving a single acting cylinder.

The spool, as shown earlier, can be open or closed centre. Although the spool valve is the most difficult to manufacture because of the close tolerances required, it is nevertheless the most popular servovalve. It is easy to achieve static pressure balance within this valve, and this is an important factor in modern high pressure systems. The valve's behaviour in mid-position can be easily arranged to be open centre, closed centre or critical centre (zero lap) by varying the amount of lap. The spool valve, however, requires a well filtered fluid, otherwise axial friction forces may build up or orifices may become choked.

In the flapper-nozzle valve, displacement of the flapper towards the nozzle makes cylinder pressure P_c increase. The flapper will lose control if it is moved further than one quarter nozzle diameter away from the nozzle. Open centre valves, like the flapper-nozzle and the jet pipe, are relatively insensitive to contaminated fluids and have a self cleaning tendency. They, however, suffer from high quiescent power loss and are hence usually only found in low power applications such as the first stages of two-stage valves. The relatively large clearances in jet pipe and in flapper-nozzle valves allow them to be manufactured more easily and cheaply than spool servovalves. Another important advantage of the jet pipe valve is its low actuating force. These forces can be high in flapper-nozzle valves (see Chapter 8).

6.2.6 Four and five-port servovalves

The spool valve shown in Fig. 6.12(a) has five ports. The jet pipe and flapper-nozzle valves could each have four external ports – supply, exhaust and two cylinder ports. Again because of their open centre characteristics, the jet pipe and flapper-nozzle valves are restricted to low power applications.

6.2.7 Flueric valves

Flueric devices such as the proportional amplifier of Fig. 1.4 and the bi-stable amplifier of Fig. 1.5 are in the main designed for signal processing and detection, but several researchers have examined the possibility of using them as power modulating valves [7, 8]. They have severe limitations, however, when used for this purpose. For example when used with oil, cavitation effects can inhibit the reattachment phenomenon in bi-stable elements and can reduce signal/noise ratios in proportional amplifiers. In addition their power recovery is poor.

There is one flueric device, however, that has the ability to handle considerable power with a reasonable efficiency. It is the vortex amplifier [9, 10], Fig. 6.13. In its simplest form it can be considered as a variable orifice. It includes, Fig. 6.13(a), a main power jet which enters a circular chamber and leaves through a hole in the centre of the chamber. (In some devices there are holes in both top and bottom plates). A control jet is placed at right angles to the incoming power jet. With no control flow the amplifier presents simple orifice resistance, the outlet pressure being determined by the pressure losses in inlet and outlet orifices. When the control jet is present the fluid in the chamber moves in a

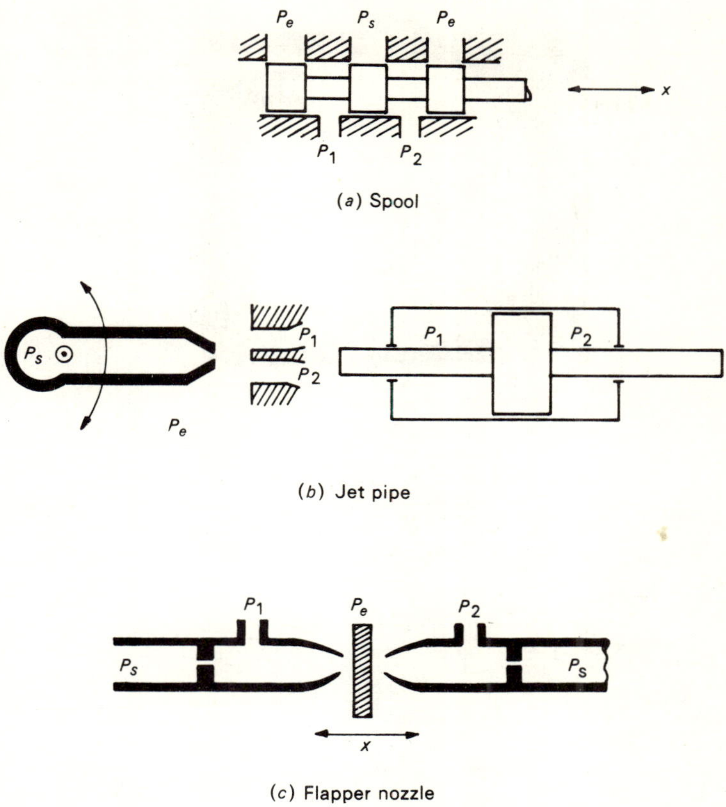

Fig. 6.12 – Five-port servovalves driving a double acting cylinder.

spiral path. The centrifugal pressure drop across the vortex chamber reduces the pressure at the outlet orifice, and hence the output flow rate decreases. The vortex amplifier has a negative gain, an increase in control flow causing a reduction of output flow. The control pressure usually has to be 20–50 per cent higher than the supply pressure – a disadvantage of vortex amplifiers. As the control flow is increased a stage is reached when the supply is completely cut off, and under these conditions the outlet flow from the device is equal to the control flow. The ratio of the maximum to the minimum output flow rate at a given supply pressure is called the **turn down ratio**. This is typically 5–10 but can, by careful design and at the expense of stability, be raised as high as 25.

The fact that the output flow rate cannot be reduced to zero may be an embarrassment is some applications, but this can be overcome by venting, Fig. 6.13(b). The receiver tube in this case is not attached directly to the body of the valve. It is separated from it by a vent. As the control flow increases the

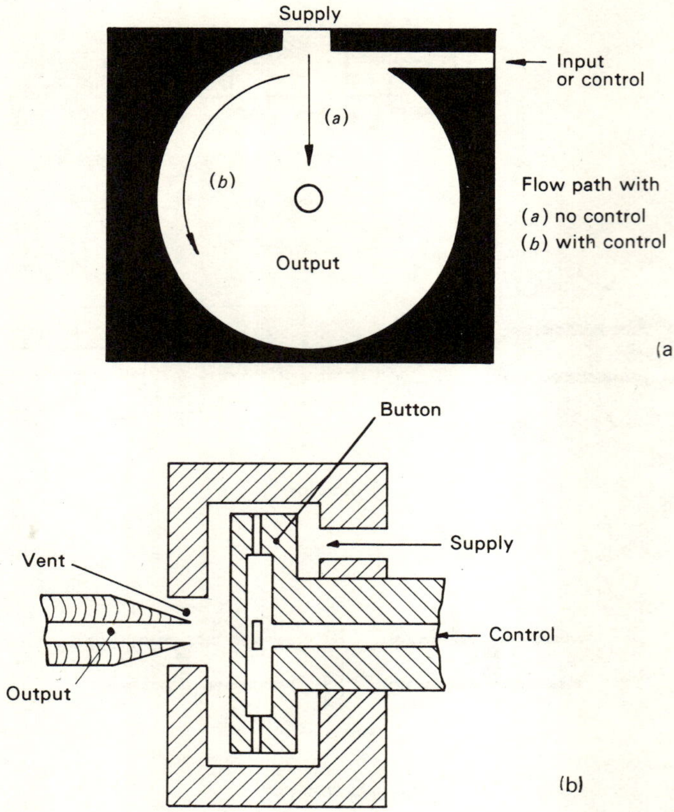

Fig. 6.13 – Vortex amplifiers.

amount of swirl in the outlet increases until ultimately the outgoing fluid spreads out in a hollow cone and does not enter the receiver tube. The figure also illustrates another design of vortex valve – the button vortex amplifier. In the simple vortex device of Fig. 6.13(a) the flow field is extremely non-uniform when there is no control flow. This is basically because there is only one supply port. As a result there is poor mixing of the supply and control flows, and indeed under some circumstances two contra-rotating vortices may form. Mixing can be improved by providing several inputs and controls around the circumference. The button configuration is extremely effective in this respect. In this case, the supply enters the vortex chamber uniformly through an annular gap in which mixing takes place. The control flow may come from the button or the vortex chamber wall.

Compared to other flueric amplifiers, the vented vortex amplifier has a high power recovery and thus has shown promise as a non-moving part servovalve.

Its use in place of the conventional spool valve offers several advantages:

1. There are no sliding spools and hence threshold and contamination effects are minimal.
2. Environmental tolerance is good; hot gases or liquid gases can be used.
3. Fewer precision machined parts are needed and tolerances are not so tight.
4. Because there are no moving parts the bandwidth can exceed 100 Hz.

In the arrangement shown in Fig. 6.14, two vented amplifiers are driven in a push-pull manner by a proportional amplifier. Increase of control signal C_1 results in a reduction of control pressure at amplifier 1 and an increase of control pressure at amplifier 2. As a result pressure P_1 increases and P_2 decreases. Note that each amplifier is basically a pilot operated three-port valve; the three ports are the supply, the vent or tank port and the outlet. The circuit is essentially the same as that of Fig. 6.6.

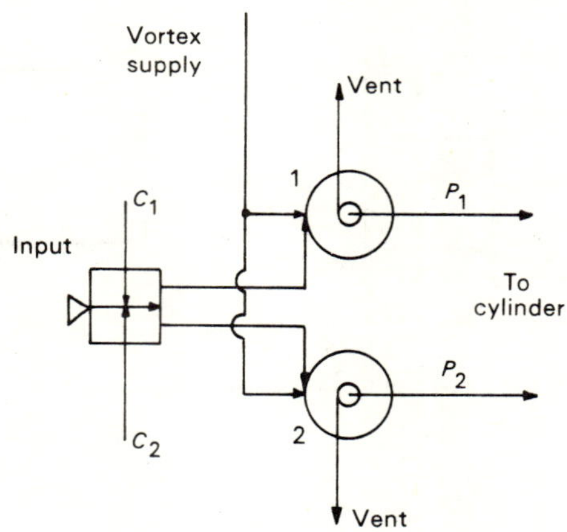

Fig. 6.14 – A servovalve using vortex amplifiers.

Vortex servovalves have been successfully applied in helicopter control systems. A case described in [11] used MIL-H-5606 oil at a pressure of about 100 bar.

6.2.8 Other directional control valves

There are two other directional control valves that do not fit neatly into the above described system of classification. These are (a) the shuttle valve and (b) the non-return valve. The shuttle valve, Fig. 6.15, provides a means for connecting two lines to one output. The higher pressure source is connected to the outlet and the two sources do not interfere with each other. It is basically a logic OR gate.

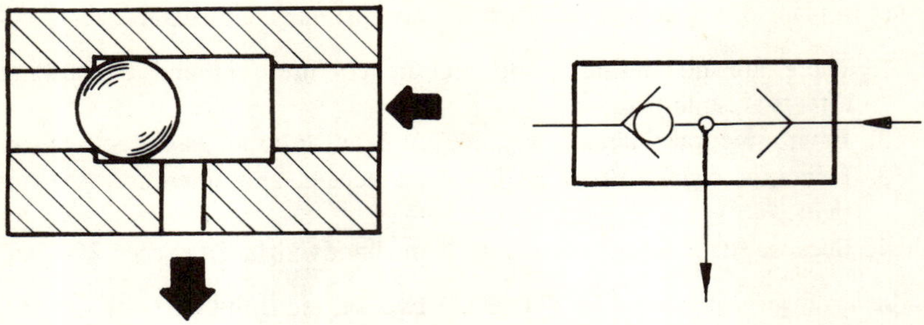

Fig. 6.15 – A shuttle valve and its symbol.

The non-return valve only allows flow in one direction. In the valve shown in Fig. 6.16 flow can pass from left to right by lifting the poppet off its seat. If the pressure on the right increases above that on the left, the poppet is forced on to its seat, stopping reverse flow.

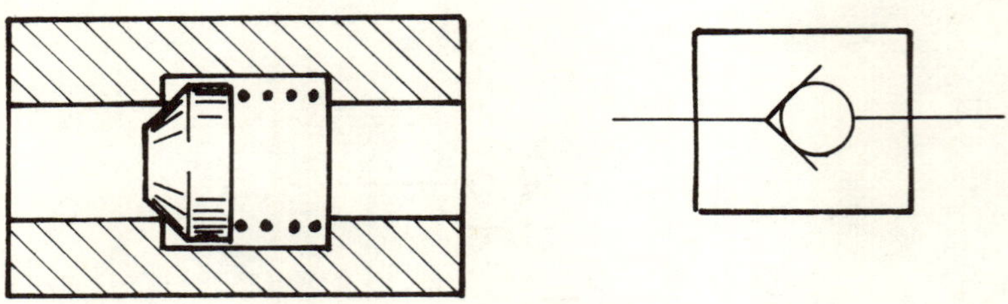

Fig. 6.16 – The check valve or non-return valve.

6.3 FLOW CONTROL

There are many cases where the speed of a piston must be fairly closely controlled. Sometimes it is necessary to make a cylinder rod retract more slowly than it extends. The required speed control can be achieved by varying the flow rate to the cylinder, using a flow control valve. Accurate speed control of pneumatic cylinders is difficult to achieve with this method because of the inherent compressibility of the air.

6.3.1 Flow control valves

The simplest of all flow controls is the needle valve, Fig. 6.17, which restricts flow in both directions. Flow to or from the cylinder is reduced by screwing down the needle.

In many systems, however, it is desirable to restrict the flow in one direction only and this can be achieved by putting a non-return valve in parallel with the restrictor. Both elements are usually combined in one valve block, Fig. 6.18.

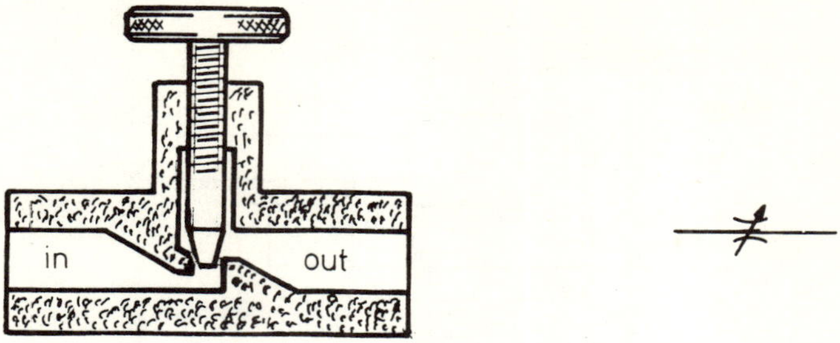

Fig. 6.17 – A needle valve and its symbol.

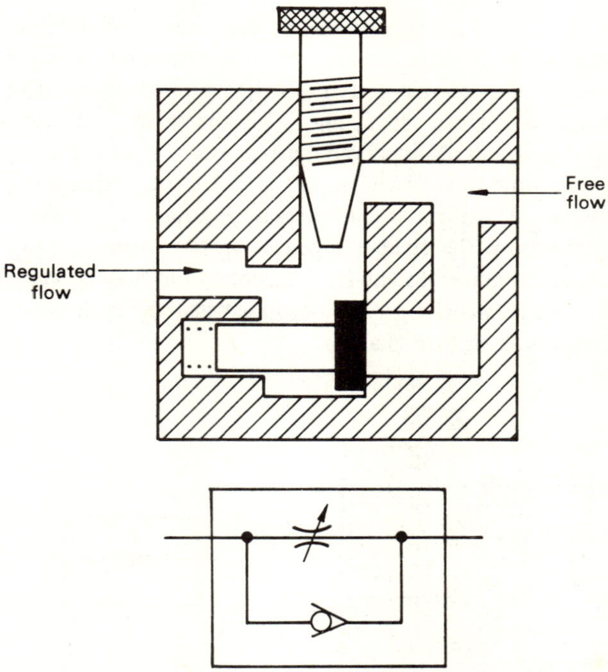

Fig. 6.18 – Valve for flow control in one direction.

Flow from left to right must find its way past the orifice, whose setting is determined by the screw. Flow from right to left forces the disc down against the spring force and allows unrestricted flow.

In pneumatic systems cylinder speed control is best achieved by installing the flow control valve in the exhaust line, so that the cylinder has to force air out through the valve. Best valve pressure drop conditions are obtained this way. This is known as **meter-out control** and has the advantage of being able to control overhauling loads. A circuit is shown in Fig. 6.19, in which the cylinder rod is controlled whilst extending but is free to retract quickly.

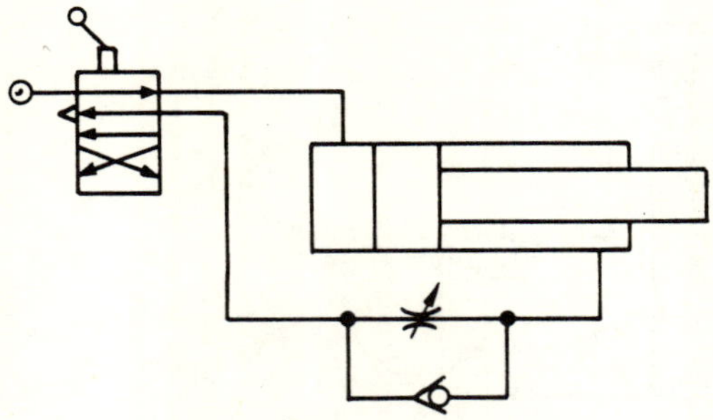

Fig. 6.19 – A meter-out speed control circuit.

Hydraulics must be used if accurate speeds are required. Unfortunately the simple restrictor has shortcomings since the flow rate through such a device is dependent both on the orifice size and the orifice pressure drop. The system load dictates the orifice pressure drop, Fig. 6.19, and hence the flow rate and rod speed are load dependent. The pressure compensated flow control valve has been designed to overcome this difficulty. In Fig. 6.20, the setting of the needle valve fixes the flow rate. If the outlet load pressure P varies, the spool moves in a direction to compensate for the change in orifice pressure drop. For example, if pressure P falls, the spool moves downwards restricting the flow Q_1 and reducing the pressure P_1. By this means the pressure drop across the needle valve is maintained constant irrespective of load changes.

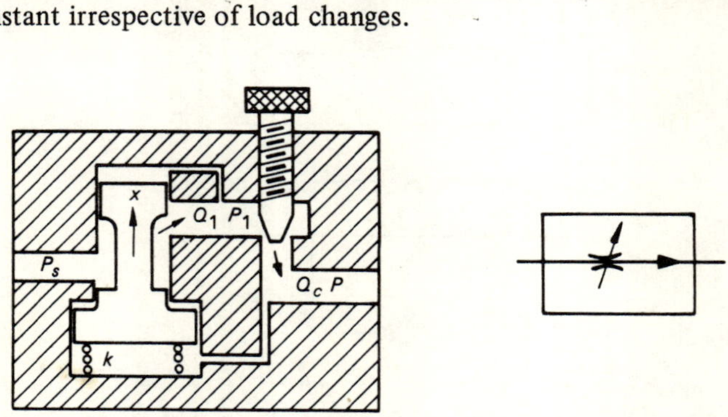

Fig. 6.20 – A pressure compensated flow control valve.

6.3.2 The quick exhaust valve

When air is exhausting from a valve-operated pneumatic cylinder it meets some resistance as it traverses the exhaust line and the valve ports. This limits the speed of the cylinder rod. If fast cylinder rod speeds are required, a quick exhaust valve is used. This exhausts the cylinder directly to the atmosphere through a large port that presents negligible resistance to flow. A quick exhaust valve and its symbol are shown in Fig. 6.21. When the cylinder is being pressurised flow passes from the inlet via the annular passage B to the cylinder. Pressure, through A, acting on the top of the diaphragm closes off the exhaust line. When the inlet line is exhausted the pressurised air in the cylinder lifts the diaphragm and allows the cylinder to exhaust rapidly.

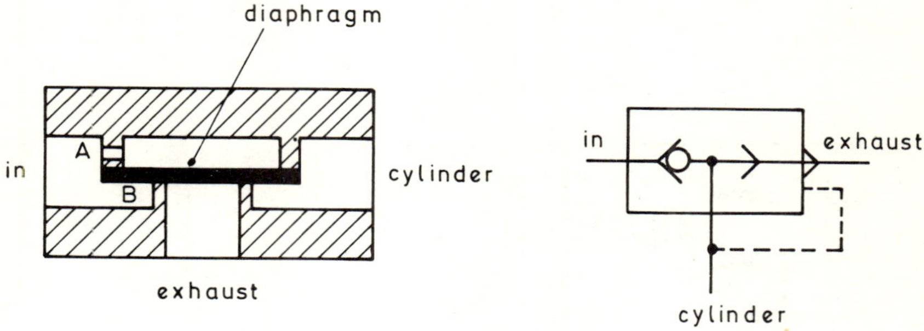

Fig. 6.21 – A quick exhaust valve.

6.4 PRESSURE CONTROL

Pressure control valves include valves for controlling pressure and valves controlled by pressure.

6.4.1 The relief valve

This protects the hydraulic circuit. If it were not for this valve, pressure might build up until a component failed or the electric motor driving the pump stalled. There are direct acting and two stage relief valves. Fig. 6.22 shows a direct acting valve. From this it can be seen that if pressure exceeds the value set by the spring, the poppet lifts and relieves the system to tank. The pressure at which the valve first lifts is referred to as the **cracking pressure**. Because of the spring, full flow pressure is greater than **cracking pressure**. This is known as **pressure override** and is an undesirable feature of direct acting values.

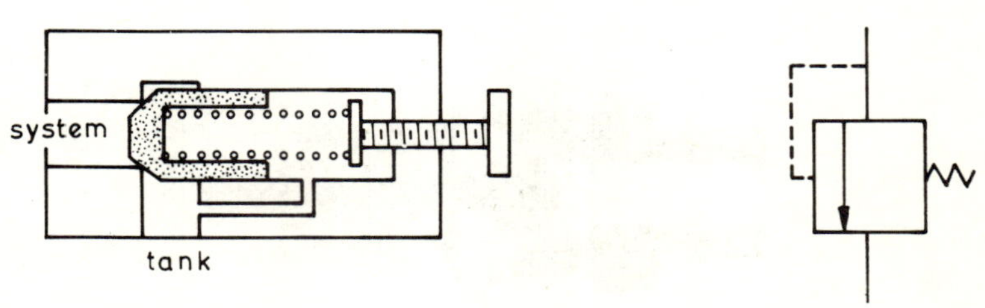

Fig. 6.22 – A single stage relief valve.

The two stage relief valve, Fig. 6.23, has practically no override. Here system pressure is fed through a hole in the main poppet to act on the pilot. When system pressure is high enough it lifts the pilot poppet off its seat, allowing flow to tank. The pressure drop across the hole in the main poppet lifts the main poppet, relieving the system to tank.

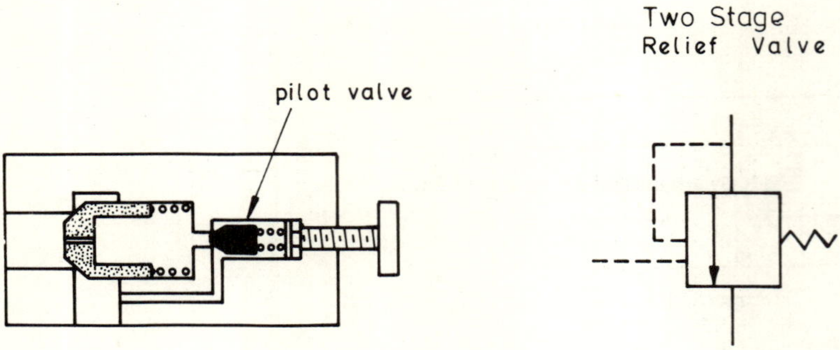

Fig. 6.23 – A two stage relief valve.

Fig. 6.24 shows two types of pneumatic relief valves. The unit shown in Fig. 6.24(a) is suitable for use as a safety device and can, for example, give automatic pressure relief on gas cylinder regulators (see section 6.4.5). The force of the spring on the seat is adjusted by means of the knurled top on the unit.

The relief valve shown in Fig. 6.24(b) is designed for use on air tanks. It works on the same principle as that of Fig. 6.24(a) but also provides a visual indication of the state of the valve.

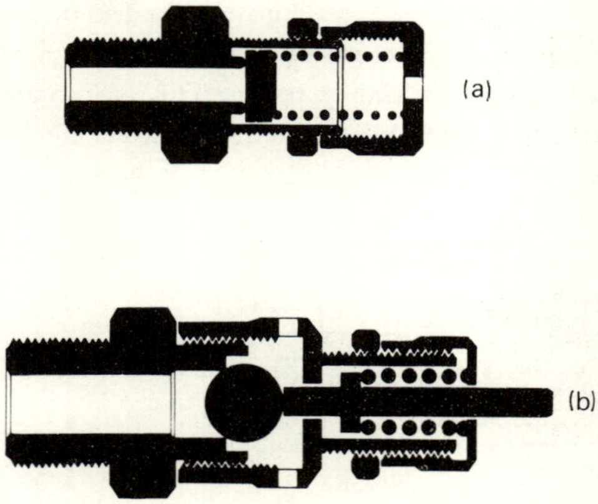

Fig. 6.24 – Pneumatic relief valves (Martonair).

6.4.2 The unloading valve

An unloading valve allows rapid discharge of oil near atmospheric pressure after system pressure has built up to a preset value, Fig. 6.25. The circuit in Fig. 6.26 illustrates the use of the unloading valve in a dual pump system. At start-up both pumps feed the system, but when a preset value of pressure is reached, pump B is unloaded. This avoids the energy loss that arises when the pump output has to flow through a relief valve at near system pressure.

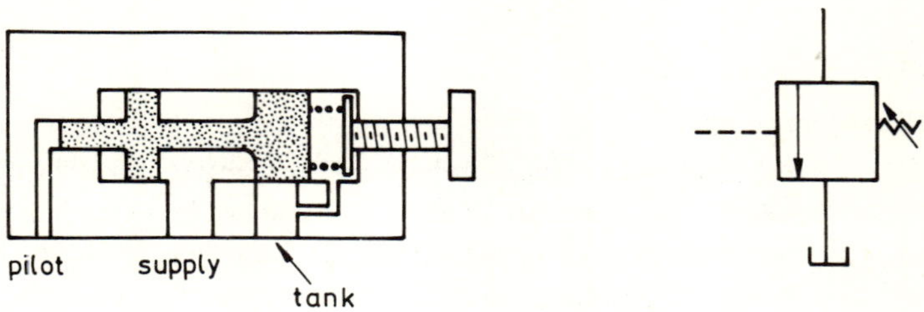

Fig. 6.25 – An unloading valve.

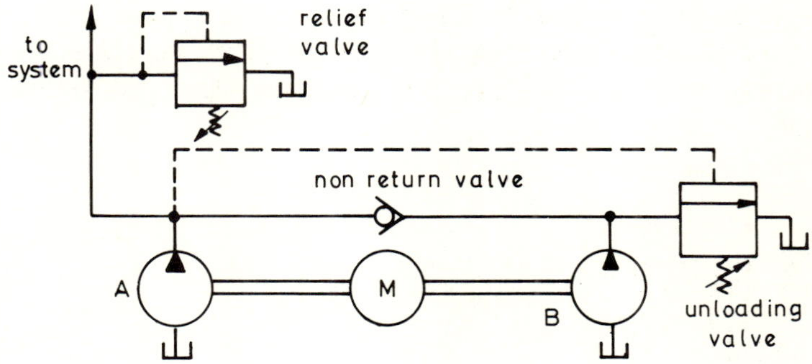

Fig. 6.26 – Use of an unloading valve in a dual pump system.

6.4.3 The pressure reducing valve

This valve limits the pressure in some part of a circuit to a value below the operating pressure of the main circuit. Fig. 6.27 shows such a valve. The output pressure acts on the left-hand end of the spool. An increase in output pressure above that preset by the spring, moves the spool to the right, restricts the flow and hence reduces the output pressure. If a relief valve were used for this purpose it would prevent further pressure build-up in the whole circuit. Note that pressure reducing valves are 'normally open' while relief valves are 'normally closed'.

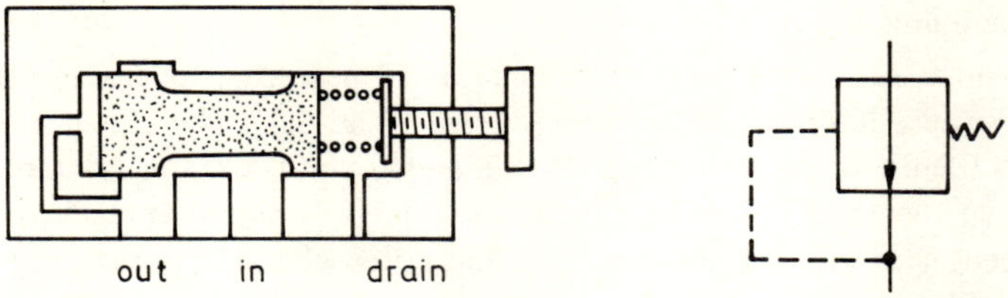

Fig. 6.27 – A pressure reducing valve.

6.4.4 The sequence valve

In circuits with more than one cylinder it is often necessary to move the cylinders in a definite sequence of movements, that is, rod one can be extended, rod two extended, rod one retracted, rod two retracted. This can be done, both in pneumatic and hydraulic systems, using sequence valves. Fig. 6.28 shows that when the inlet pressure reaches a preset value, the spool lifts, allowing flow to pass to the outlet. A circuit is shown in Fig. 6.29. When the directional control value A is actuated the supply is connected to the top of cylinder B and to the sequence valve. Cylinder rod B, which could be a clamping cylinder, extends. At the end of its stroke pressure builds up and opens the sequence valve, making cylinder rod D extend (this might be a cutter). When valve A is reset both cylinder rods retract, the flow from D bypassing the sequence valve through the check valve.

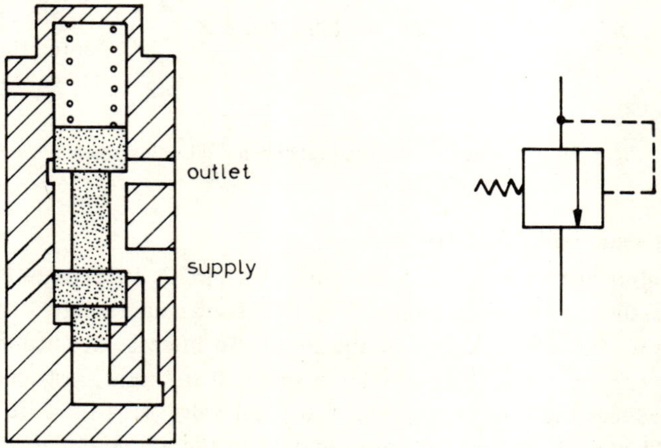

Fig. 6.28 – A sequence valve.

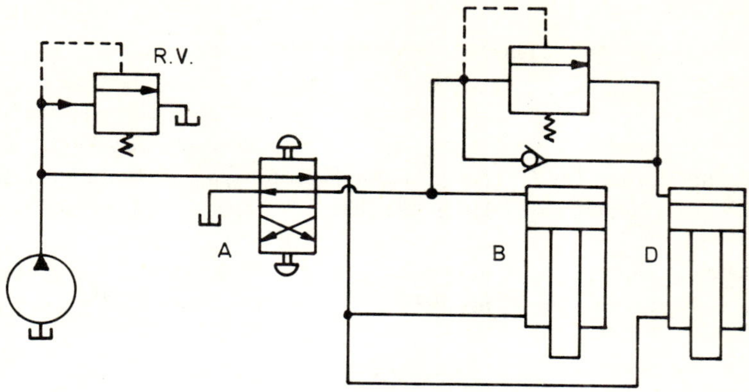

Fig. 6.29 – Circuit employing a sequence valve.

6.4.5 Pneumatic regulators

Cut in and cut out of a compressor tend to result in an unsteady supply pressure to a system. A pressure regulator is needed if a steady supply is essential. Such a valve is also useful if a pressure smaller than the main supply is needed. The majority of these valves use a flexible diaphragm, Fig. 6.30, one side of which is spring loaded and the other side is acted on by the controlled pressure. If the controlled pressure rises above the desired value, as set by the spring, the diaphagm rises, reducing the flow to the system and hence its pressure. The valve shown in Fig. 6.30(a) is 'non-relieving'. If this were controlling the pressure in a fixed volume, there would be no means of relieving the controlled pressure if it became too high. Self relieving pressure regulators are commercially available. The symbol for this valve is shown in Fig. 6.30(b).

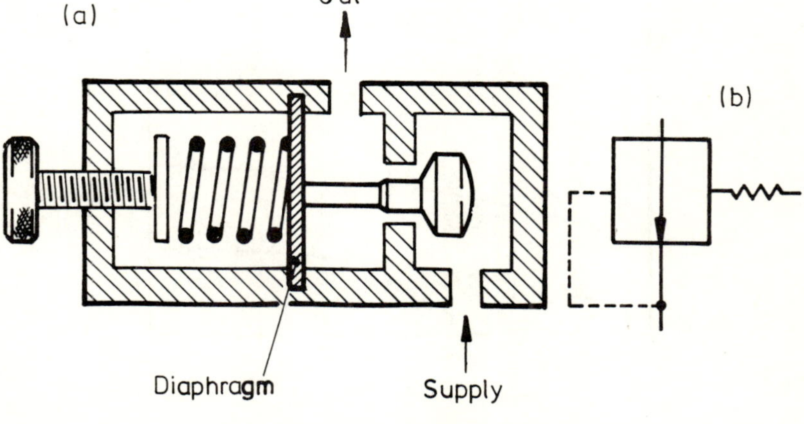

Fig. 6.30 – A non-relieving pressure regulator.

6.5 THE ELECTRIC FIELD VALVE

The properties of electrosensitive fluids were discussed in section 2.7, and it is now appropriate to show how these fluids can provide an unusual valve action. This provides a means of direct interfacing of electrical and fluid systems without the interference of the mechanical stage common in conventional valving. The first attempt at such a system is reported in [12, 13] where a vibration and shock exciter was developed and tested at the Stanford Research Institute in the early 1960s.

The electric field valve is shown in Fig. 6.31(a). It consists of five cylindrical shells forming a group of six parallel 0.75 mm gaps. A control voltage is applied

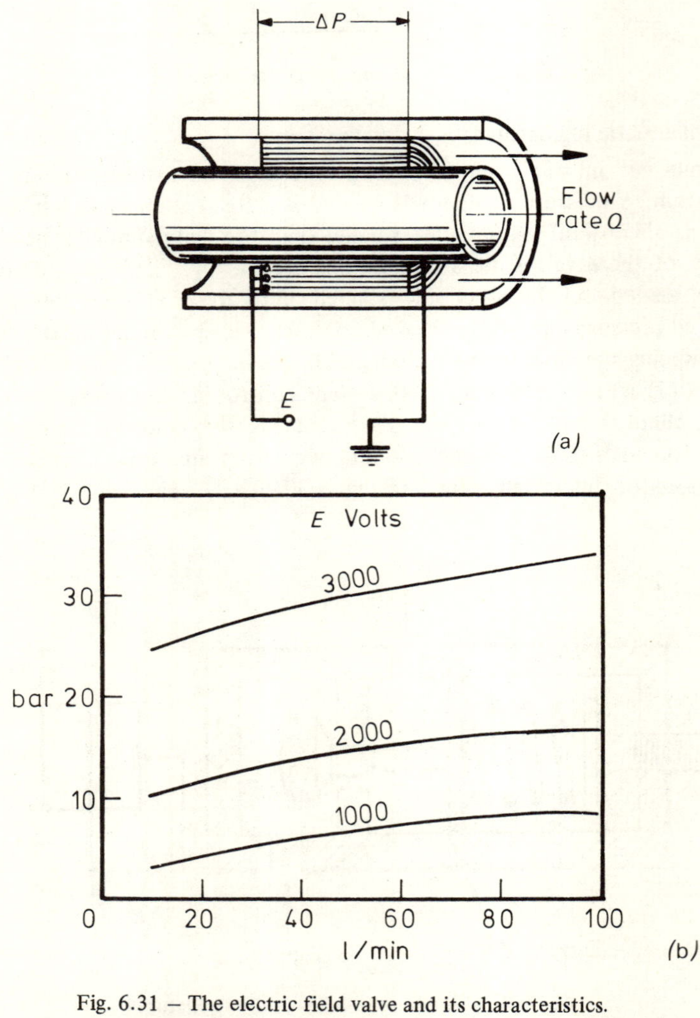

Fig. 6.31 – The electric field valve and its characteristics.

to alternative shells, and the remaining shells and the outer and inner boundary walls are grounded. The flow-pressure characteristics for the arrangement are shown in Fig. 6.31(b), and indicate the level of pressure drop that can be sustained at different electric field strengths.

The original vibrator [12] consists of a cylinder within which are assembled four electric valves around an output shaft, as shown in Fig. 6.32. The two end valves marked A and D are attached to the cylinder and function as valves only. The two centre valves B and C are mounted on the output shaft and function as both valves and pistons. The cylinder is supplied from a fixed displacement pump which provides a constant flow source. To provide vibrations, control voltages E_1 and E_2 are applied as sinusoidal waves, opposite in phase. Hence when E_1 is maximum, valves A and C try to stop the flow, but since the supply is constant flow, pressure builds up in the chamber between A and C. The pressure acting on the left face of C moves the output shaft.

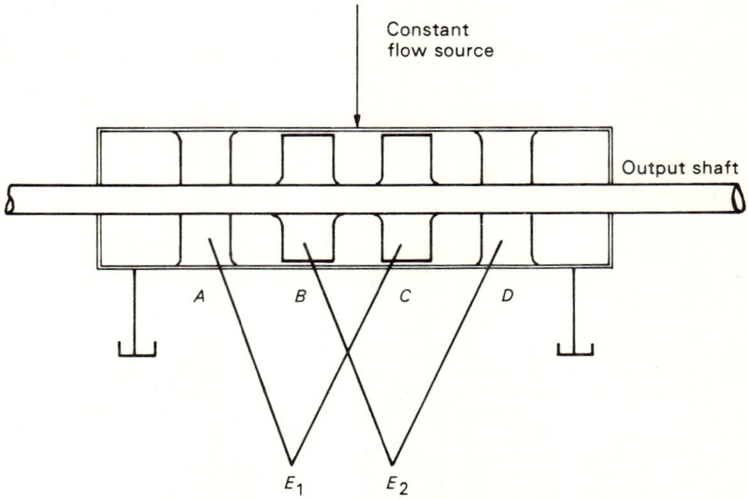

Fig. 6.32 – A vibrator using electric field valves.

The vibrator was shown to produce 14 kN static force and 10 kN peak sine wave force at 1 kHz, a frequency well beyond the useful range of conventional hydraulic vibrators. A useful output was developed at frequencies up to 2 kHz. For step inputs the response time was fast, peak load accelerations being reached in less than one millisecond.

6.5.1 Other applications of electric field valves

The relationship between the electric field strength and the stress at which flow will occur is deduced from equation (2.14) as:

$$S = \alpha V^2$$

where α = constant of proportionality dependent on the specific fluid properties.

Hence variable damping can be achieved in which the yield shear stress is proportional to the square of the field strength. In 1963 [15] this property was used to develop an automatically hardening shock absorber mount for ship applications. The electric field was generated with piezoelectric elements which have the following advantages:

1. The piezoelectric element can deliver a high voltage–low current impulse which has a good impedance match to the high resistive load of the electrosensitive fluid.
2. Because the piezoelectric element is self generating and fast responding, it can be used to sense the incident shock directly and almost immediately supply the energy required to increase the damping in the shock absorber.

The device, Fig. 6.33, consists of a moveable set of flat parallel plates which interlace with another set of parallel plates within a small tank. The performance obtained with field-off (operating as a conventional damper) and with the field

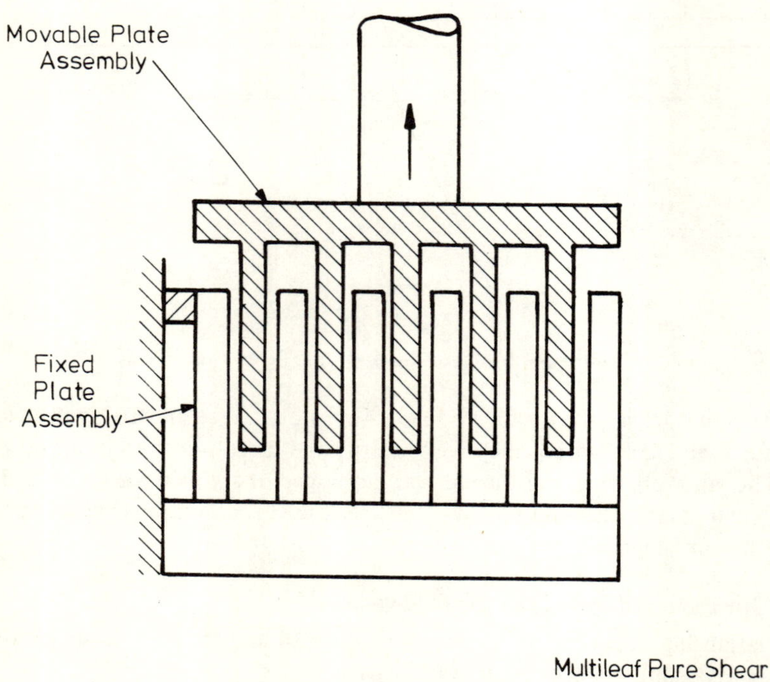

Fig. 6.33 – Automatically hardening shock absorber [15].

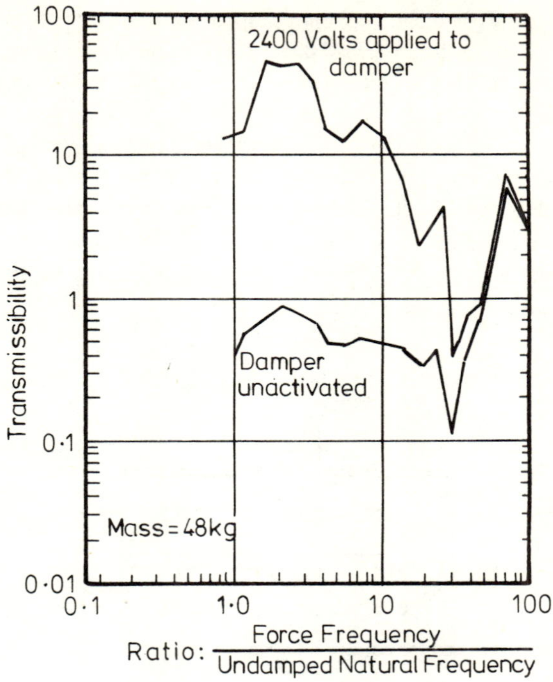

Fig. 6.34 – Performance of shock absorber [15].

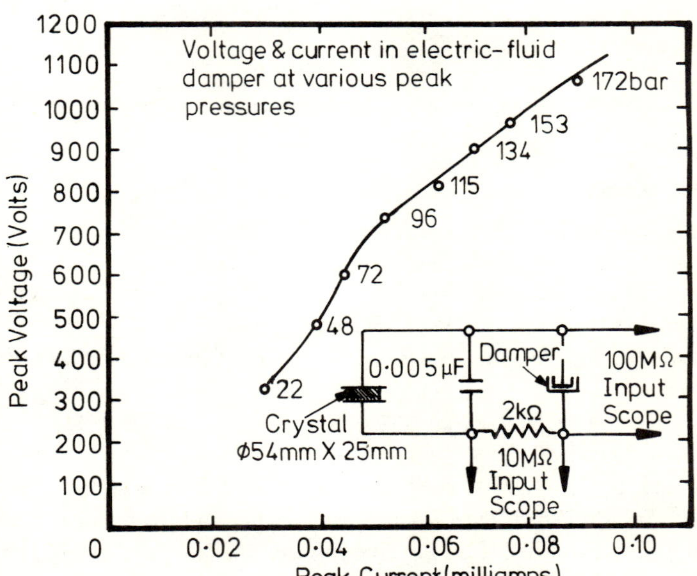

Fig. 6.35 – Voltage and current in electric-fluid damper at various peak pressures [15].

activated is shown in Fig. 6.34. The undamped natural frequency of these tests was about 7 Hz, so in the range 7 Hz to 210 Hz the shock mount behaved extremely well. Some typical results for a single crystal of barium titanate are shown in Fig. 6.35.

Another intriguing idea is the electrosensitive pumping system shown in Fig. 6.36. The signal generator provides both the power for the piston, a barium titanate crystal, and properly phased control voltage to insulated plates which form the two electric valves [16]. While this design is reported to have had limited success, a similar design [17] has developed pressures of the order of 150 mm Hg and pulsatile flows of 8 l/min.

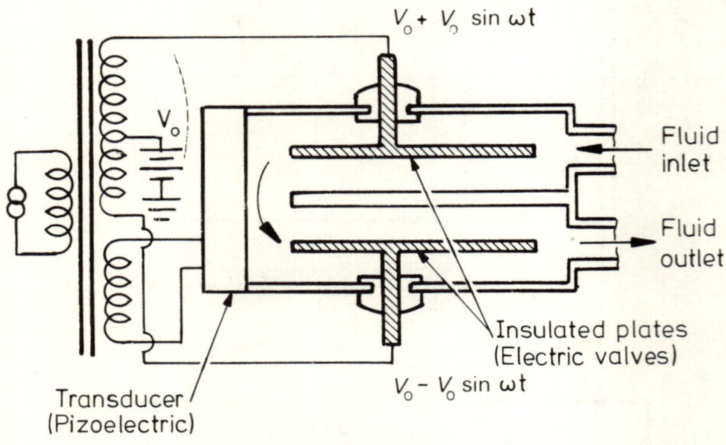

Fig. 6.36 – A pump employing electrosensitive fluids.

6.6 REFERENCES

[1] *Pneumatic Handbook*, Trade and Technical Press, England.

[2] *Hydraulic Handbook*, ibid.

[3] *Fluid Power Systems and Components – Graphical Symbols*, International Organisation for Standardisation ISO 1219/1976.

[4] McCloy, D. and Martin, H. R. (1973), *The Control of Fluid Power*, Longman.

[5] Fawcett, J. R. (1970), *Hydraulic Servomechanisms and Their Applications*, Trade and Technical Press.

[6] Merritt, H. E. (1967), *Hydraulic Control Systems*, John Wiley.

[7] Boothe, W. A. (1962), Performance evaluation of a high pressure recovery bistable fluid amplifier, *A.S.M.E. Symposium of Fluid Jet Control Devices.*

[8] McCloy, D. (1972), Oil hydraulic fluid logic, *Proc. 1972 Fluid Power International Conference.*

[9] Rivard, J. G. and Walber, J. C. (1965), A fluid state vortex hydraulic servo-valve, 21st
valve, *21st Nat. Conf. on Fluid Power*, Chicago.

[10] McCloy, D. and Stevenson, I. J. (1972), Some experiments with oil hydraulic vortex valves, *5th Cranfield Fluidics Conf.*

[11] Vann, W. D. and Honda, T. S. (1968), *Fluidic Vortex Valve Servo Actuators – an Advancement in Helicopter Flight Control Systems*, A.S.M.E. paper SP1348-690536.

[12] Eige, J. J. and Fraser, E. C. (1961), *Feasibility Study of a Vibration and Shock Exciter using Electric Field Modulation of Hydraulic Power*, W.A.D.D. Tech. Rep. 61-185, U.S.A.F. Stanford Research Institute, AD 268094, February.

[13] Sevy, R. W. (1962), *Refinement of a Hydraulic Shock-Shaker*, W.A.D.D. Tech. Rep. 62-536, U.S.A.F. Stanford Research Institute, AF 33(616)-8468, May.

[14] Martin, H. R. and Zanetel, G. (1975), Electrosensitive Fluid Characteristics Under a Pulsatory Electric Field, *A.S.M.E. Design Conf.*, paper No. 75-Dec-62, April.

[15] Oleson, S. K. and Eige, J. J. (1963), *Development of Automatically Hardening Shock Mounts*, Stanford Research Institute, Project No. PHU-4109.

[16] *Unconventional Methods for Influencing Fluid Flow*, Tech. Rep. ASD TDR 63-776, Vol. 1, Carnegie Institute of Technology, Dept. of Mech. Eng., AD 427769, 1963.

[17] Robinson, R. D., *et al.* (1975), High Frequency Piezoelectric Motor Pump for Implanted Energy Conversion, *A.C.E.M.B. Proc.*, November.

CHAPTER 7

Static characteristics of valves

7.1 INTRODUCTION

In the last chapter a variety of controlling elements and their means of operation were discussed. This chapter is concerned with the static characteristics of valves, in particular the directional control valves. The relationships among mass flow rate, pressure at the load and valve displacement are discussed. A knowledge of these characteristics allows the prediction of the steady state behaviour of the complete system, for example, what speed will the load move at for a given valve displacement and supply?

7.2 THE VARIABLE ORIFICE

The variable orifice is at the heart of most fluid power systems. Clearly then a knowledge of the orifice equations for incompressible and compressible flow is essential if static characteristics are to be calculated. These equations have already been presented in section 4.3.

7.2.1 The two-port hydraulic valve

The orifice equation for the flow of an incompressible fluid is:

$$Q = C_q a \sqrt{2(P_u - P_d)/\rho} \tag{7.1}$$

where all variables have been defined earlier (section 4.3). In a given system the dominating variables are usually the pressure drop $(P_u - P_d)$ and the orifice area. The square root term is the predominant non-linearity, but it must be remembered that even at a fixed pressure drop there are still non-linearitites present in equation (7.1). These arise because C_q is a function of Reynolds number and cavitation number. In a practical valve, Fig. 7.1, the pressure P_d and hence the pressure drop is dictated by the load.

It is common practice to present the characteristics of these valves as graphs of flow rate Q to the load, and load pressure P_d for various valve displacements.

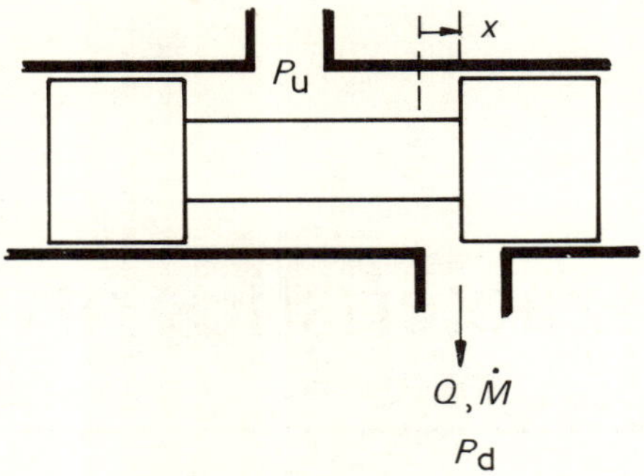

Fig. 7.1 – A two-port valve.

Equation (7.1) can be written in non-dimensional form as:

$$\bar{Q} = \bar{a}\sqrt{p_u}\,\sqrt{1-p} \tag{7.2}$$

where $p_u = P_u/P_r,\; p = P_d/P_u,\; \bar{a} = C_q a/C_{q_r} a_r,\; \bar{Q} = Q/Q_r$

and the reference quantities are:

P_r, usually the upstream or supply pressure
$C_{q_r}a_r$, usually the maximum effective orifice area
Q_r, defined as $Q_r = a_r\sqrt{2P_r/\rho}$.

Note that pressures can be either gauge or absolute provided that consistency is maintained. These valve characteristics, equation (7.2), are plotted in Fig. 7.2 assuming that P_u is constant and equal to the reference pressure P_r, that is, $p_u = 1$. As expected, maximum flow rate occurs at minimum load pressure, and there is zero flow rate (the stalled condition) when the orifice pressure drop is zero ($P_d = P_u$). Referring to Fig. 7.1, it will be seen that reverse flow occurs when $P_d > P_u$. Thus, to be precise equation (7.2) should read:

$$\bar{Q} = \bar{a}\sqrt{p_u}\,\mathrm{sgn}\,(1-p)\sqrt{(1-p)\,\mathrm{sgn}\,(1-p)} \tag{7.3}$$

It is left to the reader to determine how the plotted characteristics of Fig. 7.2 should be amended to allow for reverse flow.

The power delivered to the load can also be determined from the valve characteristics. Hydraulic power (W) at the load is the product of volumetric flow rate and delivered pressure:

$$W = QP_d \tag{7.4}$$

where P_d = pressure above ambient, that is, gauge pressure.

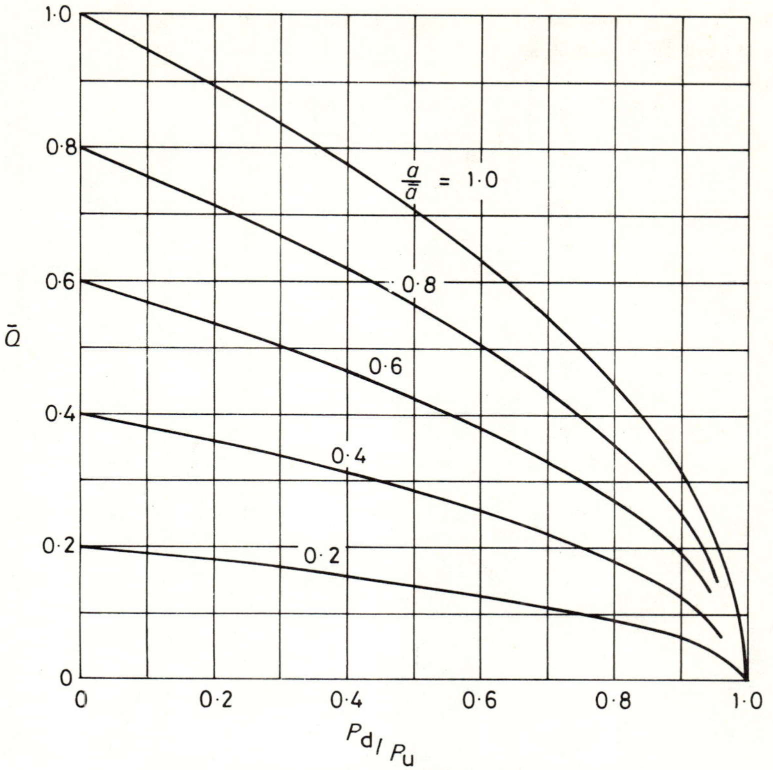

Fig. 7.2 – Valve characteristics; Hydraulic two-port valve.

Defining a reference power as $W_r = Q_r P_r$ gives the non-dimensional power as:

$$\bar{W} = \bar{Q} p p_u . \tag{7.5}$$

Substituting equation (7.2) in equation (7.5) gives:

$$\bar{W} = \bar{a} p_u^{1.5} \, p\sqrt{1-p} \tag{7.6}$$

Differentiation with respect to p shows that maximum power is transferred to the load when $p = \frac{2}{3}$. The maximum power is:

$$\bar{W}_{max} = 0.385 \, \bar{a}_{max} (p_{u_{max}})^{1.5} \tag{7.7}$$

7.2.2 The two-port pneumatic valve

Equations (4.3), (4.4) and (4.5) gave the mass flow rate through an orifice for choked and unchoked flow. They too can be presented in a more compact form if they are made non-dimensional. In this case it is necessary to define a reference mass flow rate $\dot{M}_r$ as:

$$\dot{M}_r = C_{q_r} a_r \frac{P_r}{\sqrt{T_r}} \left\{ \frac{\gamma}{R} \left(\frac{2}{\gamma + 1} \right)^{(\gamma+1)/(\gamma-1)} \right\}^{\frac{1}{2}} \tag{7.8}$$

For air this becomes:

$$\dot{M}_r = 0.0405\, C_{q_r} a_r P_r / \sqrt{T_r} \tag{7.9}$$

Thus equation (4.3) can be written:

$$\bar{\dot{M}}_r = \frac{\bar{a} p_u f(p)}{\sqrt{\bar{T}_u}} \tag{7.10}$$

where $\bar{T}_u = T_u / T_r$ and the function $f(p)$, (Fig. 7.3(a)), is, for non-choked flow:

$$f(p) = \left\{ \frac{p^{2/\gamma} - p^{(\gamma+1)/\gamma}}{\left(\frac{\gamma - 1}{2} \right) \left(\frac{2}{\gamma + 1} \right)^{(\gamma+1)/(\gamma-1)}} \right\}^{\frac{1}{2}} \tag{7.11}$$

where in general

p = ratio of the downstream to the upstream pressure.

When the flow is choked $f(p) = 1$.

The valve characteristics are plotted in Fig. 7.3(b), assuming that P_u is fixed and $P_u = P_r$, $T_u = T_r$ (Note that for compressible flow calculations it is essential that absolute pressures are used). Again reverse flow has not been considered, and this is left as an exercise for the reader.

7.3 APPLICATION OF VALVE CHARACTERISTICS

A knowledge of these characteristics enables the designer to predict steady operating conditions under different types of loading [1]. Fig. 7.4 shows several load lines superimposed on the valve characteristics for the hydraulic case; the intersections give the steady state flow rates and load pressures at different orifice areas. This is similar to the electronic engineer's method of determining the performance of amplifier-load combinations. If the cylinder load is constant (Coulomb friction) the load pressure is independent of speed and hence of volumetric flow rate, since for an incompressible fluid volumetric flow rate is the product of cylinder cross-sectional area and rod speed. Hence the load line is

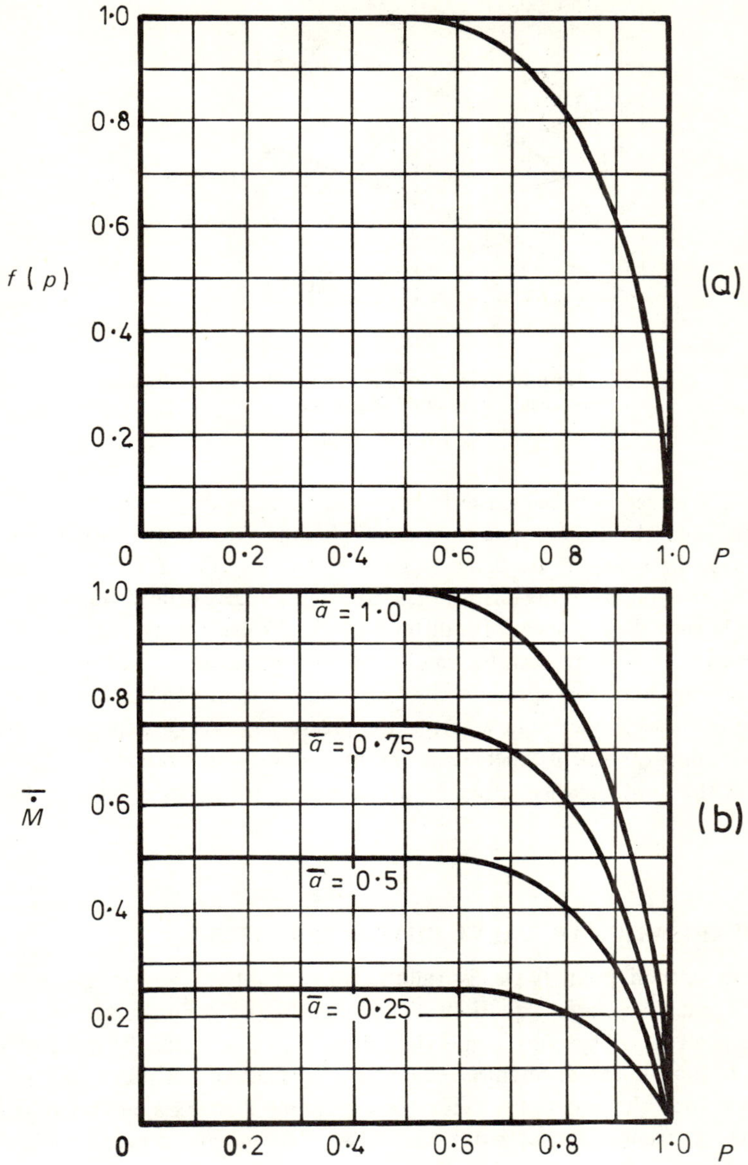

Fig. 7.3 – (a) The compressibility function and (b) valve characteristics; pneumatic two-port valve.

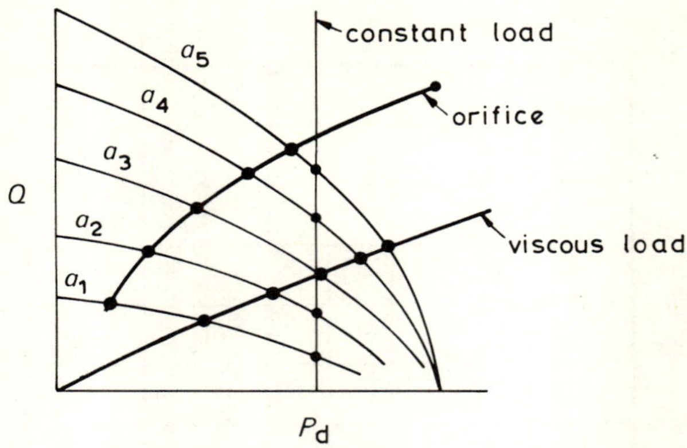

Fig. 7.4 – Superposition of load lines and valve characteristics to determine steady state operating points.

vertical. If the load is viscous, the load pressure driving the rod is proportional to velocity and the load line is then a straight line through the origin. For an orifice load, the load line is a parabola given by $Q = k\sqrt{(P_d - P_e)}$, where P_e is the pressure downstream of the orifice. The intersection gives the volumetric flow rates and in turn the rod speeds for different valve displacements.

The same procedures can be applied to pneumatic valves, but when calculating a steady state rod speed in this case it is necessary to convert the mass flow rate $\dot{M}$ to a volumetric flow rate at the pressure P_d. This is determined from the relationship $Q = \dot{M}/\rho$ where ρ is the mass density at pressure P_d. The rod velocity is then obtained as before by dividing Q by the cylinder cross-sectional area.

7.4 LINEARISATION OF VALVE CHARACTERISTICS

The above method predicts steady state performance but gives no information about dynamic response. Equations (7.2) and (7.10) are non-linear in p, and dynamic analysis of systems employing these valves accordingly involves the solution of non-linear differential equations. This usually requires the use of computers since the equations, except in the simplest of cases, are analytically intractable. Fortunately experience has shown that in fluid power servos, performance can in many cases be adequately predicted using an approximate model that has been linearised by assuming that changes in the variables are small throughout the motion. For example, consider equation (7.1):

$$Q = C_q a\sqrt{2(P_u - P_d)/\rho}\,.$$

This can be expanded as a Taylor's series about some initial point i as:

$$Q - Q_i = \delta Q = \left(\frac{\partial Q}{\partial C_q}\right)_i \delta C_q + \left(\frac{\partial Q}{\partial a}\right)_i \delta a + \left(\frac{\partial Q}{\partial P_u}\right)_i \delta P_u$$

$$+ \left(\frac{\partial Q}{\partial P_d}\right)_i \delta P_d + \left(\frac{\partial Q}{\partial \rho}\right)_i \delta \rho + \text{higher order terms.} \qquad (7.12)$$

Here δQ is an increment in Q about an initial value Q_i at P_{di}, a_i, C_{qi}, P_{ui} and ρ_i. If small changes are assumed the higher order terms may be neglected. The partial derivatives or valve coefficients are determined at the initial steady state point on the valve characteristics. They can be determined analytically from equation (7.1) or calculated from measured characteristics. Fig. 7.5 shows how this can be done by using practical results assuming P_u, C_q and ρ are fixed.

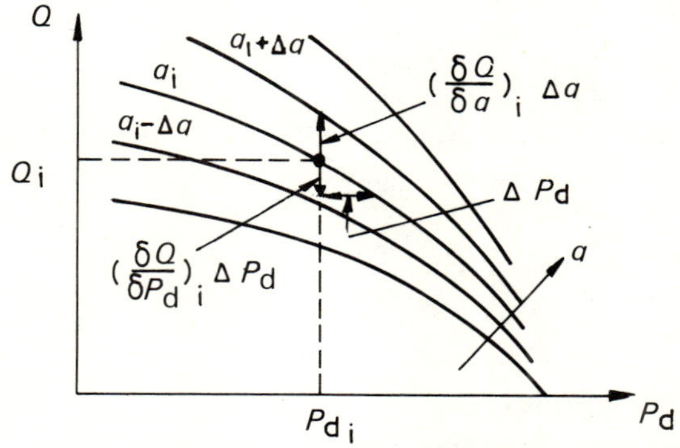

Fig. 7.5 – The linearisation of valve characteristics.

The non-dimensional relationships of equations (7.2) and (7.10) can also be linearised [3]. Putting $p_d = (P_d/P_r)$ we can write:

$$\delta \bar{Q} = \bar{Q} - \bar{Q}_i = \left(\frac{\partial \bar{Q}}{\partial \bar{a}}\right)_i \delta a + \left(\frac{\partial \bar{Q}}{\partial p_u}\right)_i \delta p_u + \left(\frac{\partial \bar{Q}}{\partial p_d}\right)_i \delta p_d \qquad (7.13)$$

the partial derivatives are evaluated as follows:

$$\left(\frac{\partial \bar{Q}}{\partial \bar{a}}\right)_i = \bar{Q}_i / \bar{a}_i$$

$$\left(\frac{\partial \bar{Q}}{\partial p_u}\right)_i = \frac{\bar{Q}_i}{2p_{ui}(1 - p_i)}$$

$$\left(\frac{\partial \bar{Q}}{\partial p_d}\right)_i = \frac{\bar{Q}_i p_i}{2(p_i - 1)p_{di}} \; .$$

For gas flow, equation (7.10) is rewritten as [2]:

$$\delta\bar{M} = \left(\frac{\partial\bar{M}}{\partial\bar{a}}\right)_i \delta\bar{a} + \left(\frac{\partial\bar{M}}{\partial p_u}\right)_i \delta p_u + \left(\frac{\partial\bar{M}}{\partial p_d}\right)_i \delta p_d + \left(\frac{\partial\bar{M}}{\partial T_u}\right)_i \delta T_u \qquad (7.14)$$

where

$$\left(\frac{\partial\bar{M}}{\partial\bar{a}}\right)_i = \frac{\bar{M}_i}{\bar{a}_i}$$

$$\left(\frac{\partial\bar{M}}{\partial T_u}\right)_i = -\frac{1}{2}\frac{\bar{M}_i}{T_{ui}}$$

$$\left(\frac{\partial\bar{M}}{\partial p_u}\right)_i = \frac{\bar{M}_i}{p_{ui}}(1+\zeta)$$

$$\left(\frac{\partial\bar{M}}{\partial p_d}\right)_i = -\zeta\frac{\bar{M}_i}{p_{di}}$$

and $\zeta = 0$ for $p_i < p_{i\ \mathrm{crit}}$ and otherwise:

$$\zeta = \frac{(\gamma-1)/2\gamma}{p_i^{(1-\gamma)/\gamma}-1} - \frac{1}{\gamma} \quad \text{(Fig. 7.6)}.$$

7.5 THREE-PORT SERVOVALVES

7.5.1 The hydraulic three-port valve

In most servosystems spool position is affected by the system error and so the variation of orifice area with spool displacement in an important factor. There are three most common porting shapes: rectangular, round and full annulus. If the ports are rectangular, the orifice area increases linearity with spool movement. This is also true of the full annulus port, which is a circumferential slot in the valve sleeve. For circular ports, however, orifice area is not linear with spool movement and the flow equations involve a non-linearity in spool displacement as well as in pressure drop. For the present it is assumed that the valve area displacement characteristic is linear so that, for example, equation (7.1) can be written:

$$Q = Kx\sqrt{(P_u - P_d)}\,. \qquad (7.15)$$

Referring to the critical centre or zero lap valve of Fig. 7.7, it can be deduced that:

For $x > 0$ $\quad Q = Kx\,\epsilon\sqrt{\epsilon(P_s - P_c)} \qquad (7.16)$

$$\epsilon = \mathrm{sgn}\,(P_s - P_c)$$

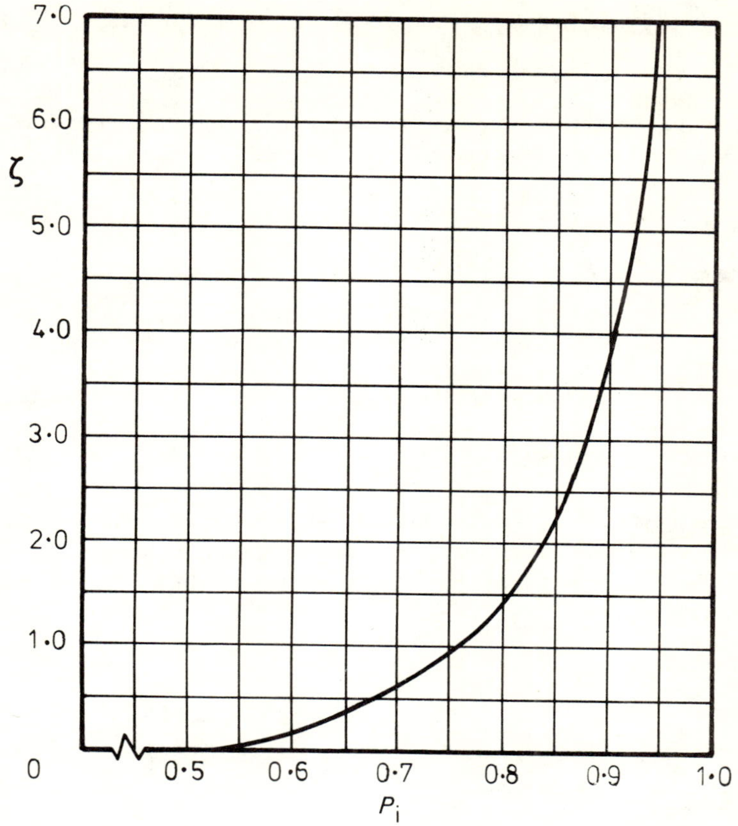

Fig. 7.6 – The function ζ used in the linearisation of pneumatic valve characteristics.

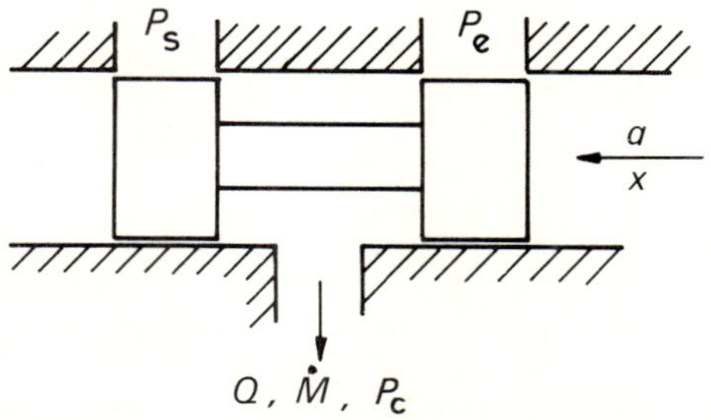

Fig. 7.7 – A three-port valve with zero lap.

For $x < 0$ $\quad Q = Kx\,\phi\sqrt{\phi(P_c - P_e)}$ (7.17)

$$\phi = \text{sgn}\,(P_c - P_e)$$

P_s, P_c and P_e are respectively supply pressure, cylinder pressure and exhaust or tank pressure. The signum terms allow for flow reversal.

Putting $\quad Q_r = Kx_r\sqrt{(P_s - P_e)/2}, P_s^* = P_s - P_e,$

$$P_c^* = P_c - P_e, p_c^* = P_c^*/P_s^*$$

leads to

For $x > 0$ $\quad \bar{Q} = \epsilon\bar{x}\sqrt{2\epsilon(1 - p_c^*)}$ (7.18)

For $x < 0$ $\quad \bar{Q} = \phi\bar{x}\sqrt{2\phi p_c^*}$ (7.19)

The cylinder pressure P_c will not in practice fall below P_v, the vapour pressure of the oil, and the lower limit on p_c^* is then p_v^*. The valve characteristics are plotted in Fig. 7.8, where p_v^* has been taken as −0.111 (which is the case when $P_s = 70$ bar, $P_e = 7$ bar, $P_v = 0$).

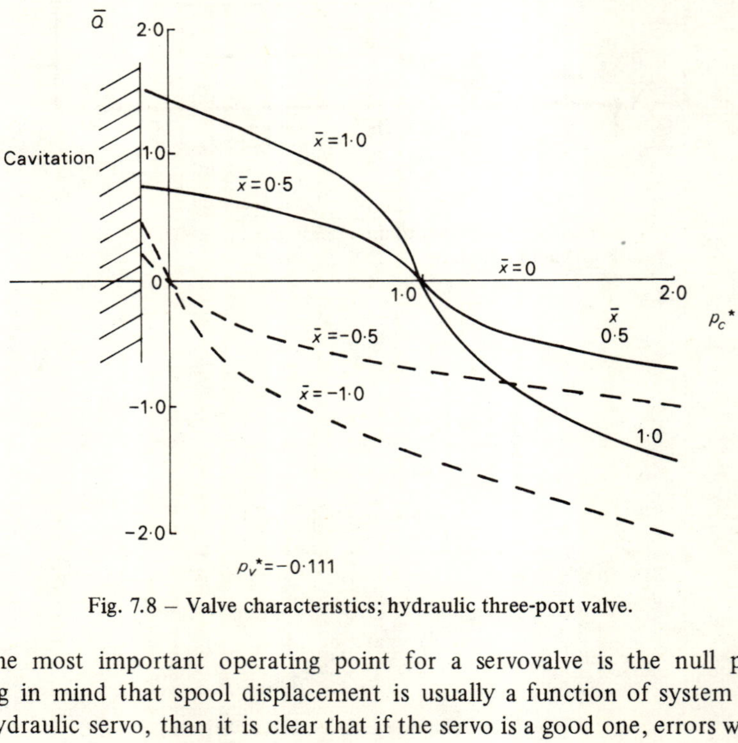

Fig. 7.8 – Valve characteristics; hydraulic three-port valve.

The most important operating point for a servovalve is the null point. Bearing in mind that spool displacement is usually a function of system error in a hydraulic servo, than it is clear that if the servo is a good one, errors will be small and the spool position should depart very little from the null position for most of its working life. Hence it is usual to linearise the valve flow about the null point. In the case discussed here, the null position is $x = 0$. For a three-port valve, the null cylinder pressure is usually chosen as $P_s^*/2$, so that there are equal potential load accelerations and decelerations. Equations (7.18) and (7.19) give, for $\bar{x} = 0$, $p_c^* = 0.5$, the **normalised flow gain** $\partial\bar{Q}/\partial\bar{x} = 1$ or **flow gain** $\partial Q/\partial x = Q_r/x_r = K\sqrt{(P_s^*/2)}$ and $\partial\bar{Q}/\partial p_c^* = 0$ or **flow pressure coefficient** $\partial Q/\partial P_c^* = 0$.

Another useful coefficient is the **pressure sensitivity** $\partial P_c^*/\partial x$, which relates to the stiffness of the valve load system [3]. This can be determined easily if the other two coefficients are known, since the theory of partial differentials gives:

$$\frac{\partial P_c^*}{\partial x} = -\frac{\partial Q}{\partial x}\bigg/\frac{\partial Q}{\partial P_c^*}$$

In this case $\partial P_c^*/\partial x = \infty$. In practice, large values of the pressure sensitivity can be achieved.

7.5.2 Underlapped hydraulic three-port valve

If the underlap U, see Fig. 7.9, is the same on supply and tank ports, the flow rate Q is:

$$Q = K(x + U)\,\epsilon\sqrt{\epsilon(P_s - P_c)} + K(x - U)\,\phi\sqrt{\phi(P_c - P_e)}$$

or

$$\bar{Q} = (\bar{x} + 1)\,\epsilon\sqrt{2\epsilon(1 - p_c^*)} + (\bar{x} - 1)\,\phi\sqrt{2\phi p_c^*}$$

where

$$Q_r = KU\sqrt{P_s^*/2}$$

and

$$x_r = U.$$

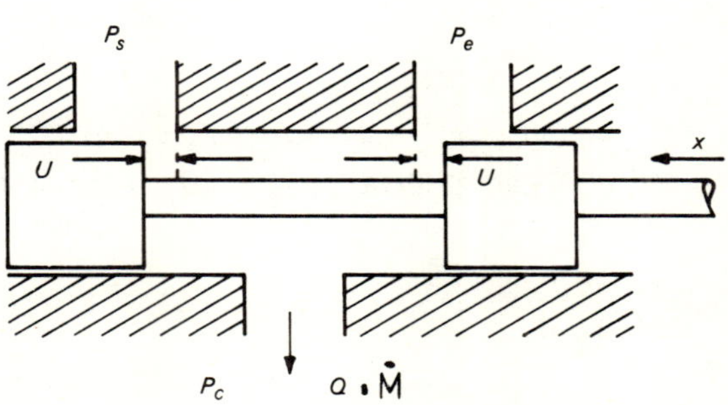

Fig. 7.9 – An underlapped three-port valve.

The null coefficients are:

$$\frac{\partial \bar{Q}}{\partial \bar{x}} = 2 \quad \text{or} \quad \frac{\partial Q}{\partial x} = 2K\sqrt{(P_s^*/2)} = 2\frac{Q_r}{U}$$

$$\frac{\partial \bar{Q}}{\partial p_c^*} = -2 \quad \text{or} \quad \frac{\partial Q}{\partial P_c^*} = -2\frac{Q_r}{P_s^*} = -KU\sqrt{(2/P_s^*)}$$

$$\frac{\partial P_c^*}{\partial x} = P_s^*/U\,.$$

Note that the flow gain is twice that for the critical centre valve. The flow pressure coefficient, which is important as far as servo stability is concerned, is no longer zero. On the other hand the stiffness has fallen. This is an open centre valve and suffers the disadvantage of a null leakage rate $\bar{Q}_l = 1.0$ or $Q_l = Q_r$.

7.5.3 Pneumatic three-port valve

Referring to Fig. 7.7 it is seen that for positive x the upstream pressure P_u is the supply pressure P_s, and the downstream pressure P_d is the cylinder pressure P_c. In developing a reference mass flow rate $\bar{\dot{M}}_r$ (equation (7.8)) it is convenient to take P_s and T_s as reference pressure and temperature respectively. Hence, for positive x, $P_u = P_s = P_r$, $P_d = P_c$, $p = p_c = P_c/P_s$.

Hence from equation (7.10):

$$\bar{\dot{M}} = \bar{x} f(p_c)\,. \tag{7.20}$$

For negative x, $P_u = P_c$, $P_d = P_e$ and $p = P_e/P_c$. It can also be assumed with reasonable accuracy that $T_u = T_s = T_d$. Hence from equation (7.10):

$$\bar{\dot{M}} = \bar{x}(P_c/P_s) f(P_e/P_c) \tag{7.21}$$

or

$$\bar{\dot{M}} = \bar{x} p_c f(p_e/p_c)\,. \tag{7.22}$$

Equations (7.20) and (7.22) are plotted in Fig. 7.10 for the case $p_e = 0.1$. Here, for negative $\bar{x}$:

$$\bar{\dot{M}} = \bar{x} p_c f\left(\frac{0.1}{p_c}\right). \tag{7.23}$$

The graphs are easily plotted by taking values of p_c, referring to Fig. 7.3 for the appropriate values of the function f and calculating $\bar{\dot{M}}$.

For positive $\bar{x}$ the characteristics can be linearised as described earlier. For negative $\bar{x}$, Fig. 7.10 shows that the mass flow rate is linear with p_c over most of the range $(f(p_e/p_c) = 1)$. Indeed it is reasonable, for small p_e, to assume linearity over the full range, in which case $\bar{\dot{M}} = \bar{x}p_c$.

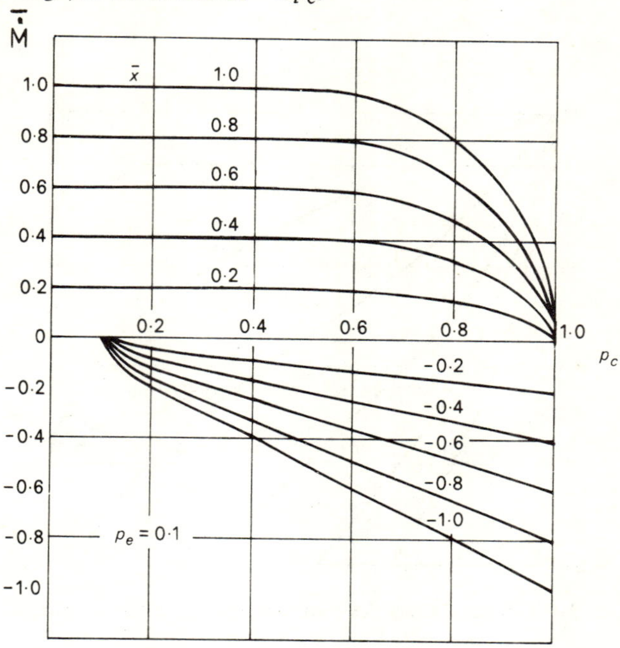

Fig. 7.10 – Valve characteristics; pneumatic three-port valve.

7.5.4 Three-port hydraulic flapper-nozzle valve

In this case, Fig. 7.11:

$$Q = Q_i - Q_o$$

$$Q = C_{qi}a_i\sqrt{2(P_s^* - P_c^*)/\rho} - C_q a\sqrt{2P_c^*/\rho} \tag{7.24}$$

where the outlet orifice area $Q = \pi dx$, d being the nozzle diameter.

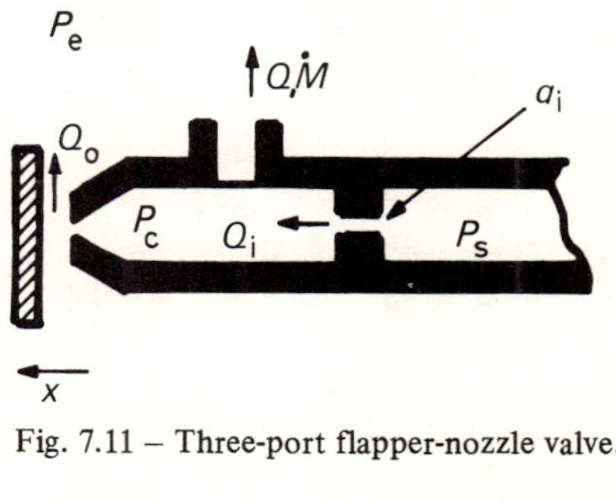

Fig. 7.11 – Three-port flapper-nozzle valve.

Define $\quad Q_r = C_{q_i} a_i \sqrt{P_s^*/\rho}\,.$

Hence $\quad \bar{Q} = \sqrt{2(1-p_c^*)} - \bar{a}\sqrt{2p_c^*} \qquad (7.25)$

with $\quad \bar{a} = C_q a / C_{q_i} a_i\,.$

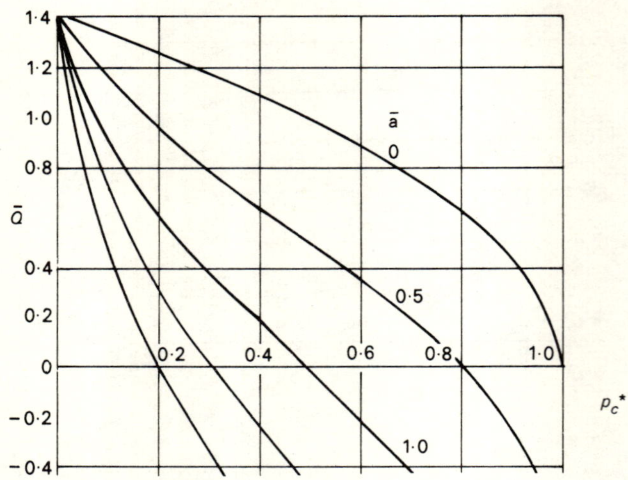

Fig. 7.12 – Valve characteristics; hydraulic three-port flapper-nozzle valve.

The valve characteristics are shown in Fig. 7.12. Flapper-nozzle valves are usually designed with $\bar{a} = 1$ and $p_c^* = 0.5$ at the null point. This gives maximum pressure sensitivity at the null point. The null coefficients are:

$$\frac{\partial \bar{Q}}{\partial \bar{a}} = -1 \quad \text{or flow gain} \quad \frac{\partial Q}{\partial x} = -C_q \pi d \sqrt{P_s^*/\rho}$$

$$\frac{\partial Q}{\partial p_c^*} = -2 \quad \text{or} \quad \frac{\partial Q}{\partial P_c^*} = -2\frac{Q_r}{P_s^*} = \frac{-2C_{q_i} a_i}{\sqrt{\rho P_s^*}}$$

Pressure sensitivity is:

$$\frac{\partial P_c^*}{\partial x} = -\frac{P_s^*}{2x_o}\frac{C_q}{C_{qi}}$$

where $\quad x_o$ = value of x at the null point.

The leakage flow rate at null, constituting a power loss, is:

$$\bar{Q}_l = 1 \quad \text{or } Q_l = Q_r = C_{q_i} a_i \sqrt{P_s^*/\rho}$$

In practice, the negative flow rates of Fig. 7.12 would be generated by fluid being forced back into the valve by an overrunning load on a cylinder.

7.5.5 Three-port pneumatic flapper-nozzle valve

Referring to Fig. 7.11 it can be seen that the mass flow rate $\dot{M}$ to the load is:

$$\dot{M} = \dot{M}_i - \dot{M}_o \, . \tag{7.26}$$

Defining $$\dot{M}_r = \left\{ \frac{\gamma}{R} \left(\frac{2}{\gamma+1} \right)^{(\gamma+1)/(\gamma-1)} \right\}^{\frac{1}{2}} C_{q_i} a_i \frac{P_s}{\sqrt{T_s}} \qquad \text{see (7.8)}$$

gives $$\bar{\dot{M}} = \bar{\dot{M}}_i - \bar{\dot{M}}_o \tag{7.27}$$

where $$\bar{\dot{M}}_i = f(P_c/P_s) = f(p_c)$$

and $$\bar{\dot{M}}_o = \bar{a} p_c f(p_e/p_c) \, . \tag{7.28}$$

Equation (7.27) is plotted in Fig. 7.13 for air flow and with $p_e = 0.1$. If $p_c = 0.5$ is chosen as the null pressure, as in the hydraulic valve, then the null orifice area ratio is 2.0 in this case. This arises because both orifices are choked in the null condition and hence the mass flow rates are proportional to the product of the orifice areas and their upstream pressures.

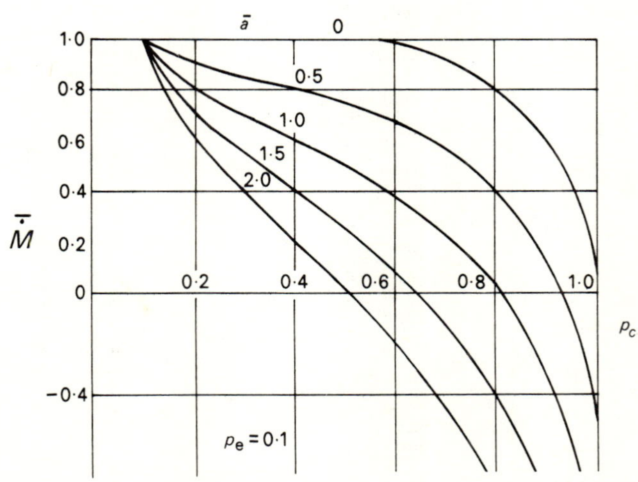

Fig. 7.13 – Valve characteristics; pneumatic three-port flapper-nozzle valve.

7.5.6 The three-port jet pipe valve

The jet pipe valve, Fig. 7.14, presents a much more difficult analytical problem than the valves discussed above. Basically the principle of operation is similar to that of the jet tube devices referred to in section 4.4, but since the receiving tube diameter in the jet pipe valve is as big, if not bigger than, the nozzle diameter, it is clear that its presence will strongly affect the flow pattern. This makes theoretical prediction difficult, and experimental determination of performance is usually necessary.

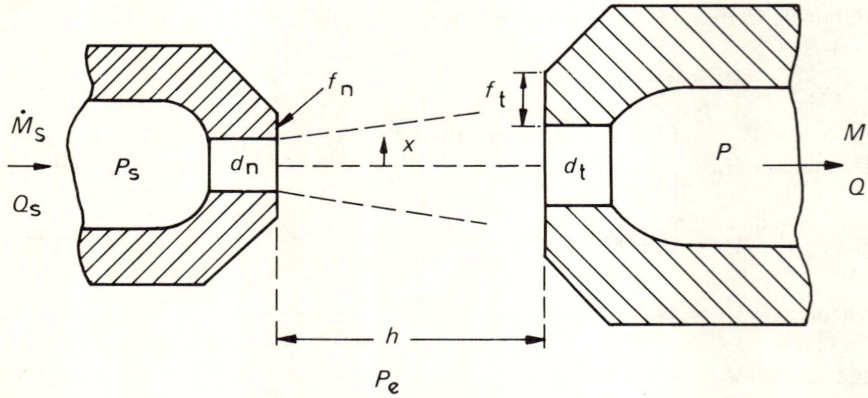

Fig. 7.14 – Three-port jet pipe valve.

Performance is usually described in terms of (a) the no-load flow recovery, (b) the blocked load pressure recovery, and (c) the energy recovery – all measured when the nozzle and tube are aligned ($x = 0$). The flow-pressure characteristics are shown qualitatively in Fig. 7.15. A comprehensive survey of jet pipe valves is

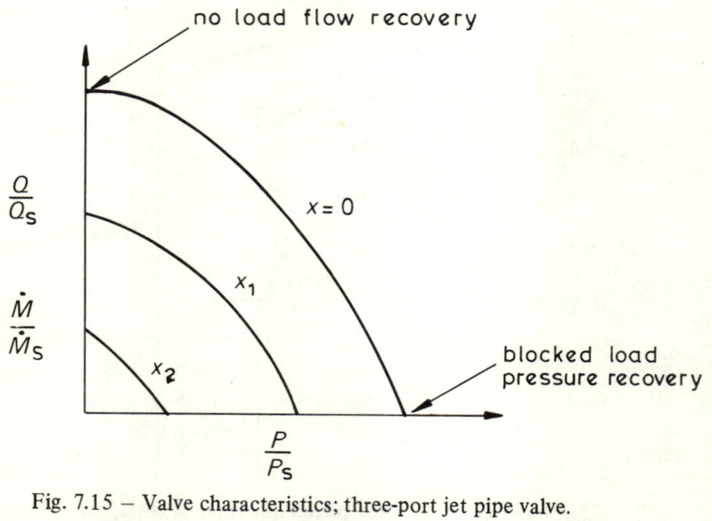

Fig. 7.15 – Valve characteristics; three-port jet pipe valve.

given in reference [4]. For an incompressible fluid the blocked load pressure recovery exceeds 97 per cent of the supply pressure when the receiver diameter $d_t < 1.4d_n$ and the axial displacement $h < 2.5d_n$. Nozzle to receiver spacings are more critical in pneumatic valves than in hydraulic valves mainly because of the rapid expansion of supersonic jets. For these cases the blocked load pressure exceeds 97 per cent when $d_t < 1.4d_n$ and $h < 0.175d_n$. Pressure oscillation can occur at some spacings.

Recovered flow rate can exceed the supply flow rate as a result of entrainment. For an incompressible fluid the no-load flow rate over a certain range of h is greater than that supplied, when $d_t > 1.2d_n$. There is an optimum spacing h, for a given (d_t/d_n), at which the recovered flow is a maximum. At this spacing the jet diameter is approximately equal to the receiver diameter. For a compressible fluid the same applies but the pressure ratio (P_e/P_s) should not be less than approximately 0.33. When driving a resistive load energy recovery is a maximum when $(d_t/d_n) = 1.4$ and $h < 3.2d_n$ for oil and $h < 0.175d_n$ for air.

7.5.7 The jet pump

Entrained flow is put to good use in the jet pump valve, Fig. 7.16. A primary jet passes through an orifice consisting of a moveable cone placed axially in an inlet hole. The jet first enters a low pressure region from which it entrains fluid.

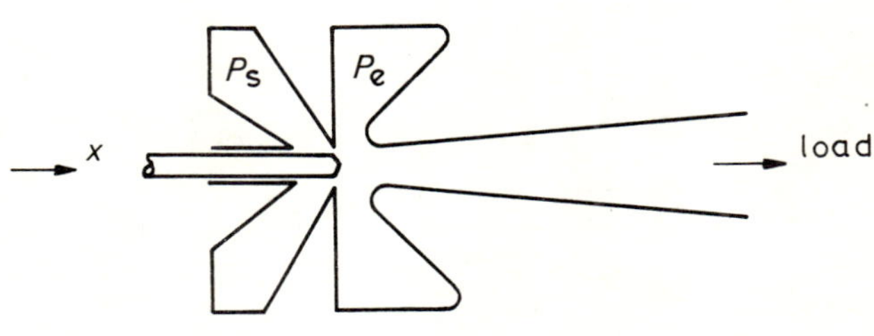

Fig. 7.16 – The jet pump valve.

It then passes through a diffusor leading to the load. Pressure, flow rate and power at the load depend on the size of the primary jet and on the amount of entrained fluid. The jet pump is claimed to have an efficiency greater than the conventional throttling valves [5]. It acts to a certain extent like a fluid transformer, converting a fluid stream of high pressure and low flow rate to one of low pressure and high flow rate. The moveable cone enables the primary jet to be varied and allows matching of the load and source characteristics with a minimum of power loss. Cavitation in the entrainment regions can impair entrainment and it may then be desirable to raise the pressure P_e in order to avoid vapourisation at the throat in hydraulic valves.

7.6 FOUR AND FIVE-PORT VALVES

These valves can control the position of double acting cylinders. Examples were shown in Fig. 6.12. They combine the functions of two three-port valves and hence their characteristics can be deduced from those derived above. The following examples illustrate this point.

7.6.1 The hydraulic five-port valve

A zero lap spool is shown in Fig. 7.17.

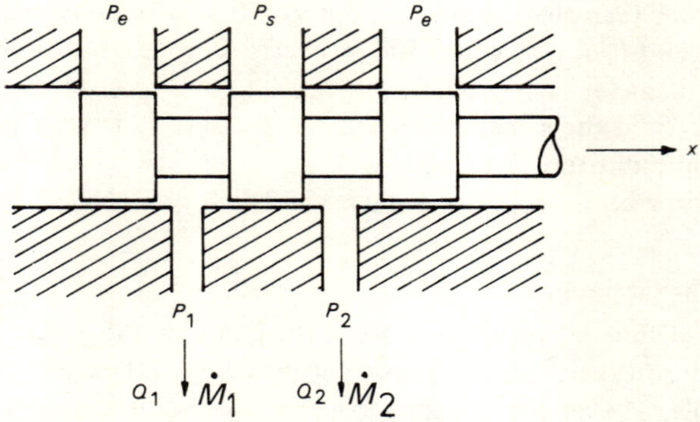

Fig. 7.17 – A five-port valve with zero lap.

For postive x:

$$\bar{Q}_1 = \epsilon_1 \bar{x}\sqrt{2\epsilon_1(1-p_1^*)}, \quad \epsilon_1 = \text{sgn}\,(1-p_1^*)$$

$$Q_2 = -\phi_2 \bar{x}\sqrt{2\phi p_2^*}, \quad \phi_2 = \text{sgn}\,p_2^*$$

For negative x:

$$\bar{Q}_1 = \phi_1 \bar{x}\sqrt{2\phi_1 p_1^*}, \quad \phi_1 = \text{sgn}\,p_1^*$$

$$\bar{Q}_2 = -\epsilon_2 \bar{x}\sqrt{2\epsilon_2(1-p_2^*)}, \quad \epsilon_2 = \text{sgn}\,(1-p_2^*)$$

where $$Q_r = Kx_r C_{q_r}\sqrt{P_s^*/2}\,.$$

The load determines the relationship between flow rates $\bar{Q}_1$ and $\bar{Q}_2$. Here steady state conditions are assumed so that for orifice loads and symmetrical double acting cylinders, $\bar{Q}_1 = -\bar{Q}_2$. Hence for positive x:

$$\epsilon_1(1-p_1^*) = \phi_2 p_2^* \tag{7.29}$$

or $$p_1^* + p_2^* = 1\,.$$

Define a load pressure $p_l = p_1^* - p_2^*$; then for positive $\boldsymbol{x}$ the flow rate is:

$$\bar{Q} = \epsilon_1 \bar{x} \sqrt{\epsilon_1 (1 - p_l)} \tag{7.30}$$

with $\epsilon_1 = \text{sgn}\,(1 - p_1^*) = \text{sgn}\,(1 - p_l)$.

The lower limit on p_1^* is p_v^* and hence since $p_l = 2p_1^* - 1$, the lower limit on p_l is $(2p_v^* - 1)$.

For negative x it is easy to show that:

$$\bar{Q} = \phi_1 \bar{x} \sqrt{\phi_1 (1 + p_l)}\,. \tag{7.31}$$

with $\phi_1 = \text{sgn}\,(1 + p_l)$.

In this case the upper limit on p_l is $(1 - 2p_v^*)$, at which point pressure P_2 falls to vapour pressure P_v.

Equations (7.30) and (7.31) are plotted in Fig. 7.18 for $\bar{x} = 0, \pm 1$ and $p_v^* = -0.111$. When $|p_l| > 1.222$ cavitation occurs in the cylinder. The null

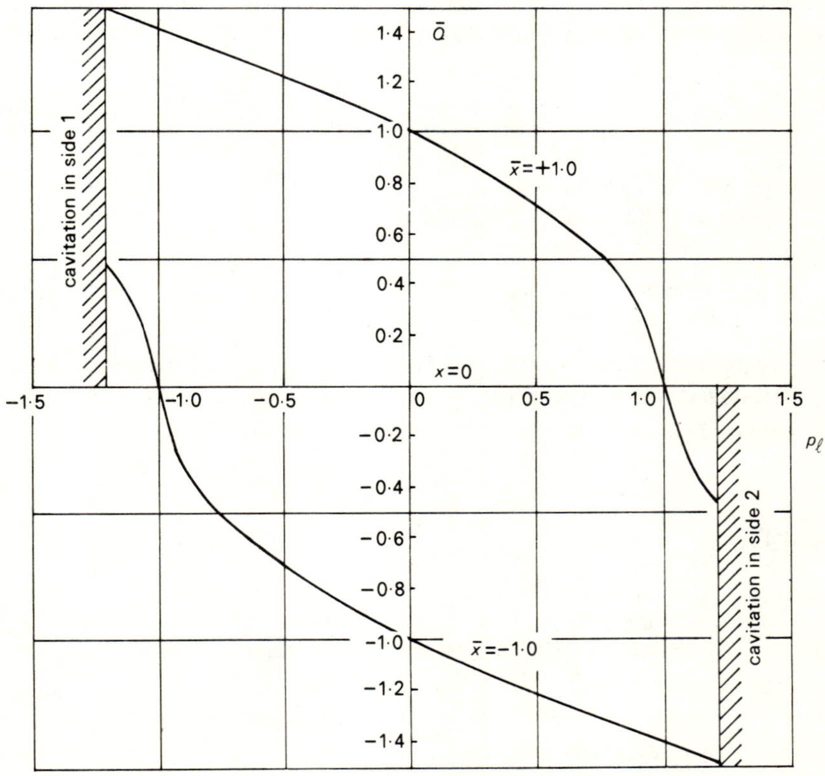

Fig. 7.18 – Valve characteristics; hydraulic five-port valve.

coefficients, evaluated at $p_1 = 0$, $\bar{x} = 0$, are $\partial Q/\partial x = K\sqrt{P_s^*}/2$, $\partial Q/\partial P_1 = 0$, and $\partial P_1/\partial x = \infty$. The piston rod velocity is Q/A where A is the effective piston area.

7.6.2 The pneumatic five-port valve

Fig. 7.17 shows that the mass flow rate passes from an upstream pressure $P_u = P_s$ to a downstream pressure $P_d = P_1$. If $P_r = P_s$ and $T_r = T_s$ equation (7.10) gives:

$$\bar{\dot{M}}_1 = \bar{a}f(P_1/P_s) = \bar{a}f(p_1)\,. \tag{7.32}$$

$\dot{M}_2$ can be calculated using P_2 as upstream pressure and P_e as downstream pressure. Equation (7.10) gives:

$$\bar{\dot{M}}_2 = -\bar{a}p_2\,f(p_e/p_2)\,. \tag{7.33}$$

The load determines the relationship between $\dot{M}_1$ and $\dot{M}_2$. For example a steady state orifice load requires that $\dot{M}_1 = -\dot{M}_2$ or that $f(p_1) = p_2\,f(p_e/p_2)$. This can be solved by trial and error using Fig. 7.3(a).

For a double acting cylinder moving at steady velocity the volumetric flow rates into and out of the cylinder are equal ($Q_1 = -Q_2$). Hence:

$$\frac{\dot{M}_1}{\rho_1} = -\frac{\dot{M}_2}{\rho_2} \quad \text{or} \quad \frac{\dot{M}_1}{\dot{M}_2} = -\frac{\rho_1}{\rho_2} = -\frac{P_1}{P_2} \tag{7.34}$$

assuming isothermal compression.

Equations (7.32) to (7.34) combine to give:

$$\frac{1}{p_1}\,f(p_1) = f(p_e/p_2) \tag{7.35}$$

This too must be solved by trial and error (a neat graphical solution is given in [1]). Having solved equation (7.35) for p_1 and p_2, the mass flow rates can be calculated. The piston rod velocity u is found as follows:

$$u = \dot{M}_1/\rho_1 A$$

and defining:

$$u_r = \dot{M}_r/\rho_r A$$

gives:

$$\bar{u} = u/u_r = \bar{\dot{M}}_1(\rho_r/\rho_1) = \bar{\dot{M}}_1(P_s/P_1) = \bar{\dot{M}}_1/p_1$$

$$\tag{7.36}$$

or

$$\bar{u} = \bar{a}f(p_1)/p_1\,.$$

The characteristics are plotted in Fig. 7.19.

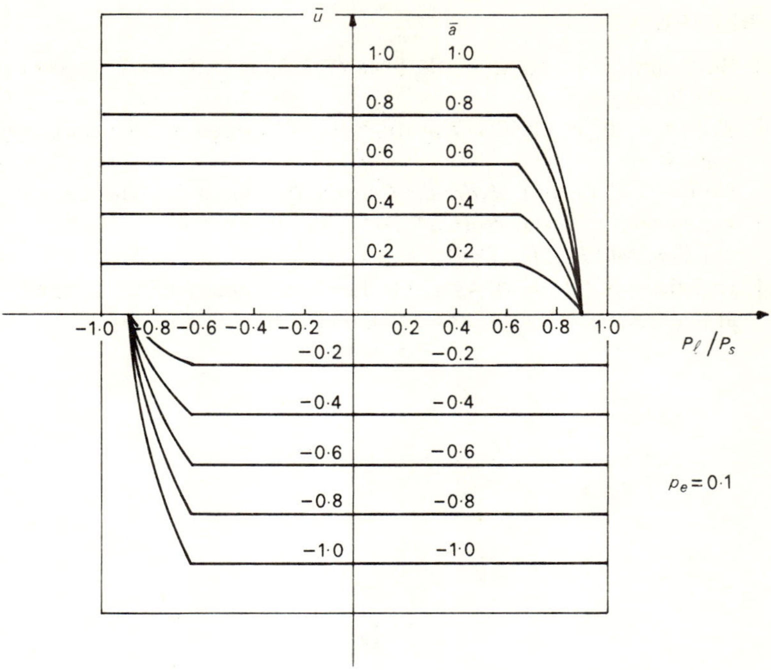

Fig. 7.19 – Valve characteristics; pneumatic five-port valve.

7.7 EXAMPLES

7.7.1

A hydraulic three-port valve controls a vertical single acting cylinder used for lifting a mass of 510 kg. The supply pressure is 71 bar, the oil density 900 kg/m^3 and the cylinder diameter is 4 cm. The maximum movement of the valve spool is limited to 1 mm, the diameter of the spool is 1 cm and the ports are annular. There is no lap. Determine (a) the value of K in equation (7.15) assuming C_q = 0.7, (b) the reference flow rate, taking $P_r = P_s$, (c) the non-dimensional steady state pressure p_c^*, (d) the non-dimensional flow rate when the spool is displaced 0.5 mm from zero, (e) the steady velocity of the load.

(*Answers*: (a) 1.037×10^{-3}; (b) 1.94 l/s; (c) 0.569; (d) 0.464; (e) 0.716 m/s.)

7.7.2

Air flows through a variable orifice from a supply pressure of 10 bar. The maximum area of orifice is 1.25 mm^2 and the flow coefficient is 0.8.(a) Linearise the mass flow equation about the steady state conditions a = 0.625 mm^2 and downstream pressure 8 bar, (b) determine the error in the mass flow predicted this way when the downstream pressure is 9 bar.

(*Answers*: (a) $\delta\bar{M} = 0.819\, \delta\bar{a} - 0.745\, \delta p$; (b) $\bar{M}$ too high by 0.264.)

7.8 REFERENCES

[1] Blackburn, J. F., Reethof, G., Shearer, J. L. (1960), *Fluid Power Control*, MIT Press.

[2] Andersen, B. W. (1967), *The Analysis and Design of Pneumatic Systems*, John Wiley.

[3] Merrit, H. E. (1967), *Hydraulic Control Systems*, John Wiley.

[4] Lichtarowicz, A. and Pearce, I. D. (1970), Characteristics of jet pipe valves, *4th Cranfield Fluidics Conf.*

[5] Dickersen, S. L.(1965), A variable fluid transformer for increasing efficiency in hydraulic control systems, Sc.D. thesis, MIT.

CHAPTER 8

Flow forces in valves

8.1 INTRODUCTION

It is desirable that the forces required to open or close a valve be as small as possible. This is particularly so for single stage servovalves where the existence of large operating forces can require the use of large and expensive torque motors. Static balance is usually achieved by the provision of a suitable number of balancing lands (see for example Figs. 6.1 and 6.3). However, even when a valve is so balanced, forces can arise due to the changes of fluid momentum within the valve. Newton's second law states that force is proportional to the rate of change of momentum. In a steady state fluid flow, force equals the net rate of outflow of momentum from a control volume. An example may make this clearer. Consider the curved pipe of varying cross section shown in Fig. 8.1. The rate of outflow of momentum in the X direction is $\rho_2 Q_2 v_2 \cos\alpha$ where ρ_2 is the mass density, Q_2 the volumetric flow and v_2 the mean velocity at station 2.

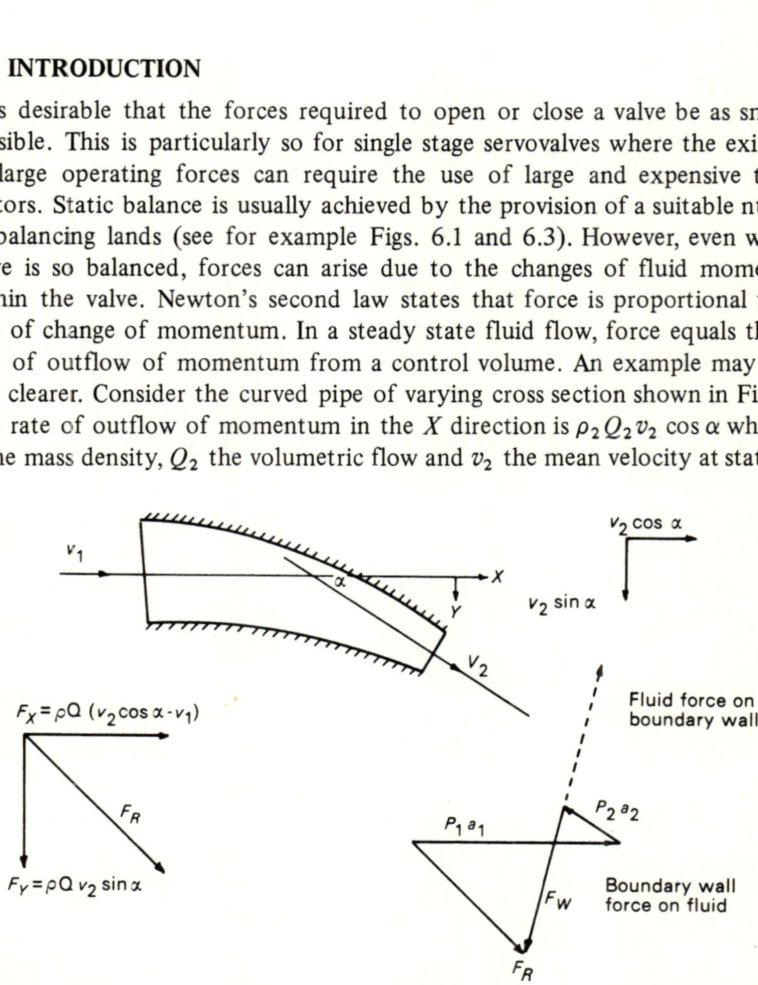

Fig. 8.1 – Forces acting on a pipe because of changes in fluid momentum.

The rate of inflow of momentum at station 1 is $\rho_1 Q_1 v_1$. Hence the force on the fluid in the X direction is:

$$F_x = \rho_2 Q_2 v_2 \cos\alpha - \rho_1 Q_1 v_1 . \tag{8.1}$$

Similarly in the y direction the force on the fluid is:

$$F_y = \rho_2 Q_2 v_2 \sin\alpha . \tag{8.2}$$

For steady flow of a liquid, $\rho_1 = \rho_2$ and $Q_1 = Q_2 = Q$

hence $$F_x = \rho Q(v_2 \cos\alpha - v_1) \tag{8.3}$$

and $$F_y = \rho Q v_2 \sin\alpha \tag{8.4}$$

In Fig. 8.1 these two forces are combined to give F_R, the resultant force that must be applied to the control volume in order to produce the given changes in momentum. Three forces combine to produce the required force F_R. These are (1) the pressure force $P_1 a_1$, (2) the pressure force $P_2 a_2$, and (3) the force F_w exerted on the fluid by the walls of the pipe. Vectorial addition determines the unknown force F_W. The reaction of the fluid on the walls of the pipe is equal and opposite to F_W. In valves this reaction force is sometimes referred to as the **Bernoulli force**, but it is more commonly called the **steady state flow force**.

8.2 STEADY STATE FLOW FORCES IN SPOOL VALVES

Fig. 8.2 shows part of a spool valve. The chamber A is pressurised and a jet of fluid enters chamber B at an angle θ: for a sharp edged orifice with no clearance between spool and sleeve, $\theta = 69^\circ$ (see section 4.5.1). The forces on the valve are found by evaluating the momentum changes in the control volume a b c d e f g h i [1]. Section ab is the vena contracta of the jet. In a conventional hydraulic valve the rate of inflow of momentum through de is much less than the rate of outflow through ab. In addition the inflow is nearly radial: thus the component of inflow momentum in the X direction may be neglected. The net rate of outflow of momentum in the X direction is $\rho Q v \cos\theta$ where the fluid velocity at ab is:

$$v = C_v \sqrt{2\Delta P/\rho} . \tag{8.5}$$

The volumetric flow rate Q is:

$$Q = a C_d \sqrt{2\Delta P/\rho} \tag{8.6}$$

where a is the orifice area. Hence the force applied to the fluid in the X direction is:

$$F_x = \rho Q v \cos\theta = 2C_d C_v a\Delta P \cos\theta . \tag{8.7}$$

Taking $C_v = 1$ and $\theta = 69°$ gives:

$$F_x = 0.72\, C_d\, a\Delta P .$$

This force will be provided by the boundary walls and by the static pressure forces discussed in the previous section. In this case the static pressure forces can be assumed to be negligible so that all of the forces in the X direction must be supplied by the spool. The reaction force on the spool tends to close the valve. The valve shown in Fig. 8.2 is pressure balanced (that is, when the orifice

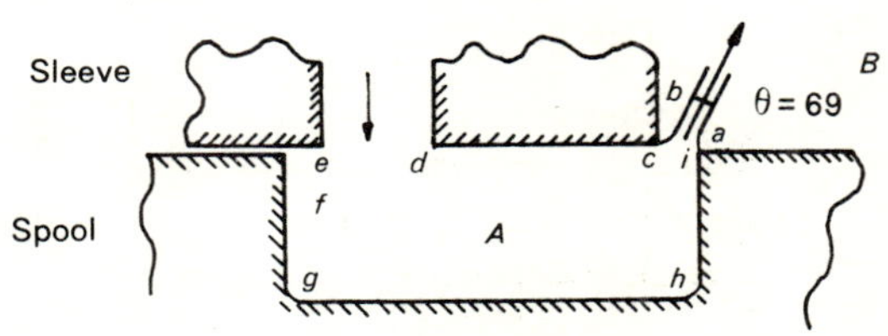

Fig. 8.2 – A spool valve chamber.

is closed there is no force acting on the valve), and in this case the only force acting on the spool is the flow force. If the upstream land fg were removed then the resultant opening force caused by the pressure on land hi could easily swamp the closing flow force.

The efflux angle θ is affected by chamber geometry, orifice clearance and sharpness (section 4.5.1).

It is worthwhile including an example at this stage in order to give an indication of the magnitude of the forces involved. If the spool displacement is 0.2 mm, the pressure drop 70 bar, $C_d = 0.7$ and the ports are circumferential in a sleeve of diameter 10 mm, then:

$$F_x = 0.72 \times 0.7 \times (\pi \times 0.01 \times 0.0002) \times 70 \times 10^5 = 22\,\text{N} .$$

Experiment has shown equation (8.7) to be satisfactory for the prediction of flow forces in hydraulic valves. For pneumatic valves recent work has brought some discrepancies to light, particularly at the larger values of spool displacement that are likely to be needed in pneumatic valves [2]. In this region of operation the pressure drop within the valve chamber, that is from de to ci in Fig. 8.2, is significant, and the analysis requires some modification (see section 8.3.3).

8.3 COMPENSATION OF FLOW FORCES IN SPOOL VALVES

These forces can be sufficiently large to limit the employment of single stage valves to low power applications. Recent papers on servo dynamics [1-7] show that flow forces reduce system gain and stability. Thus, methods of eliminating flow forces are of interest. All of these methods aim at changing the net rate of efflux of momentum from the control volume shown in Fig. 8.2.

8.3.1 Method 1. Radial hole orifices

This is illustrated in Fig. 8.3. The valve ports in this case are two small holes drilled in the sleeve in diametrically opposite positions. This ensures that there is no lateral force on the spool, that is, the momentum components $\rho Q v \sin\theta$ cancel out. Initially as the holes are being uncovered the jet angle is 69°, but as the opening increases the angle increases until at full opening it becomes 90°. At this point the net axial rate of efflux of momentum is zero and there is no flow force, Fig. 8.3(b). However, Clark [3] points out that the abrupt change in the force gradient can lead to stability problems.

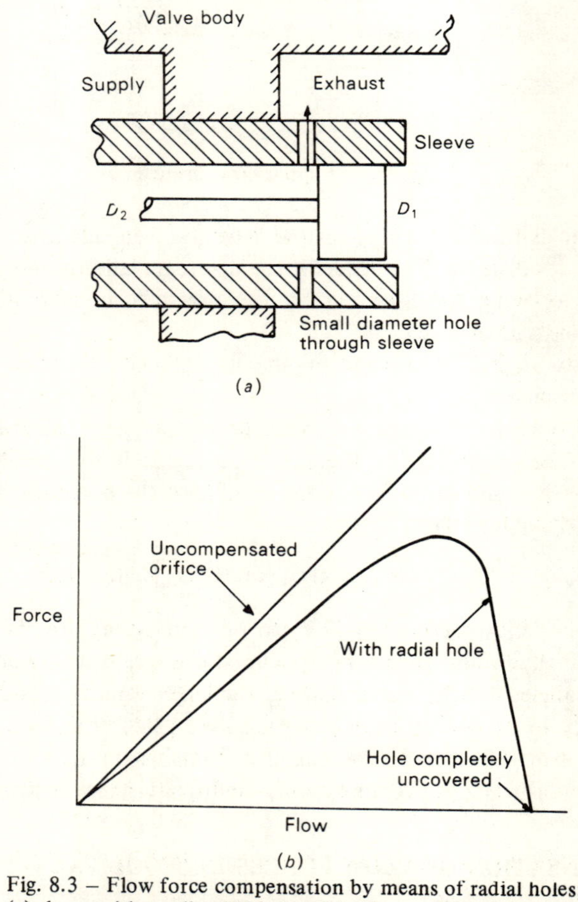

Fig. 8.3 – Flow force compensation by means of radial holes; (a) sleeve with small radial holes, (b) effect on flow force.

This can be improved if each of the radial holes is replaced by a series of smaller holes. These are arranged in a spiral so that one begins to open before its neighbour is completely closed, Fig. 8.4(a). The desired valve area displacement characteristic determines the size of these holes. Clark [3] found that an overlap $s = 0.18d$ resulted in an approximately linear area displacement characteristic. Fig. 8.4(b) shows how the flow force contributions from each hole can be made to add up to give a constant flow force, independent of flow rate.

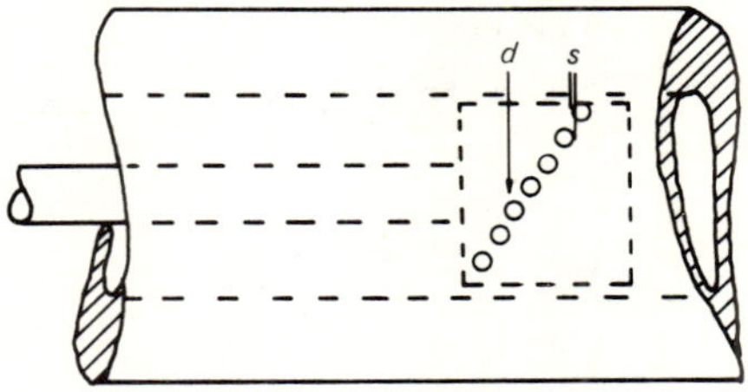

Radial holes with overlap

(a)

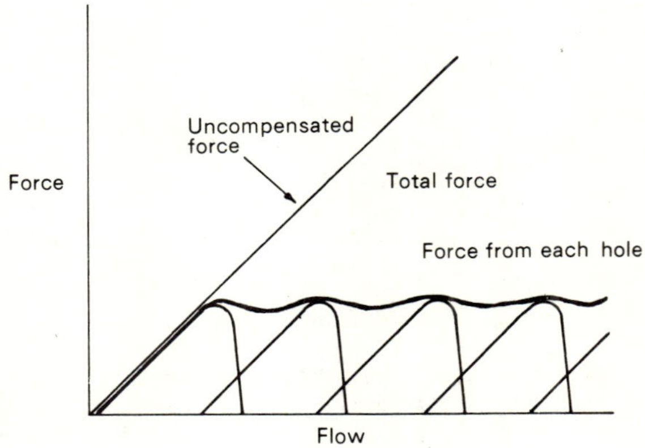

(b)

Fig. 8.4 – Use of overlapping radial holes for flow force compensation; (a) the sleeve, (b) effect on flow force.

8.3.2 Method 2. Angled port compensation

The flow force exists because there is a net rate of efflux of momentum from the control volume. In Fig. 8.2 the momentum at the outlet or orifice was the only contributor since there was no axial component of the inlet momentum. However, if the inlet port is angled as in Fig. 8.5(a) an axial component of inlet momentum will exist and this will reduce the net rate of efflux of momentum.

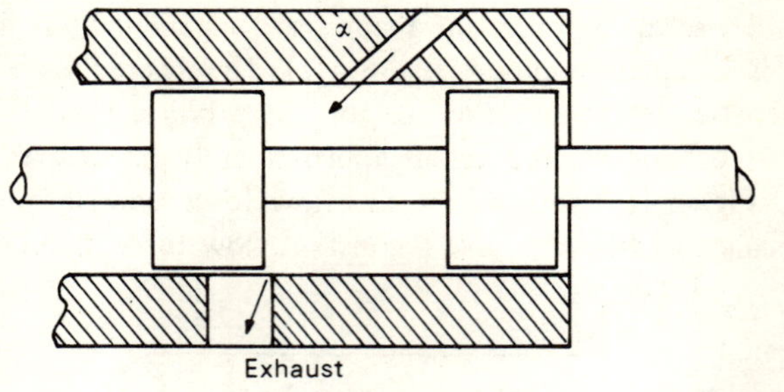

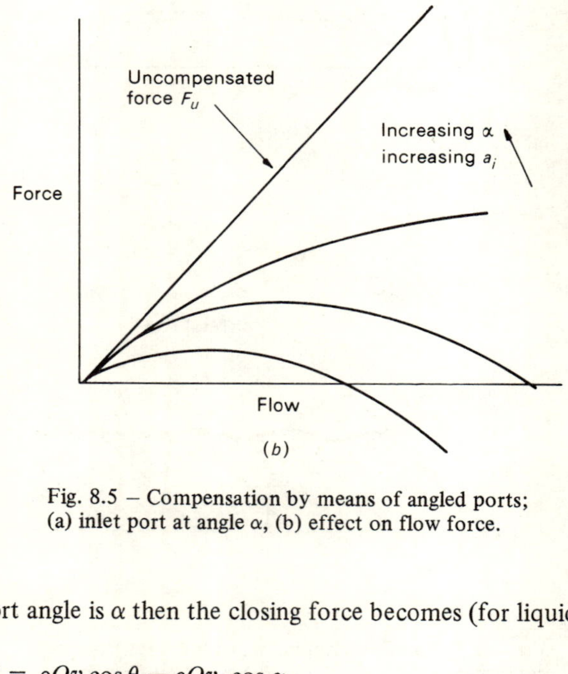

Fig. 8.5 – Compensation by means of angled ports; (a) inlet port at angle α, (b) effect on flow force.

If the inlet port angle is α then the closing force becomes (for liquid flow):

$$F = \rho Q v \cos\theta - \rho Q v_i \cos\alpha \tag{8.8}$$

where v_i is the mean velocity at the inlet port.

Hence $$F = \rho Q v \left\{\cos\theta - (v_i/v)\cos\alpha\right\}. \tag{8.9}$$

Now $$C_c a v = C_{ci} a_i v_i \tag{8.10}$$

where C_c is a contraction coefficient.

Take $C_c = 0.65$ and $C_{ci} = 0.90$. (8.11)

Hence $v_i/v = 0.72\,(a/a_i)$. (8.12)

Substitution in equation (8.9) gives:

$$F = \rho Q v \cos\theta \left\{1 - 0.72\, a \cos\alpha/(a_i \cos\theta)\right\}. \tag{8.13}$$

If $\theta = 69°$ this gives:

$$(F/F_u) = 1 - 2(a/a_i)\cos\theta \tag{8.14}$$

where F_u is the uncompensated force. Clearly the flow force is zero when $\cos\alpha = 0.5\,(a_i/a)$ or when $a = 0.5\,a_i/\cos\alpha$. For instance if $\alpha = 45°$ zero force exists when $a = 0.71\,a_i$. Since valve orifice areas are generally small this implies that the inlet area should be small too, and so this method causes an additional restriction in the valve. This can be reduced by making the inclination of the inlet port to the longitudinal axis as small as possible [4].

8.3.3 Method 3. Pressure drop compensation

In the analysis of momentum effects frictional forces in the valve chamber were neglected. However, if the annular space between spool and sleeve is small, velocity in the valve chamber will be high and a friction loss, proportional to v^2, will occur. Referring to Fig. 8.6(a), this will result in the pressure in the exhaust side of the valve at A being greater than that at B. This gives an opening force which opposes the closing force discussed above [5]. The smaller the annular area the greater the opening force. However, this cannot be carried too far since it causes choking of the flow. Lee points out that this is serious if the ratio of the wetted areas is less than 1:4. That is, if:

$$(D^2 - D_1^2)/(D^2 - D_2^2) < 1/4\,.$$

This method of compensation is sometimes referred to as the **thick stem** method.

8.3.4 Method 4. Spool profiling or spool shaping

There are two orifices in series in a five-port valve, and each contributes to the opening force. Lee and Blackburn [1] have devised a method for changing the sign of the force caused by flow from control port to exhaust port, Fig. 8.7. This is accomplished by machining tapers on parts of the valve stem and recesses in the sleeve. This is costly but it is effective and in addition it usually results in a slight lowering of the hydraulic restriction of the valve [6].

Fig. 8.7 shows a negative force port. The stem is shaped to look like a turbine bucket, and the recess in the sleeve makes the downstream chamber

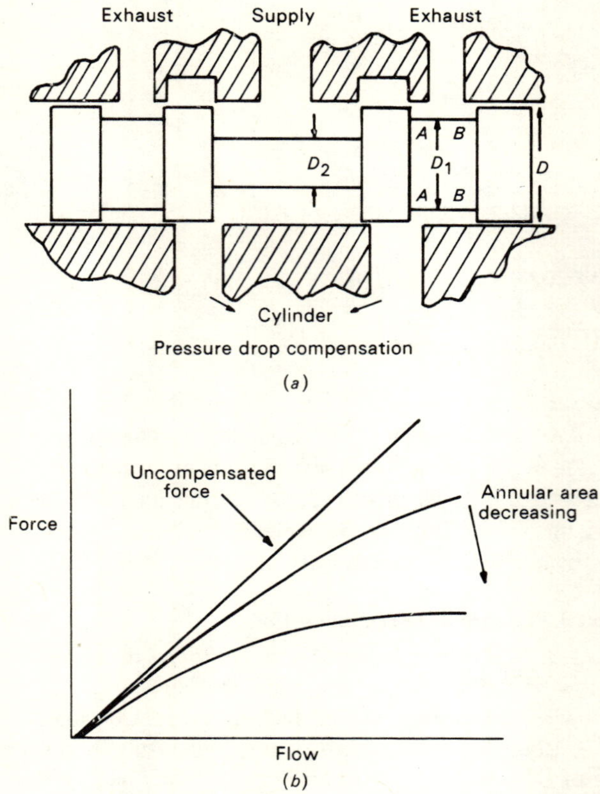

Fig. 8.6 – Use of pressure drop in the valve chamber for compensation; (a) the valve with a thick stem on the exhaust side, (b) the effect on the flow force.

larger. The jet enters at angle θ_1 and leaves at angle θ_2, and hence if the cross-section of the flow remains constant, the closing force would be:

$$F = \rho Q v(\cos\theta_1 - \cos\theta_2)\,. \tag{8.15}$$

If $\cos\,\theta_2 > \cos\,\theta_1$, that is, $\theta_2 < \theta_1$, the force will tend to open the valve. In addition to this an eddy is formed in the chamber causing an additional flow to enter the chamber at angle θ_3. This also tends to open the valve. The total force is almost impossible to predict analytically. However, the method is very effective.

Another form of spool profiling is illustrated in Fig. 8.8. In this case the intention is to increase the angle, thereby reducing the change in axial momentum. This is achieved by using a cup shaped spool face.

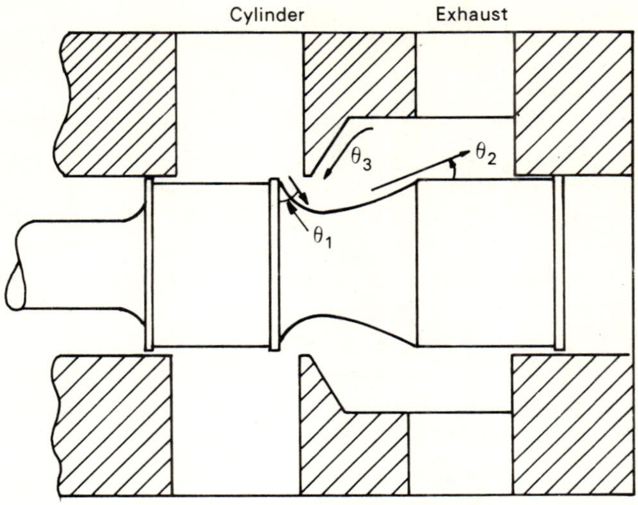

Fig. 8.7 – A negative force port.

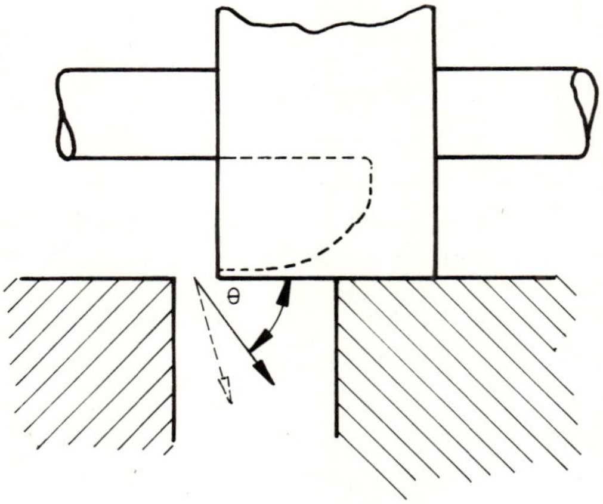

Fig. 8.8 – Use of a cup shaped piston to increase the exhaust angle.

8.4 TRANSIENT FLOW FORCES IN SPOOL VALVES

The forces so far discussed are static and their value can be calculated for a given static valve opening and flow rate. However, during unsteady or transient conditions the flow rate is not steady and the fluid accelerations involved give rise to additional flow forces. Consider the fluid slug, of length L_1, and cross-sectional area A, in the valve chamber shown in Fig. 8.9. Newton's law gives:

$$F = \text{mass} \times \text{acceleration}$$

or $$F = \rho L_1 A \frac{\mathrm{d}}{\mathrm{d}t}\left[\frac{Q_1}{A}\right] = \rho L_1 \frac{\mathrm{d}Q_1}{\mathrm{d}t}. \tag{8.16}$$

Here the fluid is assumed to be incompressible so that the volumetric flow rate through the chamber equals that through the orifice. In addition the slug is assumed to move at velocity Q_1/A. F is the force required to produce the rate of change of momentum; the reaction F_r on the spool is equal and opposite to F. For example if the flow Q_1 is increasing, the accelerating force must be in the positive x_v direction, but the reaction F_r on the spool is opposite and tends to close the valve. F_r is thus stabilising. If, however, the same analysis is applied to the outflow Q_2, it is seen that as Q_2 increases, F_r is destabilising and tends to open the valve further. Thus, in general, flow out of the valve chamber through an orifice is stabilising and flow into the valve chamber destabilising. The total closing force on the spool in Fig. 8.9 is:

$$F_{rt} = \rho L_1 \frac{\mathrm{d}Q_1}{\mathrm{d}t} - \rho L_2 \frac{\mathrm{d}Q_2}{\mathrm{d}t} \tag{8.17}$$

But $Q_1 = Q_2$ (for liquid flow).

Hence $$F_{rt} = \rho(L_1 - L_2)\frac{\mathrm{d}Q_1}{\mathrm{d}t}. \tag{8.18}$$

Thus for stability $(L_1 - L_2)$, called the **damping length** L_d, must be positive [7]. If the damping length is negative, instability can arise and the valve may 'sing'.

Since $$Q_1 = C_q a\sqrt{\frac{2}{\rho}(P_s - P_1)}$$

where $$a = kx_v,$$

then $$F_{rt} = \rho L_d k C_q \sqrt{\frac{2}{\rho}}\left\{\frac{\mathrm{d}x_v}{\mathrm{d}t}\sqrt{P_s - P_1} - \frac{x_v}{2\sqrt{P_s - P_1}}\frac{\mathrm{d}P_1}{\mathrm{d}t}\right\}. \tag{8.19}$$

The pressure rate term is usually small in comparison to the other term and can be neglected. The transient flow force can thus be written:

$$F_{rt} = \beta\sqrt{\Delta P}\frac{\mathrm{d}x_v}{\mathrm{d}t} \tag{8.20}$$

where $$\beta = \rho L_d k C_q \sqrt{2/\rho}.$$

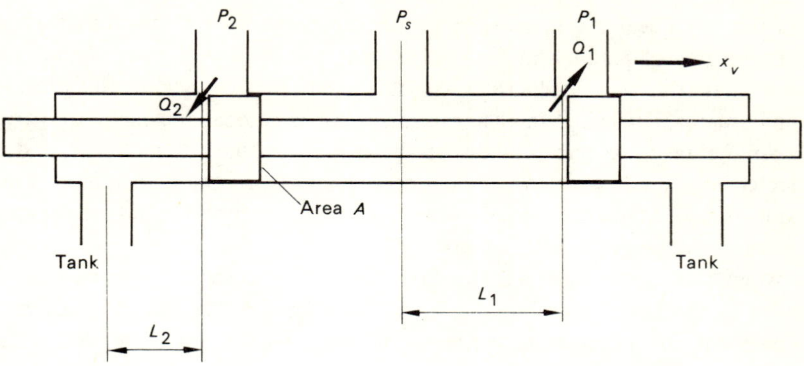

Fig. 8.9 – Supply and exhaust damping lengths, L_1 and L_2.

8.5 FLOW FORCES IN POPPET VALVES

A typical two-port poppet valve is shown in Fig. 8.10. It is provided with a balance piston which counteracts the opening force caused by the supply pressure acting on the underside of the poppet. In the closed position this opening force F_o is:

$$F_o = P_1 \left\{\frac{\pi D^2}{4}\right\} + \Delta F \tag{8.21}$$

where ΔF is an additional force caused by the pressure distribution over the

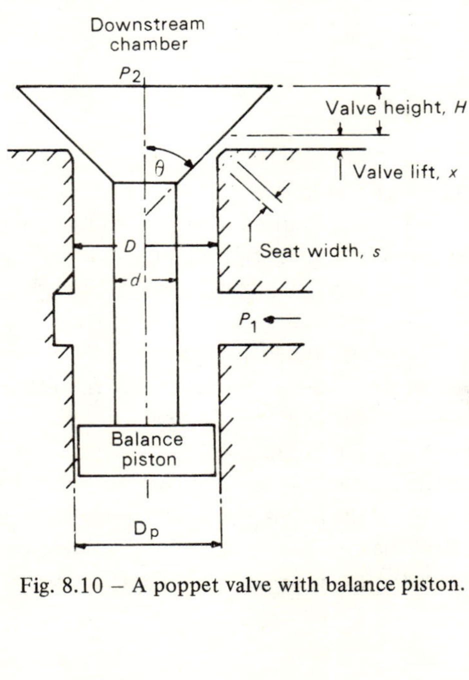

Fig. 8.10 – A poppet valve with balance piston.

when there is a seat the magnitude of ΔF is determined by the area of the seat and the degree of parallelism between poppet and seat. If the poppet angle θ is greater than the seat angle then ΔF is the product of P_1 and the projected seating area; if θ is less than the seat angle, ΔF is zero; if the faces are parallel ΔF can be taken as the product of the mean pressure, $(P_1 - P_2)/2$, and the projected seat area. A spring could be used instead of the balancing piston to provide the counteracting force, but this introduces a problem in that the spring force will increase as the valve opens.

When the valve is opened the flow forces can again be deduced from the changes in momentum within the valve. If, as in the analysis of spool valves, the momentum at the vena contracta is predominant, then the closing flow force is given by equation (8.7).

$$F = \rho Q v \cos\theta\ ,$$

where θ is the poppet angle in this case. Experiment has shown that this equation gives reasonably accurate results when the downstream chamber is large [8, 9].

For smaller chambers, vortices occur and these affect the pressure distribution on the top and sides of the poppet [10]. In such cases the downstream momentum is no longer negligible and the geometry of the chamber is important in determining whether the force is increased or decreased.

There is very little published work concerning flow forces in pneumatic poppet valves. Tsai and Cassidy [11] report experiments on a 45° poppet valve, and based on their results Anderson [12] recommends the empirical formula (for 45° poppets):

$$F = 0.97\, a \Delta P\,. \tag{8.22}$$

There are several means of reducing or compensating flow forces in poppet valves. Perhaps the most direct is to increase the poppet angle thereby reducing the change in axial momentum. If θ is increased to 90°, as in Fig. 8.11, the force is theoretically zero. Indeed it is considerably reduced in practice.

Several methods of compensation have been proposed [8]. Blackburn *et al.* [13] suggest the arrangement shown in Fig. 8.12. Two identical poppets are mounted on a common stem and adjusted to seat simultaneously. The adjustment, however, is liable to be affected by thermal expansion.

Guillon [14] proposes a method involving a negative force port similar to that discussed earlier in relation to spool valves. In Fig. 8.13, the downstream part of poppet A is so shaped that the axial momentum leaving the curved surface is greater than that at the vena contracta. An opening force is thus generated. This balances the closing forces generated in poppet B.

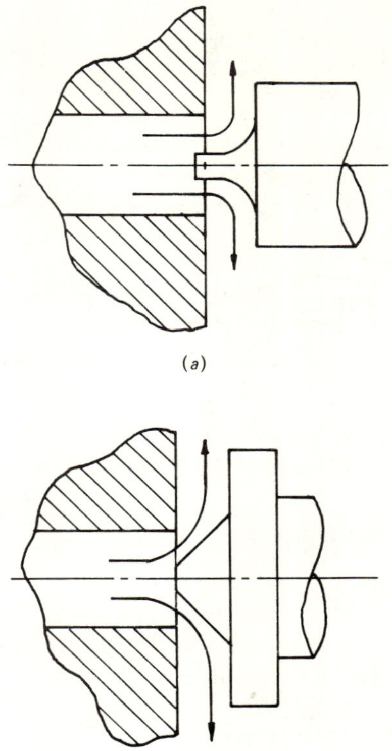

(a)

(b)

Fig. 8.11 – Flow force compensation in the poppet valve; (a) a 90° poppet, (b) another means of increasing the effective poppet angle.

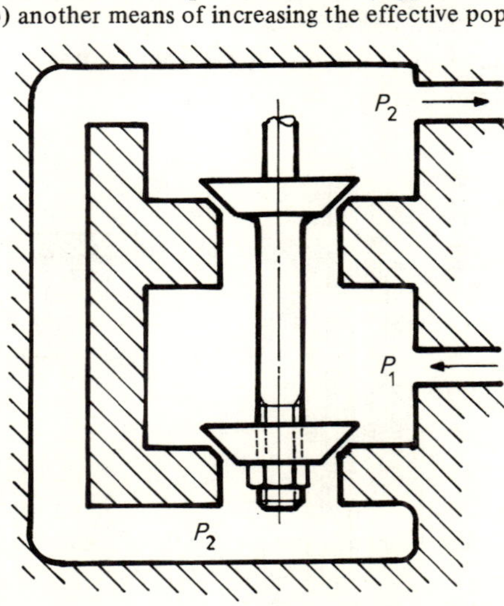

Fig. 8.12 – A twin balanced poppet valve.

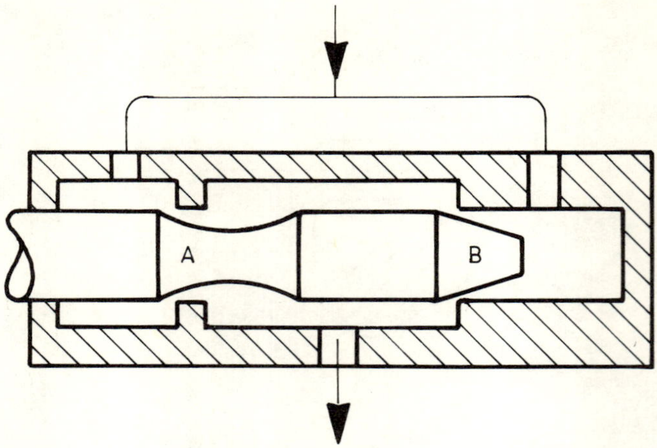

Fig. 8.13 – Balancing a poppet valve by means of a negative force port.

8.6 FORCES ON FLAPPER-NOZZLE VALVES

A sharp edged flapper-nozzle valve is shown in Fig. 8.14. If the flapper displacement x is less than one half the radius of the nozzle, the curtain area or orifice area $2\pi rx$ is less than the nozzle area and is the controlling impedance. The volumetric flow rate is given by the orifice equation:

$$Q = 2\pi rx\, C_d \sqrt{2\Delta P/\rho}$$

where $C_d = 0.6$ (see section 4.5.3). An approximate analysis of the force on the flapper can be developed, considering momentum changes [13]. With the flapper close to the nozzle, consider three sections of the flow, one far upstream of the nozzle, one at the nozzle exit, and one at the vena contracta. Bernoulli's equation applied at sections 1 and 2 gives:

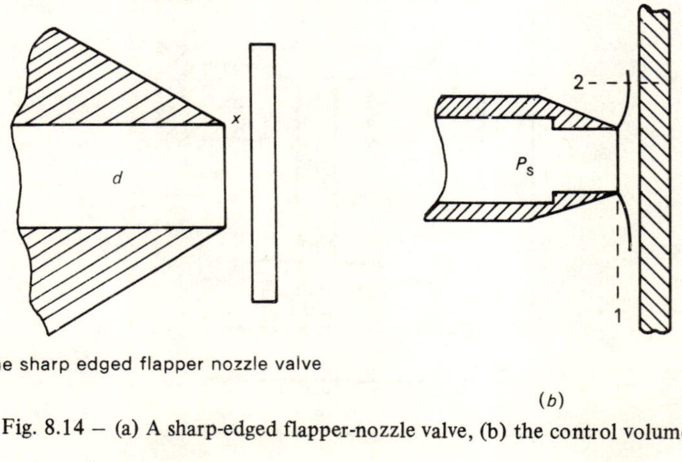

Fig. 8.14 – (a) A sharp-edged flapper-nozzle valve, (b) the control volume.

$$P_1 + \frac{1}{2}\rho v_1^2 = P_s = P_2 + \frac{1}{2}\rho v_2^2 \,.$$

For liquid flow, continuity requires that:

$$v_1 \pi r^2 = C_d v_2 \, 2\pi r x \,.$$

Momentum considerations give:

$$F = P_1 \pi r^2 + \rho Q v_1 \,.$$

Hence, neglecting P_2:

$$F = \pi r^2 P_s [1 + (2 C_d x/r)^2] \,. \tag{8.23}$$

Thus the force is initially the static pressure force, but it rises with displacement x. This result is only valid for small x/r, say $x/r < 0.2$. At greater displacements most of the assumptions used in the analysis are no longer true. For example, the velocity profile is not uniform so changes in momentum will be affected.

Pawlak [15] has examined the forces in a sharp edged pneumatic valve, and assuming isentropic flow he calculated that force would increase with displacement in the manner shown in Fig. 8.15. These results show the same trend as those for an incompressible fluid, equation (8.23), and indeed they are nearly identical if $C_d = 0.5$ in equation (8.23).

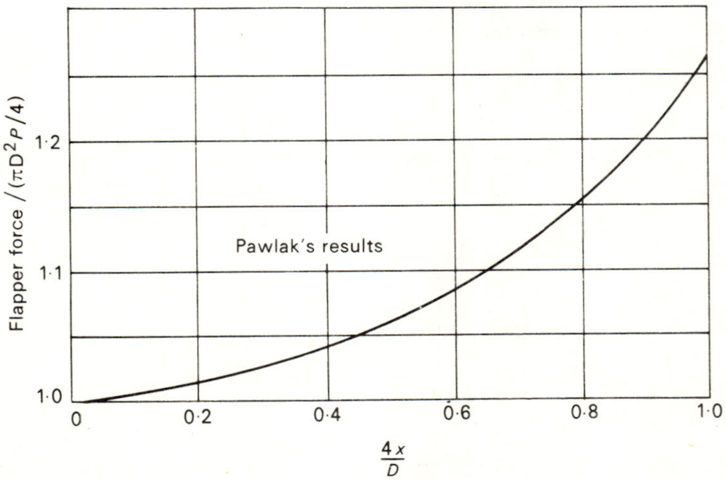

Fig. 8.15 – Flow forces in a pneumatic sharp-edged flapper-nozzle (after [15]).

When the nozzle has a flat, Fig. 8.16, the analysis becomes much more complicated. When the valve is closed the force is given by the product of the pressure and the effective nozzle area. This area will lie between the area of the nozzle πr^2 and the area of the nozzle face $\pi(r + w)^2$. Hence at the closed position:

$$\pi r^2 \Delta P < F < \pi(r + w)^2 \Delta P . \tag{8.24}$$

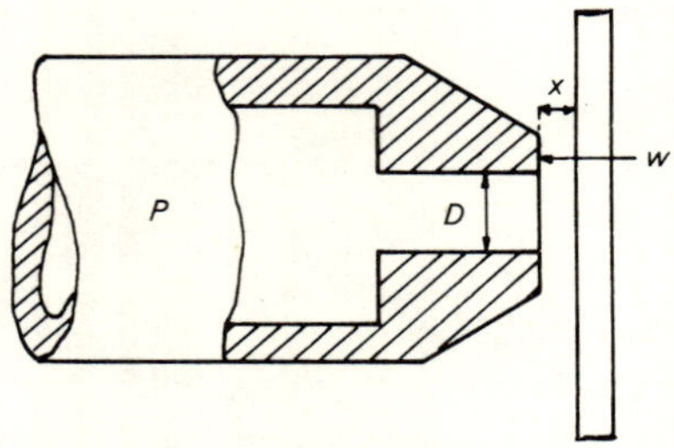

Fig. 8.16 – Flapper-nozzle valve with flat.

When the valve is opened slightly laminar flow will develop in the space between the land and flapper. If the fluid is incompressible and the flapper displacement is very small an exact solution of the Navier–Stoke's equation gives:

$$F = \frac{\pi[(r + w)^2 - r^2]\Delta P}{2 \ln [1 + (w/r)]} \tag{8.25}$$

where w is the width of the flat. The pressure distribution is given by curve (a) in Fig. 8.17. If equation (8.25) is written in the form:

$$F = C_f \pi r^2 \Delta P$$

the following variation of C_f with w can be presented.

Table 8.1.

w	0	$0.5r$	$1.0r$	$1.5r$
C_f	1.0	1.54	2.16	2.87

Hence for very small flapper displacements the flat causes an increase in force.

Pawlak has also examined the effect of nozzle flats in pneumatic flapper-nozzle valves. The continuously increasing curtain area in the radial direction

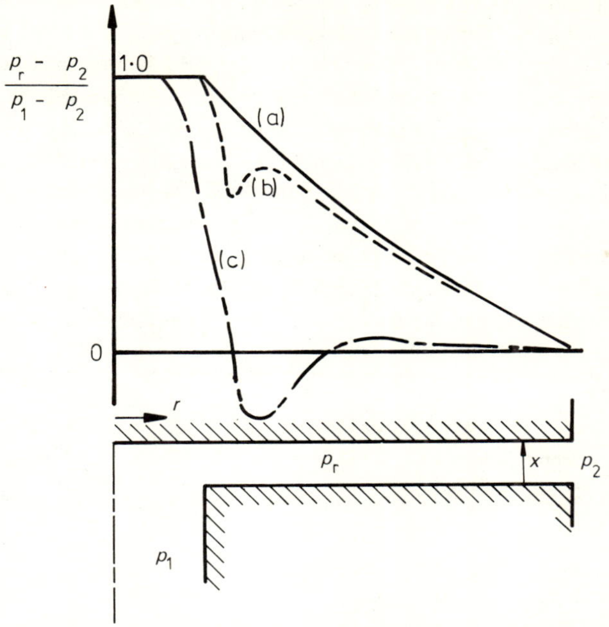

Fig. 8.17 – Effect of lift on pressure distributions in flapper-nozzle with flat; (a) small lift, laminar flow, (b) development of a separation bubble at greater lifts, (c) turbulent flow at greater lifts.

forms a diverging nozzle, and this flow can be analysed using the equations of isentropic flow with simple area change. As the curtain area increases in the radial direction, the Mach number and hence the pressure changes. Pawlak has determined the pressure distribution on the flapper and hence the force on the flapper. He examined two possibilities (a) subsonic flow, (b) supersonic flow in the passage. His results are:

Table 8.2 – Supersonic flow.

w	0	$0.5r$	$1.0r$	$1.5r$
C_f	1.0	1.308	1.519	1.691

Table 8.3 – Subsonic flow.

w	0	$0.5r$	$1.0r$	$1.5r$
C_f	1.0	2.002	3.603	5.742

Pawlak [15] found that supersonic flow occurred in all tests on his apparatus. Agreement between theory and practice was very good except at very small spacings where the effects described by equation (8.24) predominate.

Returning to Fig. 8.17, for the flow of an incompressible fluid, curve (b) illustrates that as the gap further increases a separation bubble develops at the inner corner resulting in a lowering of pressure and hence of force. If the flow becomes turbulent, curve (c), and x/w is small the reduction in force may be large enough to produce a net negative or closing force. Several investigations have reported the occurrence of these negative forces [16, 17, 18].

Fig. 8.18 shows some of Shumskii's results for the flow of kerosene [16]. There are several points worth noting; (a) the laminar flow solution is quite

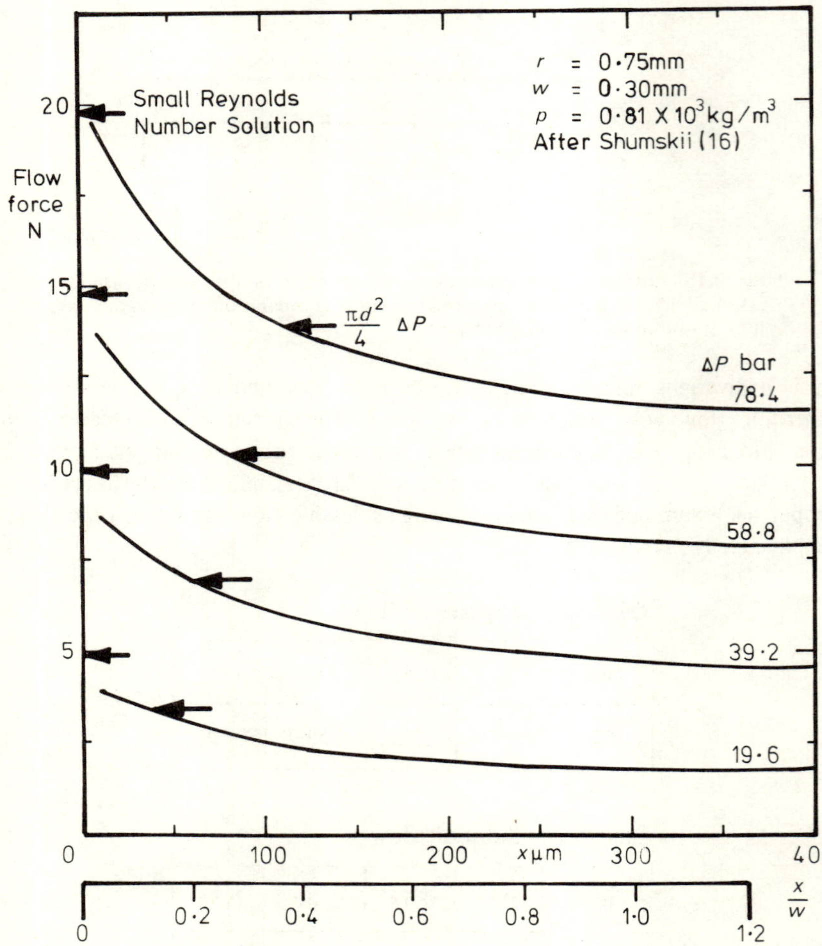

Fig. 8.18 – Measured flow forces in a hydraulic flapper-nozzle valve.

accurate, (b) at large flapper displacements the force is less than the nominal static force $\pi d^2 \Delta P/4$, as explained by the above discussion, (c) the valve operates at low values of x/w.

Burrows and Peckham [17], on the other hand, investigated a much wider range of x/w (Fig. 8.19), but on a pneumatic device. In this case the force in the closed position is given fairly accurately by the mean of the two static forces in equations (8.24). For each value of chamber pressure, the force initially falls with flapper displacement to a minimum value approximately equal to $\pi d^2 \Delta P/4$. For $x/w > 1$, it is likely that the flow is detached and the force increases in a manner similar to that predicted by Pawlak for the sharp edged orifice, Fig. 8.15. The magnitude of the increase, however, is not as great as predicted.

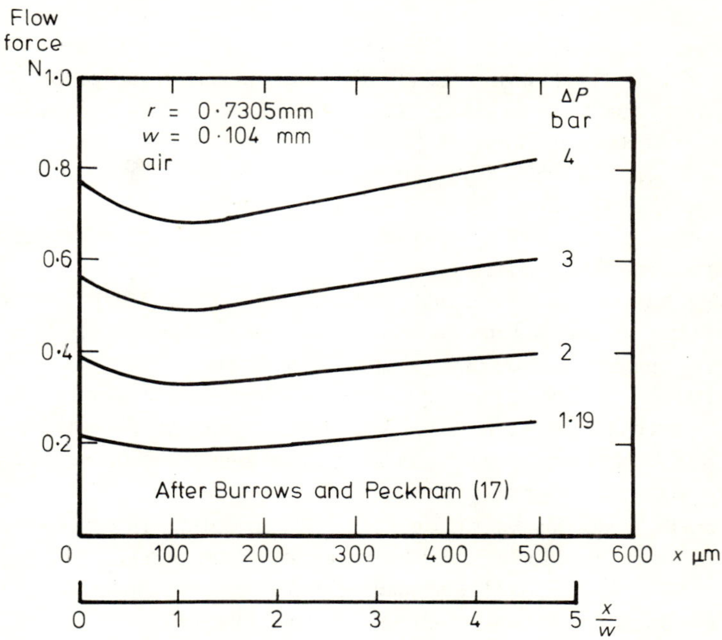

Fig. 8.19 – Measured flow forces in a pneumatic flapper-nozzle valve.

8.7 EXAMPLES

8.7.1

Oil of density 800 kg/m^3 flows at 30 m/s round a right angled elbow in a pipe of internal diameter 2 cm. The static pressure is 20 bar. Neglecting friction losses determined (a) the force required to produce the change in the momentum of the fluid, (b) the total reaction force on the pipe.

(*Answers*: (a) 319.8 N, (b) 1208.24 N.)

8.7.2

In a spool valve, like that of Fig. 8.2, the internal diameter of the sleeve is 30 mm and the stem is 10 mm thick, The radial clearance between land and sleeve is 0.5 mm. Determine (a) the jet angle θ when the spool is displaced 0.5 and 1.0 mm from the zero position (use Fig. 4.13), (b) the discharge coefficients (use Fig. 4.14), (c) the flow forces if the pressure drop is 7 bar.

(*Answers*: (a) $\theta = 90 - \alpha j = 44°, 54°$, (b) 0.74, 0.73, (c) 34.3 N, 55.3 N.)

8.8 REFERENCES

[1] Lee, S. Y. and Blackburn, J. F. (1952), Contributions to hydraulic control – axial forces on control valve pistons, *Trans. A.S.M.E.*, **74**, August.

[2] Louis, R. J. and Logan, E. (1976), Experimental investigation of flow-induced forces in pneumatic spool valves, *Trans. A.S.M.E. Jour. Dynamic Systems, Meas. and Control*, September.

[3] Clark, R. N. (1957), Compensation of steady state flow forces in spool-type hydraulic valves, *Trans. A.S.M.E.*, November.

[4] Carrington, J. E. (1966), Developments in electrohydraulic valves, *Hydraulic Pneumatic Power*, March.

[5] MacLellan, G. D. S., Mitchell, A. E., Turnbull, D. E. (1960), Flow characteristics of piston type control valves, *Proc. Symp. Recent Mech. Eng. Developments in Auto Control*, Inst. Mech. Engrs.

[6] Eynon, G. T. (1960), Developments in high-performance electromechanical servomechanisms at the Royal Aircraft Establishment, Farnborough, *Proc. Conf. Recent Mechanical Engineering Developments in Automatic Control*, **93**, Inst. Mech. Engrs. London.

[7] Lee, S. Y. and Blackburn, J. F. (1952), Contributions to hydraulic control – transient flow forces and valve instability, *Trans. A.S.M.E.*, **74**, August.

[8] Green, W. L. (1970), The poppet valve-flow force compensation, Paper 2, *Proc. 1970 Fluid Power International Conf.*, Olympia.

[9] Stone, J. A. (1960), Discharge coefficients and steady state flow forces for poppet-type valves, *Trans. A.S.M.E.*, **82**.

[10] McCloy, D. and McGuigan, R. H. (1965), *Some static and dynamic characteristics of poppet valves*, Inst. Mech. Engrs., London, April.

[11] Tsai, D. H. and Cassidy, E. C. (1961), Dynamic behaviour of simple pneumatic pressure reducer, *J. Basic Eng.*, **83**, June.

[12] Andersen, B. W. (1967), *The Analysis and Design of Pneumatic Systems*, John Wiley.

[13] Blackburn, J. F., Reethof, G., Shearer, J. L. (1960), *Fluid Power Control*, MIT Press.

[14] Guillon, M. (1969), *Hydraulic Servos System – Analysis and Design*, Butterworth.

[15] Pawlak, R. J. (1959), A Continued Study of Pneumatic Flapper Valves, M.Sc. thesis, MIT.

[16] Shumskii, N. P., The results of experimental and theoretical investigations of control devices of the nozzle-baffle type, *Pneumatic and Hydraulic Control Systems*, Pergamon Press.

[17] Burrows, C. R. and Peckham, R. G. (1977), Dynamic characteristics of a pneumatic flapper valve, *Jour. Mech, Eng. Science*, **19**, No. 3.

[18] Lichtarowicz, A. (1973), Flow and force characteristics of flapper valves, *3rd International Fluid Power Symp., B.H.R.A.*

CHAPTER 9

Some aspects of circuit design

9.1 INTRODUCTION

Later chapters will be concerned with the types of components and the ways they should be interconnected to generate a required sequence of actions by a fluid power system. This is one aspect of circuit design. In this chapter however, attention will be given more to the size of the required valve, fitting, pump or cylinder rather than the type. This enables losses and efficiencies to be calculated.

There are usually several possible solutions to most design problems, and criteria have to be established to enable a decision to be taken among these alternatives. Low cost, high reliability, high power/weight ratio, safety and easy maintenance are typical criteria for fluid power circuits. The efficient use of energy within the system is another important criterion and it will be examined later in the chapter.

In designing a system two basic types of parameter must be considered:

1. Quantities that can be determined with acceptable precision and whose values remain constant or nearly so, for example, oil stiffness.
2. Quantities that are difficult to assess, for example, seal friction and viscous damping.

In dealing with the latter the designer relies on the art of circuit design, supported by the wealth of empirical data that has accumulated over the years [3, 4, 5].

9.2 PIPE LINE DESIGN

In the selection of pipes for hydraulic circuits the following recommendations have been made:

(a) Suction lines to pumps should not carry fluid at velocities in excess of 1.5 m/s in order to reduce the possibility of cavitation at the pump.
(b) Delivery lines should not carry fluid at velocities in excess of 4.5 m/s

in order to prevent excessive shock loads on valve closure, high friction losses (that is, pressure drops greater than 5 per cent of supply pressure) and excessive heat generation.

(c) Return lines should be of larger diameter than delivery lines to avoid back pressure build-up. They should never be smaller.

General guidance in relation to pipe diameter selection, based on this information, is given in Table 9.1.

Table 9.1.

Flow rate (m^3/s)	Pipe size	
	Suction and return line dia. (N.B.) mm	Delivery line dia. (N.B.) mm
1.5×10^{-4}	15	8
3.0×10^{-4}	20	10
5.3×10^{-4}	25	15
1.3×10^{-3}	32	20
1.9×10^{-3}	40	25
3.0×10^{-3}	50	32
5.3×10^{-3}	65	40

The choice of 4.5 m/s as the limiting value of velocity in delivery lines is based on the following argument. When fluid flowing in a pipe is suddenly stopped by the rapid closing of a valve, a large pressure transient may result. The fluid adjacent to the valve is stopped initially and a pressure wave travels back to the fluid source at velocity c, given by:

$$c = \sqrt{\frac{\beta_e}{\rho}} \qquad (9.1)$$

where c = velocity of sound in the fluid
β_e = effective bulk modulus
ρ = mass density of the fluid.

When the pressure wave arrives at the end of the pipe, taking l/c seconds, l being the length of the pipe, the kinetic energy of the moving mass of fluid will be completely stored as potential energy in the fluid and in the elastic walls of the pipe. The pressure P of the compressed fluid is therefore at a maximum.

At this time a decompression wave forms and travels back to the valve. These waves continue to propagate back and forth with an associated interchange of kinetic and potential energies, until friction effects eventually dissipate the energy. At the instant of valve closure:

$$\text{Kinetic energy} = \frac{1}{2}Mv^2 = \frac{1}{2}\rho l A v^2 \tag{9.2}$$

where v = initial velocity
A = cross sectional area.

The potential energy stored in the compressed fluid will be:

$$\text{Potential energy} = \frac{1}{2}\frac{lA}{\beta_e}P^2 . \tag{9.3}$$

Hence equating (9.2) and (9.3) gives:

$$P = \rho c v . \tag{9.4}$$

If the fluid velocity is restricted to 4.5 m/s, the maximum pressure developed will be about 50 bar, which is well within the usual design safety values.

9.3 PRESSURE LOSSES IN PIPES

Determination of pressure losses in pipes is based on the d'Arcy formula for a circular pipe [1],

$$\delta p = f\frac{l}{d}\frac{\rho}{2}v^2 . \tag{9.5}$$

$$= K\rho\frac{v^2}{2}$$

where K = loss coefficient = fl/d
ρ = density of the fluid (kg/m^3)
f = friction factor
l = length of pipe (m)
d = internal diameter of pipe (m)
v = mean velocity (m/s).

The friction factor f has been shown experimentally to be a function of Reynolds numbers (Re) and of the pipe roughness. It is generally accepted for design purposes that for laminar flow ($Re < 2500$):

$$f = \frac{64}{Re} \tag{9.6}$$

and in the case of turbulent flow ($Re > 2500$):

$$f = \frac{0.332}{Re^{0.25}} \tag{9.7}$$

Table 9.2 – Standard weight steel pipe (Schedule 40).

Nominal mm	Outside mm	Inside mm	Allowable working pressure* bar
8	13.7	9.2	464
10	17.1	12.5	384
15	21.3	15.8	369
20	26.7	20.9	306
25	33.4	26.6	288
32	42.2	35.1	240
40	48.3	40.9	217
50	60.3	52.5	184

* Based on $\frac{1}{3} \times$ bursting pressure.

Data for smooth steel hydraulic pipe are listed in Table 9.2. In practice it is found more convenient to write (9.5) as:

$$\delta p = f \frac{l}{d} \frac{Q^2}{K_1} = \frac{K}{K_1} Q^2 \tag{9.8}$$

where $K_1 = (2A^2)/\rho$
Q = flow rate (m^3/s)
A = pipe cross-section area (m^2).

Values for K_1 are given in Table 9.3.

The value of Reynolds number can be determined from,

$$Re = \frac{vd}{\nu} = \frac{K}{K_1} Q^2 \tag{9.9}$$

where $K_2 = \frac{4}{\pi d}$

ν is the kinematic vicosity (m^2/s) and usually has a value about $4.0 \times 10^{-5} m^2/s$ at normal industrial working conditions. The values of K_2 for standard nominal bore pipe are also given in Table 9.3.

Table 9.3 – Pipe coefficients.

Nominal bore mm	Pipe area m^2	Coefficients	
		K_1	K_2
8	6.64×10^{-5}	1.027×10^{-11}	138
10	12.27×10^{-5}	3.506×10^{-11}	102
15	19.60×10^{-5}	8.949×10^{-11}	81
20	34.30×10^{-5}	2.740×10^{-11}	61
25	55.57×10^{-5}	7.195×10^{-11}	47
32	96.76×10^{-5}	2.181×10^{-9}	36
40	13.13×10^{-4}	4.014×10^{-9}	31
50	21.64×10^{-4}	1.090×10^{-8}	24

9.4 LOSSES THROUGH VALVES AND FITTINGS

The loss coefficients, defined by equation (9.5), can be determined for single orifices using the orifice equation:

$$Q = C_q a \sqrt{(2\delta p/\rho)} \tag{9.8}$$

derived in Chapter 4. This can be rearranged to give:

$$\delta p = \frac{1}{2}\rho V^2 \left(\frac{1}{C_q}\right)^2$$

where V = velocity through the orifice.

Hence the loss coefficient is:

$$K = 1/C_q^2 .$$

This can be applied to all the orifices discussed in Chapter 4. No similar general formulae can be established to cover all types of valves, however, since they consist of series and parallel circuits including orifice losses, changes in section, and pipe losses. In this case, the designer is very much dependent on empirical data from manufacturers and researchers. Examples of such data are given in Figs. 9.1 and 9.2 [2].

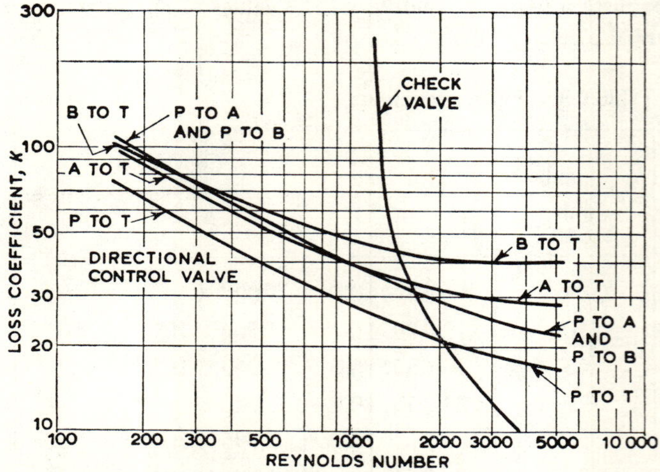

Fig. 9.1 – The loss coefficient of a directional control valve (after [2]). (Acknowledgements to the Institution of Mechanical Engineers.)

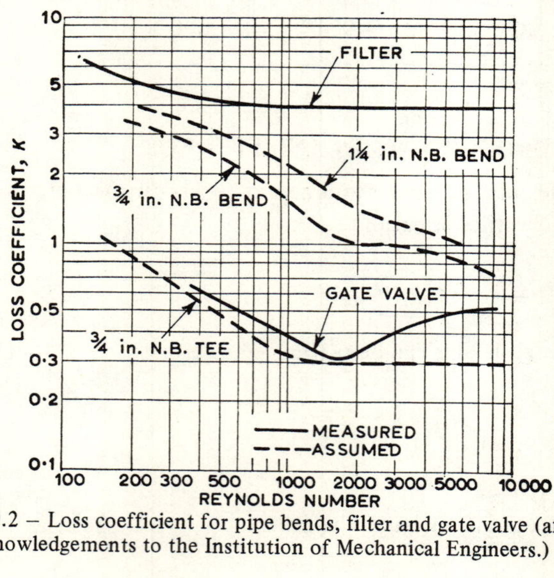

Fig. 9.2 – Loss coefficient for pipe bends, filter and gate valve (after [2]). (Acknowledgements to the Institution of Mechanical Engineers.)

Sudden contractions and expansions occur where a pipe is attached to a cylinder pump, or reservoir. The loss coefficients in these cases can be estimated from Fig. 9.3 which assumes turbulent flow [1, 7].

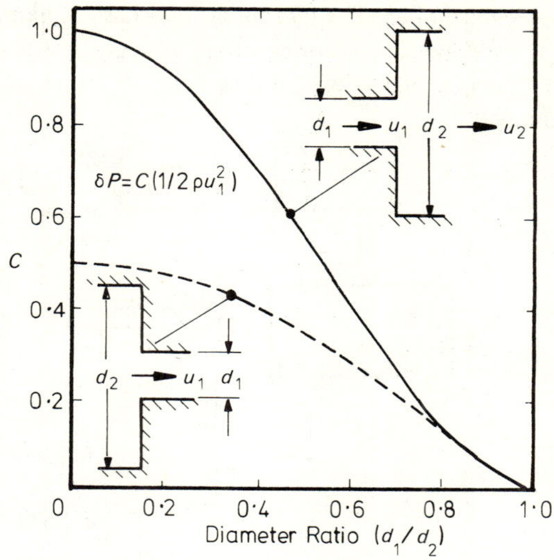

Fig. 9.3 – Loss coefficients for abrupt enlargements and contractions.

9.5 TYPICAL CIRCUIT DESIGN (after [2])

Assume that a circuit is required to move a load on the command of an operator. The functional diagram of Fig. 9.4 is the first step in the design of the required

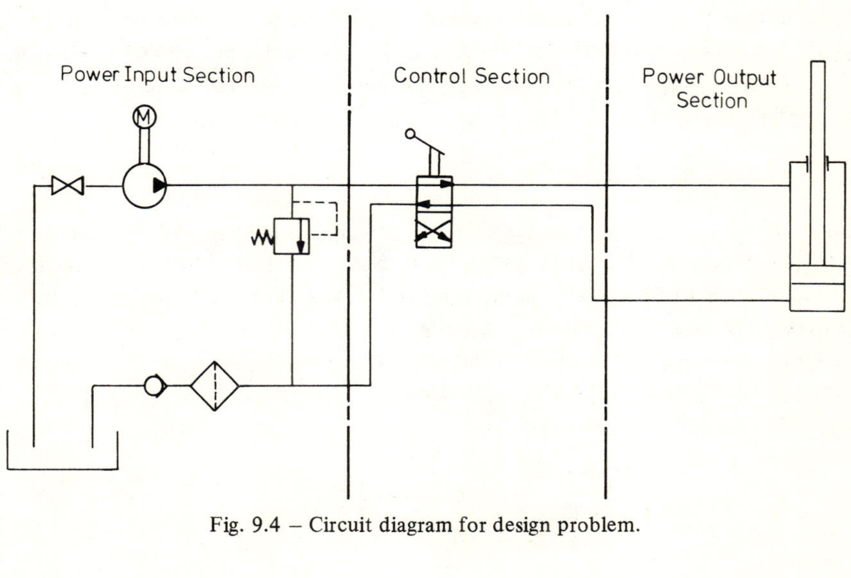

Fig. 9.4 – Circuit diagram for design problem.

system; in essence it determines the type of components required and how they should interconnect. The next step is to choose appropriate sizes of components and fittings, and for this a physical layout, Fig. 9.5, is needed together with a specification for the system. Assume that a gear pump is available with a rated delivery of 10^{-3} m^3/s at 100 bar. The hydraulic fluid has a kinematic viscosity of 4×10^{-5} m^2/s at normal working temperatures and a density of 858.2 kg/m^3. The cylinder has to move a 60 kN load on its outward stroke and the rod speed is to be about 0.15 m/s.

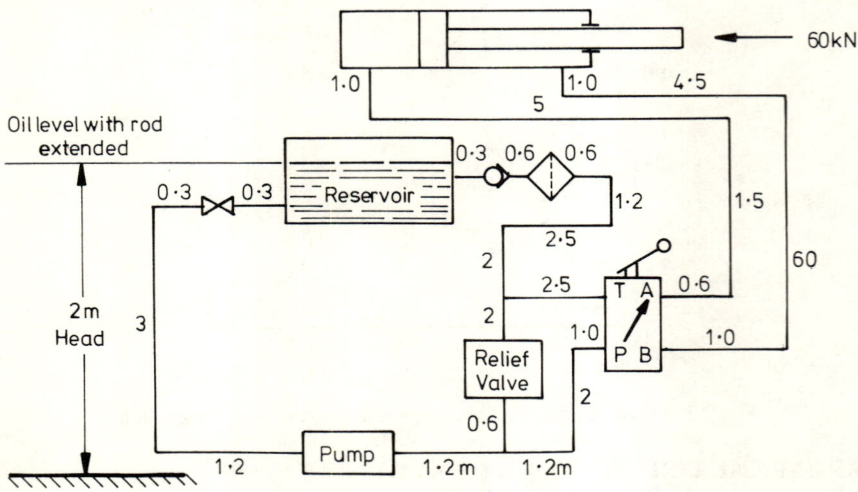

Fig. 9.5 – Physical layout for design problem.

9.5.1 Hydraulic cylinder selection

Assume that 15 per cent must be added to the force to allow for friction. From Table 9.4, which relates thrust to cylinder bore, a preliminary selection is made of a 101.6 mm bore and 50.8 mm rod. This gives a head end area of 8107 mm^2 and rod end area of 6080.0 mm^2. The steady state load equation is:

$$P_h A_h - P_r A_r = F \tag{9.9}$$

where P_h and P_r are head end and rod end gauge pressures and A_h and A_r are head end and rod end areas. In an ideal system with no losses equation (9.9) would reduce to $P_s A_h = F$ or in this case $F = 81$ kN, as shown in Table 9.4. However, in practice, losses will make P_h less than P_s, and P_r greater than tank pressure. Both of these effects will reduce the available force and thus it is essential to evaluate the losses at this stage.

The speed of extension will be:

$$v_{out} = \frac{1.0 \times 10^{-3}}{8.107 \times 10^{-3}} = 0.123 \text{ m/s}.$$

Table 9.4.

Cyl. bore or piston rod dia. (in)	Size (mm)	Area (mm^2)	Cylinder push force in Newtons at various pressures in bars										Displacement for 1 mm stroke (mm^3)
			4	6.3	10	16	25	40	63	100	160	200	
$\frac{5}{8}$	15.87	197.9	79	125	198	317	495	792	1247	1979	3107	3959	197.9
1	25.40	506.7	203	319	507	811	1267	2027	3192	5067	8107	10134	506.7
$1\frac{3}{8}$	34.93	958.0	383	604	958	1533	2395	3832	6035	9580	15328	19160	958.0
$1\frac{1}{2}$	38.10	1140.1	456	718	1140	1824	2850	4560	7183	11401	18242	22802	1140.1
$1\frac{3}{4}$	44.45	1551.8	621	978	1552	2483	3879	6207	9776	15518	24829	31036	1551.8
2	50.80	2026.9	811	1277	2027	3243	5067	8107	12769	20268	32429	40537	2026.9
$2\frac{1}{2}$	63.50	3166.9	1267	1995	3167	5067	7917	12668	19952	31669	50871	63339	3166.9
3	76.20	4560.4	1824	2873	4560	7297	11401	18242	28730	45604	72966	91208	4560.4
$3\frac{1}{4}$	82.55	5352.1	2141	3372	5352	8563	13380	21408	33718	53521	85634	107042	5352.1
$3\frac{1}{2}$	88.90	6207.2	2483	3911	6207	9931	15518	24829	39105	62072	99315	124144	6207.2
4	101.60	8107.3	3243	5108	8107	12972	20263	32429	51076	81073	129717	162147	8107.3
5	127.00	12667.7	5067	7981	12668	20268	31669	50671	79807	126677	202663	253354	12697.7
$5\frac{1}{2}$	139.70	15327.9	6131	9657	15328	24525	36320	61312	96566	153279	245247	306559	15327.9
6	152.40	18241.5	7297	11492	18242	29186	45604	72966	114922	182415	291864	364830	18241.5
7	177.80	24828.1	9931	15642	24829	39726	62072	99315	156421	248287	397260	496571	24828.7
8	203.20	32429.4	12972	20430	32429	51837	81073	129717	204305	324294	518870	648587	32429.4
$8\frac{1}{2}$	215.90	30009.7	14644	23064	36610	58576	91524	146439	230641	366097	585755	732194	36609.7
10	254.00	50670.9	20268	31923	50671	81073	126677	202683	319226	506709	810734	1013417	50670.9
12	304.80	72966.0	29186	45968	72966	116746	182415	291864	459686	729660	1167457	1459321	72966.0

For the return stroke:

$$v_{in} = \frac{8.107 \times 10^{-3}}{6.080 \times 10^{-3}} \times 0.123 = 0164\,\text{m/s}\,.$$

Both these values lie within a reasonable range of the specified speed requirement.

9.5.2 Selection of pump suction line

Table 9.1 indicates that the pipe size most suitable for the suction line, handling flows up to $1.0 \times 10^{-3}\,\text{m}^3/\text{s}$, is 32 mm nominal bore. Using equation (9.9) and Table 9.3:

$$Re = \frac{K_2 Q}{\nu} = \frac{36 \times 1.0 \times 10^{-3}}{4.0 \times 10^{-5}} = 900\,.$$

This value is within the laminar range, hence the friction factor can be estimated from equation (9.6):

$$f = \frac{64}{Re} = \frac{64}{900} = 0.0711\,.$$

From Fig. 9.8 the inlet pipe line to the pump has a length:

$$L = 0.3 + 0.3 + 3 + 1.2 = 4.8\,\text{m}\,.$$

The loss coefficient, K from equation (9.5) is given by:

$$K_{pipe} = f\frac{l}{d} = \frac{0.0711 \times 4.8}{35.1 \times 10^{-3}} = 9.72$$

where the pipe inside diameter is found in Table 9.2. Therefore, the pressure drop along the straight pipe sections of the suction line will be:

$$\delta P_{pipe} = \frac{KQ^2}{K_1} = \frac{9.72 \times (1 \times 10^{-3})^2}{2.181 \times 10^{-9}} = 4.456\,\text{kN/m}^2$$

where K_1 for 32 mm pipe is found in Table 9.3.

There are also two right-angled bends indicated in the layout. It is recommended practice that:

(a) For small pipes, $R/r = 2$, where R is the bend radius and r is the pipe radius.

(b) For larger pipes, $R/r = 4$.

These recommendations avoid excessive stresses in the region of the pipe bend. Further information on pipe bend losses can be obtained from the literature. From Fig. 9.2, for flow of Reynolds number 900, the loss coefficient is about:

$$K_{\text{bend}} = 2.3\,.$$

Hence the pressure loss due to two bends can be estimated at:

$$\delta P_{\text{bend}} = \frac{2 \times 2.3 \times (1 \times 10^{-3})^2}{2.181 \times 10^{-9}} = 2.109\,\text{kN/m}^2\,.$$

The valve in the suction line must create only a small pressure loss, so a gate valve is chosen (from Fig. 9.2) which has a loss coefficient:

$$K_{\text{gate}} = 0.4$$

assuming the same Reynolds number as the pipe. Note how small this is in comparison to the loss coefficients shown in Fig. 9.1 for directional control valves.

Hence the pressure loss, using Table 9.3, is:

$$\delta P_{\text{gate}} = \frac{KQ^2}{K_1} = \frac{0.4 \times (1 \times 10^{-3})^2}{2.181 \times 10^{-9}} = 0.183\,\text{kN/m}^2\,.$$

Finally the loss at the sudden contraction at the entrance to the suction line must be accounted for. Fig. 9.3 gives $K = 0.5$ for turbulent flow, and although the flow is laminar in this case this value will be used.

Hence $\delta P_{\text{inlet}} = 0.23\,\text{kN/m}^2$.

The total pressure loss to the pump is:

$$P = 4.456 + 2.109 + 0.183 + 0.23 = 6.978\,\text{kN/m}^2\,.$$

The layout drawing. Fig. 9.5, shows that the reservoir is sited 2 m above the pump so this will provide a pressure on the inlet side of $17.08\,\text{kN/m}^2$, since 1 m head of oil is equivalent to $8.54\,\text{kN/m}^2$.

Taking into account the loss in the suction line, the inlet pressure at the pump will be:

$$17.08 - 6.978 = 10.1\,\text{kN/m}^2$$

which is an acceptable positive pressure.

Choosing a 2 mm nominal bore pipe for the suction line would result in losses exceeding the available head, and as a result insufficient oil would flow into the pump, causing cavitation to occur.

Some pumps are designed for mounting on top of the tank, but it is generally recommended to have positive head at the pump inlet.

9.5.3 Circuit losses

The length of the delivery line to the cylinder is:

$$L_c = 1.2 + 1.2 + 2 + 1 + 0.6 + 1.5 + 5 + 1.0 = 13.5\,\text{m}\,.$$

From Table 9.1 select a 20 mm nominal bore pipe for the delivery line. Using Table 9.3 to determine the flow conditions:

$$Re = \frac{K_2 Q}{\nu} = \frac{61 \times 1 \times 10^{-3}}{4.0 \times 10^{-5}} = 1.525 \times 10^3$$

which is also within the laminar range:

$$f = \frac{64}{Re} = \frac{64}{1525} = 0.042$$

$$K_{\text{pipe}} = f\frac{L}{d} = \frac{0.042 \times 13.5}{20.9 \times 10^{-3}} = 27.13\,.$$

Now it is possible to simplify the calculation by summing loss coefficients rather than calculating pressure losses at each stage.

There are five bends of 90°, giving a total loss coefficient, from Fig. 9.2, of about:

$$K_{\text{bend}} = 5 \times 1.2 = 6.0\,.$$

The tee junction has a loss coefficient, shown in the same graph, of about:

$$K_{\text{tee}} = 0.3\,.$$

There is also a sudden expansion into the cylinder:

$$K_{\text{exp}} = 1.0 \text{ (again taking the tubulent flow figure).}$$

Hence $$K_{\text{total}} = 27.13 + 6.0 + 0.3 + 1.0 = 34.43$$

$$\delta P = \frac{K_{\text{total}}\, Q^2}{K_1} = \frac{34.43 \times (1 \times 10^{-3})^2}{2.74 \times 10^{-10}} = 121\,\text{kN/m}^2\,.$$

The characteristics for the directional valve are shown in Fig. 9.1 where in this mode the flow is from P to A. The Reynolds number through the delivery side was calculated at 1525. From Fig. 9.1 for this flow direction.

$$K_{\text{valve}} = 32$$

$$P = \frac{KQ^2}{K_1} = \frac{32 \times (1 \times 10^{-3})^2}{2.74 \times 10^{-10}} = 116.78 \text{ kN/m}^2 .$$

Therefore total losses in the delivery line to the cylinder are:

$$\delta P = 121 + 116.78 = 237.8 \text{ kN/m}^2 .$$

The pressure P_h in the head end is then (100 − 2.378) bar = 97.622 bar. The pump must also overcome the pressure losses in the return line to tank. Using 20 mm nominal bore tubing again, the line length is:

$$l = 1.0 + 4.5 + 6.0 + 1.0 + 2.5 + 2 + 2.5 + 1.2 + 0.6 + 0.6 + 0.3$$
$$= 22.2 \text{ m} .$$

Note that this line becomes a delivery line when the directional valve is moved to its other position. Because of the differential area of the cylinder piston, the flow rate will be less:

$$Q_{\text{return}} = (1 \times 10^{-3}) \times \frac{6.08 \times 10^{-3}}{8.107 \times 10^{-3}} = 7.49 \times 10^{-4} \text{ m}^3\text{/s}$$

$$Re = \frac{K_2 Q}{\nu} = \frac{61 \times 7.49 \times 10^{-4}}{4.0 \times 10^{-5}} = 1143.7$$

$$f = \frac{64}{Re} = \frac{64}{1143.7} = 0.056$$

$$K_{\text{pipe}} = f\frac{l}{d} = \frac{0.056 \times 22.2}{20.9 \times 10^{-3}} = 59.48 .$$

In addition there are five bends of 90°:

$$K_{\text{bend}} = 5 \times 1.2 = 6 .$$

One tee $K_{\text{tee}} = 0.3$.

Exit from cylinder

$$K = 0.5 .$$

Entry to tank

$$K = 1.0 .$$

Total $\quad K = 59.48 + 6 + 0.3 + 0.5 + 1.0 = 67.28 .$

$$\delta P = \frac{KQ^2}{K_1} = \frac{67.28 \times (7.49 \times 10^{-4})^2}{2.74 \times 10^{-10}} = 137.75\,\text{kN/m}^2 .$$

The directional valve has a flow path from B to T, with a Reynolds number of 1143.7. From Fig. 9.1 the loss coefficient for the valve is about:

$$K_{\text{valve}} = 45 .$$

In addition

$$K_{\text{filter}} = 4 \quad \text{and} \quad K_{\text{check valve}} = 100 .$$

Total $\quad K = 45 + 4 + 100 = 49$

$$\delta P = \frac{KQ^2}{K_1} = \frac{149 \times (7.49 \times 10^{-4})^2}{2.74 \times 10^{-10}} = 305.07\,\text{kN/m}^2 .$$

Total loss in return to tank $= (137.75 + 305.07) = 442.82\,\text{kN/m}^2$.

Hence $\quad P_r = 4.428$ bar.

It is now necessary to use equation (9.9) to check that, with these losses taken into account, there is still sufficient thrust. In this case the available thrust is 76.45 kN which exceeds the required 69 kN.

9.5.4 Circuit efficiency

The efficiency of the circuit is the ratio of the power output to the power input. The power loss in an individual component is given by the product of the volumetric flow rate and the pressure drop across the component.

Thus for the working stroke the total hydraulic losses in this circuit are:

delivery line	$2.378 \times 10^5 \times 10^{-3}$	$= 237.8$ W
return line	$4.428 \times 10^5 \times 0.749 \times 10^3$	$= 331.7$ W
	Total	$= 569.5$ W.

The output power W_o developed by the cylinder is the product of force and velocity; in this case, during the working stroke:

$$W_o = 69 \times 10^3 \times 0.1234 = 8515\,\text{W}.$$

Thus the power W_p supplied by the pump must be:

$$W_p = W_o + \text{losses}$$
$$= 8515 + 569.5 = 9084.5\,\text{W}.$$

If the pump is assumed to have an overall efficiency of 86 per cent then the power W_i supplied to the pump is:

$$W_i = 9084.5/0.86 = 10\,563.4\,\text{W};$$

The efficiency of the circuit during the working stroke is then:

$$W_o/W_i = 8515/10563.4 = 80.6\%.$$

But, since 9 kN of the cylinder thrust is required to overcome friction, the useful mechanical power is reduced by $9 \times 10^3 \times 0.1234 = 1110.6$ W. The efficiency of the system, viewed overall, is then:

$$\eta = \frac{8515 - 1110.6}{10563.4} = 70.0\%.$$

The total power degraded to heat is the sum of the hydraulic losses and the frictional losses; in this case $569.5 + 1110.6 = 1680.1$ W.

9.6 FURTHER CONSIDERATIONS OF EFFICIENCY

It is desirable, if only from economic considerations, to use energy as effectively as possible. The preceding section showed that a considerable amount of energy was degraded to heat because of hydraulic losses in pipes and fittings. These losses will be even greater in those systems that rely on throttling as the form of control. But in spite of this disadvantage, throttling is commonly used since it is a convenient way of meeting many different load requirements. Now, accepting that throttling is inefficient, it would be valuable to the designer to know how much energy is likely to be wasted in this way and if there is some way of minimising the loss. Earlier in the book, section 7.2.1, it is shown that maximum power is delivered through a single throttle (two-port valve) when the downstream pressure is two thirds of the upstream. Turning now to a four-port, servovalve driving a through-rod cylinder, Fig. 9.6, and neglecting all losses except those occurring in the valve.

Flow rate to cylinder:

$$Q_1 = kx\sqrt{(P_s - P_1)} \tag{9.10}$$

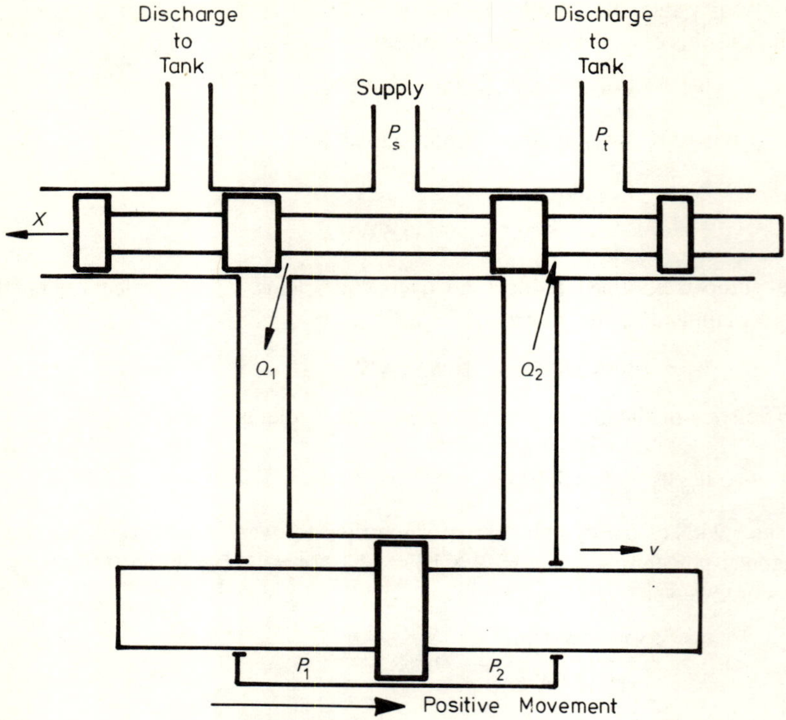

Fig. 9.6 – The hydraulic actuator.

Flow rate from cylinder:

$$Q_2 = kx\sqrt{(P_2 - P_t)} \,. \tag{9.11}$$

Neglecting compressibility effects (or assuming constant velocity v):

$$Q_1 = Q_2 = Av \tag{9.12}$$

where A = effective piston area.

Thus equations (9.10), (9.11) and (9.12) give:

$$P_2 = P_s + P_t - P_1 \,. \tag{9.13}$$

Now considering the load F,

$$(P_1 - P_2)\,A = P_l A = F \tag{9.14}$$

where P_l = load pressure.

From equations (9.13) and (9.14):

$$P_1 = 0.5\,(P_s + P_t + P_l)\,. \tag{9.15}$$

Hence substituting in equation (9.10):

$$Q_1 = \frac{kx}{\sqrt{2}}\sqrt{[(P_s - P_t) - P_1]}\,. \tag{9.16}$$

This is plotted in Fig. 9.7.

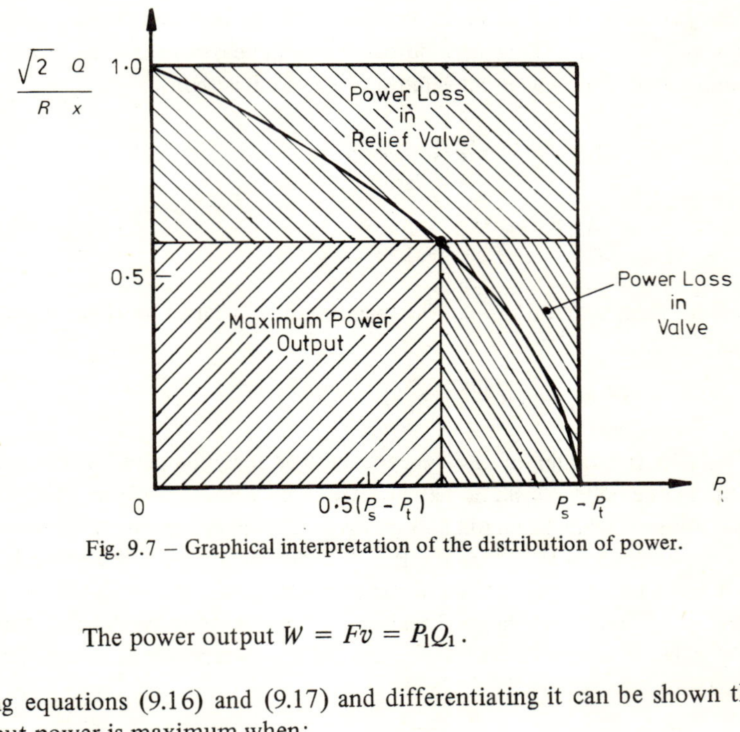

Fig. 9.7 – Graphical interpretation of the distribution of power.

$$\text{The power output } W = Fv = P_l Q_1\,. \tag{9.17}$$

Using equations (9.16) and (9.17) and differentiating it can be shown that the output power is maximum when:

$$P_l = \frac{2}{3}(P_s - P_t)\,. \tag{9.18}$$

At this condition the remaining one third of the total available pressure drop occurs in the metering orifices of the valve. Because of symmetry there is a pressure drop of $(P_s - P_t)/6$ over both inlet and outlet orifices. The maximum output power $\hat{W}$ is:

$$\hat{W} = \frac{2}{3\sqrt{3}}\frac{k\hat{x}}{\sqrt{2}}(P_s - P_t)^{1.5} = 0.272\,k\hat{x}\,(P_s - P_t)^{1.5} \tag{9.19}$$

where $\hat{x}$ = maximum valve movement.

Since power is the product of flow rate and load pressure it can also be interpreted as an area under the valve characteristics of Fig. 9.7. Maximum power occurs when the area of the rectangle QP_l is a maximum. In order to calculate the efficiency of the system it is necessary to compare this power output with the power supplied to the system. This is determined by pump capacity and relief valve setting. If the pump is matched to the valve so that the pump can just meet the maximum demand then the required pump capacity is $Q = kx\sqrt{[(P_s - P_t)/2]}$. The supply pressure P_s will be determined by the stalling load requirement. The power delivered to the system is then:

$$W_d = (P_s - P_t)\,k\hat{x}\sqrt{[(P_s - P_t)/2]}$$

$$= \frac{k\hat{x}}{\sqrt{2}}(P_s - P_t)^{1.5}\,. \tag{9.20}$$

Thus under these circumstances the maximum efficiency is (from equations (9.19) and (9.20)):

$$\eta = \hat{W}/W_d = 2/(3\sqrt{3}) = 38.49\%.$$

Where has the remainder of the power gone to? It is dissipated in the control valve and in the relief valve. It has already been shown that one third of the available pressure drop occurs in the control valve at optimum conditions. The relevant power loss is indicated on Fig. 9.7. As far as the relief valve is concerned it must be appreciated that at the optimum point on Fig. 9.7 the pump is supplying more fluid than the load requires; the excess must be diverted through the relief valve at supply pressure. The resultant power loss is also illustrated on Fig. 9.7. The power balance is:

Power input	$= 0.7071\,k\hat{x}\,(P_s - P_t)^{1.5}$
Power output	$= 0.2720\,k\hat{x}\,(P_s - P_t)^{1.5}$
Loss in relief valve	$= 0.2989\,k\hat{x}\,(P_s - P_t)^{1.5}$
Loss in control valve	$= 0.1352\,k\hat{x}\,(P_s - P_t)^{1.5}$.

The losses in the relief valve and the required input power could be reduced by employing a pressure compensated pump whose output reduces with increasing pressure.

9.7 TIME VARYING LOADS

The above analysis assumed that the load was constant and optimum. This is rarely so in practice, and it is instructive to examine the effects of load variations. To illustrate this point a comparison will be made of two methods of speed control: meter-in and bleed-off [6]. The two circuits are shown in Fig. 9.8. It is assumed that the losses in the metering valves and in the relief valves are the predominant ones so that others may be neglected. The cylinder dimensions are:

$$\begin{aligned} &\text{stroke} &&= 0.8\,\text{m} \\ &\text{blank end area } A_1 &&= 20 \times 10^{-4}\,\text{m}^2 \\ &\text{rod end area } A_2 &&= 12 \times 10^{-4}\,\text{m}^2\,. \end{aligned}$$

The load requirement is:

Period 1: rod extends for 40 cm at 0.2 m/s against 1 kN load.

Period 2: rod extends for 20 cm at 0.1 m/s against 10 kN load.

Period 3: rod extends for 20 cm at 0.05 m/s against 20 kN load.

Period 4: rod dwells extended for 4 seconds against 20 kN load.

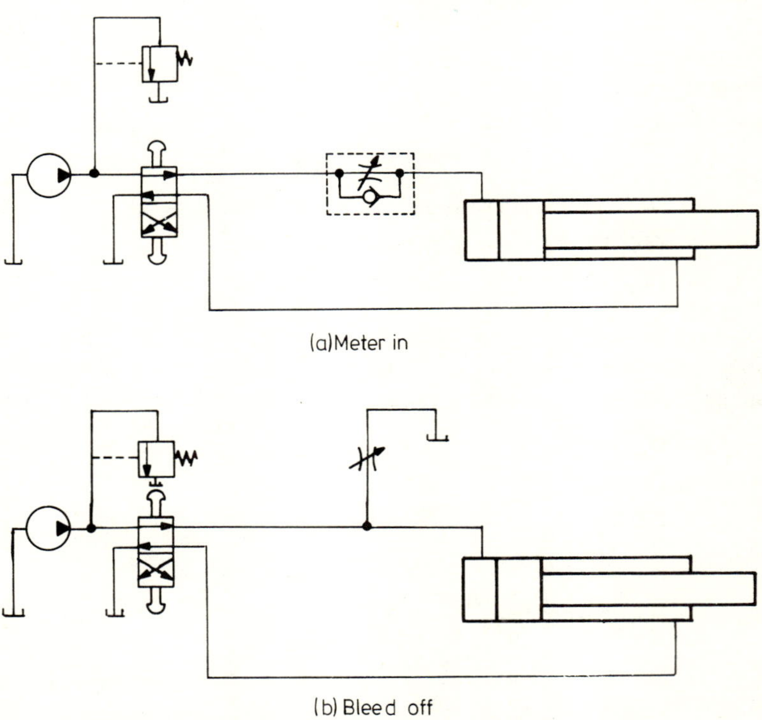

Fig. 9.8 – (a) Meter-in speed control, (b) bleed-off speed control.

The times required are:

Period 1: 2 s
Period 2: 2 s
Period 3: 4 s
Period 4: 4 s
Total 12 s .

The pressures required in the blank end are:

Period 1: $10^3/A_1 = 5$ bar gauge
Period 2: $10^4/A_1 = 50$ bar
Period 3: $2 \times 10^4/A_1 = 100$ bar
Period 4: $2 \times 10^4/A_1 = 100$ bar .

$$\begin{aligned}\text{Required pump capacity} &= A_1 \times \text{maximum velocity}\\ &= 20 \times 10^{-4} \times 0.2\\ &= 4 \times 10^{-4}\,\text{m}^3/\text{s} = 0.4\,\text{l/s}.\end{aligned}$$

Required relief valve setting (allowing 10 per cent) = 110 bar.

9.7.1 Meter-in

Hence hydraulic energy input is:

$$\begin{aligned}
&\text{Period 1: } (4 \times 10^{-4} \times 5 \times 10^5)\,2 &&= 0.40\,\text{kJ}\\
&\text{Period 2: } (4 \times 10^{-4} \times 1.1 \times 10^5)\,2 &&= 8.80\,\text{kJ}\\
&\text{Period 3: } (4 \times 10^{-4} \times 1.1 \times 10^5)\,4 &&= 17.60\,\text{kJ}\\
&\text{Period 4: } (4 \times 10^{-4} \times 1.1 \times 10^5)\,4 &&= 17.60\,\text{kJ}\\
&\text{Total} &&= 44.40\,\text{kJ}.
\end{aligned}$$

The energy ouput is force $\times$ velocity $\times$ time:

$$\begin{aligned}
&\text{Period 1: } 0.2 \times 10^3 \times 2 &&= 0.4\,\text{kJ}\\
&\text{Period 2: } 0.1 \times 10^4 \times 2 &&= 2.0\,\text{kJ}\\
&\text{Period 3: } 0.05 \times 2 \times 10^4 \times 4 &&= 4.0\,\text{kJ}\\
&\text{Period 4: } 0 \times 2 \times 10^4 \times 4 &&= 0.0\,\text{kJ}\\
&\text{Total} &&= 6.4\,\text{kJ}.
\end{aligned}$$

the efficiency is $6.4/44.4 = 14.4\%$
the average output power $= 6.4 \times 10^3/12 = 533$ W
the input power $= 44.40 \times 10^3/12 = 3.7$ kW .

The energy losses can be catalogued as follows:

In the relief valve: flow rate × pressure drop × time

Period 1:	no loss	=	0.00 kJ
Period 2:	$(4-2)10^{-4} \times 110 \times 10^5 \times 2$	=	4.40 kJ
Period 3:	$(4-1)10^{-4} \times 110 \times 10^5 \times 4$	=	13.20 kJ
Period 4:	$(4-0)10^{-4} \times 110 \times 10^5 \times 4$	=	17.60 kJ
	Total	=	35.20 kJ .

In the metering orifice:

Period 1:	valve by-passed	=	0.00 kJ
Period 2:	$2 \times 10^{-4} \times 60 \times 10^5 \times 2$	=	2.40 kJ
Period 3:	$1 \times 10^{-4} \times 10 \times 10^5 \times 4$	=	0.40 kJ
Period 4:	no loss	=	0.00 kJ
	Total	=	2.80 kJ .

9.7.2 Bleed-off

In the meter-in case the majority of energy loss occurred in the relief valve. This can be reduced by using bleed-off. In this case the input requirement is:

Period 1:	$(4 \times 10^{-4} \times 5 \times 10^5)2$	=	0.40 kJ
Period 2:	$(4 \times 10^{-4} \times 50 \times 10^5)2$	=	4.00 kJ
Period 3:	$(4 \times 10^{-4} \times 100 \times 10^5)4$	=	16.00 kJ
Period 4:	$(4 \times 10^{-4} \times 100 \times 10^5)4$	=	16.00 kJ
	Total	=	36.40 kJ

Here relief valve losses are reduced to zero since pump delivery not required by the load is diverted by the metering valve. The losses in the metering valve are:

Period 1:	no flow	=	0.00 kJ
Period 2:	$(4-2)10^{-4} \times 50 \times 10^5 \times 2$	=	2.00 kJ
Period 3:	$(4-1)10^{-4} \times 100 \times 10^5 \times 4$	=	12.00 kJ
Period 4:	$(4-0)10^{-4} \times 100 \times 10^5 \times 4$	=	16.00 kJ
	Total	=	30.00 kJ

The output power is, as before, 6.4 kJ. Hence the efficiency in this case is:

$$\text{efficiency} = 6.4/36.4 = 17.6\%.$$

9.8 USE OF ACCUMULATORS TO REDUCE ENERGY REQUIREMENTS

The above examples show that the requirement to have a pump large enough to meet maximum output velocities can result in energy wastage in valves at low output velocities. An accumulator can be used to assist the pump to meet maximum velocity requirements. Thus a smaller pump can be used thereby reducing input power requirements and energy losses.

9.8.1 Flow averaging

As an example of the application of an accumulator, consider a system in which the hydraulic cylinder requires the flow rates shown in Fig. 9.9(a). The cycle of operation consists of a slow extension, a delay, a fast retraction and a further delay. The pressure supplied to the cylinder must not fall below 50 bar during the cycle.

Without the use of an accumulator it would be necessary to use a pump capable of meeting the maximum demand, that is, 3 l/s. With an accumulator in the circuit the pump need only supply the average flow rate over a cycle. In this case the average flow rate Q_a is 0.8 l/s.

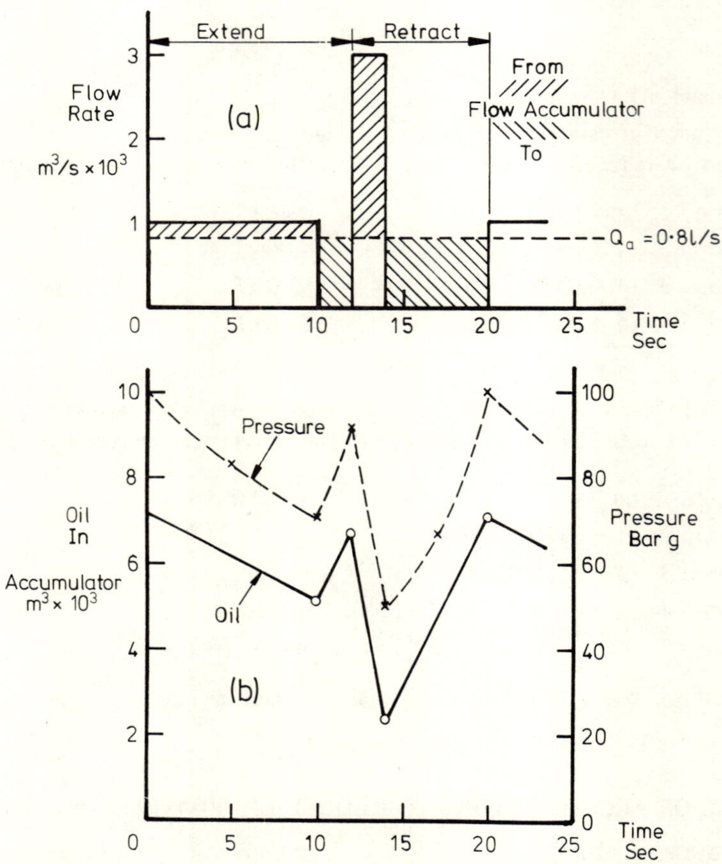

Fig. 9.9 – (a) Flow rate demand to meet loading requirements, (b) variation with time of oil volume and pressure in accumulator.

Knowing this, it is now possible to calculate the variation with time of the oil volume in the accumulator. Assuming that this is v_o at the start of the cycle, Fig. 9.9 (b) is then used to give the following:

Time (s)	Oil volume (m^3)	
0	v_o	$= v_o$
10	$v_o - (0.2 \times 10^{-3} \times 10)$	$= v_o - 2 \times 10^{-3}$
12	$(v_o - 2 \times 10^{-3}) + (0.8 \times 10^{-3} \times 2)$	$= v_o - 0.4 \times 10^{-3}$
14	$(v_o - 0.4 \times 10^{-3}) - (2.2 \times 10^{-3} \times 2)$	$= v_o - 4.8 \times 10^{-3}$
20	$(v_o - 4.8 \times 10^{-3}) + (0.8 \times 10^{-3} \times 6)$	$= v_o$

Thus the oil volume is maximum at the beginning and end of the cycle and is a minimum after 14 seconds, that is, after the fast retraction. The change in volume is $4.8 \times 10^{-2}\,m^3$.

The next task is to determine the size of the accumulator needed. Fig. 9.10 illustrates the three basic states of the accumulator (1) precharge, (2) maximum oil volume and maximum pressure during the cycle, (3) minimum oil volume and minimum oil pressure during the cycle. The following information is available:

$$V_3 - V_2 = 4.8 \times 10^{-3}\,m^3 \tag{9.21}$$

$$P_3 = 50 \text{ bar gauge.}$$

It is common practice to set the precharge pressure at 10 bar less than the minimum pressure during the cycle. Thus:

$$P_1 = 40 \text{ bar gauge.}$$

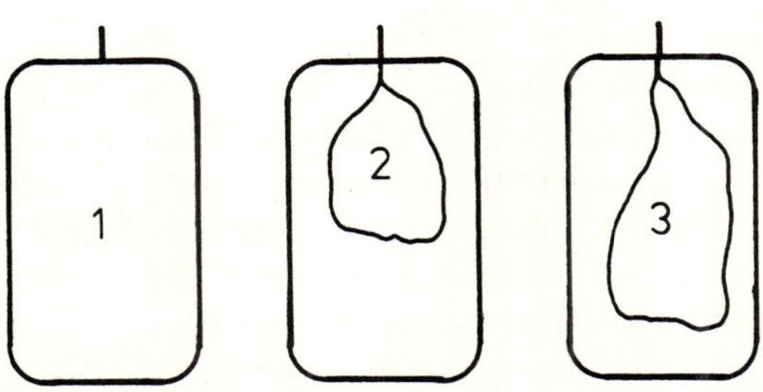

Fig. 9.10 – Different accumulator states (a) the precharge condition, (b) maximum oil volume and pressure during the working cycle, (c) minimum oil volume and pressure during the working cycle.

It is assumed that the maximum working pressure for the pump is 100 bar gauge. Thus:

$$P_2 = 100 \text{ bar gauge.}$$

The objective is to determine V_1 which is the accumulator size. If isothermal conditions are assumed then:

$$P_1V_1 = P_2V_2 = P_3V_3 \tag{9.22}$$

where the pressures are absolute.

It is easy to show that:

$$V_3 = 9.7 \times 10^{-3}\,\text{m}^3$$

$$V_2 = 4.9 \times 10^{-3}\,\text{m}^3$$

and $$V_1 = \text{accumulator size} = 12.07 \times 10^{-3}\,\text{m}^3 .$$

The volume of oil v_o in the accumulator at the start of the cycle can now be calculated. It is $V_1 - V_2 = 7.17 \times 10^{-3}\,\text{m}^3$. This allows Fig. 9.9(b) to be completed, showing the quasi-static variations of oil volume throughout the cycle. Pressures can be calculated using equation (9.22). The power W_p required from the pump is:

$$W_p = PQ = 100 \times 10^5 \times 0.8 \times 10^{-3} = 8\,\text{kW}.$$

If there were no accumulator the power required would be:

$$W_p = PQ = 50 \times 10^5 \times 3 \times 10^{-3} = 15\,\text{kW}$$

which is considerably greater.

9.9 EXAMPLES

9.9.1

A hydraulic circuit with accumulator employs a 150 mm diameter cylinder with a 100 mm diameter rod to crush a car body into bale size. The cylinder has a stroke of 1.83 m and the minimum crushing pressure is 172 bar abs. With a preload pressure of 35 bar abs. the rod extends in 20 s and retracts in 40 s. The time between crushing operations is 6 mins, and the maximum pump pressure is 207 bar abs. Assuming that the temperature in the accumulator remains constant,

(a) find the capacity of the accumulator

(b) compare the power required with that required if an accumulator were not used.

(*Answers*: (a) 0.94 m^3; (b) 2.893 kW c.f. 27.81 kW.)

9.9.2

The suction line to a hydraulic pump is 1 m long. Neglecting all losses except those due to pipe friction, determine the minimum diameter of the pipe if cavitation is not to occur at the pump inlet. The pump draws 2 l/s of an oil with a density 870 kg/m3

ensity

a density 870 kg/m^3 and kinematic viscosity 75×10^{-6} m^2/s.

(*Answer*: 15.05 mm.)

9.10 REFERENCES

[1] Streeter, V. L. (1966), *Fluid Mechanics*, McGraw-Hill.

[2] Hamilton, J. D. (1961), Steady state flow characteristics of fluid power components, *Conf. Oil Hydraulic Power*, paper 19, Inst. Mech. Engrs., London.

[3] Westway, C. R. and Loomis, A. W. (ed.) (1977), *Cameron Hydraulic Data*, Ingersoll-Rand, Woodcliffe Lake, N.J.

[4] Yeaple, F. D. (1966), *Hydraulic and Pneumatic Power Control*, McGraw-Hill.

[5] *Design Engineers Handbook for Fluid Power* (1972), Parker-Hannifin.

[6] Bradbury, F. (1972), *Hydraulic Systems and Maintenance*, Iliffe.

[7] Walshaw, A. C. and Jobson, D. A. (1972), *Mechanics of Fluids*, Longman.

CHAPTER 10

Dynamic characteristics of circuits

10.1 INTRODUCTION

One of the most neglected aspects of circuit design is performance under dynamic conditions. Chapter 9 gave an insight into the design of pipeworks under steady flow conditions, so that an efficient transfer of power could be achieved.

This required an understanding of the resistive characteristics of components and piping. However, the system also has energy storage characteristics in the form of fluid and mechanical mass, together with structural and fluid elasticity. While the resistive, that is energy dissipative elements, behave fairly well independent of excitation frequency, the other two storage elements are very sensitive to the frequency content of the transmitted power.

These elements become important in such areas as:

(a) Excitation of model resonance.
(b) Standing wave phenomena resulting in pipework damage.
(c) Radiation of noise.

There are strong analogies that can be drawn between electrical and fluid power systems under these conditions, allowing the fluid power engineer to take advantage of the well established theories and techniques of the electrical engineer.

10.2 FLOW NETWORKS

The use of flow networks arises from the consideration of physical events that result from a given interconnection of fluid power components [8]. The behaviour of the components can be approximately characterised by a set of flows and pressures associated with input and output ports, along with the constraints that relate these flows and pressures [2]. Further constraints are introduced by their interconnection. This can be approximately expressed by borrowing the two Kirchhoff laws from electrical engineering.

Any interconnection of elements and sources is called a **network**. The elements are the valves and cylinders; the sources are the pumps and compressors and the network interconnections are the pipework. Flow and pressure distributions within such a network must satisfy three basic constraints:

1. The flow-pressure relationship for the elements. This is the equation relating the input to the output across a device.
2. The equivalent of Kirchhoff's current law. If q_i represents the flow through the ith pipe of the network, then at any junction of pipes:

$$\Sigma q_i = 0 . \tag{10.1}$$

3. The equivalent of Kirchhoff's voltage law. If δP_i is the pressure drop across the ith pipe of the network then around a closed circuit:

$$\Sigma \delta P_i = 0 . \tag{10.2}$$

Kirchhoff's flow law (10.1) states that the algebraic sum of the flows entering a connection point (or alternatively the sum of the flows leaving a connection point) is zero. A **connection point** or **junction** is a point in the flow network at which one or more elements or sources terminate. For example, for the situation shown in Fig. 10.1(a):

$$q_1 - q_2 - q_3 - q_4 = 0 . \tag{10.3}$$

Kirchhoff's pressure law states that the algebraic sum of the pressure drops around a closed path is zero. This is illustrated in Fig. 10.1(b) where the pressure drops in a clockwise direction around the closed path abcda are given by:

$$\delta p_1 - \delta p_2 + \delta p_3 + \delta p_4 = 0 \tag{10.4}$$

where, for example, δp_1 represents the pressure drop between a and b.

Calculations of the head loss and flow distribution in networks generally fall into two groups:

(a) Those simple layouts, where fittings other than the straight pipe play a minor role.

(b) Systems of varied compexity in which elements other than straight pipes becomes of much greater importance.

It is possible to estimate the head loss along uniform straight pipes with an error not exceeding 10 per cent. However in many circuits these losses form only part of the total circuit resistance. Information on the head loss through fittings, valves, and other elements is not readily available. Errors in the published data of up to 30 per cent can be expected, so that a much higher degree of uncertainty must be experienced when dealing with circuit network problems of this type.

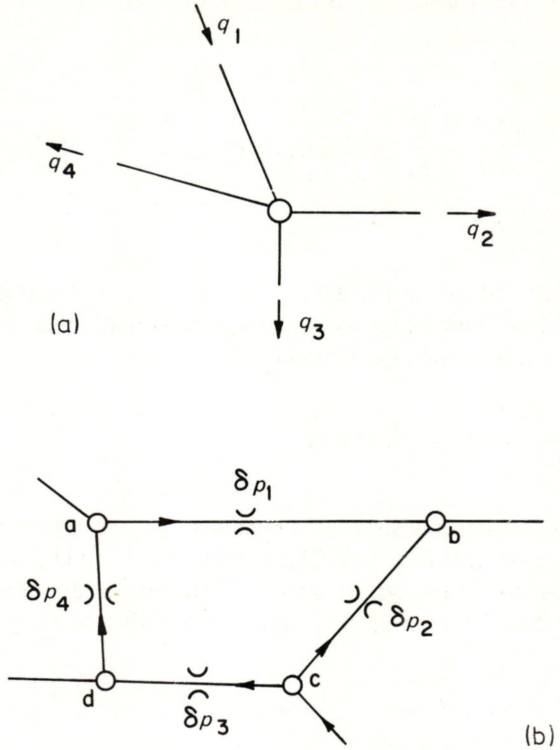

Fig. 10.1 – (a) Flow law, (b) pressure law.

10.3 FLOW NETWORKS FOR STEADY CONDITIONS

In general, under steady flow conditions, the effect of fluid resistance is characterised by a pressure drop in the direction of flow. Hence, the analogy of Ohm's law in electrical engineering will be:

$$\delta P = R_{\mathrm{h}} Q^{\mathrm{n}} . \tag{10.5}$$

From basic fluid mechanics [1] it can be shown for laminar flow conditions in a circular pipe that:

$$n = 1.0$$

and $$R_{\mathrm{h}} = \frac{128 \mu l}{\pi d^4} .$$

For the turbulent flow case, the analogy is not quite as satisfactory. Here:

$$n = 2.0$$

$$C = \frac{2f\rho l}{A^2 d} = \frac{K}{K_1} \text{ (see equation (9.8))}$$

and $$\delta p = CQ^2. \tag{10.6}$$

This relationship is shown in Fig. 10.2, where it can be seen that the resistance is only linear within limited regions. A good approximation in this case can be obtained by differentiation of equation (10.6):

$$\frac{\mathrm{d}(\delta P)}{\mathrm{d}Q} = R_{\mathrm{h}} = 2\surd(C\delta P) \tag{10.7}$$

In most physical systems, resistance is related to power loss from the system in the form of irretrievable heat loss. This implies that in fluid systems the power delivered to the inlet of the system is greater than that produced at the outlet.

Hence, if the inlet pressure to a pipe is p_1 and the outlet p_2, the power loss will be:

$$W = (p_1 - p_2)Q$$

or $$\mathrm{W(Watts)} = \delta P(\mathrm{N/m^2}) \times Q(\mathrm{m^3/s})\,. \tag{10.8}$$

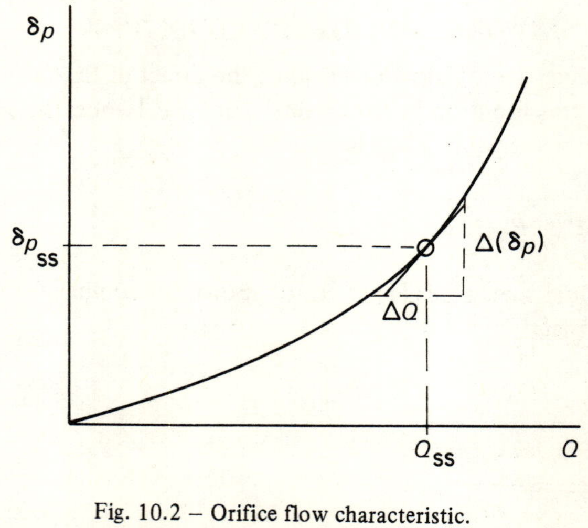

Fig. 10.2 – Orifice flow characteristic.

10.4 LINEAR RESISTANCE NETWORKS

In Fig. 10.3(b), (c) and (d) three different pipe networks are shown. Each is composed of three separate pipes, A, B, C of known $\delta p - Q$ characteristics. For simplicity, it is assumed that in all the networks the three pipes have the characteristics as shown in Fig. 10.3(a). Using the information developed in the previous section it is desired to find the pressure drop in each case, if a flow is specified.

For the case shown in Fig. 10.3(b):

$$P_1 - P_2 = R_A Q$$

$$P_2 - P_3 = R_B Q$$

$$P_3 - P_4 = R_C Q$$

since the flow is common to each tube. Hence,

$$P_1 - P_4 = Q(R_A + R_B + R_C)\,.$$

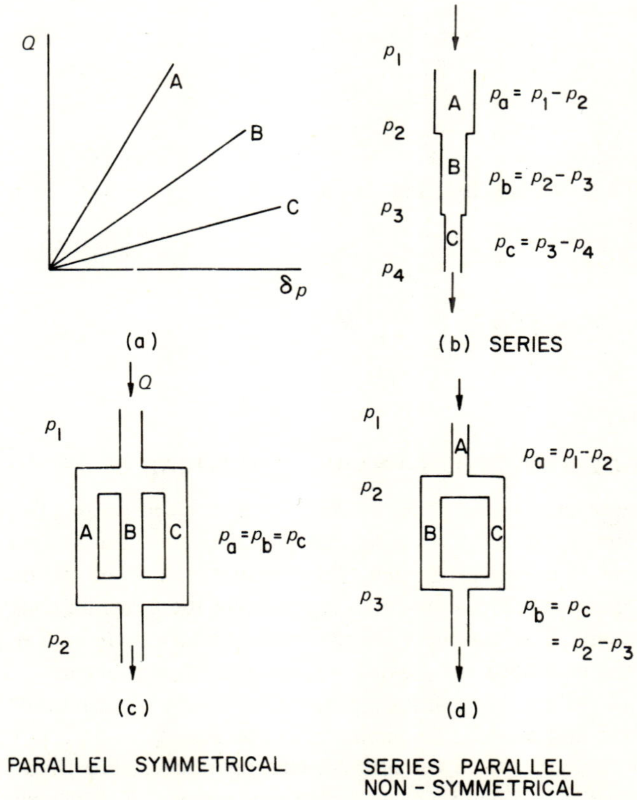

Fig. 10.3 – Network types.

This would represent a series network.

In the case of Fig. 10.3(c), the resistances are in a parallel configuration, and the flow through each leg is different.

$$(p_1 - p_2) = R_A Q_A$$
$$(p_1 - p_2) = R_B Q_B$$
$$(p_1 - p_2) = R_C Q_C .$$

Now at the junction,

$$Q = Q_A + Q_B + Q_C$$
$$= \frac{(p_1 - p_2)}{R_A} + \frac{(p_1 - p_2)}{R_B} + \frac{(p_1 - p_2)}{R_C} .$$

Hence:

$$p_1 - p_2 = \left[\frac{1}{\dfrac{1}{R_A} + \dfrac{1}{R_B} + \dfrac{1}{R_C}}\right] Q .$$

It is not difficult to show that Fig. 10.3(d) is a combination of the other two arrangements:

$$p_1 - p_2 = \left\{R_A + \frac{1}{\dfrac{1}{R_B} + \dfrac{1}{R_C}}\right\} Q .$$

10.5 FLOW NETWORKS FOR DYNAMIC CONDITIONS

In order to meet the requirements dictated by the specification, the engineer must have some means of estimating the parameters of the system, and its behaviour under dynamic conditions. This requires 'sizing' of the dimensions under steady state conditions and then formulating the dynamic equations for further assessment. The 'goodness' of the mathematical model will depend on how detailed the equations are. Having reached this stage it becomes clear that many types of system (mechanical, pneumatic, electric, etc.) result in very similar equations, and on this basis analogies can be drawn.

Because hydraulic and pneumatic systems operate using a fluid medium, strong analogies [1] can be drawn with electrical circuits. This allows many of the techniques used by electrical engineers to be used and often gives a clearer understanding of the problems. This approach can be developed further on the **analogue** computer, where active rather than passive networks can be developed. Care should be taken, however, as there are limitations on the analogies that can be drawn between the flow of fluid and the flow of electrical current [5].

The types of elements to be considered have corresponding electrical analogies. For example, the behaviour of the connecting tubes, or pipes for a network of flow paths is analogous to that of an electrical network. Volume rate of flow (Q) of the working fluid is analogous to the electric current flow (I), and fluid pressure (P) is analogous to electrical voltage (V).

The working fluid may be either a liquid or a gas, and many of the elements may be non-linear, particularly in the case of pneumatic elements. Hence, direct linear analogies may often be limited to a small range.

Fluid systems can be considered as a collection of interconnected basic elements. The differential equations which represent them are obtained by the application of the laws of networks and junctions, as used in section 10.2. Hence it is possible to construct an electrical network analogous to a fluid circuit, starting from a knowledge of the electrical components that are analogous to the basic fluid elements.

One basic link between all physical systems is energy or rate of doing work. For example, the mechanical equivalent of heat is the link between mechanical and thermal systems.

10.5.1 Fluid resistance

The effect of a fluid resistance is characterised by a pressure drop in the direction of the flow. This is analogous to the drop in voltage across an electrical resistance. Again care has to be taken with this analogy, since Ohm's law is a linear relationship and a fluid restrictor will only exhibit linear characteristics under laminar flow conditions. The relationships are shown in Table 10.1.

There are four basic types of fluid resistance commonly encountered:

Porous plug
Laminar tube flow
Turbulent pipe flow
Orifice

For pressure drops small compared with the average pressure of the system, the characteristics of these elements can be taken as identical for liquid or gas flows. This allows compressibility effects to be ignored. The fluid resistance R_h is measured in fluid ohms.

10.5.2 Fluid inductance

The equivalent of electrical inductance is **fluid inertance** which is a measure of inertia forces developed when a fluid is accelerated in a pipe or passage. It can be determined experimentally by measuring the rate of change of flow rate in a fluid line, when a given pressure difference is slowly applied across its ends. If the change is too rapid, compressibility effects cannot be ignored. The relationship is shown in Table 10.1, where the fluid inductance L_h is measured in fluid Henries.

10.5.3 Fluid capacitance

A fluid element is which potential energy is stored may be thought of as a fluid capacitor, equivalent to an electrical capacitor. The capacitance can be determined experimentally by recording the amount of fluid that must be slowly introduced into a pipe line to produce a unit rise in pressure.

Energy storage can take the form of compression of the liquid, or elasticity of the cylinder or piping walls. The relationships are shown Table 10.1, where the fluid capacitance C_h is measured in fluid Farads.

In general, therefore R_h, L_h and C_h can be termed the fluid impedances of a fluid system, and hence the general relationship is:

$$\text{Pressure} = \text{hydraulic impedance} \times \text{flow rate}$$

$$P = Z_h \times Q\,. \tag{10.9}$$

Under dynamic conditions, the relationships can be treated in the same way as normal electrical relationships, in which the basic frequency forms are obtained by replacing the operator $\mathrm{d}/\mathrm{d}t$ by ωj.

Most components can be thought of as two-port devices, in the sense that energy can enter at one end, and exit at the other. In this respect therefore, there can be hydraulic impedance values allocated to the input and the output of a component. At each port there will exist a pressure and flow, the product of these two variables determining the power.

10.6 FLUID IMPEDANCE

The fluid or hydraulic impedance relationship was given in (10.9) and can be thought of as the fluid power version of Ohm's law. Impedance in the electrical sense can be a composite term of resistance, capacitance, and inductance. Fluid systems can be modelled in the same way and, like the electrical circuit, will present a different impendance value for each frequency of excitation.

The individual fluid element shown in Table 10.1 will have distinct impedance/frequency characteristics as summarised in Fig. 10.4. There are several levels of modelling the behaviour of fluid flow under dynamic conditions in pipe work. These are:

1. Pure lumped parameter models in which a distributive system such as a pipe filled with oil is treated as a single lump.

2. Multi-lumped parameter models in which the pipe is treated as a series of inter-connected small lumps.
3. Distributed parameter models requiring the solution of the wave equation.

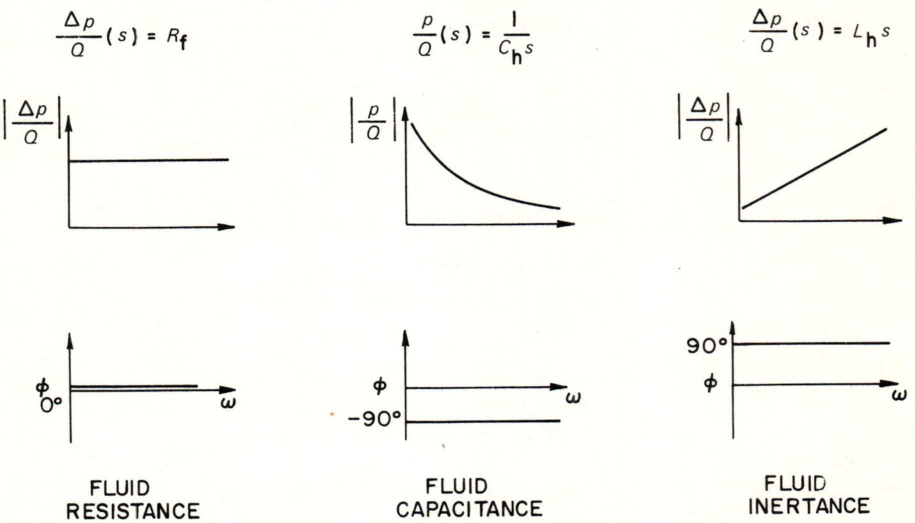

Fig. 10.4 – Impedance characteristics of basic element.

The problem with the simple approach defined in subsection 1 above is that it will not demonstrate the time it takes for a wave to propagate along a pipe line. Hence standing wave phenomena such as 'water hammer' cannot be modelled [6].

Consider the section of pipe shown in Fig. 10.5(a) of length L, this can be considered to offer resistance R_L to the following oil; it can offer inductive impedance due to the effort required to accelerate and decelerate the oil, and can offer capacitive impedance due to the elasticity of the oil and the elasticity of the pipe.

The electrical analogue of this pipe section is shown in Fig. 10.5(c) where it can be seen that:

$$\text{Input impedance} = \frac{P_u}{Q_u} = \left(R_h + L_h s + \frac{1}{C_h s}\right). \quad (10.10)$$

$$\text{Output impedance} = \frac{P_d}{Q_d} = \frac{1}{C_h s}. \quad (10.11)$$

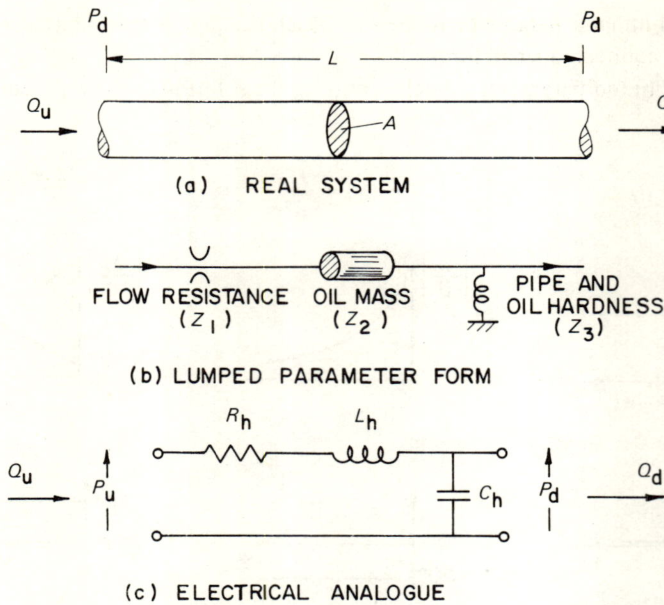

Fig. 10.5 – Lumped parameter model of pipe line.

If we can assume that $Q_u = Q_d$ for the pipe section, then the pipe transfer function will be:

$$\frac{\text{Output impedance}}{\text{Input impedance}} = \frac{P_d}{P_u} = \frac{1}{L_h C_h s^2 + R_h C_h s + 1}\,. \tag{10.12}$$

The behaviour of the impedances with increasing excitation frequency can be determined by making the substitution $s = \omega j$. Hence (10.10) becomes:

$$Z_u = R_h + \left\{ L_h \omega - \frac{1}{C_h \omega}\, j\,. \right\} \tag{10.13}$$

The input impedance will have a minimum value, that is offer least resistance to sinusoidal flow, when the frequency is:

$$\omega = \omega_n = \frac{1}{\sqrt{(L_h C_h)}} \tag{10.14}$$

which will in fact be the fundamental vibration mode (natural frequency) for that particular section of pipe.

Equation (10.12) is the classical form of the second order differential equation for a vibrating system and can be compared with the general form:

$$\frac{\text{Output}}{\text{Input}} = \frac{1}{\dfrac{1}{\omega_n^2}s^2 + \dfrac{2c}{\omega_n}s + 1} . \qquad (10.15)$$

Hence it can be seen by comparing (10.15) with (10.14), that:

$$\omega_n = \frac{1}{\sqrt{(L_h C_h)}}; \qquad c = \frac{R_h}{2}\sqrt{\frac{C_h}{L_h}} . \qquad (10.16)$$

Suppose the pipe shown in Fig. 10.5 has the following parameters:

10 mm N.B. steel pipe

$I.D.$	$= 12.5$ mm	P_u	$= 100$ bar
$O.D.$	$= 17.1$ mm	E_{steel}	$= 210\ \text{GN/m}^2$
t	$= 2.3$ mm	ρ_{oil}	$= 858.2\ \text{kg/m}^3$
L	$= 3$ m	μ_{oil}	$= 4 \times 10^{-2}\ \text{Ns/m}^2$

For steel pipe where the internal and outside diameters are approximately equal, the bulk modulus [7] is:

$$\beta_{steel} = \frac{tE}{d} = \frac{2.3 \times 210 \times 10^9}{17.1} = 2.824 \times 10^{10}\ \text{N/m}^2 .$$

For oil in a hydraulic circuit, a bulk modulus of $1.3 \times 10^9\ \text{N/m}^2$ is not an unreasonable assumption since experiment shows there will be about 2 per cent trapped air in the oil.

The equivalent bulk modulus for a pipe containing oil will be:

$$\frac{1}{\beta_e} = \frac{1}{2.824 \times 10^{10}} + \frac{1}{1.3 \times 10^9}$$

$$\beta_e = 1.242 \times 10^9\ \text{N/m}^2 .$$

The capacitive, inductive, and resistive components can be determined from Table 10.1.

$$C_h = \frac{AL}{\beta_e} = \frac{12.27 \times 10^{-5} \times 3}{1.242 \times 10^9} = 2.963 \times 10^{-13}\ \text{m}^4\text{s}^2/\text{kg}$$

TYPE	ELEMENT	RELATIONSHIP	ELECTRICAL ANALOG	ELEMENT	RELATIONSHIP
RESTRICTOR	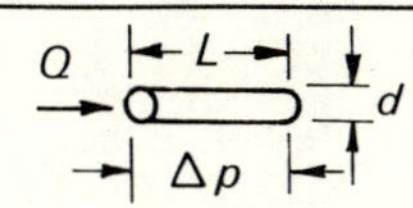	LAMINAR FLOW $\Delta p = R_h Q$ $R_h = \frac{128\mu L}{\pi d4}$	RESISTOR	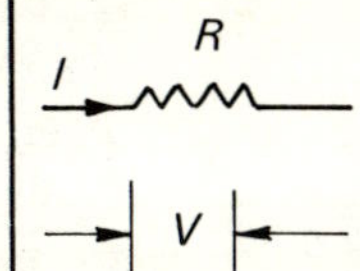	$V = RI$ OHMS LAW
	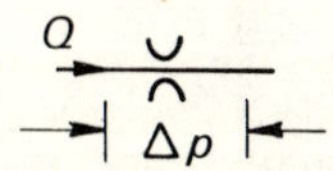	TURBULENT FLOW $\Delta p = R_h Q\|Q\|$ $R_h = \frac{\phi}{2Cd^2a^2}$			
INDUCTANCE	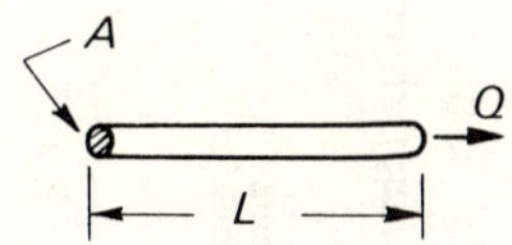	$\Delta p = L_h \frac{dQ}{dt}$ $L_h = \frac{\phi L}{A}$	INDUCTOR	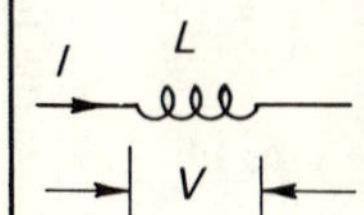	$V = L \frac{dI}{dt}$
CAPACITANCE	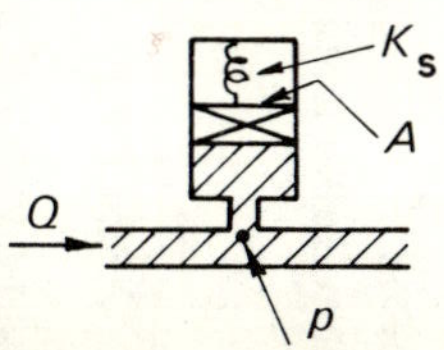	$\frac{dp}{dt} = \frac{1}{C_h} Q$ $C_h = \frac{A^2}{K_s}$ LIQUID SPRING $C_h = \frac{AL}{B}$	CAPACITOR	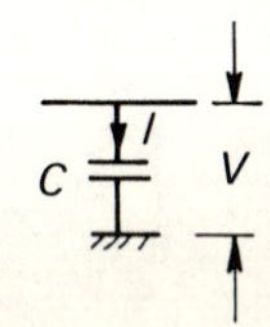	$V = \frac{1}{C}\int I\,dt$

Table 10.1.

$$L_h = \frac{\rho L}{A} = \frac{858.2 \times 3}{12.27 \times 10^{-5}} = 2.098 \times 10^7 \text{ kg/m}^4$$

and assuming laminar flow:

$$R_h = \frac{128\mu l}{\pi d^4} = \frac{128 \times 4 \times 10^{-2} \times 3}{3.14 \times (12.5 \times 10^{-3})^4}$$

$$= 2.003 \times 10^8 \text{ kg/m}^2\text{s}\,.$$

Substituting these values into equation (10.13) gives:

$$Z_u = 2.003 \times 10^8 + \left\{2.098 \times 10^7\,\omega - \frac{1}{2.963 \times 10^{-13}\,\omega}\right\} j\,.$$

The solution of this relationship over the frequency range 20–100 Hz is given in Table 10.2.

Table 10.2

f	ω	$L_h\omega(\times 10^{-9})$	$\frac{1}{C_h\omega}(\times 10^{-9})$	$\lvert Z_u \rvert(\times 10^{-9})$	$\phi°$
20	125.6	2.635	26.87	24.23	−89.5
30	188.4	3.952	17.91	13.95	−89.1
40	251.2	5.270	13.43	8.162	−88.5
50	314.0	6.587	10.74	4.157	−87.2
70	439.6	9.222	7.677	1.555	+82.6
80	502.4	10.54	6.717	3.858	+86.9
90	565.2	11.86	5.971	5.892	+88.1
100	628.0	13.17	5.374	7.796	+88.5

The minimum value of input impedance occurs at:

$$\omega_n = \frac{1}{\sqrt{(L_h C_h)}} = \frac{1}{\sqrt{(2.098 \times 10^7 \times 2.963 \times 10^{-13})}}$$

$$= 401.08 \text{ rad/s} = 63.86 \text{ Hz}\,.$$

Plots of the amplitude and phase of both the input and output impedances are shown in Fig. 10.6. The input impedance is dominated by the capacitive components up to the natural frequency (stiffness controlled) and by the inertance of the oil above the natural frequency (mass controlled). Note the rapid change in phase angle at the natural frequency. This is characteristic of very low damping in a resonant system. At low frequencies the pressure lags the flow and this lag reduces until at the natural frequency the pressure and flow are in phase. At high

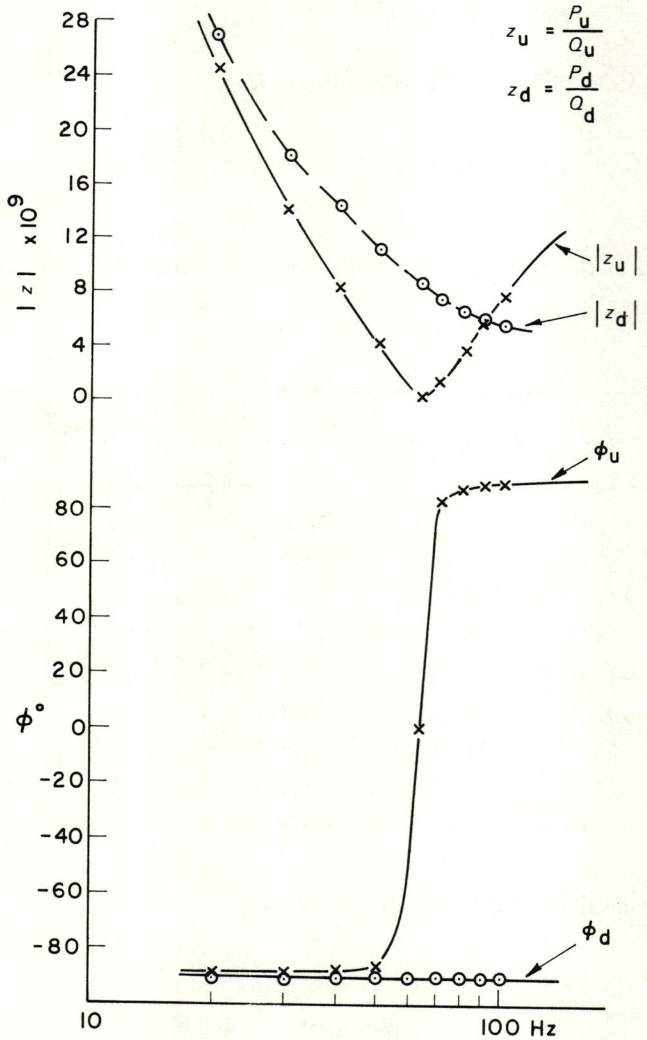

Fig. 10.6 – Input and output impedance plots for a pipe containing oil.

frequencies the pressure leads the flow. Compare these results with Fig. 10.4. From equation (10.16), the damping ratio for this typical length of hydraulic pipe is:

$$c = \frac{2.003 \times 10^8}{2}\sqrt{\left(\frac{2.963 \times 10^{-13}}{2.098 \times 10^{-7}}\right)} = 0.012$$

confirming that the damping is low.

The solution of the transfer function, equation (10.15), for both the time domain and frequency domain is well documented in the literature [3, 4]. Fig. 10.7 shows typical plots for a low damping condition.

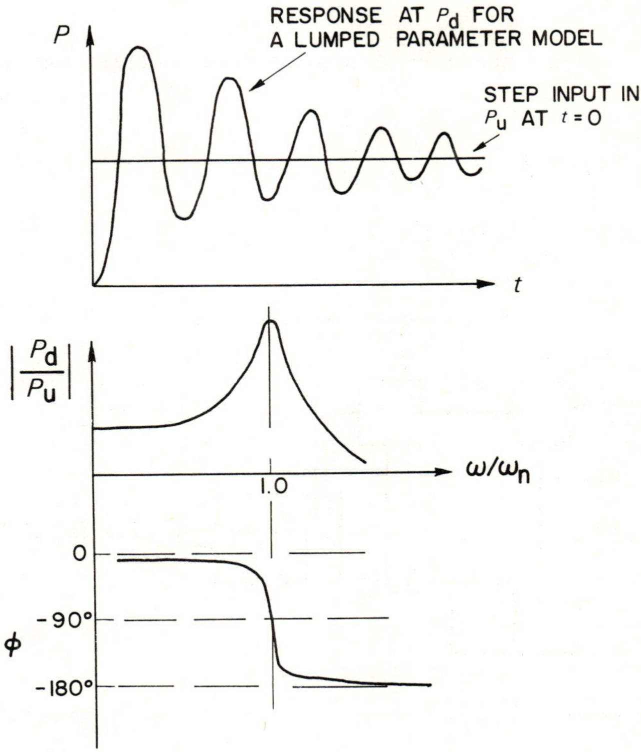

Fig. 10.7 – Frequency and transient response of a typical low damped system.

10.7 IMPEDANCE APPROACH APPLIED TO SYSTEMS

Consider a simple pump circuit and the electrical equivalent shown in Fig. 10.8. The mass and spring elements are referred to in the electrical diagram as inductive and capacitive elements. For the hydraulic system, the force F applied by the piston gives:

$$F = M\frac{d^2x}{dt^2} + kx = M\frac{dv}{dt} + kx\ .$$

Dividing by the piston area A, and rearranging into the impedance form:

$$\begin{aligned} P &= \frac{M}{A^2}\frac{dQ}{dt} + \frac{k}{A^2}\int Q\,dt \\ &= L_h\frac{dQ}{dt} + \frac{1}{C_h}\int Q\,dt\ . \end{aligned}$$

This equation can now be converted into its frequency domain form by letting $s = \omega j$ again:

$$\begin{aligned} \frac{P}{Q}(\omega j) &= L_h\omega j - \frac{j}{C_h\omega} \\ &= j\left\{\frac{\omega^2 L_h C_h - 1}{\omega C_h}\right\}\ . \end{aligned} \qquad (10.17)$$

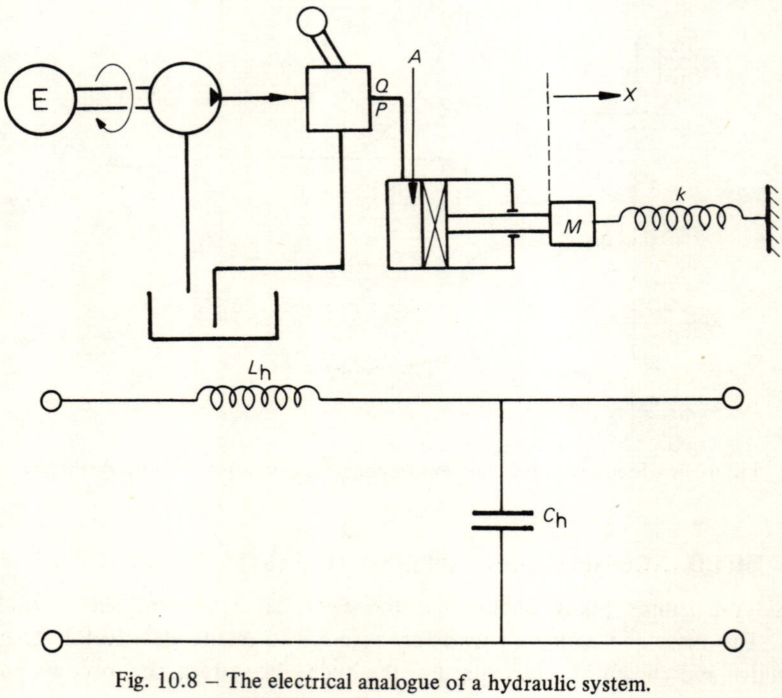

Fig. 10.8 – The electrical analogue of a hydraulic system.

This is a similar impedance characteristic to that shown in Fig. 10.6. Hence at the natural frequency, the pump will see a greatly reduced circuit input impedance and would not be required to deliver as much power as it would have at other frequencies of operation. This phenomenon is also discussed in Chapter 12.

The variation of circuit impedance with frequency has become more significant in recent years with the introduction of fluid logic devices and alternating flow hydraulics. The way energy propagates through the oil and its surrounding mechanical structure is also important to the noise control engineer, especially when pipe lengths are such that standing waves occur.

The lumped parameter approach discussed earlier is valid when the time required for a pressure wave to travel the length of the line is short with respect to the period of the highest frequency wave that is to be transmitted. This approach is certainly not suitable for step or impulse responses for example, since the functions contain all frequencies. In other words, single lumped parameter models are only useful for frequency response characteristics over frequency ranges considerably below the natural frequency of the line. A much more accurate and general approach to line dynamics can be achieved using a distributed parameter model.

In the case of liquid transmission lines, the rapid closing of a valve causes fluid to be stopped suddenly, and a pressure wave travels back to the 'incompressible' fluid source at velocity c, given by:

$$c = \sqrt{(\beta_e/\rho)} \tag{10.18}$$

where c = velocity of sound in the fluid (1365.5 m/s for oil)
β_e = effective bulk modulus
ρ = mass density of fluid.

When the pressure wave reaches the source at the upstream end of the pipe (in L/c seconds), then the kinetic energy of the moving mass of fluid has been completely stored as potential energy in the elastic pipe and fluid. The pressure P_c of the compressed fluid is a maximum. At this point a decompression wave forms and travels back to the valve. These waves continue to travel up and down the pipe with associated interchanges of kinetic and potential energies until friction expends all the energy. This type of behaviour can only be studied effectively using a distributive parameter model. The results of this approach are very important when analysing the propagation of noise through fluid power circuits, or the transmission of pulses between fluidic digital components.

10.8 THE WAVE EQUATION AND DISTRIBUTED PARAMETER MODELS

Consider an element of fluid in a pipe, Fig. 10.9. The net mass flow per unit time through the element, dx, must equal the time rate increase of mass within that space, hence:

$$\rho v A - \rho A\left(v + \frac{\partial v}{\partial x}\,\mathrm{d}x\right) = \frac{\partial}{\partial t}(\rho A)\mathrm{d}x$$

where v = average velocity at a cross-section of pipe,

or
$$\frac{\partial}{\partial t}(\rho A)\mathrm{d}x = -\frac{\partial}{\partial x}(\rho v A)\mathrm{d}x\,. \qquad (10.19)$$

Completing the differentiation of (10.19), gives,

$$A\frac{\partial \rho}{\partial t} + \rho\frac{\partial A}{\partial t} = -\left(vA\frac{\partial \rho}{\partial x} + \rho A\frac{\partial v}{\partial x} + \rho v\frac{\partial A}{\partial x}\,.\right) \qquad (10.20)$$

For a compressible fluid such as hydraulic oil, changes in density are determined from the bulk modulus:

$$\mathrm{d}\rho = \frac{\rho}{\beta}\,\mathrm{d}P\,. \qquad (10.21)$$

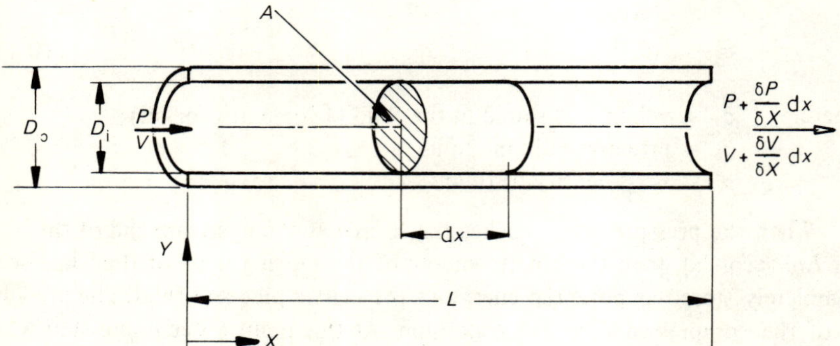

Fig. 10.9 – Element of fluid in a pipe.

Differentiating (10.21) with respect to both time and distance,

$$\frac{\partial \rho}{\partial t} = \frac{\rho}{\beta}\frac{\partial P}{\partial t} \qquad (10.22)$$

and
$$\frac{\partial \rho}{\partial x} = \frac{\rho}{\beta}\frac{\partial P}{\partial x}\,. \qquad (10.23)$$

Combine (10.20) with (10.22):

$$-\left(vA\frac{\partial\rho}{\partial x}+\rho A\frac{\partial v}{\partial x}+\rho V\frac{\partial A}{\partial x}\right)=\frac{A\rho}{\beta}\frac{\partial P}{\partial t}+\rho\frac{\partial A}{\partial t} \tag{10.24}$$

and also (10.24) with (10.23),

$$-\left(\frac{vA\rho}{\beta}\frac{\partial P}{\partial x}+\rho A\frac{\partial v}{\partial x}+\rho v\frac{\partial A}{\partial x}\right)=\frac{A\rho}{\beta}\frac{\partial P}{\partial t}+\rho\frac{\partial A}{\partial t}\,. \tag{10.25}$$

Now consider a thin-walled elastic tube. The change in cross-sectional area due to a pressure increase is given by:

$$\mathrm{d}A=\frac{A}{E(D_\mathrm{o}/D_\mathrm{i}-1)}\,\mathrm{d}P\,. \tag{10.26}$$

Combine (10.25) and (10.26):

$$-\left\{\frac{v}{\beta}\frac{\partial P}{\partial x}+\frac{v}{E((D_\mathrm{o}/D_\mathrm{i})-1)}\frac{\partial P}{\partial x}+\frac{\partial v}{\partial x}\right\}=\frac{1}{\beta}\frac{\partial P}{\partial x}+\frac{1}{E(D_\mathrm{o}/D_\mathrm{i}-1)}\frac{\mathrm{d}P}{\mathrm{d}t}\,. \tag{10.27}$$

If the equivalent bulk modulus for the pipe and the oil contained in it is defined as β_e where:

$$\frac{1}{\beta_\mathrm{e}}=\frac{1}{\beta}+\frac{1}{E(D_\mathrm{o}/D_\mathrm{i}-1)}\,. \tag{10.28}$$

Then (10.27) becomes:

$$-\left\{\frac{v}{\beta_\mathrm{e}}\frac{\partial P}{\partial x}+\frac{\partial v}{\partial x}\right\}=\frac{1}{\beta_\mathrm{e}}\frac{\partial P}{\partial t}\,. \tag{10.29}$$

The differential force acting across the element $\mathrm{d}x$ will be:

$$P-\left(P+\frac{\partial P}{\partial x}\,\mathrm{d}x\right)=\frac{F}{A}$$

hence:

$$F=-A\frac{\partial P}{\partial x}\,\mathrm{d}x\,. \tag{10.30}$$

The rate of change of momentum (M) for the element of fluid will be:

$$F = \frac{\mathrm{d}}{\mathrm{d}t}(M) = \frac{\partial}{\partial x}(\rho A v^2)\mathrm{d}x + \frac{\partial}{\partial t}(\rho v A)\mathrm{d}x. \tag{10.31}$$

Note that the effects of friction have been ignored, (see later). Substituting from (10.22), (10.23) and (10.28):

$$-\left(1 + \frac{v^2 \rho}{\beta_e}\right)\frac{\partial P}{\partial x} = \rho \frac{\partial v}{\partial t} + 2V\rho \frac{\partial v}{\partial x} + \frac{v\rho}{\beta_e}\frac{\partial P}{\partial t}. \tag{10.32}$$

For most fluid pumping applications the flow velocity v is small with respect to the wave propagation velocity, that is, velocity of sound in the fluid.

Hence (10.32) reduces to:

$$-\frac{\partial P}{\partial x} = \frac{\rho}{A}\frac{\partial Q}{\partial t} \tag{10.33}$$

where Q is the volumetric flow rate. This is called the **momentum equation.** Applying the same assumption to (10.29):

$$-\frac{\partial Q}{\partial x} = \frac{A}{\beta_e}\frac{\partial P}{\partial t} \tag{10.34}$$

which is known as the **continuity equation.**

Comparing the coefficients of these two equations with Table 10.1 it can be concluded that:

$$\frac{\rho}{A} = \text{inductance per unit length of pipe}$$

$$\frac{A}{\beta_e} = \text{capacitance per unit length of pipe.}$$

If the forces due to friction are of the same magnitude as the inertia and pressure forces, then the momentum equation (10.33) becomes:

$$-\frac{\partial P}{\partial x} = \frac{\rho}{A}\frac{\partial Q}{\partial t} + K_e Q \tag{10.35}$$

where $K_e = R_h/L$ = resistance per unit length of pipe.

Further differentiation of (10.33) and (10. 34) results in two forms of the one-dimensional wave equation:

$$\frac{\partial^2 P}{\partial x^2} = \frac{1}{c^2}\frac{\partial^2 P}{\partial t^2} \tag{10.36}$$

$$\frac{\partial^2 Q}{\partial x^2} = \frac{1}{c^2}\frac{\partial^2 Q}{\partial t^2} \tag{10.37}$$

where $c = \sqrt{(\beta_e/\rho)}$, the velocity of sound.

Equations (10.33) and (10.34) can be written in a more general form:

$$\frac{\partial P}{\partial x} = -Z(s)Q \tag{10.38}$$

$$\frac{\partial Q}{\partial x} = -Y(s)P. \tag{10.39}$$

Since the line possesses the general properties of inertance, capacitance and resistance, the series impedance and shunt admittance are given by:

$$Z(s) = L_h s + R_h$$

$$Y(s) = C_h s$$

where L_h = inertance per unit length of line
C_h = capacitance per unit length of line
R_h = resistance per unit length of line
s = operator d/dt.

By further partial differentiation of (10.38) and substitution:

$$\frac{\partial^2 P}{\partial x^2} = -Z(s)\frac{\partial Q}{\partial x} = Z(s)Y(s)Q$$

or

$$P = \frac{1}{Z(s)Y(s)}\frac{\partial^2 P}{\partial x^2} \tag{10.40}$$

also

$$Q = \frac{1}{Z(s)Y(s)}\frac{\partial^2 Q}{\partial x^2}. \tag{10.41}$$

The product of impedance and admittance, common to both these equations, is a complex number known as the **propagation constant** γ, where:

$$\gamma^2 = Z(s)Y(s) = \frac{As}{\beta_e}\left(\frac{\rho}{A}s + K_e\right)$$

$$= C_h s(L_h s + R_h)\,. \qquad (10.42)$$

The propagation constant is usually expressed in terms of unit line length.

The characteristic impedance of the pipe is defined as:

$$Z_c = \sqrt{\frac{Z(s)}{Y(s)}} = \sqrt{\left\{\frac{A_s}{\beta_e}\Big/\left(\frac{\rho}{A}s + K_e\right)\right\}}$$

This is equivalent to the hydraulic impedance of an infinitely long line. It can be seen that both γ and Z_c are dependent on the physical properties of the pipe and the frequency of excitation.

10.9 SOLUTION OF THE WAVE EQUATIONS FOR HARMONIC EXCITATION

For steady forced oscillation of the fluid in a pipe, after transients have died away, the pressure and flow behaviour can be expressed by:

$$Q = \hat{Q}\,e^{j\omega t} \qquad (10.44)$$

$$P = \hat{P}\,e^{j\omega t} \qquad (10.45)$$

where $\hat{P}$ and $\hat{Q}$ are also functions of x, the position at which they are measured along the pipe. These could be the ripples in flow and pressure propagated from the outlet of a piston pump.

Assuming these to be solutions of (10.35) and (10.34):

$$\frac{\partial \hat{Q}}{\partial x}e^{j\omega t} = -\frac{A}{\beta_e}\,j\omega\hat{P}\,e^{j\omega t} \qquad (10.46)$$

and

$$\frac{\partial \hat{P}}{\partial x}e^{j\omega t} = -\left(\frac{\rho}{A}\hat{Q}\,\omega j + K_e\hat{Q}\right)e^{j\omega t}\,. \qquad (10.47)$$

Differentiating and using (10.42):

$$\frac{\partial^2 \hat{P}}{\partial x^2} - \gamma^2 \hat{P} = 0$$

$$\frac{\partial^2 \hat{Q}}{\partial x^2} - \gamma^2 \hat{Q} = 0\,.$$

The general solutions for these partial differential equations at a particular instant in time will be:

$$\hat{P}(x) = \hat{P}_\text{i}\,\text{e}^{-\gamma x} + \hat{P}_\text{r}\,\text{e}^{+\gamma x} \tag{10.48}$$

$$\hat{Q}(x) = \hat{Q}_\text{i}\,\text{e}^{-\gamma x} + \hat{Q}_\text{r}\,\text{e}^{+\gamma x}\,. \tag{10.49}$$

The first term in these equations represents an incident wave travelling from the source down the pipe, and the second a reflected wave returning from the far end of the pipe.

From equation (10.9), the hydraulic impedance at any position along a pipe, subjected to harmonic excitation, will be,

$$Z(x) = \frac{\hat{P}(x)}{\hat{Q}(x)} = \frac{\hat{P}_\text{i}\,\text{e}^{-\gamma x} + \hat{P}_\text{r}\,\text{e}^{\gamma x}}{\hat{Q}_\text{i}\,\text{e}^{-\gamma x} + \hat{Q}_\text{r}\,\text{e}^{\gamma x}}\,. \tag{10.50}$$

The relationships between $\hat{P}_\text{i}$ and $\hat{Q}_\text{i}$, and $\hat{P}_\text{r}$ and $\hat{Q}_\text{r}$ can be found by substitution of the solutions $\hat{Q} = \hat{Q}_\text{i}\,\text{e}^{-\gamma x}$ and $\hat{P}_\text{i} = \hat{P}_\text{i}\,\text{e}^{-\gamma x}$ into equation (10.34) giving:

$$-\frac{\partial}{\partial x}(\hat{Q}_\text{i}\,\text{e}^{-\gamma x}) = \frac{A}{\beta_\text{e}}\frac{\partial}{\partial t}(\hat{P}_\text{i}\,\text{e}^{-\gamma x})$$

and from (10.42) and (10.43):

$$\hat{Q}_\text{i} = \frac{A}{\beta_\text{e}\gamma}\frac{\partial \hat{P}_\text{i}}{\partial t} = \frac{A\omega j}{\beta_\text{e}\gamma}\hat{P}_\text{i}$$

$$= \frac{1}{Z_\text{c}}\hat{P}_\text{i}\,. \tag{10.51}$$

Similarly for the other solutions $\hat{Q} = \hat{Q}_\text{r}\,\text{e}^{\gamma x}$ and $\hat{P} = \hat{P}_\text{r}\,\text{e}^{\gamma x}$

$$\hat{Q} = -\frac{1}{Z_\text{c}}\hat{P}_\text{r}\,. \tag{10.52}$$

If these values are substituted into (10.50), the exponential function for impedance is expressed as:

$$Z(x) = \frac{\hat{P}(x)}{\hat{Q}(x)} = \frac{Z_c(\hat{P}_i\,e^{-\gamma x} + \hat{P}_r\,e^{\gamma x})}{\hat{P}_i\,e^{-\gamma x} - \hat{P}_r\,e^{\gamma x}} \quad . \tag{10.53}$$

Two values of this impedance expression are of particular interest; the input impedance to the line at $x = 0$, see Fig. 10.9, and the termination impedance at $x = L$.

At $x = 0$, equating (10.48) and (10.49) gives:

$$\hat{P}(0) = \hat{P}_i + \hat{P}_r \tag{10.54}$$

$$\hat{Q}(0) = \frac{1}{Z_c}(\hat{P}_i - \hat{P}_r)\,, \tag{10.55}$$

or solving for $\hat{P}_i$ and $\hat{P}_r$:

$$\hat{P}_i = \frac{1}{2}[\hat{P}(0) + Z_c\hat{Q}(0)] \tag{10.56}$$

$$\hat{P}_r = \frac{1}{2}[\hat{P}(0) - Z_c\hat{Q}(0)]\,. \tag{10.57}$$

At any point along the pipe, from (10.48), (10.56) and (10.57):

$$\hat{P}(x) = \frac{1}{2}[\hat{P}(0) + Z_c\hat{Q}(0)]\,e^{-\gamma x} + \frac{1}{2}[\hat{P}(0) - Z_c\hat{Q}(0)]\,e^{+\gamma x}$$

$$= \hat{P}(0)\,\frac{e^{-\gamma x} + e^{+\gamma x}}{2} + Z_c\hat{Q}(0)\,\frac{e^{-\gamma x} - e^{+\gamma x}}{2}$$

$$= \hat{P}(0)\cosh\gamma x - Z_c\hat{Q}(0)\sinh\gamma x \tag{10.58}$$

and by a similar process: (10.58)

$$\hat{Q}(x) = \frac{\hat{P}(0)}{Z_c}\sinh\gamma x + \hat{Q}(0)\cosh\gamma x\,. \tag{10.59}$$

These relationships result in:

$$Z(L) = \frac{\hat{P}(L)}{\hat{Q}(L)} = \frac{\hat{P}(0)\cosh\gamma L - Z_c\hat{Q}(0)\sinh\gamma L}{\hat{Q}(0)\cosh\gamma L - \dfrac{\hat{P}(0)}{Z_c}\sinh\gamma L}$$

or

$$Z(L) = \frac{Z(0) - Z_c\tanh\gamma L}{1 - \dfrac{Z(0)}{Z_c}\tanh\gamma L}. \tag{10.60}$$

10.10 MATCHED AND UNMATCHED TRANSMISSION LINES

The physical realisation of equation (10.48) is two interfering pressure waves, one travelling down the pipe (incident) and the other reflected back up the pipe [6]. To ensure an efficient transfer of energy, the reflected wave should not occur.

Hence (10.48) reduces to:

$$\hat{P}(x) = \hat{P}_i\,e^{-\gamma x}. \tag{10.61}$$

This condition will exist if the line is terminated in the characteristic impedance $Z(L) = Z_c$, and is the most efficient way to operate a pipe line.

On the other hand open circuit, $Z(L) = \infty$, and short circuit, $Z(L) = 0$, lines are mismatched. In these cases reflections can occur and under certain conditions standing waves will result. In addition to reducing the power transmission efficiency, standing waves cause noise and vibration problems. These will be discussed in Chapter 18 on fluid power system noise. In practice the noise source is at the beginning of a pipe line, for example a piston pump will propagate ripples down a hydraulic line. It is therefore more convenient to express the pressure and flow at any point along the line in terms of the source values.

Equation (10.53) can be separated and expressed in the form:

$$\hat{P}(x) = \hat{P}_i\,e^{-\gamma x} + \hat{P}_r\,e^{+\gamma x} \tag{10.62}$$

$$\hat{Q}(x) = \frac{\hat{P}_i}{Z_c}\,e^{-\gamma x} - \frac{\hat{P}_r}{Z_c}\,e^{+\gamma x}. \tag{10.63}$$

At the source, $x = 0$, and the pump will have an output performance defined by $P(0) = P(0)\,e^{jwt}$. By substitution of these boundary conditions:

$$\hat{P}(0) = \hat{P}_i + \hat{P}_r \tag{10.64}$$

$$\hat{Q}(0) = \frac{\hat{P}_i}{Z_c} - \frac{\hat{P}_r}{Z_c} \tag{10.65}$$

At the termination, $x = L$, and (10.53) gives:

$$Z(L) = \frac{\hat{P}(L)}{\hat{Q}(L)} = \frac{\hat{P}_i e^{-\gamma L} + \hat{P}_r e^{+\gamma L}}{\frac{\hat{P}_i}{Z_c} e^{-\gamma L} - \frac{\hat{P}_r}{Z_c} e^{+\gamma L}}$$

and by rearrangement:

$$\frac{\hat{P}_i}{\hat{P}_r} = -e^{2\gamma L} \frac{1 + \frac{Z(L)}{Z_c}}{1 - \frac{Z(L)}{Z_c}} \tag{10.66}$$

Now (10.64) can be rewritten as:

$$\hat{P}(0) = \hat{P}_i + \hat{P}_i \frac{\hat{P}_r}{\hat{P}_i}$$

and using (10.66) gives:

$$\hat{P}_r = \frac{\hat{P}(0)}{1 - e^{2\gamma L} \frac{(Z_c + Z(L))}{(Z_c + Z(L))}} = \frac{-\hat{P}(0) e^{-\gamma L} (Z_c - Z(L))}{2[Z_c \sinh \gamma L + Z(L) \cosh \gamma L]} \tag{10.67}$$

$$\hat{P}_i = \hat{P}_r \frac{\hat{P}_i}{\hat{P}_r} = \frac{\hat{P}(0) e^{\gamma L} (Z_c + Z(L))}{2[Z_c \sinh \gamma L + Z(L) \cosh \gamma L]} \tag{10.68}$$

Equation (10.67) and (10.68) can now be substituted into (10.62):

$$\hat{P}(x) = \frac{\hat{P}(0)[Z_c \sinh \gamma(L - x) + Z(L) \cosh \gamma(L - x)]}{Z_c \sinh \gamma L + Z(L) \cosh \gamma L} \tag{10.69}$$

where $\hat{P}(0) = \hat{P}(0) e^{j\omega t} = \hat{P}(0)(\cos \omega t + j \sin \omega t)$.

By a similar process it can be shown that:

$$\hat{Q}(x) = \frac{\hat{P}(0)[Z_c \cosh \gamma(L - x) + Z(L) \sinh \gamma(L - x)]}{Z_c[Z_c \sinh \gamma L + Z(L) \cosh \gamma L]} . \tag{10.70}$$

10.11 PERFORMANCE CHARACTERISTICS OF A PUMP EXCITING A BLOCKED LINE

As an example of the application of the theory developed in the previous sections, consider a nine piston pump operating at 1800 rev/min. Pressure pulse ripples are propagated down the hydraulic line, arbitrarily chosen to be 5.04 m long. At the end of the line is a control valve which is closed. It is required to establish the general performance of the circuit and to determine if the line length is such that there will be no excessive vibrations in the circuit.

From equations (10.69) and (10.70), the impedance anywhere along the line can be established:

$$Z(x) = \frac{\hat{P}(x)}{\hat{Q}(x)} = \frac{Z_c [Z_c \sinh \gamma(L-x) + Z(L) \cosh \gamma(L-x)]}{Z_c \cosh \gamma(L-x) + Z(L) \sinh \gamma(L-x)} \,. \quad (10.71)$$

Since the pump is driving fluid against a closed valve at the end of the line, there will be zero flow at this point, hence:

$$Z(L) = \frac{\hat{P}(L)}{\hat{Q}(L)} = \frac{\hat{P}(L)}{0} = \infty \,. \quad (10.72)$$

If the valve were open then the load impedance would be:

$$Z(L) = \frac{\hat{P}(L)}{\hat{Q}(L)} = \frac{0}{\hat{Q}(L)} = 0$$

while for a matched line:

$$Z(L) = Z_c \,.$$

From (10.69) the pressure at any point along the line under blocked load conditions, will be:

$$\frac{\hat{P}(x)}{\hat{P}(0)} = \frac{\cosh \gamma (L-x)}{\cosh \gamma L} = \frac{\cos (\omega/c)(L-x)}{\cos (\omega L/c)} \quad (10.73)$$

and for a matched line:

$$\frac{\hat{P}(x)}{\hat{P}(0)} = \frac{\sinh \gamma (L-x) + \cosh \gamma (L-x)}{\sinh \gamma L + \cosh \gamma L} = e^{-\gamma x} \,. \quad (10.74)$$

Similarly from (10.70), the flow at any point along the line under blocked load conditions will be:

$$\frac{\hat{Q}(x)}{\hat{P}(0)} Z_c = \frac{\sinh \gamma (L-x)}{\cosh \gamma L} = \frac{j \sin (\omega/c)(L-x)}{\cos (\omega/c) L} \tag{10.75}$$

and for a matched line:

$$\frac{\hat{Q}(x)}{\hat{P}(0)} Z_c = e^{-\gamma x} . \tag{10.76}$$

Equations (10.73) and (10.75) indicate that for the mismatched line the incident and reflected waves reinforce each other to establish a standing wave in the oil. This will translate itself into a violent vibration of the pipe under certain ill-chosen conditions. On the other hand, equations (10.74) and (10.76) show that under matched conditions no standing wave pattern is generated, hence producing no vibrational problems. It is not always possible to design for matched conditions, so it is therefore necessary to ensure that line lengths under mismatched conditions are chosen to minimise the standing wave. From equation (10.71), under blocked load conditions, the impedance at any point in the line is:

$$Z(x) = \frac{Z_c \cosh \gamma (L-x)}{\sinh \gamma (L-x)} . \tag{10.77}$$

It is evident from Fig. 10.6 that flow resistance in oil systems does not play a significant role under pulsatile flow conditions, and therefore it is reasonable to assume that the propagation constant γ per unit length of pipe as defined in equation (10.42) can be reduced to:

$$\gamma = \sqrt{(C_h L_h)}\, j\omega = \omega j \sqrt{(\gamma/\beta)} = \frac{\omega}{c} j ;$$

substitution in (10.77) gives:

$$Z(x) = Z_c \frac{\cosh\{(\omega/c) j (L-x)\}}{\sinh\{(\omega/c) j (L-x)\}} \tag{10.78}$$

but $\cosh j\theta = \cos \theta$ and $\sinh j\theta = j \sin \theta$, so that (10.78) can be rewritten as the source impedance:

$$Z(0) = -jZ_c \cot (\omega L/c) . \tag{10.79}$$

Resonance occurs in the pipe for values of L which give zero values of impedance at the source. Hence $Z(0) = 0$ for:

$$\frac{\omega L}{c} = \frac{\pi}{2}, \frac{3\pi}{2}, \frac{5\pi}{2} \text{ etc.}$$

The lowest frequency (fundamental) producing resonance in this length of pipe is:

$$f_0 = \frac{c}{4L} = \frac{1365.5}{4 \times 5.04} = 67.73 \text{ Hz}.$$

The fundamental pumping frequency is:

$$f_p = 9 \times \frac{1800}{60} = 270 \text{ Hz},$$

which will be close to the pipe resonance at 203 Hz. A clearer insight into the situation can be seen, if the ratio of load pressure to input pressure is plotted against selected line lengths. For a blocked line, (10.73) gives for $x = L$:

$$\frac{\hat{P}(L)}{\hat{P}(0)} = \frac{\cosh \gamma (L - L)}{\cosh \gamma L} = \frac{1}{\cos (\omega L/c)}. \tag{10.80}$$

These results for lines of different lengths are plotted in Fig. 10.10, for a pump

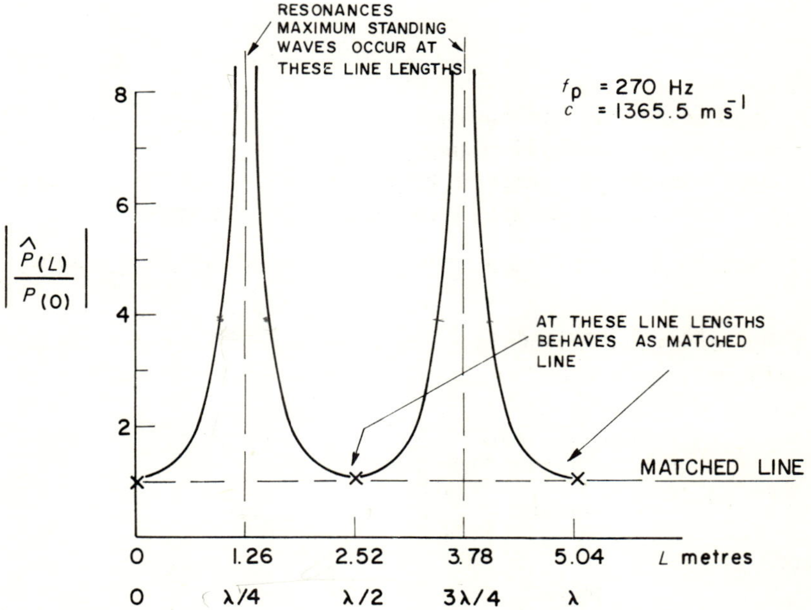

Fig. 10.10 – Pressure magnitude for lines of different length.

ripple frequency of 270 Hz. This graph shows that for line lengths of $L = \frac{1}{4}\lambda$, $\frac{3}{4}\lambda$, etc., pressure resonance occurs and maximum standing waves are established in the pipe. For the line lengths of $L = 0, \frac{1}{2}\lambda, \lambda$, etc., the pressure ratio becomes unity, which corresponds to a matched line, see Fig. 10.11.

The pressure and flow distributions along the specific line under discussion, can be plotted from equations (10.73) through to (10.76). These are shown, together with the impedance distribution, in Fig. 10.11.

10.12 EXAMPLES

1. A 20 m long pipe, 20 mm diameter, is filled with hydraulic oil. One end of the pipe is connected to large tank and the other end is fitted with a piston. The piston is excited at 25 Hz by a vibrator which oscillates with an amplitude of 10 mm.

   If the flow is considered frictionless, find (a) the hydraulic impedance and (b) the magnitude of the pressure oscillations at the piston end. Take the oil density as 858 kg/m^3 and bulk modulus as 1.39×10^9 N/m^2.

   (*Answers*: (a) 1.498×10^8 kg/m^4 s; (b) 0.12 bar.)

2. A pipe line 0.813 mm diameter and 30.5 cm long contains fluid of density 1106 kg/m^3. The viscosity of the fluid is 4×10^{-2} Ns/m^2. Assuming the flow in the pipe to be laminar and that it varies sinusoidally such that:

   $$Q = \hat{Q} \sin \omega t ,$$

   then:

   (a) Compute the fluid resistance.
   (b) Compute the fluid inertance.
   (c) Determine the range of frequencies over which the pipe can be modelled as a pure resistance, within 10 per cent error.

   (*Answers*: (a) 1.14×10^{12} kg/m^{2}s; (b) 6.5×10^8 kg/m^4; (c) $\omega < 128$ Hz.)

3. A pipe of length l has a fixed orifice at the downstream end. If the orifice coefficient is k, show that:

   $$|Q(l)| = \frac{\hat{Q}(0)}{\left\{\cos^2 \gamma l + \dfrac{\sin^2 \gamma l}{k^2 Z_c^2}\right\}^{\frac{1}{2}}}$$

   and

   $$\phi = \tan^{-1}\left(\frac{\tan \gamma l}{k Z_c}\right).$$

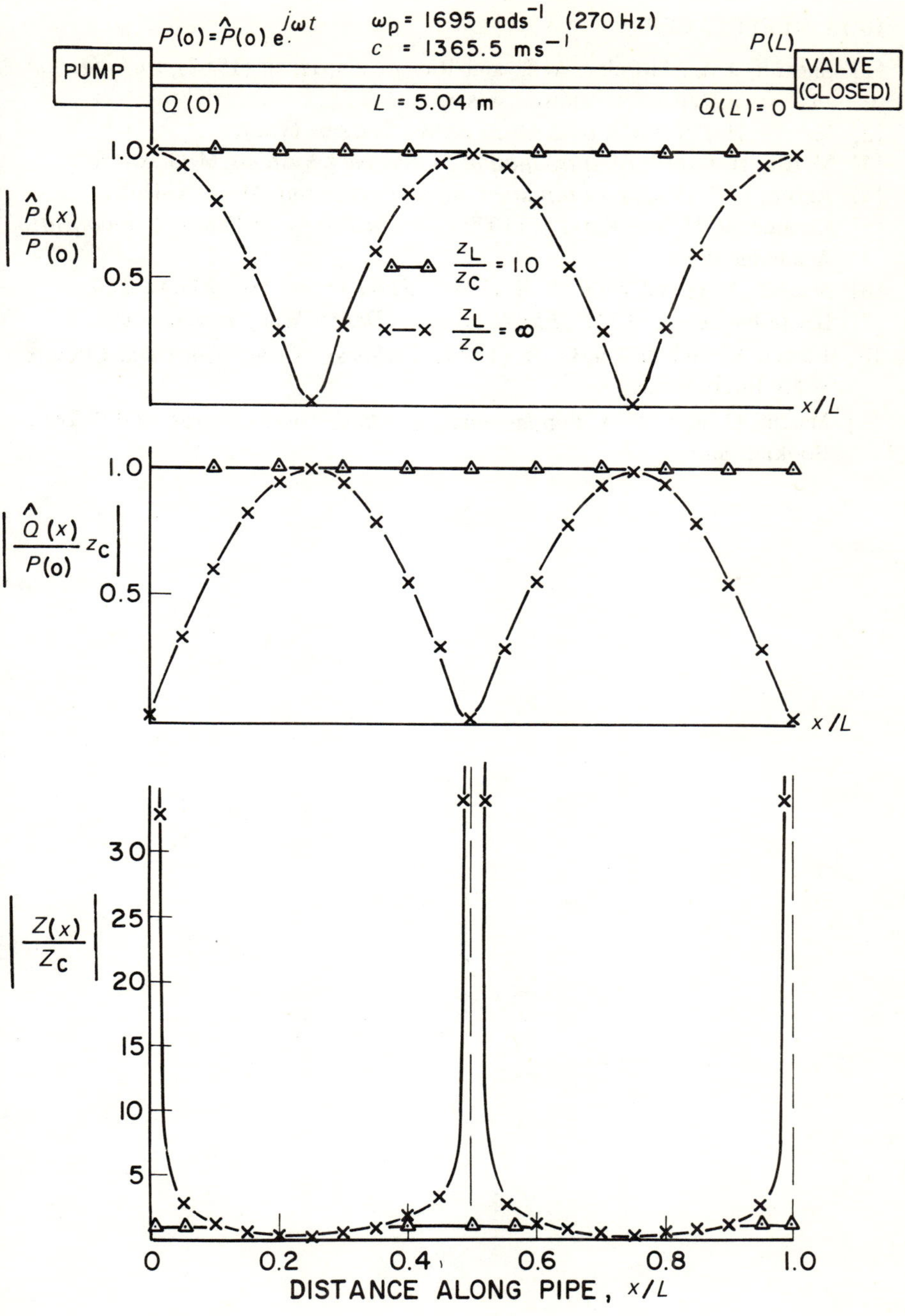

Fig. 10.11 – Performance of a blocked line.

10.13 REFERENCES

[1] Shearer, J. L., Murphy, A. J., and Richardson, H. H. (1967), *Introduction to System Dynamics*, Addison Wesley.

[2] Merritt, H. E. (1967), *Hydraulic Control Systems*, Wiley.

[3] Martin, H. R. (1968), *Introduction to Feedback Systems*, McGraw-Hill.

[4] Raven, F. H. (1961), *Automatic Control Engineering*, McGraw-Hill.

[5] Kirshner, J. M. and Katz, S. (1975), *Design Theory of Fluidic Components*, Academic Press.

[6] Streeter, V. L. and Wylie, E. B. (1978), *Fluid Transients*, McGraw-Hill.

[7] Belsterling, C. A. (1971), *Fluidic Systems Design*, Wiley Interscience.

[8] Foster, K. and Parker, G. A. (1970), *Fluidics – Components and Circuits*, Wiley Interscience.

[9] Martin, H. R. (1977), *Fundamentals of Fluid Power Control*, AUEW Tech. Booklet, Surrey.

CHAPTER 11

Hydraulic servomechanisms

11.1 PRECISION CONTROL SYSTEMS IN HYDRAULICS

Generally speaking, the hydraulic servomotor provides fast response, high force and short stroke characteristics which are satisfactory for many fields of application. In the aircraft industry this type of actuator is used for power operated controls, autostabilisers and autopilots. In general industry it finds favour in speed control systems for prime movers, and in the operation of process control valves.

The machine tool industry requires control systems that are able to produce high accuracy with good finish. The machine must be rigid and must be able to produce a good standard of repeatability. The introduction of automatic control techniques and hydraulic servos has brought vast improvements, and even further enhancement can be achieved by using computer control. These methods, however, could not be fully effective without the introduction of the special techniques developed in engineering measurements, so that accurate positioning to 1 μm or better can be obtained.

Many machine tools use hydraulic systems for both the basic drive mechanism and the control of the cutting tool. In general the advantages of hydraulic components are:

Ease of accurate control of work table position and velocity.
Good stiffness characteristics.
Zero backlash.
Rapid response to change in speed or direction.
Low rate of wear.

Good stiffness characteristics are especially important in the light of recent developments in automatic machine tool control and metal cutting; these emphasised the need for rigid, backlash free drives with an adequate response to control demands.

Less sophisticated machine tool control systems use direct input signals to the control valve from a template. For more complex metal cutting problems numerical control is commonly used.

The effects of compressibility of the oil and the contribution to stiffness by the mechanical structure associated with the simple hydraulic servomotor are often not fully appreciated, especially if the servomotor is accelerating a large mass. Furthermore, the output may be subjected to external forces, these introducing signals into the loop via the servomotor output. Some idea of the effect of these conditions on performance can be gathered from a linearised analysis of the system.

11.2 SIMPLE MODEL OF A HYDRAULIC SERVO

The hydraulic servomechanism in its basic form is shown in Fig. 11.1(a). When lightly loaded it acts as an integrator, since the flow rate, controlled by the spool valve position, produces a rate of movement of the output piston rod.

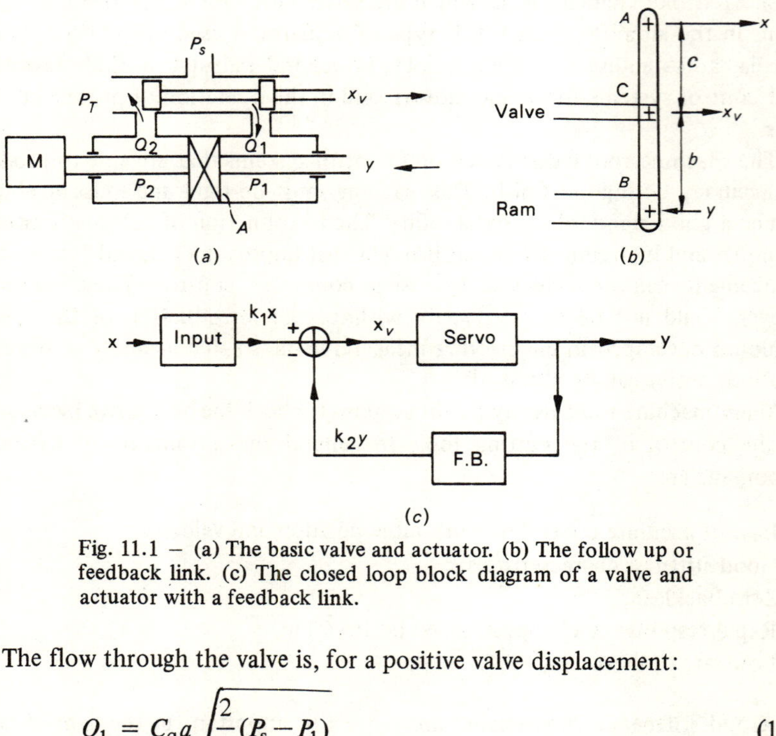

Fig. 11.1 – (a) The basic valve and actuator. (b) The follow up or feedback link. (c) The closed loop block diagram of a valve and actuator with a feedback link.

The flow through the valve is, for a positive valve displacement:

$$Q_1 = C_q a \sqrt{\frac{2}{\rho}(P_s - P_1)} \tag{11.1}$$

and:

$$Q_2 = C_q a \sqrt{\frac{2}{\rho}(P_2 - P_t)}\,. \tag{11.2}$$

Normally, the tank pressure is assumed atmospheric, and the flow rates Q_1 and Q_2 are assumed equal. The effects of compressibility are also ignored in this simple analysis. Hence mean flow rate Q is given by:

$$Q = \frac{Q_1 + Q_2}{2} = \frac{C_q a}{2}\sqrt{\frac{2}{\rho}}\,[\sqrt{(P_s - P_1)} + \sqrt{P_2}]\,; \tag{11.3}$$

but $P_s - P_1 = P_2$ from equations (11.1) and (11.2), and therefore:

$$Q = C_q a\sqrt{\frac{2}{\rho}}\,[\sqrt{(P_s - P_1)}]\,. \tag{11.4}$$

Now if the load pressure is defined as:

$$P_L = P_1 - P_2$$

then

$$P_1 = P_L + P_2$$

$$= P_L + (P_s - P_1)$$

and therefore:

$$P_1 = \frac{1}{2}(P_L + P_s)\,.$$

This allows the pressure drop across the valve to be expressed in terms of supply pressure and load pressure:

$$P_s - P_1 = P_s - \frac{1}{2}P_L - \frac{1}{2}P_s = \frac{1}{2}(P_s - P_L)$$

so (11.4) becomes:

$$Q = C_q a\sqrt{\frac{2}{\rho}}\sqrt{\frac{P_s - P_L}{2}}\,. \tag{11.5}$$

If the load at the output is small, then the load pressure is small and $P_1 = P_2 \simeq \frac{1}{2}P_s$. This allows (11.5) to be written:

$$Q = A\frac{\mathrm{d}y}{\mathrm{d}t} = K_1 x_v\sqrt{\frac{P_s}{2}}$$

where

$$K_1 x_v = C_q a\sqrt{\frac{2}{\rho}}$$

or

$$y = \frac{K_1}{A}\sqrt{\frac{P_s}{2}}\int x_v \mathrm{d}t\,. \tag{11.6}$$

In most cases position control of the output is desired, and a simple way to achieve this is to introduce a follow-up lever as shown in Fig. 11.1(b). A movement at the input x is initially pivoted about B, causing a displacement of the spool at C. The spool displacement allows flow into the cylinder, making the rod move in a direction to cancel the spool movement. During this movement the lever pivot point is at A. The block diagram is shown in Fig. 11.1(c).

Consider similar triangles in the lever geometery:

Pivot about B:

$$k_1 x = \frac{b}{c+b}x$$

Pivot about A:

$$k_2 y = \frac{c}{c+b}y\,.$$

Hence:

$$x_v = k_1 x - k_2 y = \frac{b}{c+b}x - \frac{c}{c+b}y\,, \tag{11.7}$$

but:

$$x_v = \frac{A}{K\sqrt{(P_s/2)}}\frac{\mathrm{d}y}{\mathrm{d}t}$$

$$\frac{A}{K_1\sqrt{(P_s/2)}}\frac{\mathrm{d}y}{\mathrm{d}t} + \frac{c}{c+b}y = \frac{b}{c+b}x$$

$$\frac{y}{x} = \frac{b/c}{1+\left\{\dfrac{c+b}{c}\,\dfrac{A}{K_1\sqrt{(P_s/2)}}\right\}s} \tag{11.8}$$

where $s \equiv \mathrm{d}/\mathrm{d}t$ operator.

It is easily seen from this transfer function that $y = (b/c)x$ in the steady state. This is a position control. Dynamically the system is a simple lag with:

$$\text{Gain} = b/c$$

$$\text{Time constant} = \frac{c+b}{c}\,\frac{A}{K_1\sqrt{(P_s/2)}}\,.$$

11.3 EFFECTS OF OIL COMPRESSIBILITY

A more realistic model of a hydraulic servomotor can be developed, taking into account compressibility and the inertia of the load. In this case the load pressure acting across the piston in Fig. 11.1 can become large during acceleration and retardation periods. Under these conditions oil compressibility must be taken into account since it can be shown to have a destabilising effect. It will be seen that the equations describing the system are non-linear in nature, and it is necessary to linearise them by considering small perturbations about an initial value. This approach was first introduced by N. F. Harpur [1].

From Fig. 11.1(a) it is seen that flow rate into the cylinder is:

$$Q_1 = f(x_v, P_s - P_1)\,. \tag{11.9}$$

In other words the flow rate into the cylinder is a function of valve displacement and pressure drop across the control orifice; we are not concerned with the exact relationship.

Similarly, the flow rate out of the other side of the cylinder is given by:

$$Q_2 = f(x_v, P_2 - P_T) \tag{11.10}$$

where P_T is normally zero relative to atmospheric pressure.

Consideration of the flow relationship through the cylinder must take into account the compressibility of the working fluid. Therefore if the bulk modulus of the fluid is defined as:

$$\beta = -V\frac{dP}{dV} \tag{11.11}$$

then it can be shown that the flow rate into the driving side is:

$$Q_1 = \frac{V_1}{\beta_1}\frac{dP_1}{dt} + A\frac{dy}{dt}\,. \tag{11.12}$$

Equation (11.12) can be visualised physically as:

$$\text{Total flow} = \begin{bmatrix}\text{Flow due to}\\\text{compressibility}\end{bmatrix} + \begin{bmatrix}\text{Flow due to}\\\text{piston movement}\end{bmatrix}.$$

Similarly

$$Q_2 = -\frac{V_2}{\beta_2}\frac{dP_2}{dt} + A\frac{dy}{dt}\,. \tag{11.13}$$

Equations (11.12) and (11.13) can now be considered in terms of small perturbations about an initial steady state, giving:

$$\delta Q_1 = A\frac{\mathrm{d}}{\mathrm{d}t}\delta y + \frac{V_1}{\beta_1}\frac{\mathrm{d}}{\mathrm{d}t}\delta P_1$$

and

$$\delta Q_2 = A\frac{\mathrm{d}}{\mathrm{d}t}\delta y - \frac{V_2}{\beta_2}\frac{\mathrm{d}}{\mathrm{d}t}\delta P_2 \,.$$

Hence the average flow:

$$\delta Q = \frac{\delta Q_1 + \delta Q_2}{2} = \frac{1}{2}\left\{\frac{V_1}{\beta_1}\frac{\mathrm{d}}{\mathrm{d}t}\delta P_1 - \frac{V_2}{\beta_2}\frac{\mathrm{d}}{\mathrm{d}t}\delta P_2 + 2A\frac{\mathrm{d}}{\mathrm{d}t}\delta y\right\}.$$

If V is the total swept volume of the cylinder, then at the centre position:

$$V_1 = V_2 = V/2$$

and hence considering small perturbations about this piston position gives:

$$\delta Q = A\frac{\mathrm{d}}{\mathrm{d}t}\delta y + \frac{V}{4\beta}\frac{\mathrm{d}}{\mathrm{d}t}(\delta P_1 - \delta P_2) \,. \tag{11.14}$$

This assumes a common value for bulk modulus through the system. The most interesting case, from a stability point of view, is when the external load is mainly inertia and internal or external friction is ignored. This means that any system damping is due entirely to the control valve characteristics.

The load equation is:

$$P_1 - P_2 = \frac{M}{A}\frac{\mathrm{d}^2 y}{\mathrm{d}t^2}$$

or in terms of perturbations:

$$\delta P_1 - \delta P_2 = \frac{M}{A}\frac{\mathrm{d}^2}{\mathrm{d}t^2}\delta y \,. \tag{11.15}$$

The non-linear relationship associated with flow through the valve, equations (11.1) and (11.2), are linearised by applying Taylor's formula for two variables, neglecting all powers above the first.

$$\delta Q_1 = \frac{\partial Q_1}{\partial x_v}\,\delta x_v + \frac{\partial Q_1}{\partial P_1}\,\delta P_1 \tag{(11.16)}$$

$$\delta Q_2 = \frac{\partial Q_2}{\partial x_v}\,\delta x_v + \frac{\partial Q_2}{\partial P_2}\,\delta P_2 \tag{11.17}$$

where $Q_1 = Q_s + \delta Q_1$; $x_v = x_{vs} + \delta x_v$ etc., and the partial derivatives are evaluated at the particular initial steady state (P_{1s}, P_{2s}, x_{vs}, Q_s). In other words the flow is made up of a contribution due to valve displacement, and a contribution due to pressure change.

These equations are now linear although their coefficients are functions of the steady state conditions at which the system is being studied. In the steady state the flow rates into and out of the cylinder are equal, hence:

$$\frac{\partial Q_1}{\partial x_v} = \frac{\partial Q_2}{\partial x_v} = C_x \tag{11.18}$$

$$-\frac{\partial Q_1}{\partial P_1} = \frac{\partial Q_2}{\partial P_2} = C_p\,. \tag{11.19}$$

This is demonstrated in Fig. 11.2. Realistic evaluation of these constants is difficult (see section 7.4):

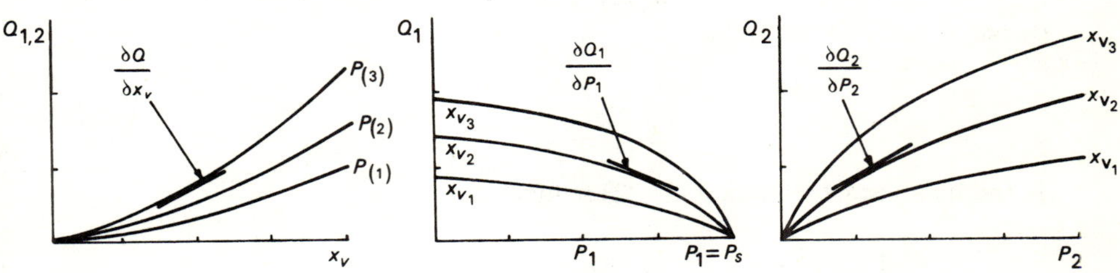

Fig. 11.2 – Static characteristics of the critical centre four-port valve of Fig. 11.1.

The average flow through the valve can therefore be determined from equations (11.16) and (11.17):

$$\delta Q = C_x \delta x_v - \frac{C_p}{2}(\delta P_1 - \delta P_2)\,. \tag{11.20}$$

Equating (11.14) and (11.20):

$$C_x\delta x_v - \frac{C_p}{2}(\delta P_1 - \delta P_2) = \frac{V}{4\beta}\frac{d}{dt}(\delta P_1 - \delta P_2) + A\frac{d}{dt}\delta y \tag{11.21}$$

and introducing the load term from equation (11.15):

$$C_x\delta x_v - \frac{C_p}{2}\frac{M}{A}\frac{d^2}{dt^2}\delta y = \frac{VM}{4\beta A}\frac{d^3}{dt^3}\delta y + A\frac{d}{dt}\delta y\,. \tag{11.22}$$

Equation (11.22) is the forward path equation for the servomotor. If now a feedback to the valve is arranged, then:

$$\delta x_v = \delta x_i - \delta y$$

where δx_i = incremental input movement.

Combining with equation (11.22) the transfer operator for the closed loop condition becomes:

$$\delta y = \frac{C_x/A}{\dfrac{VM}{4\beta A^2}s^3 + \dfrac{C_pM}{2A^2}s^2 + s + \dfrac{C_x}{A}}\,\delta x_i\,. \tag{11.23}$$

The combined stiffness of the two oil columns, each of length half the stroke, is easily shown to be:

$$k_o = \frac{4\beta A^2}{V}\,. \tag{11.24}$$

In the steady state, equation (11.22) reduces to:

$$\frac{d}{dt}\delta y = \frac{C_x}{A}\delta x_v = \frac{1}{A}\frac{\partial Q}{\partial x_v}\delta x_v$$

giving the open path gain relationship between the piston rate and the spool valve displacement.

The second term in the denominator of equation (11.23) can be written:

$$\frac{C_pM}{2A^2} = \frac{C_x}{A}M\frac{C_p}{2AC_x} = \frac{KM}{k_v}$$

where $K = C_x/A$

$k_v = 2A(C_x/C_p)$, the valve stiffness (section 7.5.1).

The closed loop equation (11.23) can now be rewritten in a more general form:

$$\delta y = \frac{K}{\frac{M}{k_o}s^3 + \frac{KM}{k_v}s^2 + s + K}\,\delta x_i\,. \tag{11.25}$$

The open path transfer operator from (11.23) then becomes:

$$\delta y = \frac{K}{s\left\{\frac{M}{k_o}s^2 + \frac{KM}{k_v}s + 1\right\}}\,\delta x_v\,. \tag{11.26}$$

The effect of compressibility is to introduce a quadratic lag term to the basic integration of equation (11.6). The system can therefore be subject to unstable oscillations.

Equation (11.26) can be written in the more familiar form:

$$\delta y = \frac{K}{s\left\{\frac{1}{\omega_n^2}s^2 + \frac{2c}{\omega_n}s + 1\right\}}\,\delta x_v \tag{11.27}$$

where $\omega_n^2 = k_o/M = 4\beta A^2/MV$ and c is the damping ratio.

If there are long pipe runs between the components, the inertia of the fluid in the pipes must be considered in addition to the load. Consider a piston of area A acted on by a pressure P_1 and driving a mass M with acceleration $\mathrm{d}^2y/\mathrm{d}t^2$. The fluid is exhausted through a pipe of area A_p with a mass m of fluid in the pipe. The fluid in the pipe accelerates at $\mathrm{d}^2z/\mathrm{d}t^2 = (A/A_p)\mathrm{d}^2y/\mathrm{d}t^2$, if compressibility is neglected. Hence, if P_2 is the exhaust pressure,

$$A(P_1 - P_2) = M\frac{\mathrm{d}^2y}{\mathrm{d}t^2}\,.$$

The force applied to the fluid in the cylinder to accelerate the fluid in the pipe will be:

$$P_2A_p = m\frac{\mathrm{d}^2z}{\mathrm{d}t^2} = m(A/A_p)\frac{\mathrm{d}^2y}{\mathrm{d}t^2} \tag{11.28}$$

and therefore:

$$P_1A = M + \frac{A}{A_p}^2 m\,\frac{\mathrm{d}^2y}{\mathrm{d}t^2}\,. \tag{11.29}$$

This can considerably increase the effective inertia seen by the piston.

11.4 STABILITY CONSIDERATIONS

Applying the Routh–Hurwitz stability criterion to the characteristic equation of the transfer function (11.25), gives the relationship for stability as:

$$\frac{M}{k_o} K < \frac{KM}{k_v}$$

or

$$k_o > k_v \, .$$

In other words the stiffness of the oil column must be greater than the effective valve stiffness if stability is to be satisfactory. Comparison of equations (11.26) and (11.27) produces the relationship:

$$\text{Damping ratio } c = \frac{KM}{k_v}\frac{\omega_n}{2} = \frac{K\sqrt{(Mk_o)}}{2k_v} \, .$$

This shows that the stiffness of the valve, k_v, must be reduced to improve stability. This is more effective than increasing the oil column stiffness. In practice the damping term will include friction terms and the structural stiffness.

An increase of valve gain C_x will also decrease stability since:

$$k_v = 2A\frac{C_k}{C_p} = \frac{2A^2}{C_p}K \, .$$

It should be noted, however, that the valve stiffness, k_v, is rather complex to assess, since, as can be seen from Fig. 11.2, C_p and C_x vary considerably during working conditions. Intelligent assessment of the results is therefore required, (see sections 7.5.1, 11.8 and 11.12).

It has been shown that the flow rate through the valve is:

$$Q = K_1 x_v \sqrt{(P_s - P_1)} = K_1 x_v \sqrt{P_2} \, .$$

Differentiating this equation with respect to pressure:

$$\frac{\partial Q}{\partial P_{1,2}} = C_p = \frac{K_1 x_v}{2\sqrt{(P_s - P_1)}} = \frac{K_1 x_v}{2\sqrt{P_2}}$$

and with respect to valve displacement:

$$\frac{\partial Q}{\partial x_v} = C_x = K_1\sqrt{(P_s - P_1)} = K_1\sqrt{P_2} \, .$$

But $P_1 - P_2 = P_L$ and hence:

$$C_p = \frac{K_1 x_v}{\sqrt{2(P_s - P_L)}} \;;\quad C_x = K_1 \sqrt{\frac{P_s - P_L}{2}}$$

Therefore

$$k_v = 2A \frac{C_x}{C_p} = K_1 \sqrt{\frac{P_s - P_L}{2}} \; \frac{\sqrt{2(P_s - P_L)}}{K_1 x_v} \; 2A$$

$$= \frac{2A(P_s - P_L)}{x_v} .$$

Now if, as is usual in a servo, the valve is operating about $x_v = 0$, the valve stiffness $k_v = \infty$ and the stability criterion predicts instability. In practice the situation is not quite so drastic as there is generally some form of leakage present, and certainly other forms of damping. Underlap can be used to improve stability by ensuring that $x_v \neq 0$.

11.5 DETERMINATION OF OPTIMUM RESPONSE

While use of the Routh-Hurwitz stability criterion in the previous section determined the conditions for stability, there is no guide to the relationship between k_v and k_o to maintain a staisfactory transient response. This can easily be assessed as follows.

Making the substitution $s = j\omega$ in equation (11.26):

$$\frac{\delta y}{\delta x_v}(j\omega) = \frac{K}{j\omega \left\{ 1 - \frac{M}{k_o}\omega^2 + \frac{KM}{k_v}\omega j \right\}}$$

The phase angle ϕ from this equation is given by:

$$\tan\phi = \frac{\omega - (M/k_o)\omega^3}{-\omega^2 MK/k_v}$$

If the system is excited at the natural frequency ω_n, then the numerator becomes:

$$\omega_n - \frac{M}{k_o}\omega_n^3 = \sqrt{\frac{k_o}{M}} - \left\{ \frac{M}{k_o} \times \frac{k_o}{M} \times \sqrt{\frac{k_o}{M}} \right\} = 0.$$

and hence, $\phi = -180°$.

Under this condition the amplitude ratio is given by:

$$\left|\frac{\delta y}{\delta x_v}\right| = \frac{K}{\sqrt{\left\{\left(\frac{\omega_n^2 KM}{k_v}\right)^2 + \left(\omega_n - \frac{M}{k_o}\omega_n^3\right)^2\right\}}}$$

$$= \frac{K}{\sqrt{\left\{\left(\frac{k_o KM}{Mk_v}\right)^2 + \left[\sqrt{\frac{k_o}{M}}\left(1 - \frac{M}{k_o}\frac{k_o}{M}\right)\right]^2\right\}}} = \frac{k_v}{k_o} \cdot$$

Hence, as shown in Fig. 11.3, the gain margin must be:

$$\text{Gain margin} = 1 - k_v/k_o \ .$$

This agrees with the Routh-Hurwitz condition – if $k_v = k_o$, the gain margin is zero and the system is unstable.

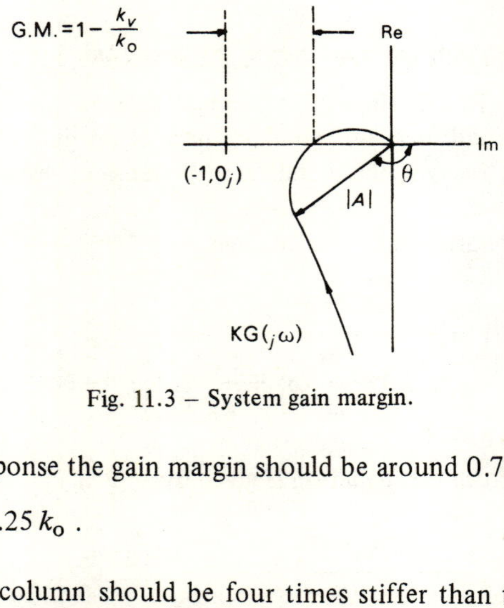

Fig. 11.3 – System gain margin.

For a good response the gain margin should be around 0.75 and hence:

$$k_v = 0.25\,k_o \ . \tag{11.30}$$

Therefore, the oil column should be four times stiffer than the valve stiffness, but this is not always possible in practice, and some form of artificial stabilizing loop may be needed.

To determine an optimum value of the gain K, a phase margin of 45° is chosen as consistent with the gain margin. It has been shown by Nightingale [2] that for this condition:

$$K = 1.2\sqrt{\left\{\frac{k_v}{M}\left(1 - \frac{k_v}{2k_o}\right)\right\}}.$$

11.6 RESPONSE CHARACTERISTICS

The closed loop equation can be written:

$$\frac{\delta y}{\delta x_i}(s) = \frac{K\omega_n^2}{s^3 + 2c\omega_n s^2 + \omega_n^2 s + K\omega_n^2}$$

giving the characteristic equation:

$$\frac{K\omega_n^2}{s(s^2 + 2c\omega_n s + \omega_n^2)} = -1. \tag{11.31}$$

Three types of transient response can result, depending on whether the roots are real or complex. A study of the patterns that result from parameter variation can rapidly be assessed by using the root locus method. The transient response can be assessed by examining the dominant root or roots, which are positioned nearest the imaginary axis.

For $c > 1.0$, the open loop poles of (11.31) all lie on the real axis, Fig. 11.4.

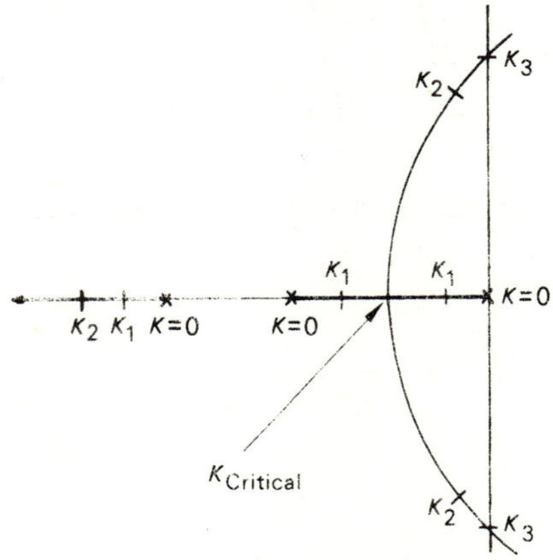

Fig. 11.4 – Root locus for equation (11.31) when damping ratio $c > 1.0$.

The transient response, Fig. 11.5, is exponential and shows no oscillation until the critical gain $K = K_{critical}$ is reached. Up to this condition the system is sluggish and the response time is dominated by the lag term with the biggest time constant.

If now the damping ratio is reduced so that $c < 1.0$, the open loop poles are complex and the transient response is never purely exponential. The root locus for this case is shown in Fig. 11.6. It can be seen from transient behaviour that at lower gains (K_1 for example), the simple lag term dominates making the transient response unacceptable, Fig. 11.7. Ideally, a servomechanism with this type of characteristic equation should have $c > 1.0$ and the transient response adjusted by means of the gain K. However, an inherent damping ratio $c > 1.0$ is very rare in practice.

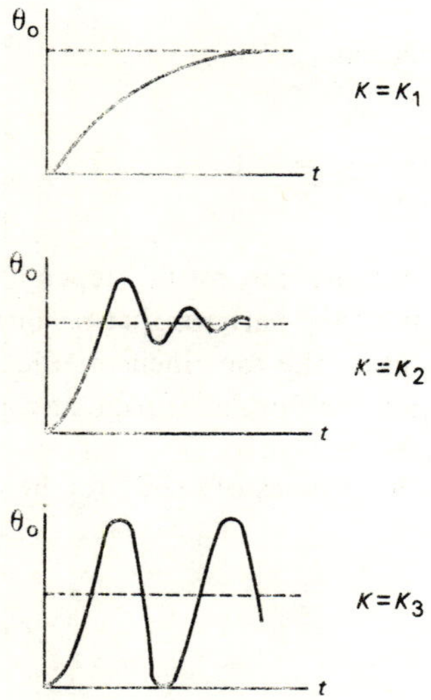

Fig. 11.5 – Step response for pole locations indicated on Fig. 11.4.

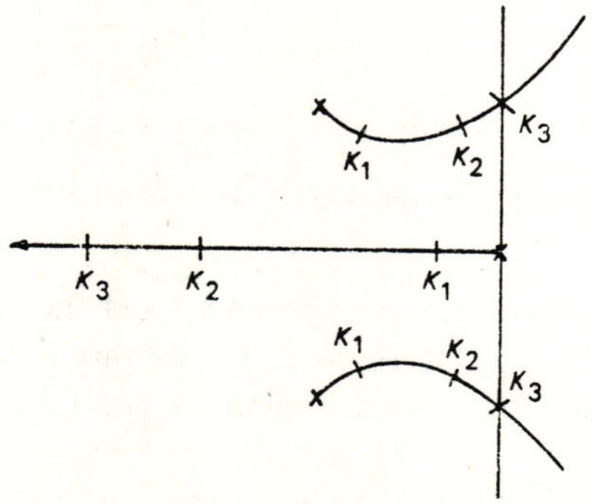

Fig. 11.6 – Root locus for equation (11.31) when damping ratio $c < 1.0$.

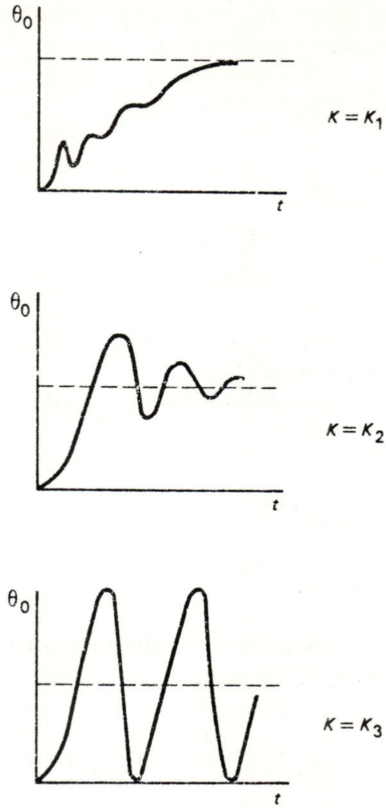

Fig. 11.7 – Step response for pole locations indicated on Fig. 11.6.

11.7 EFFECTS OF LOAD DISTURBANCES

Since it has been indicated that rigidity is of importance in many industrial systems, it is worthwhile examining the dynamic stiffness of a hydraulic servo. The stiffness of importance here is that which resists forces applied at the output when the input is fixed. It will be shown that if this applied force is oscillated harmonically, the stiffness presented to it by the servo loop varies considerably with frequency.

From equation (11.21) for a hydraulic servomotor,

$$C_x \delta x_v - \frac{C_p}{2}(\delta P_1 - \delta P_2) = \frac{V}{4\beta}\frac{d}{dt}(\delta P_1 - \delta P_2) + A\frac{d}{dt}\delta y \qquad (11.32)$$

and for closed loop control:

$$\delta x_v = \delta x_i - \delta y \, .$$

If the input is held fixed at $x_i = 0$, and if the output is forced harmonically by an external force F then since:

$$\delta F = -(\delta P_1 - \delta P_2)A$$

equation (11.32) becomes:

$$-C_x \delta y + \frac{C_p}{2}\frac{\delta F}{A} = -\frac{V}{4\beta A}\frac{d}{dt}\delta F + A\frac{d}{dt}\delta y \,. \tag{11.33}$$

Note that the sign of the force is negative since it is a force applied to the system and not one supplied by the system.

For a harmonic excitation the derivatives in (11.33) can be replaced by $j\omega$ and hence:

$$\frac{\delta F}{\delta y}(\omega j) = \frac{C_x + A\omega j}{\dfrac{C_p}{2A} + \dfrac{V}{4\beta A}\omega j} \tag{11.34}$$

where $\delta F/\delta y$ is a complex number called the **dynamic stiffness.**

Rationalising (11.34) to a single complex form gives:

$$\frac{\delta F}{\delta y} = \frac{2A}{C_p^2 + \left(\dfrac{V\omega}{2\beta}\right)^2}\left\{ C_x C_p + \frac{VA}{2\beta}\omega^2 + \left(C_p A - \frac{C_x V}{2\beta}\right)\omega j \right\} . \tag{11.35}$$

If $C_p/C_x = V/2\beta A$, the imaginary part of (11.35) becomes zero and the servo excited in this manner will become unstable. This is identical to the results shown in section 11.4.

For very low frequencies of operation (11.35) reduces to:

$$\frac{\delta F}{\delta y} \approx 2A\frac{C_x}{C_p} = k_v \,. \tag{11.36}$$

Care must be exercised in extending this result to zero frequency since the valve might be slightly overlapped and the resultant stiffness would be that of the oil locked in the cylinder.

For very high frequencies, on the other hand, equation (11.35) reduces to:

$$\frac{\delta F}{\delta y} = \frac{4\beta A^2}{V} = k_o \tag{11.37}$$

which is just the stiffness of the oil column in the cylinder.

The condition for stability for the servo was shown in section 11.4 to be $k_o/k_v > 1.0$, which implies that the lower the valve stiffness the more stable the unit. This, however, conflicts with the static stiffness requirement given in (11.36). To obtain both these conditions at an acceptable level would require the use of additional feedback loops.

11.8 DESIGNING A HYDRAULIC SERVO

Much development time and expense can be saved by an appreciation at the design stage of the effect of a low natural frequency on servo performance. The natural frequency of the system is closely related to the bandwidth, which in turn determines the speed of response of the system. Hence the higher the natural frequency, the faster the system response. Since the natural frequency is dependent on both inertia and stiffness, both of these should be kept to an optimum value. However, even an intensive design effort will be limited by the stiffness-weight characteristics of the materials available and by the hydraulic fluid bulk modulus.

In a response test on a servo system, frequencies of the same order as the resonant frequencies contribute large angles of lag, and any attempt to correct this by increasing the servo gain only accentuates the resonant peak. However, if the resonant frequency can be kept outside the working bandwidth the situation is much improved.

The problem of designing for a high natural frequency must be approached simultaneously from several different aspects. The structure itself must have minimum mass, a feature which also helps to minimise the kinetic energies to be dissipated in the event of a runaway failure. In addition the structure should be designed for maximum stiffness; in this respect box type structures sometimes offer greatest possibilities.

In order to increase the stiffness of the oil column, the actuator should be designed for maximum piston area and minimum stroke.

An example of the assessment of stiffness for the arrangement shown in Fig. 11.9 can now be developed. It is assumed that the push rods and rocker arm of the servo motor shown in Fig. 11.8, are stiff and the main contribution to flexibility is from the output shaft.

The stiffness of a shaft in torsion is given by:

$$\frac{T}{\theta} = \frac{GI}{l}$$

where
T = applied torque
θ = angular deflection
G = modulus of rigidity (85 GN/m^2)
I = polar moment of inertia
l = lengths of the sections.

For the first section of Fig. 11.8:

$$I_1 = \frac{\pi}{2}\left(\frac{d}{2}\right)^4 = \frac{3.14}{2} \times (0.0095)^4 = 1.278 \times 10^{-8}\,\text{m}^4$$

where d = section diameter.

Hence:

$$\left(\frac{T}{\theta}\right)_1 = \frac{85 \times 10^9 \times 1.278 \times 10^{-8}}{0.11} = 9.875 \times 10^3\ \text{Nm/rad}\,.$$

Also for the second section:

$$I_2 = \frac{3.14}{2} \times (0.013)^4 = 4.484 \times 10^{-8}\,\text{m}^4$$

hence:

$$\left(\frac{T}{\theta}\right)_2 = \frac{85 \times 10^9 \times 4.484 \times 10^{-8}}{0.05} = 7.622 \times 10^4\ \text{Nm/rad}\,.$$

Since these two stiffnesses are in series, the overall mechanical stiffness is:

$$\left(\frac{T}{\theta}\right)_\text{T} = \left\{\left(\frac{\theta}{T}\right)_1 + \left(\frac{\theta}{T}\right)_2\right\}^{-1} = 8.743 \times 10^3\ \text{Nm/rad}\,.$$

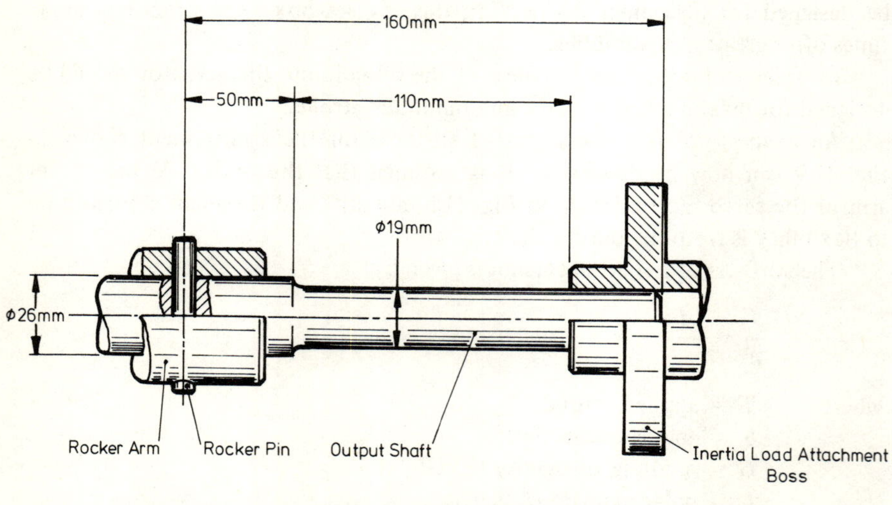

Fig. 11.8 – Mechanical connection between actuator and load.

The oil stiffness can be assessed using the dimensions shown for the piston arrangement of Fig. 11.9.

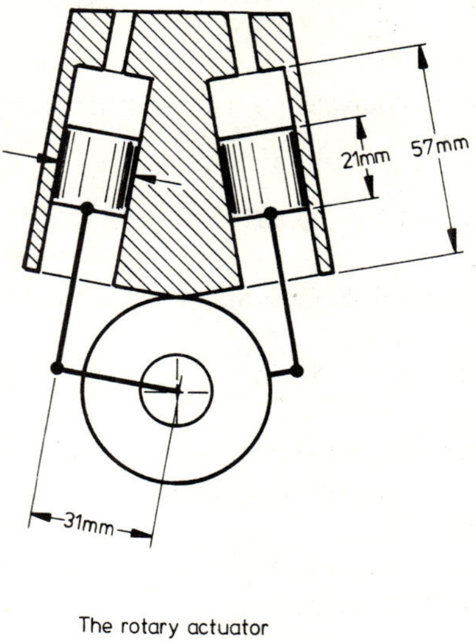

Fig. 11.9 – The rotary actuator.

The volume of the piston is:

$$0.021 \times (0.022)^2 \times \frac{3.14}{4} = 7.978 \times 10^{-6}\,\mathrm{m}^3\,,$$

and the volume of the cylinder is:

$$(0.022)^2 \times 0.057 \times \frac{3.14}{4} = 2.166 \times 10^{-5}\,\mathrm{m}^3\,.$$

The resultant swept volume of the cylinder is therefore:

$$2.166 \times 10^{-5} - 7.978 \times 10^{-6} = 1.368 \times 10^{-5}\,\mathrm{m}^3\,.$$

The oil stiffness k_o is given by (11.24):

$$k_o = \frac{4\beta A^2}{V}\,.$$

Taking a value of $\beta = 1.38\ \text{GN/m}^2$ for industrial hydraulic oil:

$$k_o = \frac{4 \times 1.38 \times 10^9 \times 1.44 \times 10^{-7}}{1.368 \times 10^{-5}} = 5.81 \times 10^7\ \text{N/m}\,.$$

Converting this into an equivalent torsional stiffness by introducing the torque arm of 0.031 m:

$$k_o = 5.81 \times 10^7 \times (0.031)^2 = 5.58 \times 10^4\ \text{Nm/rad}\,.$$

Since the oil and structural stiffness are in series, the overall stiffness is given by:

$$k_T = 7.558 \times 10^3\ \text{Nm/rad}\,.$$

Oil volume in the passages between the control valve and the pistons have been ignored, since the volumes of these passages should be very small compared with the swept volume.

In this particular example, the load inertia is $0.9744\ \text{kgm}^2$; hence the natural frequency can be estimated as:

$$f_n = \frac{1}{2\pi}\sqrt{\frac{7.558 \times 10^3}{0.9744}} = 14.02\ \text{Hz}\,.$$

If the natural frequency were based solely on the oil stiffness, then its value would be significantly higher, at 38.1 Hz. This would lead the designer to think he had a much faster response (wider bandwidth) than actually existed. Hence the stiffness of the mechanical structure can have a significant effect on the final performance.

11.9 PERFORMANCE ASSESSMENT

For this particular system the specification is given as follows:

Supply pressure = 120 bar
Output load inertia = $0.9744\ \text{kgm}^2$
Required steady state rate = 600 deg/s
Required response ±0.75 deg at 15 Hz
Viscous load = 0.569 Nms/rad
Load torque = 20.3 Nm.

From these data the engineer can assess the required natural frequency, piston sizes, and information regarding the choice of the control valve. The process of assessment must follow a number of paths; the results are then compared in order to choose the correct system parameters.

The following method can be used to determine the piston area in relation to the frequency response requirements. Assuming sinusoidal excitation of the output, then:

$$\theta_o = \hat{\theta}_o \sin \omega t$$

where $\hat{\theta}_o$ is the maximum output amplitude required at a frequency of ω rad/s. Double differentiation of this equation will produce the expression for output acceleration:

$$\frac{d^2\theta_o}{dt^2} = -\hat{\theta}_o \omega^2 \sin \omega t$$

giving:

$$\text{Maximum acceleration} = -\hat{\theta}_o \omega^2 \, .$$

From the specification the values for $\hat{\theta}_o$ and ω can be substituted, and hence the maximum acceleration required to meet the performance can be determined:

$$\left(\frac{d^2\theta_o}{dt^2}\right)_{max} = 0.75 \times (2 \times \pi \times 15)^2 = 6655 \text{ deg/s}^2 \, .$$

Since the inertia of the output load is known, the necessary torque required to be produced by the actuator can be calculated.

$$\text{Torque} = J \frac{d^2\theta_o}{dt^2} = \frac{0.9744 \times 6655\pi}{180} = 113.2 \text{ Nm} \, .$$

To this figure an additional 10 per cent is added as an estimate of internal friction, seals, etc., hence:

$$\text{Required torque} = 124.5 \text{ Nm} \, .$$

The supply pressure is given as 120 bar, and therefore the piston area can be calculated:

$$\text{Piston area required} = \frac{124.5}{0.031} \times \frac{1}{12 \times 10^6} = 3.35 \times 10^{-4} \text{ m}^2 \, .$$

This calculation assumes that all the supply pressure is available for acceleration. The actual piston area from Fig. 11.9 is 3.799×10^{-4} m^2 so there is plenty in hand.

Having considered the conditions necessary to produce adequate acceleration to satisfy the frequency response, it is now necessary to look at the conditions required to satisfy the velocity requirements.

At steady velocity the force to be overcome is:

$$\text{Force} = \frac{20.3 + 0.569 \times 600/57.3}{0.031} = 847.0\,\text{N}\,.$$

For most efficient work conditions it can be shown that two-thirds of the supply pressure should be used in this case (section 9.6); hence:

$$\text{Piston area required} = \frac{847}{12 \times 10^6 \times \dfrac{2}{3}} = 1.06 \times 10^{-4}\,\text{m}^2\,.$$

Clearly the acceleration requirement provides sufficient piston area.

The next step is to determine the maximum flow requirement from the steady rate condition so that the valve can be designed. For the steady rate condition:

$$\frac{\mathrm{d}\theta_o}{\mathrm{d}t} = 600\ \text{deg/s} = 10.42\ \text{rad/s}\,.$$

The linear piston rate will be:

$$\frac{\mathrm{d}x}{\mathrm{d}t} = r\frac{\mathrm{d}\theta_o}{\mathrm{d}t} = 0.031 \times 10.42 = 0.323\,\text{m/s}$$

and hence the maximum flow rate must be:

$$Q_{\text{max}} = A\frac{\mathrm{d}x}{\mathrm{d}t} = 0.323 \times 3.35 \times 10^{-4} = 1.08 \times 10^{-4}\,\text{m}^3/\text{s}\,.$$

The maximum oscillatory rate required is:

$$\left(\frac{\mathrm{d}\theta_o}{\mathrm{d}t}\right)_{\text{max}} = 0.75 \times 2 \times 3.14 \times 15 = 70.65\ \text{deg/s}$$

$$\left(\frac{\mathrm{d}x}{\mathrm{d}t}\right)_{\text{max}} = 0.031 \times \frac{70.7}{57.3} = 3.82 \times 10^{-2}\,\text{m/s}$$

$$Q_{\text{max}} = 3.82 \times 10^{-2} \times 3.799 \times 10^{-4} = 1.45 \times 10^{-5}\,\text{m}^3/\text{s}\,.$$

This figure of flow rate is much less than that required by the steady rate condition.

11.10 CHOICE OF CONTROL VALVE

The simplest single stage valve for controlling both the direction of the flow and the flow rate is the three-port valve, but this will control the flow to only one side of the load, and a differential area cylinder must be used, Fig. 11.10. A more common form of the single stage valve is the five-port valve of Fig. 11.11.

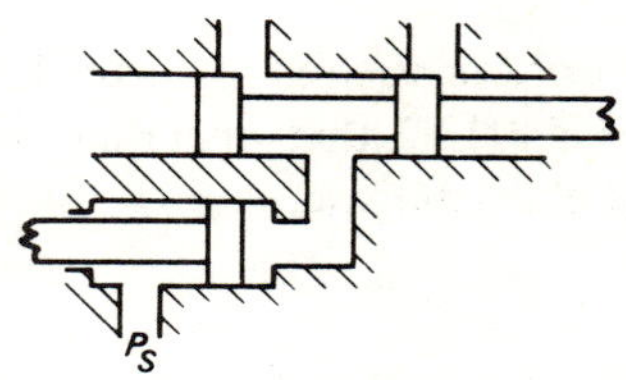

Fig. 11.10 – Three-port valve and single acting cylinder.

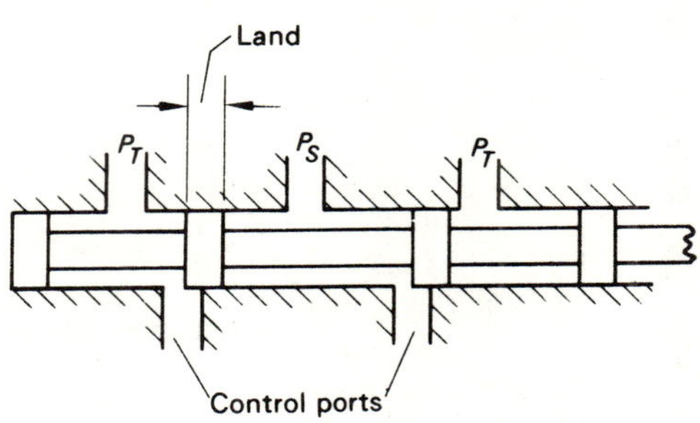

Fig. 11.11 – Five-port servovalve.

The valve is made up of a spool, the sliding portion and a liner or sleeve. Two sets of orifices are bored symmetrically around the sleeve, together with three other sets of holes to provide passages for the inlet pressure and the outlet exhaust to tank. The orifices must be symmetrically arranged around the sleeve. There should not be less than two although three orifices are preferable. This is to ensure that the pressure is distributed evenly around the spool, reducing stiction effects.

Fluid is metered through the control valve ports when the spool uncovers a segment of the orifice. This uncovered area is non-linear with respect to spool displacement (see section 4.3.2).

An example of this is shown in Fig. 11.12 in which the uncovered area is calculated from the relationship:

$$a = \frac{d^2}{8}(\theta - \sin\theta)$$

$$\cos\frac{\theta}{2} = \frac{d - 2x_v}{d}$$

where a = exposed orifice area
d = orifice diameter
x_v = spool displacement.

In this case there are three orifices, each of diameter 1.93 mm.

It can be seen from Fig. 11.12 that operating the valve at larger displacements helps to linearise the characteristic.

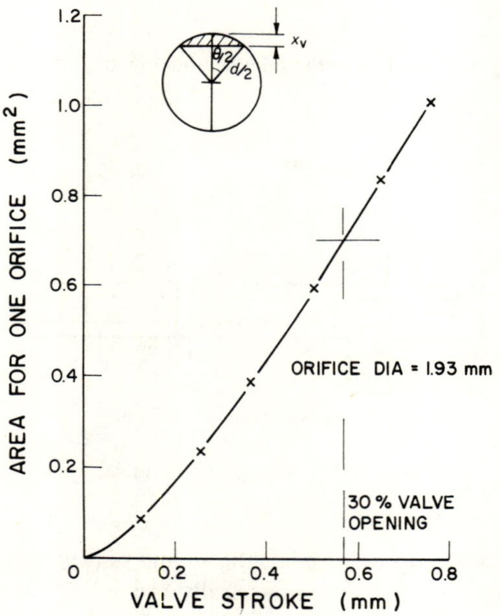

Fig. 11.12 – Area-stroke characteristics of a segmented orifice.

The section of the spool which controls the uncovering of the orifice area is termed the **land**. A valve has **underlap** or **overlap** if the land is smaller or larger respectively than the orifice diameter, when the spool is central.

An overlapped valve saves fluid loss when the spool is central but tends to reduce the accuracy of the servo. Its use also tends to reduce servo stability.

An underlapped valve gives much more sensitive control at the expense of fluid loss. Its use tends to improve the servo stability. Although in most cases the ideal valve would have zero lap, in practice this is difficult to achieve without the use of very expensive production techniques. Typical flow characterisitics are shown in Fig. 11.13.

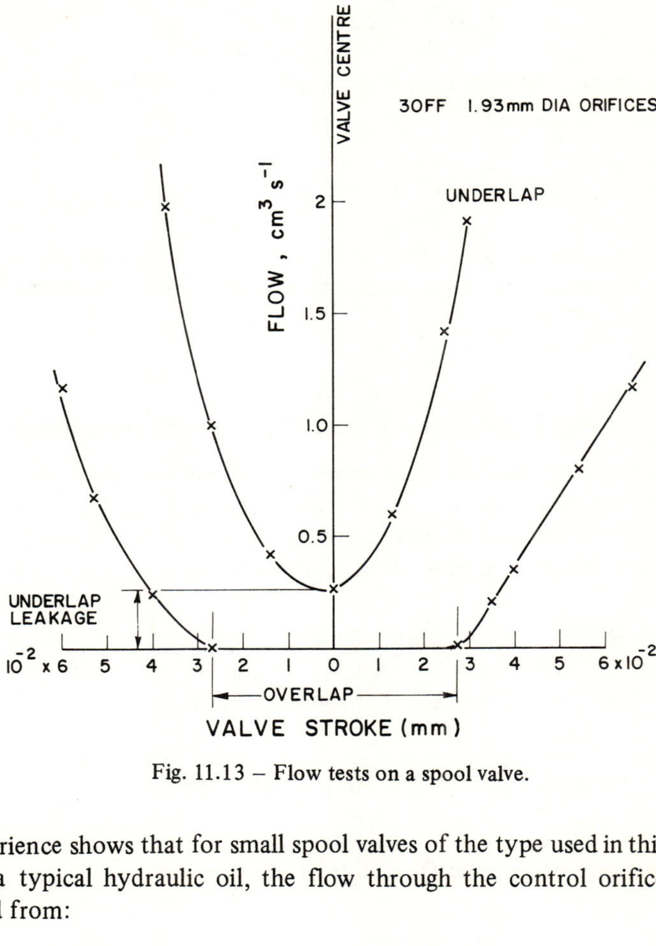

Fig. 11.13 – Flow tests on a spool valve.

Experience shows that for small spool valves of the type used in this example and for a typical hydraulic oil, the flow through the control orifices can be calculated from:

$$Q = 3.12 \times 10^{-2} a\sqrt{(P_s - P_1)}\,.$$

In the example, the maximum flow required is 1.08×10^{-4} m^3/s and for optimum conditions (section 9.6) the pressure drop across one set of orifices is:

$$P_s - P_1 = \frac{1}{3} \times \frac{1}{2} \times 12 \times 10^6 = 20 \text{ bar}\,.$$

This takes into account the maximum efficiency criterion, that is, $\frac{1}{6}P_s$ drop across each of the valve's two orifice sets, leaving $\frac{2}{3}P_s$ for driving the load. Hence the total required orifice area is:

$$a_T = \frac{1.08 \times 10^{-4}}{3.12 \times 10^{-2} \times \sqrt{(2.0 \times 10^6)}} = 0.82\,\text{mm}^2 .$$

Assuming three orifices are used in each control port set then, for each orifice:

$$a = \frac{1}{3} \times 2.45 \times 10^{-6} = 0.82\,\text{mm}^2.$$

This area should be available when the orifice opening is about 0.3 to 0.4 times the diameter. Fig. 11.13 shows that three 1.93 mm diameter orifices are adequate.

11.11 A METHOD FOR EVALUATING FREQUENCY RESPONSE

It was necessary in section 11.3 to introduce a linearising process which resulted in the pressure gain term C_p and the flow gain term C_x. Fig. 11.2 shows that these two parameters vary considerably in value, and it is difficult to select a reasonable set of values in order to estimate the frequency response at the design stage.

In this section a semi-empirical mathematical model is proposed to overcome these difficulties for the designer [6]. From Fig. 11.1(a) and equation (11.4), the load flow Q_L through the valve is given by:

$$Q_L = \frac{1}{2} C_q a \sqrt{\frac{2}{\rho}} \left\{ \sqrt{(P_s - P_1)} + \sqrt{(P_2 - P_t)} \right\} \tag{11.38}$$

which was shown in equation (11.5) to reduce to:

$$Q_L = C_q a \sqrt{\frac{2}{\rho}} \sqrt{\frac{P_s - P_L}{2}} . \tag{11.39}$$

In section 11.3 it was shown that when compressibility is taken into account, the load flow Q_L through the cylinder is given by:

$$Q_L = A \frac{dy}{dt} + \frac{V}{4\beta} \frac{d}{dt} (P_1 - P_2) . \tag{11.40}$$

Finally, the load equation is:

$$P_L = P_1 - P_2 = \frac{M}{A}\frac{d^2y}{dt^2} \, . \tag{11.41}$$

Combining equations (11.39), (11.40) and (11.41) gives the open loop relationship:

$$C_q a \sqrt{\frac{2}{\rho}}\sqrt{\frac{P_s}{2}}\sqrt{\left(1 - \frac{M}{P_s A}\frac{d^2y}{dt^2}\right)} = A\left\{\frac{dy}{dt} + \frac{VM}{4\beta A^2}\frac{d^3y}{dt^3}\right\} . \tag{11.42}$$

This equation can be made more general in application, and easier to solve, if it is converted into a non-dimensional form by defining:

Non-dimensional time: $\tau = \omega_n t$

Reference displacement: $\hat{y} = \dfrac{C_q \hat{a}\sqrt{\dfrac{P_s}{2}}\sqrt{\dfrac{2}{\rho}}}{A\omega_n} = y/y^*$

Inertia load factor: $\alpha = \dfrac{MC_q \hat{a}\,\dfrac{P_s}{2}\,\dfrac{2}{\rho}\,\omega_n}{P_s A^2}$

Reference valve area: $\hat{a} = a/a^*$.

Substituting these values into (11.42) gives the open loop equation in non-dimensional form:

$$x_v^* \left(1 - \alpha\frac{d^2y^*}{d\tau^2}\right) = \frac{d^3y^*}{d\tau^3} + \frac{dy^*}{d\tau} \, . \tag{11.43}$$

where $a^* = x_v^{*n}$ and $n = 1.0$. The factor n is different for round, square, and annular orifices [8].

The relationship between the load factor α and the inherent valve damping can be more easily understood if equation (11.43) is approximated by a linear relationship. Expanding the left-hand side of the equation and neglecting terms greater than first order, gives:

$$x_v^* \left(1 - \frac{1}{2}\alpha\frac{d^2y^*}{d\tau^2}\right) = \frac{d^3y^*}{d\tau^3} + \frac{dy^*}{d\tau} \, . \tag{11.44}$$

This can now be rearranged into a more familiar form, comparable to equation (11.26):

$$y^* = \frac{1}{s\left(s^2 + \dfrac{1}{2}x_v^*\alpha s + 1\right)} x_v^* . \tag{11.45}$$

Inspection of the denominator shows that the damping is a function of the magnitude of the input displacement. This explains why, in practical test results, the frequency response curves of a hydraulic servo change depending on the size of the input signal.

The left-hand side of equation (11.44) is an approximation to the load flow. It can be written in the more general form:

$$Q_L^* = x_v^*(1 - K_L P_L^*), \tag{11.46}$$

Since $P_L^* = \alpha d^2y^*/d\tau^2$ and the coefficient of $\frac{1}{2}$ is replaced by a general constant K_L in (11.46). The characteristics of this equation, in comparison with the true equation, that is equation (11.39), are shown in Fig. 11.14, where it can be seen that $K_L = \frac{1}{2}$ overestimates the true curve and $K_L = 1.0$ underestimates the true curve.

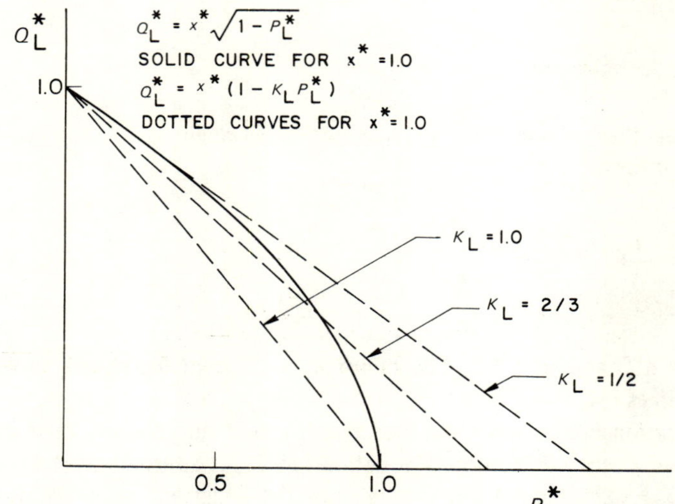

Fig. 11.14 – Estimation of damping characteristics.

A better model can be obtained if it is assumed that the value of K_L is such that the average error must be zero over the range $1.0 < P_L^* < 0$:

$$\int_0^{1.0} (\text{error})\,dP_L^* = 0\,. \tag{11.47}$$

Let the error be defined as:

$$\text{error} = (1 - P_L^*)^{\frac{1}{2}} - (1 - K_L P_L^*)\,.$$

The from (11.47):

$$\int_0^{1.0} [(1 - P_L^*)^{\frac{1}{2}} - (1 - K_L P_L^*)]\,dP_L^* = 0$$

giving:

$$K_L = \frac{2}{3}.$$

It can be seen from Fig. 11.14 that this gives a better estimation to the true load flow-pressure characteristic. Equation (11.45) now becomes:

$$y^* = \frac{1}{s\left(s^2 + \frac{2}{3}\alpha x_v^* s + 1\right)} x_v^* \tag{11.48}$$

indicating the damping ratio estimated at $\frac{1}{3}x_v^*\alpha$.

To estimate the frequency response, let:

$$x_v^* = \hat{x}_v^* \sin \omega^* \tau$$

which yields, for the $x_v^* = 1.0$ case:

$$|y^*| = \frac{1}{\omega^* \sqrt{\left\{\left(1 - \omega^{*2}\right)^2 + \left(\frac{2}{3} x_v^* \alpha \omega^*\right)^2\right\}}} \tag{11.49(a)}$$

with a phase lag of:

$$\phi = \tan^{-1}\left\{\frac{\frac{2}{3} x_v^* \alpha \omega^*}{1 - \omega^{*2}}\right\} + \frac{\pi}{2}. \tag{11.49(b)}$$

These results are shown in Fig. 11.15, where a frequency response calculated from (11.49(a) and (b)), is compared with a digital computer solution using the true equation given by (11.43). Two inertia load cases are shown.

The approximate model defined by equations (11.49(a) and (b)) has the following advantages over Harpur's original model:

(a) Frequency responses can be hand-calculated for a family of servo systems.
(b) The problems of trying to estimate realistic values for C_x and C_p are simplified.
(c) The estimate of the inherent damping due to fluid flow through the valve is more realistic.
(d) The effect of valve displacement on the system response can be estimated.

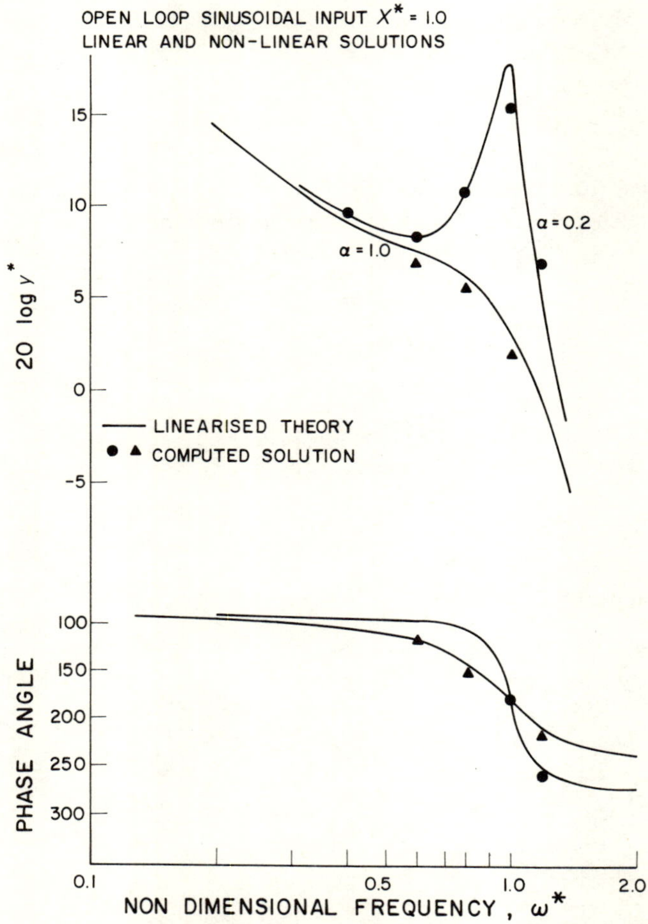

Fig. 11.15 – Comparison of linearised theory and computer solution.

11.12 CAVITATION IN THE HYDRAULIC ACTUATOR

The analysis discussed so far assumes that the inflow to the cylinder and the outflow are identical. This is not always the case. For example, with large inertia loading conditions, high transient pressures may be developed in the cylinder as the load is decelerated. On the inflow side of the cylinder the closing valve is not able to supply enough fluid to keep that side full, and cavities start to form. The load is, in fact, overrunning. Hence the outflow side of the cylinder has a high peak pressure and the inflow side has been reduced to the vapour pressure of the oil. The presence of the large peaking pressures may give rise to fatigue failure, and the presence of cavities in the fluid can result in implosion damage. Pressure peaks of up to five times the supply pressure have been recorded experimentally.

The presence of valve overlap [11] which is sometimes purposely employed to reduce leakage across the valve, or may be present owing to manufacturing tolerances, can increase the pressure peaks considerably. On the other hand, the compressibility of the fluid helps to reduce the magnitude of the peaks. It can be shown that the worst condition occurs when the valve is within the lap region and the load is decelerating. The fluid is then trapped in the cylinder by the valve lap and absorbs the kinetic energy of the moving mass.

The conditions that give rise to cavitation can be estimated by developing the linear equations for the actuator.

Equation (11.22) can be written:

$$\delta x_v = a\frac{d^3}{dt^3}\delta y + b\frac{d^2}{dt^2}\delta y + f\frac{d}{dt}\delta y\,. \tag{11.50}$$

This third-order differential equation relates the actuator and valve displacements. However, it is required to find the conditions for cavitation, and to do this, the relationship between actuator pressure and valve movement is required. Since the inflow and outflow are assumed equal up to cavitation:

$$Q_1 = Q_2\,.$$

Hence, from equations (11.16) and (11.17):

$$\delta Q_1 = C_x\delta x_v - C_p\delta P_1$$

$$\delta Q_2 = C_x\delta x_v + C_p\delta P_2$$

then, from Fig. 11.2:

$$\delta P_1 + \delta P_2 = 0 \tag{11.51}$$

and hence from (11.15):

$$\delta P_1 = \frac{M}{2A}\frac{d^2}{dt^2}\delta y\,. \tag{11.52}$$

Combining equations (11.52) and (11.50) and introducing the operator s:

$$\frac{2A}{M}\frac{\delta P_1}{\delta x_v} = \frac{s}{as^2 + bs + f}\,. \tag{11.53}$$

The natural frequency of the load and oil column system is:

$$\omega_n^{\,2} = \frac{f}{a} = \frac{4\beta A^2}{MV}\,.$$

If the valve is oscillated at frequency ω and amplitude $\delta\hat{x}_v$, it can be shown that the amplitude of the pressure waveform from equation (11.53) will be:

$$\delta P_1 = \left\{\frac{2A^2}{MK_1\omega_n(P_s - P_1)}\right\}^{-1}\frac{\omega^*\delta\hat{x}_v}{\sqrt{\{(1-\omega^{*2})^2 + 4c^2\omega^{*2}\}}} \tag{11.54}$$

where $\omega^* = \omega/\omega_n$,

and the damping ratio:

$$c = \frac{MC_p\omega_n}{4A^2} = \frac{MK\omega_n x_{vr}}{8A^2\sqrt{(P_s - P_1)}}\,. \tag{11.55}$$

Let $\delta P_1^* = \delta\hat{P}_1/P_s$, then equation (11.53) reduces to:

$$\delta P^* = \alpha\sqrt{\left\{\frac{1}{2}\left(1 - \frac{P_{1r}}{P_s}\right)\right\}}\left\{\frac{\omega^*}{\sqrt{\{(1-\omega^{*2})^2 + 4c^2\omega^{*2}\}}}\right\} \tag{11.56}$$

where the load factor:

$$\alpha = \frac{MK_1\omega_n\delta\hat{x}_v\sqrt{(P_s/2)}}{P_sA^2}$$

and subscript r refers to the steady state condition.

Considering the particular case $x_{vr} = 0$, that is, the valve oscillating about the centre position:

$$P_{1r} = [P_s/2] \, .$$

From equation (11.55), $c = 0$, then:

$$\delta P_1^* = \frac{\alpha\omega^*}{2|1-\omega^{*2}|} \, . \qquad (11.57)$$

When the valve is oscillating at the natural frequency, equation (11.57) indicates that the actuator pressure amplitude is infinite. At both zero and infinite excitation frequency the pressure δP_1 is zero. Cavitation arises when δP_1 exceeds $\frac{1}{2}P_s$. For this case equation (11.51) shows that δP_2 will be less than $-\frac{1}{2}P_s$ and the total pressure on the exhaust side of the actuator is:

$$P_{2r} + \delta P_2 = \tfrac{1}{2}P_s - \tfrac{1}{2}P_s = 0 \, .$$

The critical value of α for cavitation occurrence is found from (11.57) letting $\delta\hat{P}_1^* = 0.5$:

$$\alpha_c = \frac{|1-\omega^{*2}|}{\omega^*} \, . \qquad (11.58)$$

These results are shown in Fig. 11.16 (heavy line).

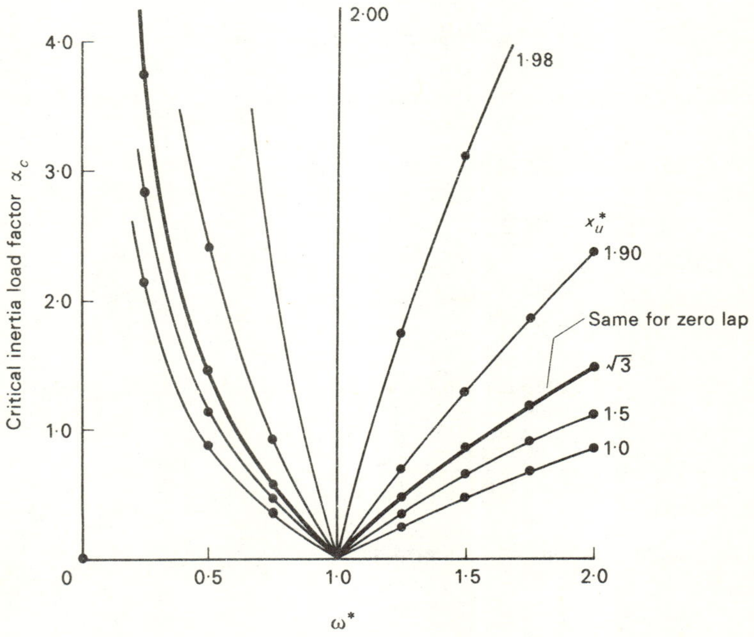

Fig. 11.16 – Critical load factor α_c, above which cavitation occurs.

It is clear from the results that it is difficult to avoid cavitation in the region of the natural frequency. This result is confirmed in the more detailed analysis in Chapter 12, where it is shown that damping tends to improve the situation.

If the fluid in considered to be incompressible, the bulk modulus β is infinite and hence α must be reduced, since the natural frequency is infinite:

$$\alpha_i = \alpha\omega^* = \frac{MK_1\,\delta x\,\sqrt{(P_s/2)}\,\omega}{P_s A^2}$$

and equation (11.58) becomes:

$$\frac{\alpha_{ic}}{\omega^*} = \frac{1-\omega^{*2}}{\omega^*}$$

$$\alpha_{ic} = 1.0 \text{ since } \omega^* = 0\,.$$

11.13 EFFECT OF UNDERLAP ON CAVITATION

For the underlapped valve, shown in Fig. 11.17, operating within the lap, x_v is less than the underlap x_u:

$$Q_1 = K_1(x_v + x_u)\sqrt{(P_s - P_1)} - K_1(x_u - x_v)\sqrt{P_1}\,. \tag{11.59}$$

For small oscillations about valve central position, $x_{vr} = 0$ and $P_{1r} = P_s/2$, giving:

$$Q_1 = C_x\delta x_v - C_p\delta P_1$$

where

$$C_x = 2K_1\sqrt{(P_s/2)}$$

$$C_p = \frac{-K_1 x_u}{\sqrt{(P_s/2)}}\,.$$

From these values, the pressure amplitude can be found as before, giving:

$$\delta\widehat{P}_1^* = \frac{\alpha\omega^*}{\sqrt{\{(1-\omega^{*2})^2 + 4c^2\omega^{*2}\}}}\,. \tag{11.60}$$

It is easily shown that the damping ratio is:

$$c = \frac{MC_p\omega_n}{4A^2} = \frac{\alpha}{2}x_u^*$$

where the non-dimensional underlap $x_u^* = \hat{x}_u/\delta x_v$. For operation within the underlap $x_u^* > 1$ and equation (11.60) becomes:

$$\delta \hat{P}_1^* = \frac{\alpha \omega^*}{\sqrt{\{(1-\omega^{*2})^2 + \alpha^2 x_u^{*2} \omega^{*2}\}}}. \tag{11.61}$$

At first glance one would expect this equation to reduce to equation (11.57) when the underlap x_u^* is reduced to zero. This is not the case since it should be borne in mind that for the case considered, x_u^* must exceed unity.

The pressure amplitudes of equations (11.57) and (11.61) are compared in Fig. 11.17. Underlap is effective in reducing the pressure peaks in the region of the natural frequency. When $\hat{P}_1^*$ exceeds unity, $\hat{P}_2^*$ becomes negative and cavitation occurs.

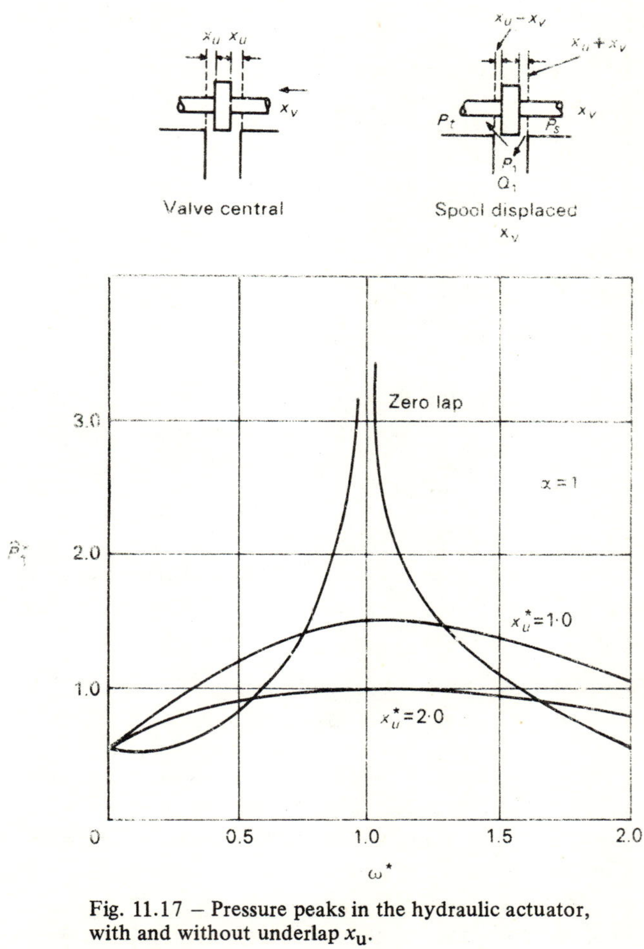

Fig. 11.17 – Pressure peaks in the hydraulic actuator, with and without underlap x_u.

The critical load factor α_c can also be calculated. If cavitation occurs when $\delta\widehat{P}_1^* = 0.5$, then:

$$\alpha_c = \frac{0.5\sqrt{\{(1-\omega^{*2})^2 + \alpha_c^2 x_u^{*2}\omega^{*2}\}}}{\omega^*}. \tag{11.62}$$

These results are also shown in Fig. 11.16. If $x_u^* > 2$ cavitation does not occur at any frequency. It is interesting to note that the boundary for $x_u^* = \sqrt{3}$ coincides with that for zero lap.

11.14 EXAMPLES

1. A valve controlled hydraulic cylinder with a stroke of 150 mm and useful piston area of 645 mm^2 is used to control the velocity of a mass of 25 kg. There is viscous friction present defined by the coefficient 2.6×10^3 Ns/m.

   (a) Determine the natural frequency of the system when using oil of density 858 kg/m^3.

   The following parameters were measured experimentally:

   $$C_p = 9600 \text{ mm}^5/\text{Ns}$$

   $$\beta_{oil} = 1400 \text{ N/mm}^2.$$

   (b) Determine the damping ratio of the system.

   (*Answers*: (a) $f_n = 111$ Hz; (b) $c = 0.17$.)

2. A specification for a hydraulic control system requires that the output of the servo shall be able to reach an acceleration of 40 rad/s^2 when the output amplitude is 5 deg, If the load inertia is 6.74 kg/m^2, and if an error of 0.5 radians gives a torque output of 337 Nm, determine:

   (a) If the specification is met.
   (b) What alterations are required to maintain the specification?
   (c) What is the value of the damping coefficient required to keep the overshoot to 10 per cent?

   (*Answers*: (a) No, $\omega_n = 10$ rad/s and should be at least 21.44 rad/s; (b) increase K; (c) damping coefficient increased to 171.68 Ns/m.)

3. The tool of a copy lathe is positioned using a valve-controlled hydraulic cylinder . There is a mechanical feedback linkage between the cylinder and the valve such that the spool displacement is equal to the positional error. The valve spool is 5 mm in diameter, and the valve discharge coefficient is 0.6. Assume an annular control orifice.

   The fluid has a specific gravity of 0.8 and the supply pressure is 19.6 bar.

If the piston diameter is 100 mm, and the feed rate is 5 mm/s, determine:

(a) The diametrical error on a workpiece which is being machined to a tape of 1 mm/mm.

If the mechanical feedback is now replaced by a solenoid drive to the spool, estimate:

(b) The solenoid force required to operate the cylinder at 20 Hz, and at an amplitude of 0.5 mm.

Take the spool mass as 0.05 kg.

(*Answers*: (a) 0.0857 mm; (b) 0.8495 N.)

11.15 REFERENCES

[1] Harpur, N. F. (1953), Some Design Considerations of Hydraulic Servos of Jack Type, *Proc. of a Conf. on Hydraulic Servomechanisms*, Inst. Mech. Engrs. (London), February.

[2] Nightingale, J. M. (1956), Hydraulic Servo Fundamentals, *Machine Design.*

[3] Blackburn, J. F. *et al.* (1960), *Fluid Power*, M.I.T. Press.

[4] Martin, H. R. (1977), *Fundamentals of Fluid Power Control*, A.U.E.W.-T.A.S.S., Surrey, England.

[5] Merritt, H. E. (1967), *Hydraulic Control Systems*, Wiley.

[6] Martin, H. R. and Keating, T. (1974), Calculating Hydraulic Servo Freguency Response. *Hydraulics and Pneumatics Mag.*, July.

[7] Martin, H. R. and Keating, T. (1971), Response Testing of a Hydraulic Servomechanism with Mechanical Feedback. *B.H.R.A 2nd Int. Conf. on Fluid Power.* University of Surrey.

[8] Montgomery, J. and Martin, H. R. (1972), *The Effect of Valve Area Gain on the Performance of the Hydraulic Servomechanism*, A.S.M.E. Winter Meeting, November, paper No. 72WA/Flcs-7.

[9] Keating, T. and Martin, H. R. (1973), Mathematical Models for the Design of Hydraulic Actuators, *Trans. I.S.A.*, **12**.

[10] Martin, H. R. and Carrington, J. (1964), Designing the Servo Drive for a Rocking Table, *Hydraulic and Pneumatics Mag.*, July.

[11] McCloy, D. (1965), Pressure Peaks in the Hydraulic Actuator, *Trans. A.S.M.E., J. Basic Eng.*

CHAPTER 12

Electro-hydraulic servomechanisms

12.1 SINGLE STAGE HYDRAULIC VALVE

The increasing demands from modern control systems necessitate greater flexibility between controlling elements. This has resulted in the combining of electrical signalling with fluid power actuators. The need to change from electrical to hydraulic control at some point in the loop has led to the development of an electric torque motor to actuate the hydraulic servo valve. The torque motor receives its inputs from an electronic amplifier.

A block diagram of a typical servo loop is shown in Fig. 12.1. The output torque T_o of the motor is a function of two variables.

$$T_o = K_g \Delta i - K_s \theta_r \tag{12.1}$$

where K_g = torque motor gain (Nm/A)
Δi = input differential current (A)
K_s = torque motor stiffness (Nm/rad)
θ_r = torque motor shaft deflection (rad).

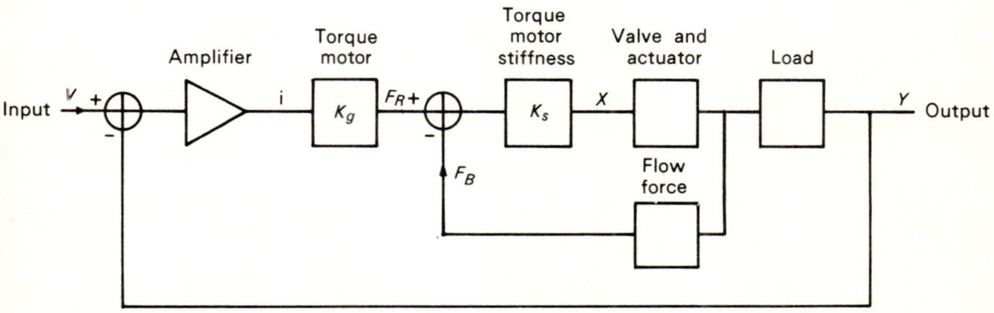

Fig. 12.1 – Block diagram of an electro-hydraulic servo.

This characteristic complicates the problem, since the valve movement is determined by the interaction of the torque motor force F_R which tries to open the valve, and the flow force which tries to close it. This interaction will be discussed in more detail later. On the other hand, the servo amplifier can deal with a variety of signals, such as those generated by electrical compensation networks.

The torque motor is an important component in the electro-hydraulic servomotor, since it converts the error signal into mechanical movement at the spool valve. This is achieved by the use of a simple permanent magnet motor with limited rotational movements. In this type of torque motor the differential current across coils produces a resultant torque and movement. The characteristic of this device is therefore dependent on two variables, and the output torque T_o is given by equation (12.1).

A plot of some typical characteristics is shown in Fig. 12.2. It is seen that for any particular input differential current, the torque motor shaft position is

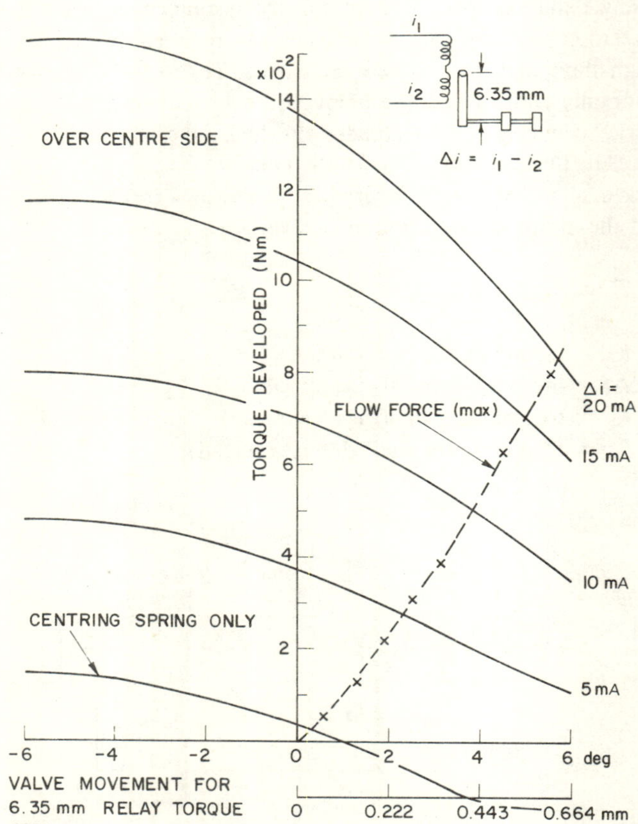

Fig. 12.2 – Torque motor characteristics.

dependent on the external load at the valve assembly. The external loading T_L consists of flow forces, friction and centring spring forces.

In practice there would be typically 10 mA present in each coil of the torque motor so the $\Delta i = 5$ mA curve in Fig. 12.2 means:

$$\Delta i = i_1 - i_2 = (12.5 - 7.5)\,\text{mA} = 5\,\text{mA}$$

but

$$i_1 + i_2 = \text{const.}$$

If the torque motor arm is held at centre, this differential current would develop 3.7×10^{-2} Nm of torque. If the torque arm were forced in the wrong direction to $-6°$, then 4.8×10^{-2} Nm would oppose this effort. If the torque motor is left free, it would hit the $+6°$ stop and produce 1.15×10^{-2} Nm of torque against the stop.

To see the interaction with the valve, the maximum flow force developed across a spool with two sets of 3×1.930 mm diameter control orifices (see Chapter 11) is plotted. The spool valve is assumed to have a pressure drop of 140 bar across the two sets of control orifices. This can limit the valve opening for any particular current differential.

An approximation to the dynamic performance of the torque motor can be found by summing the torques acting on such a system. Apart from having to accelerate the rotor of inertia J, the torque developed by the action of a differential current must overcome the viscous damping term with coefficient f, the relay stiffness and the external load torque.

Hence

$$J\frac{\mathrm{d}^2\theta_r}{\mathrm{d}t^2} + f\frac{\mathrm{d}\theta_r}{\mathrm{d}t} + k_s\theta_r + T_L = k_g i$$

giving a transfer function, if $T_L = 0$, of:

$$\theta_r = \left[\frac{k_g/J}{s^2 + (f/J)s + k_s/J}\right] i\,. \tag{12.2}$$

A typical torque motor would have the following parameters:

Rotor inertia (J)	$= 5.054 \times 10^{-7}$ kgm^2
Damping coefficient (f)	$= 1.356 \times 10^{-4}$ Nm s/rad
Gain (k_g)	$= 13.82$ Nm/A
Stiffness (k_s)	$= 2.657 \times 10^{-1}$ Nm/rad
Back e.m.f. (k_e)	$= 0.46$ V.s/rad
Coil resistance (R)	$= 10\,k\Omega$ each coil
Coil inductance (L)	$= 10$ H at 750 Hz .

Hence, the natural frequency of the torque motor is:

$$f_n = \frac{1}{2\pi}\sqrt{\frac{k_s}{J}} = \frac{1}{2\pi}\sqrt{\frac{2.657 \times 10^{-1}}{5.054 \times 10^{-7}}} = 115.45 \text{ Hz}$$

and the damping ratio:

$$c = \frac{f}{J} \times \frac{1}{2\omega_n} = \frac{f}{2\sqrt{(Jk_s)}} = 0.185.$$

The bandwidth of this component would not normally limit the response of most hydraulic systems.

Although the torque motor and amplifier have acceptable bandwidths as separate components, it is often the case that in combination severe limitations on the overall frequency response can occur if matching of the components is ignored.

The amplifier is effectively a voltage generator driving current through its own impedance and the combined impedance, resistive and inductive, of the torque coils. Thus the amplifier must drive the current against the back e.m.f. developed in the torque motor. Owing to the magnitude and phase relationship of the back e.m.f. this voltage increases with frequency and amplitude, tending to cancel the operating voltage of the servo amplifier. This results in attenuation of the output. High impedance output values and current feedback in the servo amplifier are effective methods of overcoming these problems.

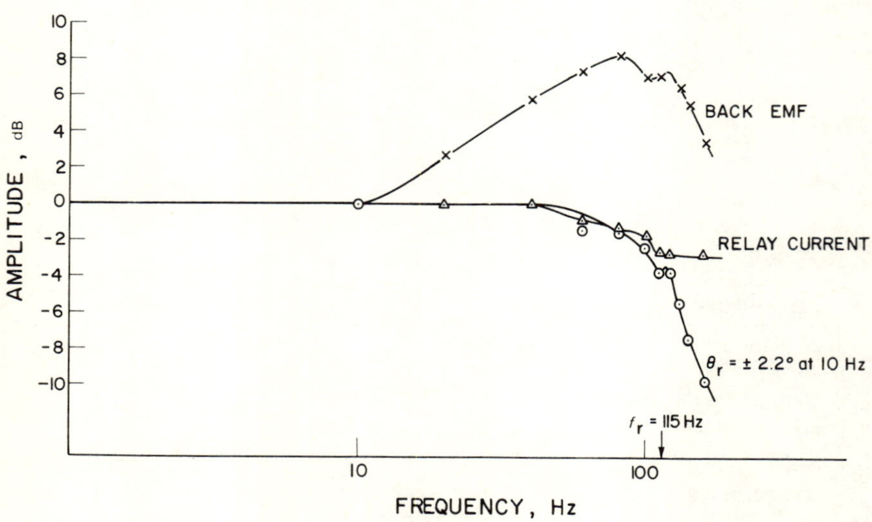

Fig. 12.3 – Servo amplifier – relay frequency response.

If V_R is defined as the voltage available to drive the relay coils, and V_B the e.m.f. generated, then the useful voltage at the coils at any particular frequency is:

$$V_R - V_B = V_R - k_e\dot{\theta}_r = (R + L\omega j)\,i\,.$$

Typically, at 100 Hz, the generated back e.m.f. is 290 V, with 300 V maximum available at the amplifier output. Typical test results are shown in Fig. 12.3.

The effect of loading due to flow forces is also shown in Fig. 12.2. This will be further discussed later in the chapter.

12.2 TWO STAGE ELECTRO-HYDRAULIC VALVE

When large fluid flow rates are required it is usually necessary to use two stages in the control valve arrangement. Basically, this is necessary in order to maintain a satisfactory response at high flow rates. Loss of response in single stage valves at large flow rates, is due to high spool inertia and large flow forces, resulting in the need for a large torque motor. Feedback loops are often provided at each stage, making them more flexible.

A typical arrangement in conjunction with a cylinder is shown in diagrammatic form in Fig. 12.4. It consists of a pilot spool valve, having an electrical feedback path to the servo amplifier. The output from the servo amplifier actuates the torque motor attached to the pilot spool. The movement of the pilot spool varies the differential pressure acting on the ends of the main spool. This is clearly more effective for dealing with large forces. The output flow rate from the main spool is used to drive the piston, which is also provided with a feedback loop. Simplified transfer functions can be developed for the system shown in Fig. 12.4 as follows:

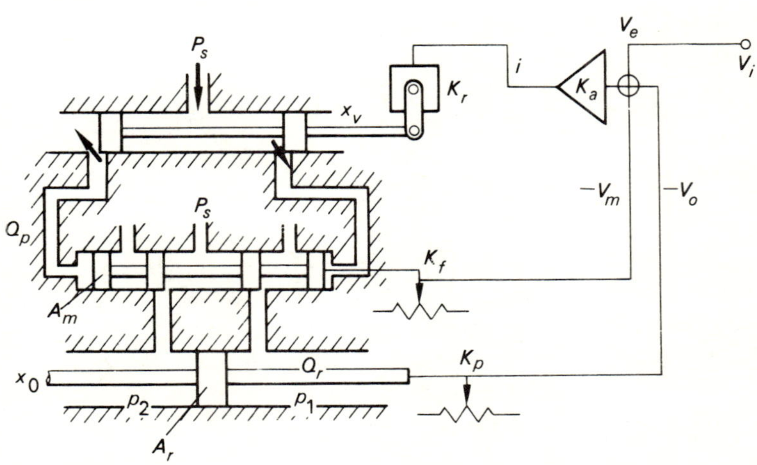

Fig. 12.4 – Two-stage valve and actuator with feedback.

For the pilot spool:

$$Q_p = K_1 x_v \sqrt{(P_s/2)} . \tag{12.3}$$

For the main spool:

$$Q_r = K_1' x_m \sqrt{(P_s/2)} . \tag{12.4}$$

An interaction occurs between the two stages; it can be described by:

$$Q_p = A_m \frac{dx_m}{dt} . \tag{12.5}$$

From equations (12.3) and (12.4):

$$x_v = \frac{Q_p}{K_1 \sqrt{(P_s/2)}} = \frac{A_m}{K_1 \sqrt{(P_s/2)}} \frac{dx_m}{dt} \tag{12.6}$$

and from (12.4):

$$x_m = \frac{Q_r}{K_1' \sqrt{(P_s/2)}} . \tag{12.7}$$

Differentiating equation (12.7) gives:

$$\frac{dx_m}{dt} = \frac{1}{K' \sqrt{(P_s/2)}} \frac{dQ_r}{dt} . \tag{12.8}$$

Substituting (12.8) in (12.6) gives:

$$x_v = \frac{2A_m}{P_s K_1 K_1'} \frac{dQ_r}{dt} .$$

Hence,

$$Q_r = \frac{K_1 K_1' P_s}{2A_m} \int x_v \, dt = A_r \frac{dx_o}{dt} . \tag{12.9}$$

The output acceleration d^2x_o/dt^2 is the proportional to the input x_v. This is an unsatisfactory form of control giving rise to stability problems.

In order to improve the control of the system, feedback from the main spool to the servo amplifier is required. This is represented by the voltage V_m, a function of the main spool displacement.

Now:

$$x_v = V_e K_a K_r \tag{12.10}$$

and

$$V_m = K_f x_m \tag{12.11}$$

where K_f = the gain of the spool transducer.

Now:

$$V_e = V_i - V_m \,. \tag{12.12}$$

Equating (12.6) and (12.10) gives:

$$K_a K_r V_e = \frac{A_m}{K_1 \sqrt{(P_s/2)}} \frac{dx_m}{dt} \,. \tag{12.13}$$

From (12.12) and (12.11) with (12.13):

$$\begin{aligned} V_e &= \frac{A_m}{K_a K_r K_1 \sqrt{(P_s/2)}} \frac{dx_m}{dt} \\ &= V_i - K_f x_m \,. \end{aligned}$$

The transfer function is, therefore:

$$\begin{aligned} \frac{x_m}{V_i}(s) &= \frac{1}{\left\{ \dfrac{A_m}{K_1 K_a K_r \sqrt{(P_s/2)}} s + K_f \right\}} \\ &= \frac{1}{K_f(1 + T_c s)} \end{aligned} \tag{12.14}$$

where $T_c = A_m / K_1 K_a K_f K_r \sqrt{(P_s/2)}$.

But from equation (12.7) substituted in (12.14):

$$Q_r(s) = \left[\frac{K_1' \sqrt{(P_s/2)}}{K_f (1 + T_c s)} \right] V_i \,. \tag{12.15}$$

For a step change in V_i the flow rate to the cylinder reaches a steady state exponentially, Fig. 12.5. In this steady state the piston rod moves at constant velocity instead of at a constant acceleration as indicated by equation (12.9).

In order to maintain position control of the load a voltage V_o is fed back from the piston rod.

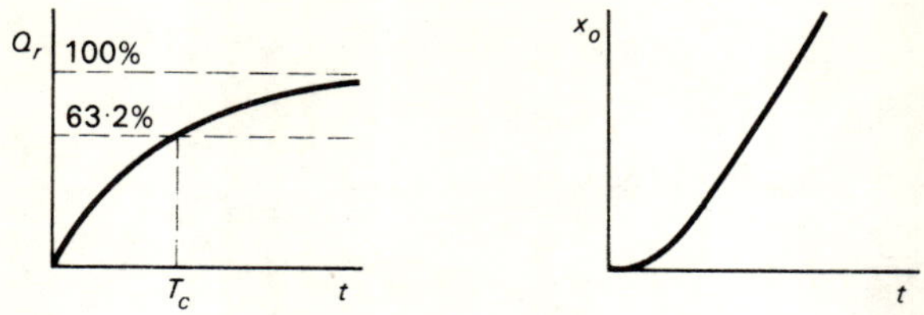

Fig. 12.5 – Effect of feedback from main spool valve.

The new arrangement is shown in the block diagram of Fig. 12.6. From equation (12.15), referring to Fig. 12.4 where V_i becomes V_e, owing to the introduction of an additional feedback loop:

$$Q_r = \frac{K_1'\sqrt{(P_s/2)}}{K_f(1+T_c s)} V_e = A_r \frac{dx_o}{dt} \ . \qquad (12.16)$$

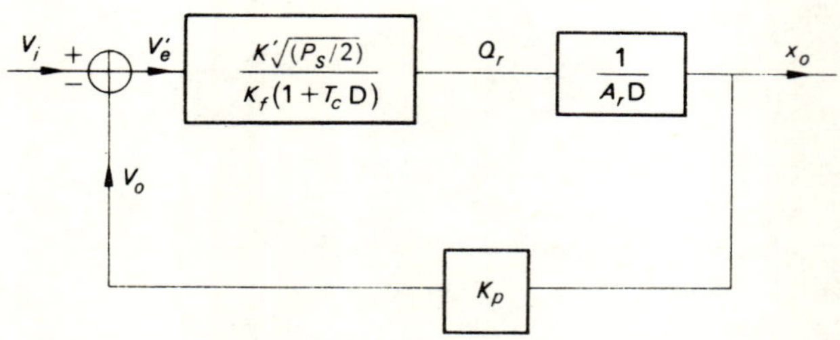

Fig. 12.6 – Block diagram showing position feedback from the actuator.

But

$$V_e' = V_i - V_o = V_i - K_p x_o \ . \qquad (12.17)$$

Combining (12.16) and (12.17):

$$A_r \frac{dx_o}{dt} = \frac{K_1'\sqrt{(P_s/2)}}{K_f(1+T_c s)} (V_1 - K_p x_o)$$

therefore

$$K_1'\sqrt{(P_s/2)}\, V_i = K_f A_r \frac{dx_o}{dt} + K_f A_r T_c \frac{d^2 x_o}{dt^2} + K_1' K_p \sqrt{(P_s/2)} x_o$$

or

$$K_1'\sqrt{(P_s/2)}\, V_i(s) = [K_f A_r T_c s^2 + K_f A_r s + K_1' K_p \sqrt{(P_s/2)}]\, x_o(s) \ .$$

The transfer function is therefore:

$$\frac{x_o}{V_i}(s) = \frac{K_1'\sqrt{(P_s/2)}}{K_f A_r T_c s^2 + K_f A_r s + K_1' K_p \sqrt{(P_s/2)}}'$$

$$= \frac{1/K_p}{\dfrac{K_f A_r T_c}{K_1' K_p \sqrt{(P_s/2)}} s^2 + \dfrac{K_f A_r}{K_1' K_p \sqrt{(P_s/2)}} s + 1} . \tag{12.18}$$

There is now a direct relationship between x_o and V_i in the steady state given by:

$$x_o = \frac{1}{K_p} V_i$$

and hence position control is achieved. Notice that the transfer function (12.18) is a quadratic, indicating that the system can exhibit continuous oscillations if:

$$\frac{K_f A_r}{K_1' K_p \sqrt{(P_s/2)}} \leqslant 0 ,$$

the frequency of the resulting oscillations being given by:

$$f_n = \frac{1}{2\pi} \frac{K_1' K_p \sqrt{(P_s/2)}}{K_f A_r T_c} \text{ Hz} .$$

If the cylinder were moving a load dominated by inertia than the quadratic term due to compressibility would need to be included in equation (12.18), raising the order to four and introducing a possible maximum lag of 360°. Care must be taken, therefore, to ensure that stability is maintained when combining commercially bought valves with existing plant.

When optimising the loops in such systems one should allow sufficient margin for wear and deterioration of the system; for example damping due to friction in a system usually reduces with wear. It is therefore good practice to provide adjustment of loop gains to help maintain performance.

12.3 GENERAL ANALYSIS OF THE ELECTRO-HYDRAULIC SERVO

The analytical development in Chapter 11 was basically a linearised approach. In this chapter a more detailed analysis is considered and solutions are obtained using one of the computer simulation techniques.

Considering the actuator arrangement shown in Fig. 12.7 and assuming the valve to have zero lap, the flow rates into and out of the actuator are:

For $x_v > 0$:

$$Q_1 = \mathrm{sgn}\,(P_s - P_1) K_1 x_v \sqrt{(|P_s - P_1|)}$$

$$Q_2 = K_1 x_v \sqrt{P_2} \tag{12.19}$$

For $x_v < 0$:

$$Q_1 = K_1 x_v \sqrt{P_1}$$

$$Q_2 = \mathrm{sgn}\,(P_s - P_2) K_1 x_v \sqrt{(|P_s - P_2|)}\,.$$

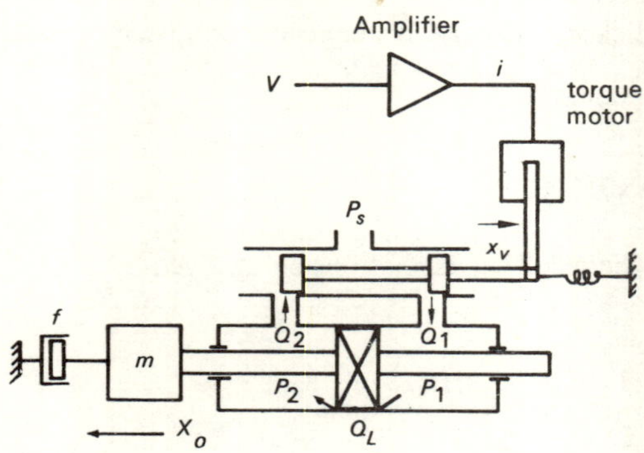

Fig. 12.7 – Basic open loop system.

Note that on a full simulation it is necessary to consider negative spool movements and the possibility of reverse flow [1, 2, 3]. This was not essential in the earlier analysis. These equations are made non-dimensional by letting $x^* = x_v/\hat{x}_v$, $p = P/P_s$, $Q^* = Q/\hat{Q}$ with $Q = A\hat{u}$ and $\hat{u} = K_1\hat{x}_v\sqrt{(P_s/2)}\,/A$. Hence:

$x_v^* > 0$:

$$Q_1^* = \sqrt{2}x_v^*\,\mathrm{sgn}\,(1 - p_1)\sqrt{(|1 - p_1|)}$$

$$Q_2^* = \sqrt{2}x_v^*\sqrt{p_2} \tag{12.20}$$

$x_v^* < 0$:

$$Q_1^* = \sqrt{2}x_v^*\sqrt{p_1}$$

$$Q_2 = \sqrt{2}x_v^*\,\mathrm{sgn}\,(1 - p_2)\sqrt{(|1 - p_1|)}\,.$$

The continuity equations (11.12) and (11.13) can also be made non-dimensional. In addition to the above reference quantities non-dimensional time is defined as $\tau = \omega t$ where ω is the driving frequency. Equation (11.12) becomes:

$$Q_1^* = u^* + \frac{V}{2\beta}\omega P_s \frac{dp_1}{d\tau}\frac{1}{Kx_v\sqrt{(P_s/2)}} \tag{12.21}$$

or

$$Q_1^* = u^* + \frac{2\omega^*}{\alpha}\frac{dp_1}{d\tau} \tag{12.22}$$

where the inertia load factor α is:

$$\alpha = \frac{\text{maximum inertia load at the natural frequency}}{\text{stalled load}}$$

or

$$\alpha = \frac{M\omega_n K\hat{x}_v\sqrt{(P_s/2)}}{P_s A^2}$$

and $\omega^* = \omega/\omega_n$ where ω_n is the mass and oil column natural frequency $\sqrt{(4\beta A^2/MV)}$ (at the rod mid-position).

By a similar argument equation (11.13) reduces to:

$$Q_2^* = u^* - \frac{2\omega^*}{\alpha}\frac{dp_2}{d\tau}\,. \tag{12.23}$$

If the load consists of inertia M and viscous friction f then the load equation is:

$$P_L = P_1 - P_2 = \frac{M}{A}\frac{d^2x_o}{dt^2} + \frac{f}{A}\frac{dx_o}{dt} \tag{12.24}$$

or

$$p_L = p_1 - p_2 = a\omega^*\frac{du^*}{d\tau} + \gamma u^* \tag{12.25}$$

where the viscous load factor γ is:

$$\gamma = \frac{\text{maximum viscous force}}{\text{stall force}} = \frac{fK_1\hat{x}_v}{P_s A^2}\sqrt{(P_s/2)}\,. \tag{12.26}$$

12.4 SIMPLE CASE

If cavitation is not allowed to occur there is no reversal of flow, and a simplified equation results. Since under this condition the flow into the cylinder approximately equals the flow out at all times, equations (12.21) can be combined to give:

For $x^* > 0$:

$$Q_L^* = \sqrt{2} x_v^* \sqrt{(1-p_1)} \tag{12.27}$$

and for $x^* < 0$:

$$Q_L^* = \sqrt{2} x_v^* \sqrt{p_1}$$

where Q_L^* = average load flow through the valve.

From equations (12.24) and (12.25), the flow into the cylinder is given by:

$$Q_L^* = u^* + \frac{1}{\alpha}\omega^* \frac{dp_L}{d\tau}. \tag{12.28}$$

But since inflow and outflow are equal:

$$p_1 + p_2 = 1.0 ,$$

also:

$$p_L = p_1 - p_2 .$$

Hence:

$$p_1 = \tfrac{1}{2}(1 + p_L). \tag{12.29}$$

Hence from equations (12.25) through to (12.29):

$$x_v^* \left\{1 - \left(\alpha\omega^* \frac{du^*}{d\tau} + \gamma u^*\right) \frac{x_v^*}{|x_v^*|}\right\}^{\frac{1}{2}} = u^* + \frac{\gamma}{\alpha}\omega^* \frac{du^*}{d\tau} + (\omega^*)^2 \frac{d^2u^*}{d\tau^2}. \tag{12.30}$$

It has been established earlier that the flow force F_β in the negative x direction is given by:

For $x_v^* > 0$:

$$F_\beta = \sqrt{2\rho}\,[Q_1\sqrt{(P_s - P_1)} + Q_2\sqrt{P_s}]\cos\theta . \tag{12.31}$$

For $x_v^* < 0$:

$$F_\beta = -\sqrt{2\rho}\,[Q_1\sqrt{P_1} + Q_2\sqrt{(P_s - P_2)}]\cos\theta . \tag{12.32}$$

It can be seen from Fig. 12.1 that this force opposes the torque motor output so that the valve displacement will be determined by:

$$x_v = K_s(F_R - F_\beta)\,.$$

Since the inflow and outflow are equal, equation (12.31) can be written:

$x_v^* > 0$:

$$F_\beta = 2\sqrt{2\rho}\,[Q_1\sqrt{(P_s - P_1)}]\cos\theta$$

$x_v^* < 0$:

$$F_\beta = -2\sqrt{2\rho}\,[Q_1\sqrt{P_1}]\cos\theta\,. \tag{12.33}$$

If the reference flow force $\hat{F}_\beta = 2\sqrt{\rho}\cos\theta\, K_1\hat{x}_v P_s/\sqrt{2}$ then equation (12.33) becomes:

$$F_\beta^* = x_v^*\left\{1 - \alpha\frac{du^*}{d\tau} + \gamma u^*\frac{x_v^*}{|x_v^*|}\right\}. \tag{12.34}$$

The valve displacement for a sinusoidal electrical input signal will be:

$$x_v = K_s(K_g\hat{V}\sin\omega t - F_\beta)$$

assuming for simplicity that K_g includes the amplifer gain. Hence:

$$x_v^* = \sin\tau - \delta F_\beta^* \tag{12.35}$$

where the flow force factor δ is:

$$\frac{2\sqrt{\rho}\cos\theta\, K_1\hat{x}_v P_s}{\sqrt{2}K_g V}\,.$$

and

$$\hat{x}_v = K_s K_g\hat{V}\,.$$

The system equations to be solved are then:

$$x_v^*\left\{1 - \alpha\omega^*\frac{du^*}{d\tau} + \gamma u^*\frac{x_v^*}{|x_v^*|}\right\}^{\frac{1}{2}} = u^* + \frac{\gamma\omega^*}{\alpha}\frac{du^*}{d\tau} + \omega^{*2}\frac{d^2u^*}{d\tau^2} \tag{12.36}$$

$$F_\beta^* = x_v^*\left\{1 - \left(\alpha\frac{du^*}{d\tau} + \gamma u^*\right)\frac{x_v^*}{|x_v^*|}\right\} \tag{12.37}$$

$$x_v^* = \sin\tau - \delta F_\beta^* \tag{12.38}$$

12.5 RESULTS FROM COMPUTER STUDY

Although this analysis does not allow results to be recorded when the actuator is cavitating, it is possible to determine the cavitation boundaries. This is achieved by increasing the load factor α until p_L becomes equal to unity. Results from this examination [1, 2, 3, 4, 5, 6] are shown in Fig. 12.8 and should be compared with those shown in Fig. 11.16. Also included in this figure is the effect of viscous friction on the cavitation boundary. It is seen that a little friction is quite effective particularly at the natural frequency.

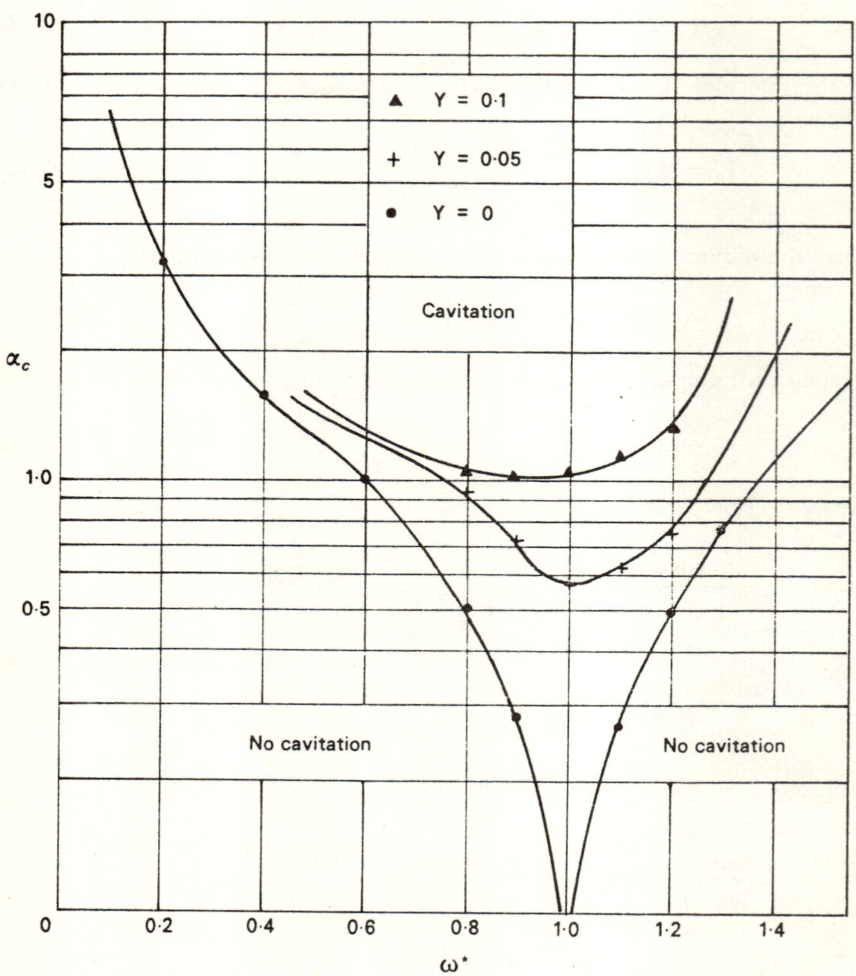

Fig. 12.8 – Effect of viscous friction γ on the cavitation boundary.

A similar study of the effects of leakage gives results very similar to those of Fig. 12.8, except that the coefficients are 50 per cent lower in magnitude. The sensitivity of the cavitation boundary to small changes in leakage factor is to be expected, since the use of an orifice bleed is one of the established methods of reducing cylinder pressure peaks.

The leakage factor ϕ is defined as follows. Since the leakage past the piston can be assumed to be laminar:

$$Q_e = l(P_1 - P_2)$$

and therefore equations (11.12) and (11.13) become:

$$Q_1 = A\frac{dx_o}{dt} + \frac{V}{2\beta}\frac{dP_1}{dt} + l(P_1 - P_2) \tag{12.39}$$

$$Q_2 = A\frac{dx_o}{dt} - \frac{V}{2\beta}\frac{dP_2}{dt} + l(P_1 - P_2)\,. \tag{12.40}$$

Using the same process of non-dimensionalising:

$$Q_1^* = u^* + \frac{2}{\alpha}\omega^*\frac{dp_1}{d\tau} + \frac{Ml\omega_n}{A^2}\frac{P_s A^2}{MK_1\hat{x}_v\omega_n\sqrt{(P_s/2)}}(p_1 - p_2)\,.$$

Define the leakage factor as:

$$\phi = \frac{Ml\omega_n}{A^2}\,.$$

Then (12.39) and (12.40) become:

$$Q_1^* = \left\{2\,\frac{u^*}{2} + \frac{1}{\alpha}\omega^*\frac{dp_1}{d\tau} + \frac{\phi}{2\alpha}(p_1 - p_2)\right\}$$

$$Q_2^* = \left\{2\,\frac{u^*}{2} - \frac{1}{\alpha}\omega^*\frac{dp_2}{d\tau} + \frac{\phi}{2\alpha}(p_1 - p_2)\right\}$$

and therefore equation (12.36) becomes:

$$x_v^*\left\{1 - \alpha\omega^*\frac{du^*}{d\tau} + \gamma u^*\frac{x_v^*}{|x_v^*|}\right\}^{\frac{1}{2}}$$

$$= \left(1 + \gamma\frac{\phi}{\alpha}\right)u^* + \left(\phi + \frac{\gamma}{\alpha}\right)\omega^*\frac{du^*}{d\tau} + (\omega^*)^2\frac{d^2u^*}{d\tau^2}\,. \tag{12.41}$$

The effects of viscous damping and leakage on the open loop frequency responses are shown in Figs. 12.9 and 12.10. It should be noted that since the characteristic equation of the system is third order, velocity feedback and viscous loading do not have the same effect.

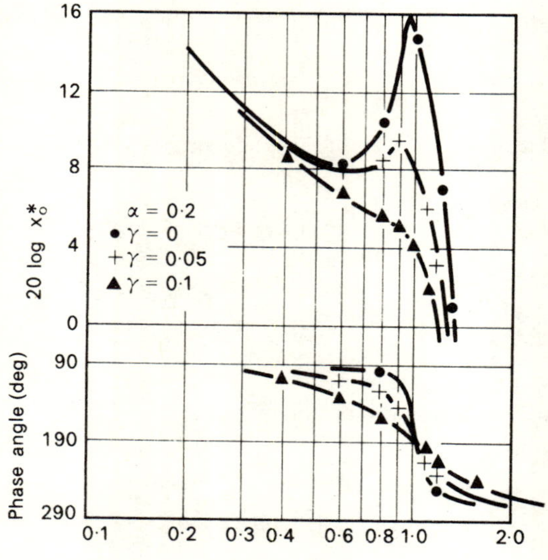

Fig. 12.9 – Effect of viscous friction on open loop response.

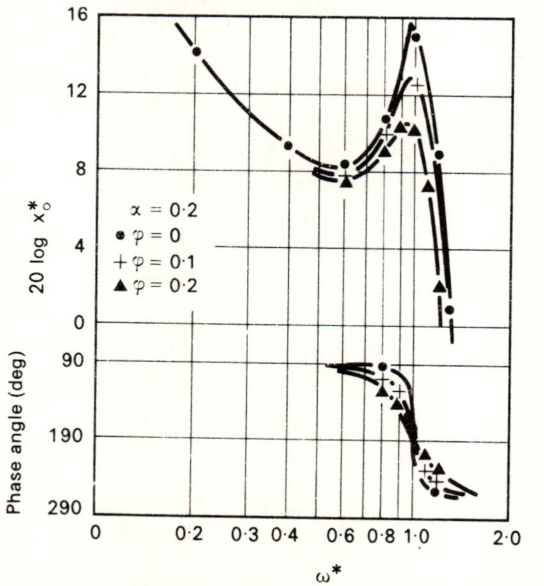

Fig. 12.10 – Effect of leakage on open loop response.

12.6 EFFECTS OF FLOW FORCES

The presence of flow forces has two main effects on the dynamic behaviour of the servo loop [3, 7, 8]:

(a) The open loop static gain of the servo is reduced.
(b) The stability of the servo is affected.

It was shown earlier that the flow force opposes the torque motor driving force, hence affecting the valve position. Equation (12.34) clearly shows that the flow force varies with the load pressure in the actuator. These effects will be fed back directly to the control valve; the torque motor is in fact working into a frequency and amplitude dependent load. As the forcing frequency approaches the natural frequency determined by the output mass and the oil column stiffness, the valve flow decreases. At the natural frequency the inertia load is simply bouncing on the oil column and the sum of the displacement flow and compressibility flow is very small. This is demonstrated in the results shown in Fig. 12.11. The explanation for the decrease in flow can be seen from the graphical analysis shown in Fig. 12.12.

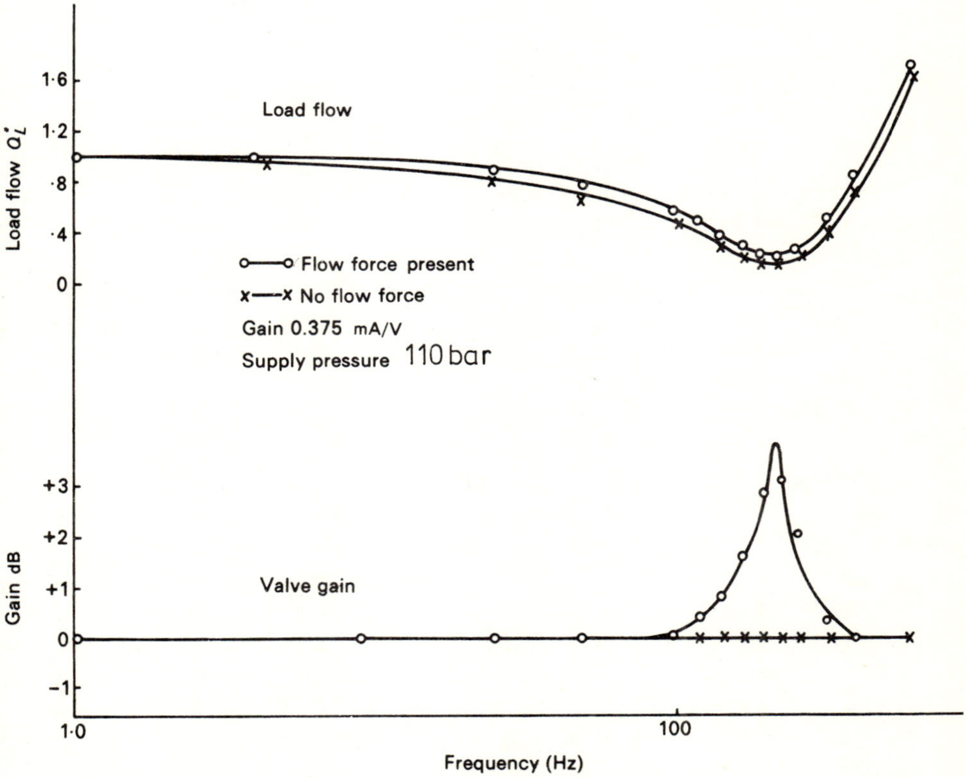

Fig. 12.11 – Experimental results showing the behaviour of the load flow and the valve gain.

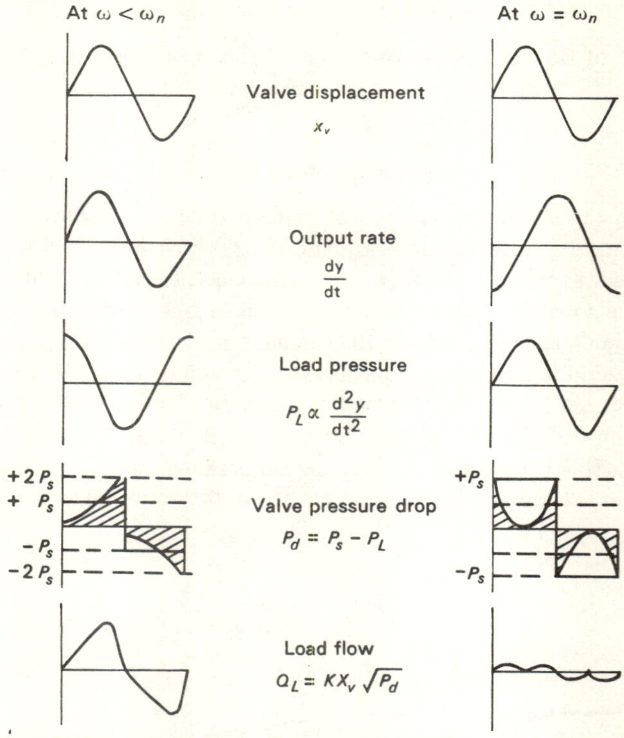

Fig. 12.12 – (a) Experimental results showing behaviour of load flow and valve gain. (b) Behaviour of load flow under dynamic conditions.

Since there is very little flow through the valve the flow force is small and the valve displacement is very sensitive to torque motor current, making the forward loop gain high in this region. In other words the gain of the servo is a function of frequency, being largest in the region of the natural frequency; this is the worst possible condition with regards to loop stability. Results demonstrating the effect of flow forces are shown in Fig. 12.13.

It is seen that although the overall gain level is reduced owing to the first effect mentioned above, the resonant amplitude is greatly increased in the region of the natural frequency.

In Fig. 12.14, the value of the flow force coefficient required to induce instability even in the open loop condition is defined as δ_c. This condition is modified depending on the amount of viscous damping (γ) present. With the smaller inertia loads, viscous damping has a pronounced effect. From these results it is apparent that the two most important factors to be considered are the type of loading and the position of the natural frequency of the oil column and inertia relative to the operational bandwidth. If the natural frequency is well beyond this bandwidth, a reduction in open loop gain is the main effect

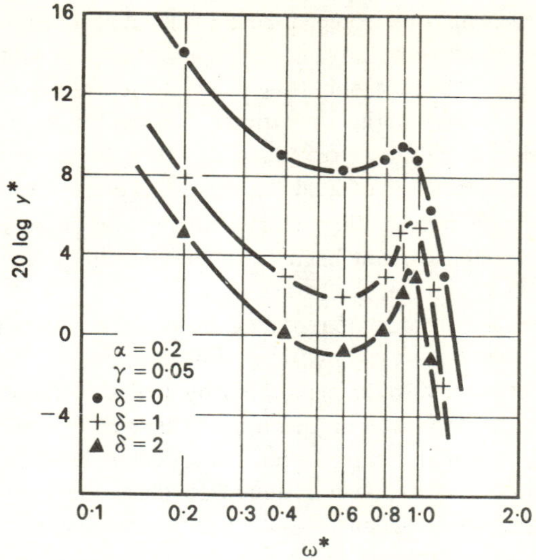

Fig. 12.13 – Effect of flow force on open loop response.

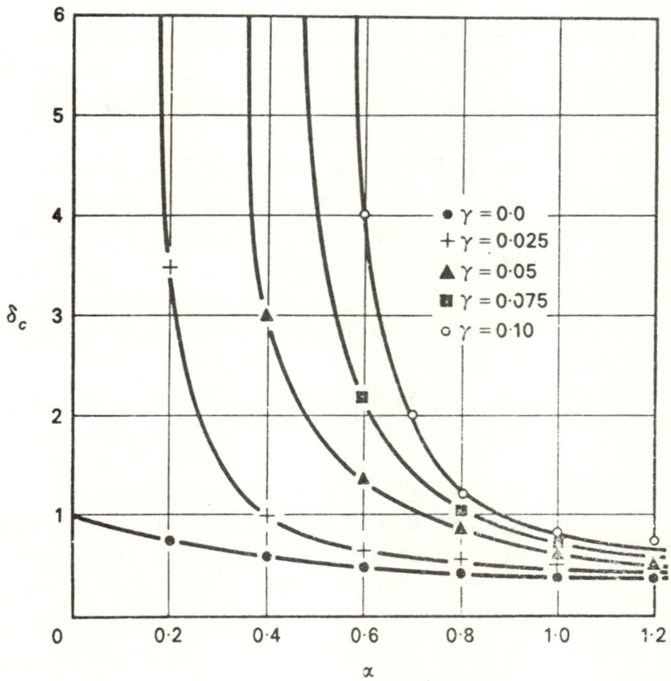

Fig. 12.14 – The critical flow force factor δ_c.

of the flow force. In this case power is being wasted in driving an extra load on the valve; it may therefore be advantageous, especially in aircraft [9, 10] and missile applications, to reduce the flow force. The results show that, if the natural frequency lies within the operational bandwidth, it may be necessary, from considerations of stability, to reduce the flow force. If the load on the servo is mainly of a viscous or of a spring nature, the stability of the system is not so drastically affected as when inertia load predominates. In fact with a spring load, a case which has not been considered in this analysis, the Bernoulli force can be used to advantage.

There are several methods of reducing flow force effects (see Chapter 8). The flow forces may be greatly reduced, for example, by shaping the spool and sleeve of the control valve as elaborated by Lee and Blackburn and Eynon [8]. This has the disadvantage of being an expensive operation, especially if a high degree of flow force compensation is desired. It has the advantage, however, that the torque motor power required to drive the control valve can be kept within reasonable limits, and this, as has been mentioned before, may be of importance in missile applications.

The valve reaction force may be swamped by a spring load. This may be achieved by increasing the torque motor stiffness or by fitting a valve-centralising spring. By such means, the compliance K_s and hence the flow force factor δ are reduced.

Another method of swamping the reaction force is to use a more powerful torque motor. This increases the relay amplifier gain K_g and hence reduces the flow factor δ. This method may be limited by difficulty in obtaining good response rates from the larger torque motors.

The destabilising effect of the flow force can be reduced by the introduction of a viscous load, perhaps in the form of a dashpot or by the use of an orifice. The presence of a leak tends to reduce the positional accuracy of the system, but the increase in gain thus made possible may more than compensate for this. Such damping effects can of course be obtained by using electrical networks and transducers. For instance, acceleration feedback can be approximated by using transducers to measure the load pressure.

If flow forces are completely eliminated, the electro-hydraulic gain, the flow per unit error voltage, is high. This feature is undesirable because most of the non-linearities are in this part of the loop, and it may mean that the electrical gain, relay driving voltage per unit error in load displacement, must be reduced. This can cause loss of accuracy in a positional system. In general, it is preferable to keep the electro-hydraulic gain as low as possible because the electrical gain is much more flexible. Hence one of the best methods of improving system performance is to remove some of the frequency-dependent flow force by partial shaping, and to reduce the electro-hydraulic gain by fitting a spring.

It should be borne in mind that, even if flow force effects are eliminated, an undesirable resonance peak can still occur when α is small, and the introduction of artificial damping may be necessary.

12.7 MORE EXACT SIMULATION

When cavitation occurs it can no longer be assumed that the inflow and outflow are equal. The symmetry has been destroyed and the flow through each valve port must be treated separately. The equations to be solved are, for positive spool movements:

Inflow:

$$\sin\tau\sqrt{\frac{1}{2}(1-p_1)} = \frac{u^*}{2} + \frac{1}{\alpha}\omega^*\frac{\mathrm{d}p_1}{\mathrm{d}\tau} + \frac{\phi}{2\alpha}(p_1 - p_2)\,. \tag{12.42}$$

Outflow:

$$\sin\tau\sqrt{\frac{1}{2}p_2} = \frac{u^*}{2} - \frac{1}{\alpha}\omega^*\frac{\mathrm{d}p_1}{\mathrm{d}\tau} + \frac{\phi}{2\alpha}(p_1 - p_2) \tag{12.43}$$

Load:

$$p_1 - p_2 = \alpha\omega^*\frac{\mathrm{d}u^*}{\mathrm{d}\tau} + \gamma u^* + \eta\frac{u^*}{|u^*|} \tag{12.44}$$

where η is the seal friction coefficient. The equations for negative x also follow from (12.20).

It is necessary to obtain a clear physical picture of the mechanism of cavitation before any simplifying assumptions are made. If, as in Fig. 12.15, the cylinder pressure falls to zero at some point in the stroke but recovers before the valve crosses centre, no cavitation bubble will form. The side of the cylinder on which the zero pressure exists is connected to exhaust, and so no outflow,

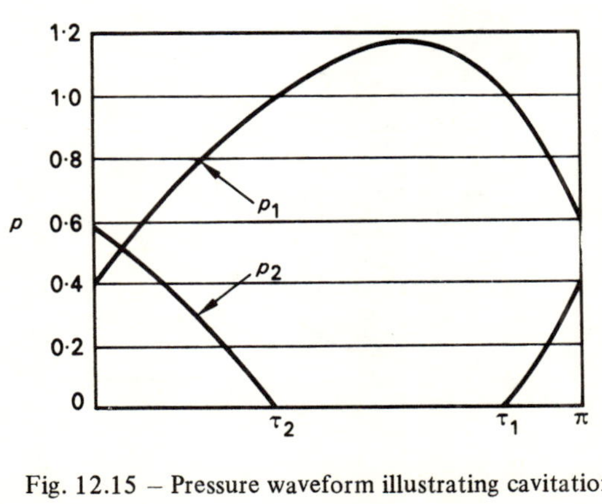

Fig. 12.15 – Pressure waveform illustrating cavitation.

except that due to piston movement, takes place. This side remains full of oil and so no cavitation bubble forms. Theoretically, the attainment of zero pressure will cause the oil to vapourise, but there is no experimental evidence to show that vaporisation does in fact occur. When the pressure on the high-pressure side drops below system pressure, the low-pressure side will begin to increase and continuity of flow is maintained. Inflow and outflow are equal and $p_1 + p_2 = 1$. However, during the time that the pressure on the low-pressure side is zero (and the pressure on the high-pressure side exceeds P_s) equations (12.42) and (12.44) describe the dynamics.

A cavitation bubble can form if the pressure on one side of the cylinder is zero when the valve crosses centre. In the case illustrated in Fig. 12.16, the pressure on that side of the cylinder connected to the supply falls to zero and does not recover before the valve crosses centre. During the interval τ_1 to π the pressure drop across the inlet port is P_s and this does not provide sufficient flow to keep the inlet side full. The piston breaks away from the fluid and a bubble forms. When the valve crosses centre that side of the cylinder containing the bubble is now connected to the exhaust. The pressure in this side cannot rise until the bubble has been repaired at time τ_2. Hence from time 0 to τ_2 equations (12.42) and (12.44) are used, from time τ_2 to τ_1 equations (12.42), (12.43) and (12.44) are used, and from τ_1 to π equations (12.43) and (12.44) are used.

To simplify the analysis it is assumed that the bubble fills instantaneously and so p_2 begins to rise as soon as p_1 falls below unity. This is in effect assuming equal inflow and outflow when neither pressure is zero. Thus when $p_1 > 1$, equations (12.42) and (12.44) are used, when $p_2 > 1$ equations (12.43) and (12.44) are used, and when p_1 and p_2 are both less than one, equations (12.42), (12.43) and (12.44) are used.

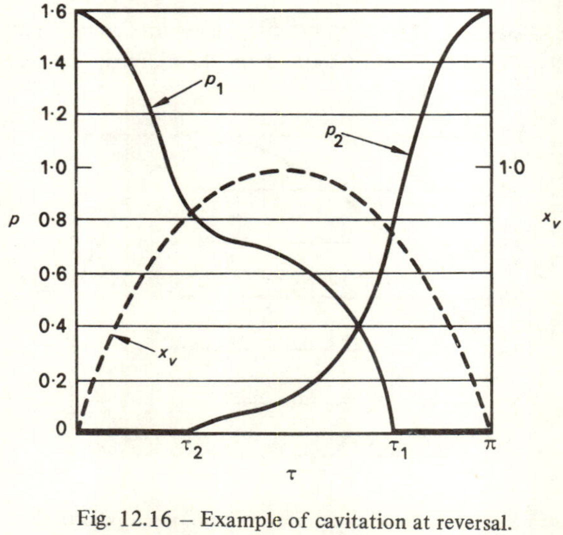

Fig. 12.16 – Example of cavitation at reversal.

The assumption that the bubble fills instantaneously simplifies the analysis considerably. However, the reader interested in a more exact treatment is referred to [14, 15]. With reference to Fig. 12.17, cavitation bubbles exist in the region marked 'cavitation at reversal'. It is seen that the values of inertia load factor required to produce this condition are relatively large and it is unlikely that such values would be achieved in the closed loop condition, except perhaps near the natural frequency. Calculations have shown that the time required to fill these cavities is relatively small and so the assumption of instantaneous filling is not unrealistic.

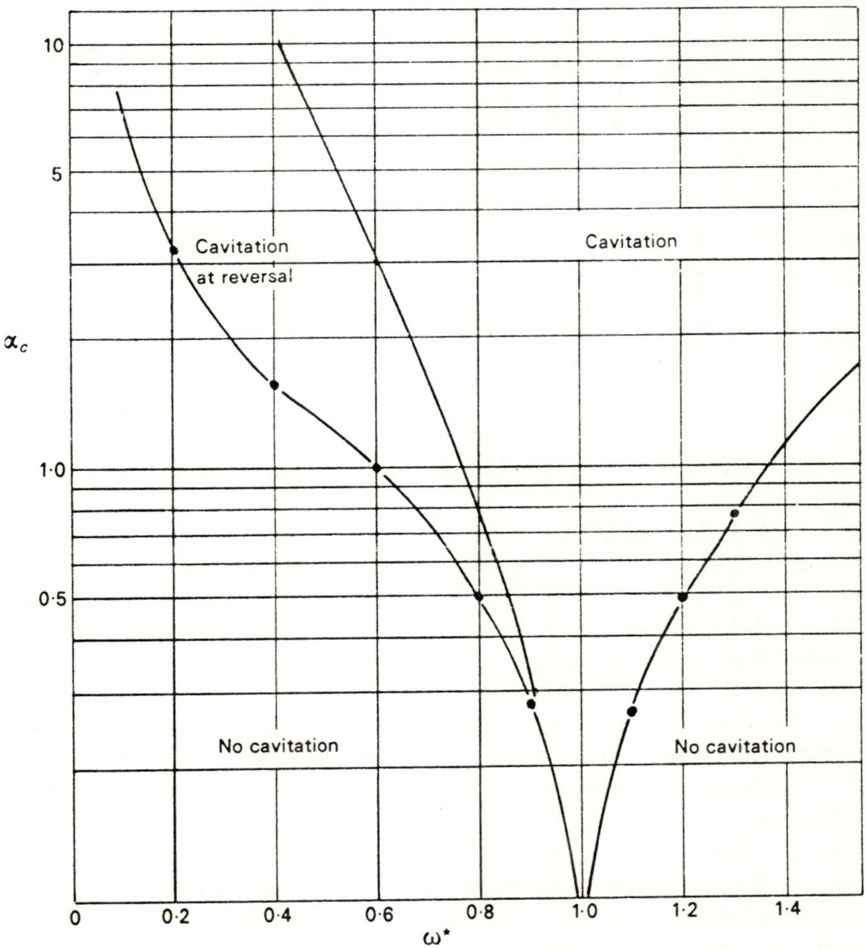

Fig. 12.17 – Cavitation boundaries.

These conditions are used to investigate the effect of large inertia loads (Fig. 12.18). Increasing α has a strong stabilising effect owing to the valve flows present. It is interesting to note that if Bernoulli forces are non-existent, the increasing α is stabilising as shown. If flow forces are present, however, valve flow no longer stabilises but destabilises, and large α is destabilising.

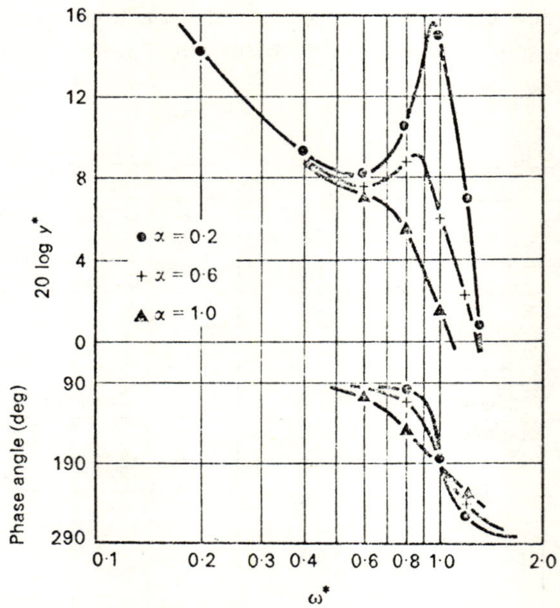

Fig. 12.18 – Effect of inertia load factor on the open loop response.

12.8 SMALL SIGNAL RESPONSES

The analysis of a linear servo system, excluding leakage effects, indicates that there are no 'threshold' conditions. The system responds to the smallest input signals and disturbances, whilst the time taken for similar movements is constant irrespective of the overall amplitude scale of movement.

In practice, however, a hydraulic servo has non-linearities which detract from this ideal, and experimental results show that as the input amplitude is reduced, the output waveform is not in exact proportion to the input signal. An example of this is shown in Fig. 12.19, which is an experimental result. If the input is reduced still further, eventually a stage is reached when there is no response at the output for a finite disturbance at the input. This condition is termed the **threshold region**. As the input amplitude is increased, the system operates more definitely and approaches the behaviour exhibited by the linear servo. As the input is increased still further, a stage is ultimately reached where saturation occurs. This will be the result of one or more of the system components reaching the limit of their output.

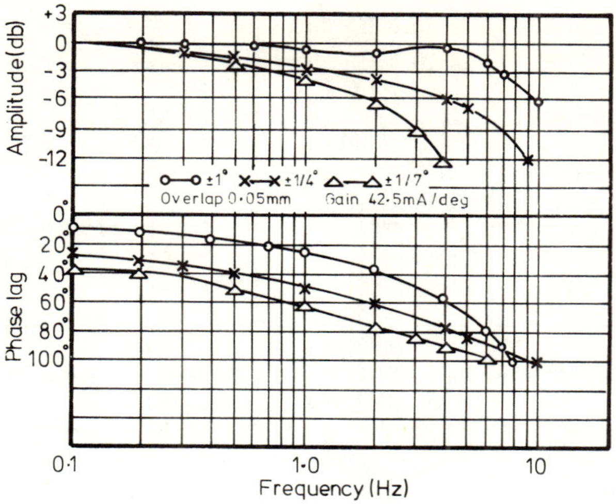

Fig. 12.19 – Experimental results showing variation of servo loop response with input amplitude.

Electro-hydraulic servo systems are often used for applications that require a fast response to cater for an emergency or extreme condition. However, it is very probable that only a small fraction of their life will be spent under these conditions, and a good deal of this time will be used during the test and development period. Much the greater part of the life of the servo will be spent responding to small signals, and in fact much of this time will be spent around the threshold region. The threshold region may be considered as the dead zone of a servomechanism, and investigations [11, 16] reveal that there are two main characteristics:

Static dead zone, defined as the maximum signal to which the system fails to respond after being quiescent for a finite period.

Dynamic dead zone, defined as the maximum signal to which the system fails to respond, whilst the input has some form of movement.

Some experimental results are shown in Figs. 12.20 and 12.21.

Static dead zone appears to be due mainly to friction either of the piston seals or of the hydraulic valve, and develops with time as lubricant is squeezed from between rubbing surfaces. The pressure which elastomeric seals exert on the cylinder walls ultimately exudes the lubricant, with a consequent increase in the coefficient of friction, which reduces again after the first movement causes relubrication; dead zone from this cause is of more significance if there is a bleed orifice, or if underlapped spool valves are used.

Hydraulic spool valves are normally 'pressure balanced' to avoid large operating forces, but even so quite large frictional forces can develop, of the

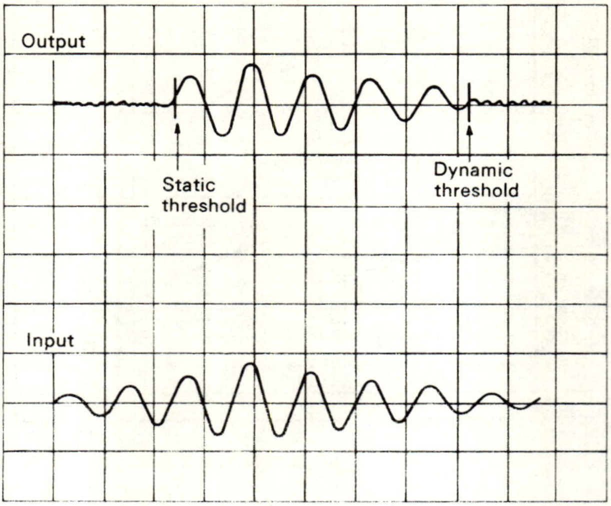

Fig. 12.20 – Threshold characteristics.

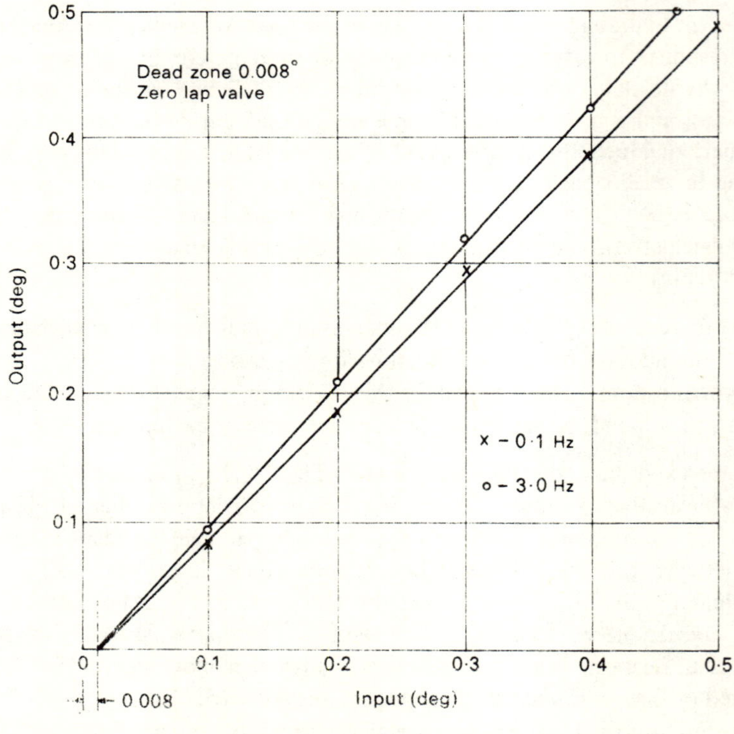

Fig. 12.21 – Dynamic threshold results – servo output v input.

order of 0–20 N with a miniature valve, if it is quiescent for a length of time. This is because side forces are present on the valve owing to unbalanced pressure gradients around the periphery of the spool. These side forces ultimately squeeze out the fluid in the clearance between the spool and sleeve, and a high coefficient of friction again results. Inadequate filtration will also contribute greatly to increased spool valve friction.

It is fortunate that the causes of static dead zone are not inherently additive, that is, almost all the error signal necessary to break the valve friction becomes available to break the seal friction when the valve has moved. Since static dead zone is mainly a function of friction it is obviously a very erratic phenomenon, as attempts at measurement soon reveal.

Dynamic dead zone is usually only about 30 per cent of the static dead zone, and is the combined effect of any valve overlap and backlash, friction, hysteresis and resolution of the various components in the system. Friction in this case will be of a more viscous nature. Continuous efforts are made to reduce the dynamic dead zone, yet in the last analysis a certain amount of dead zone is essential. With round-ported overlapped and semi-overlapped valves, analytical treatment concludes that the servomechanism is unstable with the valve in the neutral position. The fact that these servomechanisms are stable is due to the friction in the system. Although the servomechanism is nominally in the dead zone it may often be observed to drift slowly and erratically with zero input signal within the dead zone until a position is found at which it can stay long enough to establish a static dead zone condition.

Predictions of the magnitude of dead zone are difficult, because the ultimate gain of the system may not be known, and it is impracticable to control the equivalent dead zone of the components, except to ensure that it does not exceed a certain limit. Thus, unless very extensive testing of components has been done, any calculation will present the worst case, the final measured results possibly being up to three or four times better.

12.9 METHODS OF IMPROVING PERFORMANCE

Having accepted the fact that the threshold condition is a real problem in practice, it is now necessary to investigate a method of improving the situation. This must fall into two general approaches:

(i) In addition to good basic design, improve the design and manufacturing tolerances of the components.
(ii) Alter the system arrangement, possibly with additional loops.

To achieve satisfactory performance the first of these approaches is not always the most economic since development should be aimed at widening tolerances in order to reduce manufacturing costs. Hence the second approach would seem to offer the more economic method. An example of this is shown in Fig. 12.22

where the improvement achieved by using small amplitude high frequency excitation of the control valve is evident. This may not always be an acceptable solution.

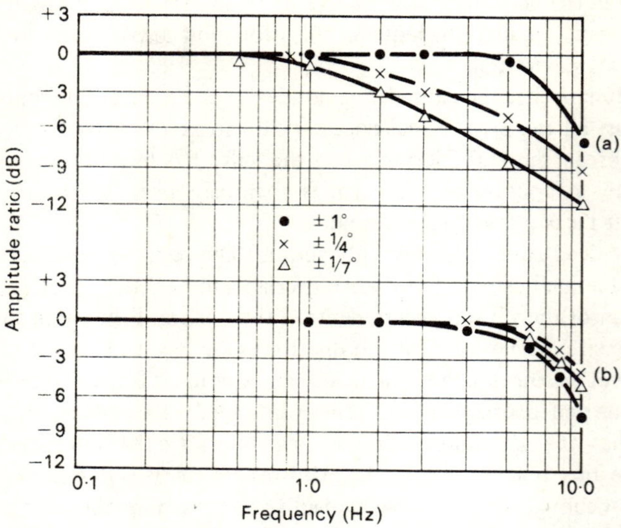

Fig. 12.22 – Effect of dither on amplitude response. Zero lap, gain 42.5 mA/deg. (a) No dither. (b) 175 Hz dither – for different input amplitudes.

Consider the system shown in Fig. 12.23. In this case the threshold (T) occurs in the forward loop path. The transfer function will be:

$$\theta_o = [K_A][T][KG(s)]\theta_e$$

and for the feedback loop:

$$\theta_e = \theta_i - \theta_o' = \theta_i - K_f\theta_o \, .$$

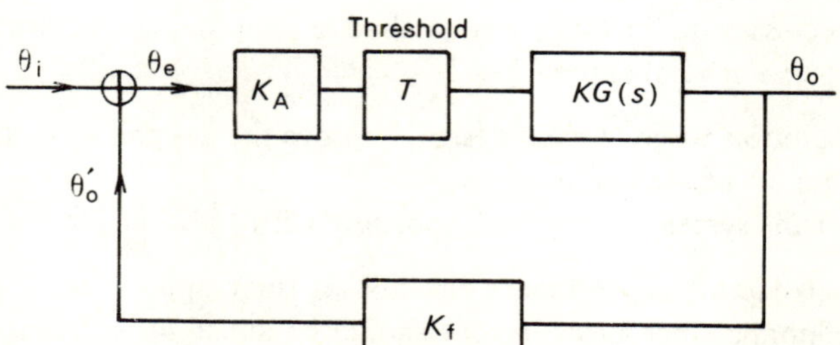

Fig. 12.23 – Closed loop with threshold included.

These two relationships combine to give the closed loop transfer function:

$$\frac{\theta_o}{\theta_i} = \frac{[K_A][T][KG(s)]}{1 + [K_f][K_A][T][KG(s)]} .$$

It has been a well known practice to reduce the threshold by increasing the forward loop gain K_A so that T makes a smaller contribution to the overall gain.

If the characteristic for T were a dead zone, for example valve overlap as, shown in Fig. 12.24, the characteristic could be approximated by two gains, K_1 and K_2. The threshold could then be described by:

$$|T| = [K_1] \text{ for } \theta > +h \text{ and } \theta < -h$$
$$= [K_2] \text{ for } -h < \theta < h .$$

By increasing the amplifier gain K_A, an attempt is made to make the effect of K_2 small, but in doing so K_1 is also affected, so that the system gain is greatly increased outside the limits of $\pm h$. This will result in an alteration of the system stability and performance, and hence some form of additional shaping loop would be required to return the system to its original performance. A phase lag network, for example, would help to attenuate the increased gain at the higher frequencies. Clearly, this is a sledge-hammer approach. Ideally a gain increase in the threshold region only is required, and this to contribute no phase lag. This may be possible with non-linear networks.

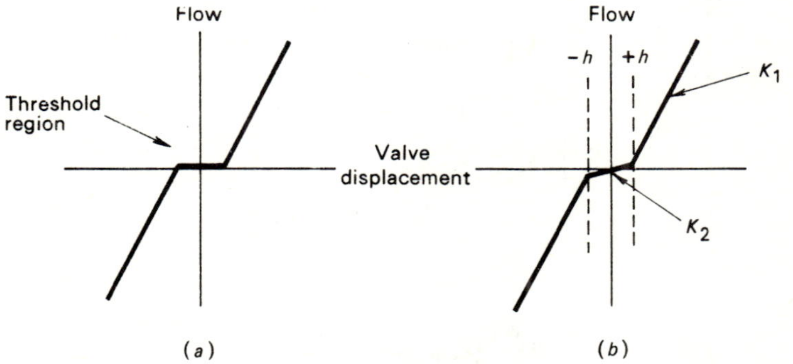

Fig. 12.24 – Approximations to the threshold characteristics.

Some practical results of the interrelationship between attempts to improve threshold and the resultant effects on performance are shown in Table 12.1. The type of output waveform distortion experienced is shown in Fig. 12.25, where the effect of the inertia load bouncing on the oil column is clearly seen.

Table 12.1.

Run	Threshold (deg)	Waveform	Gain (mA/deg)	Lap condition (mm)	Bleed diameter (mm)	Velocity feed-back	Acceler-ation feed-back
1	0.085	Poor	25.6	Overlap 0.0254	–	–	–
2	0.29	Much better	25.6	Overlap 0.0254	0.343	–	–
3	0.12	Better	50.0	Overlap 0.0254	0.343	Yes	–
4	0.088	Worse	25.8	Overlap 0.0254	–	Yes	–
5	0.044	Worse	57.5	Overlap 0.0254	0.152	Yes	–
6	0.05	Much better	25.0	Overlap 0.0254	–	–	Yes
7	0.008	Much better	42.5 (175 Hz dither)	Zero	–	–	Yes

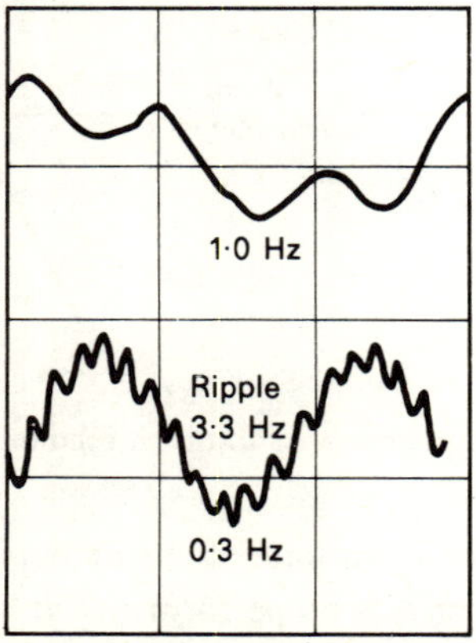

Fig. 12.25 – Output waveform distortion, showing effect of inertia load bouncing on the oil column.

12.10 COMPUTER STUDY OF SMALL AMPLITUDE RESPONSE

The flow through the control ports of the valve has been described by:

$$Q = K_1 a \sqrt{(P_s - P_L)}$$

where a is the uncovered orifice area. This equation is only approximately true since the discharge coefficient will vary at small openings and the orifice area displacement curve will be non-linear. Some comparisons are drawn in Fig.12.26.

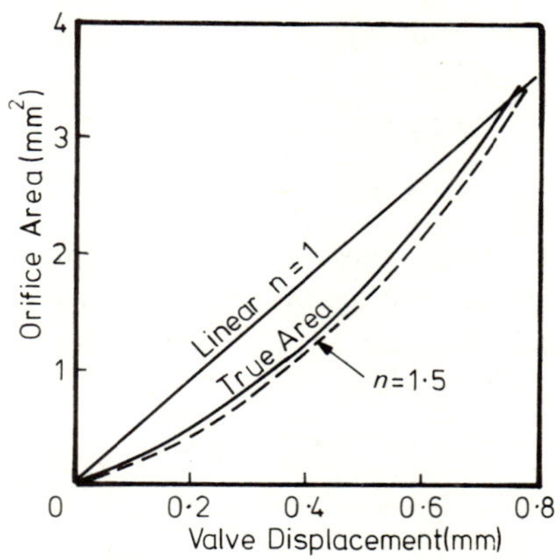

Fig. 12. 26 – Comparison between true and simulated orifice openings.

Using the equations developed in section 12.3, and assuming that cavitation is unlikely to occur under small amplitude response conditions, then the open loop condition is:

$$a^* \left\{1 - \left(\alpha\omega^* \frac{du^*}{d\tau} + \gamma u^* + \eta \frac{u^*}{|u^*|}\right)\right\}^{\frac{1}{2}}$$

$$= \omega^{*2} \frac{d^2u^*}{d\tau^2} + \left(\frac{\gamma}{\alpha} + \phi\right)\omega^* \frac{du^*}{d\tau} + \left(1 + \frac{\phi\gamma}{\alpha}\right)u^* + \frac{\phi}{\alpha}\eta \frac{u^*}{|u^*|} \quad . \qquad (12.45)$$

When non-dimensional area is used, the inertia load factor is defined slightly differently:

$$\alpha = \frac{MK_1 \hat{a}\sqrt{(P_s/2)}\,\omega_n}{P_s A^2}$$

where $\quad a^* = a/\hat{a}.$

It is seen from Fig.12.26 that the most realistic area characteristic is given by:

$$a = K_2 x_v^n$$

where $n = 3/2$.

$n = 1$ will be the usual linear relationship.

For a sinusoidal input, let:

$$x_v = T\hat{x}_v \sin \tau$$

where T = some factor by which the input is altered and describes the input amplitude selected.

Hence:

$$a^* = \frac{K_2 x_v^n}{a} = \frac{K_2 (T\hat{x}_v \sin \tau)^n}{K_2 x_v^n} = (T \sin \tau)^n \ . \tag{12.46}$$

It can be seen [12] from considerations of simple relationships that the minimum requirement to achieve a threshold condition is that a leakage path must be present except in the case of valve overlap. Without a leakage path any opening of the valve will allow full system pressure to act on the piston and hence

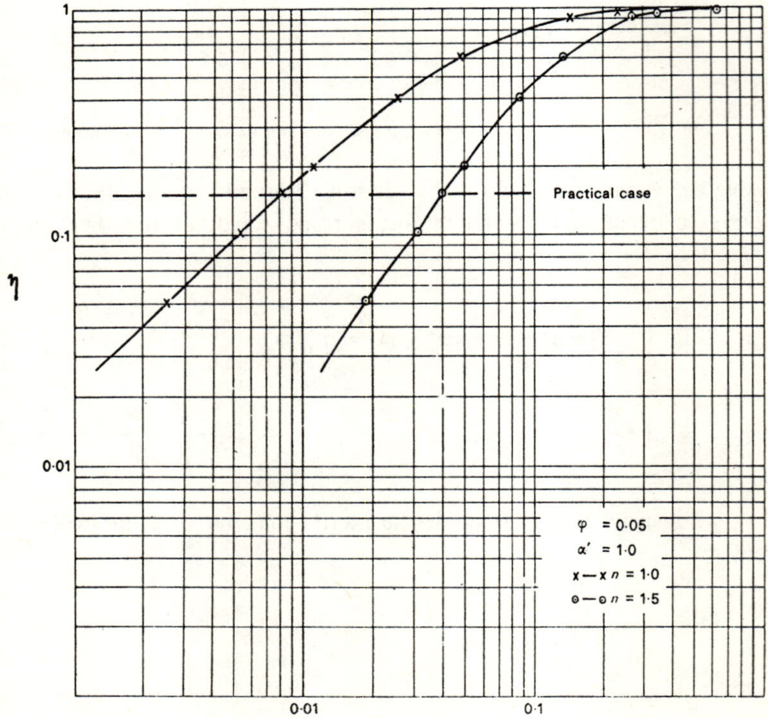

Fig. 12.27 – Theoretical variation of absolute threshold with seal friction.

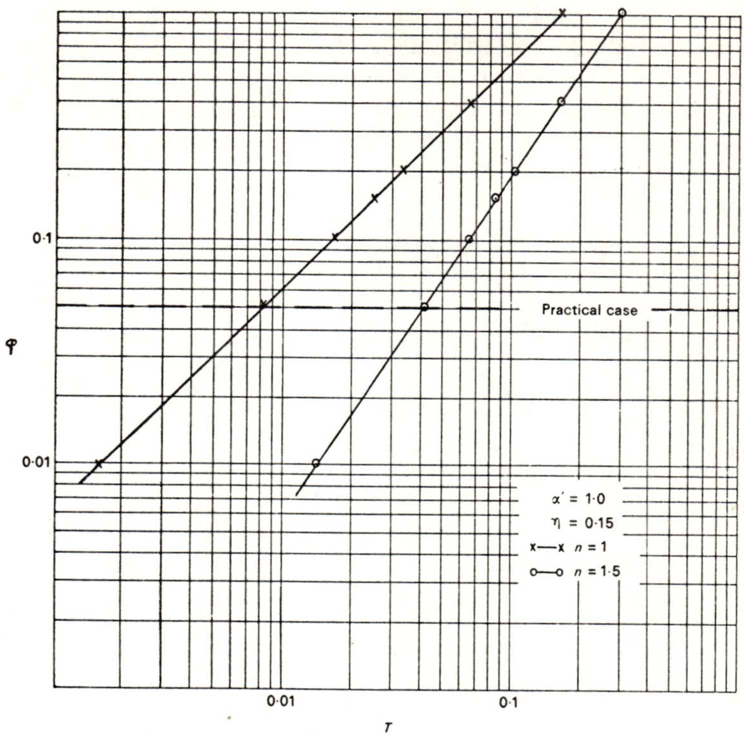

Fig. 12.28 – Theoretical variation of absolute threshold with leakage.

overcome any resistance. However, a real system will have a leakage path present. Before the piston can move flow must pass through the leakage path present. to establish a pressure sufficient to overcome the load.

The theoretical absolute threshold value can be obtained from equation (12.45) if it is assumed that all output motion ceases, giving u^*, $\mathrm{d}u^*/\mathrm{d}\tau$ and $\mathrm{d}^2u^*/\mathrm{d}\tau^2$ all zero. Hence:

$$T = \left\{\frac{\phi\eta}{\alpha\sqrt{(1-\eta)}}\right\}^{1/n} . \tag{12.47}$$

From this equation it is seen that only leakage and seal friction affect the final steady state conditions. These results are shown in Figs. 12.27 and 12.28. This shows clearly the necessity to assume a more realistic valve area-displacement characteristic.

In the case of seal friction it can be seen that the threshold value is infinite when the seal friction reaches 100 per cent. In other words, the seal friction force becomes equal to the maximum available force.

The computer result shows that at the smaller amplitudes ($T < 0.0173$, 1.7%) the system becomes oscillatory, as shown in Fig. 12.29. A stage is reached when a small input will cause the system to oscillate at its natural frequency. It seems likely that at these small amplitudes, conditions can be such that the inherent damping contribution from the servovalve is not able to keep the system stable.

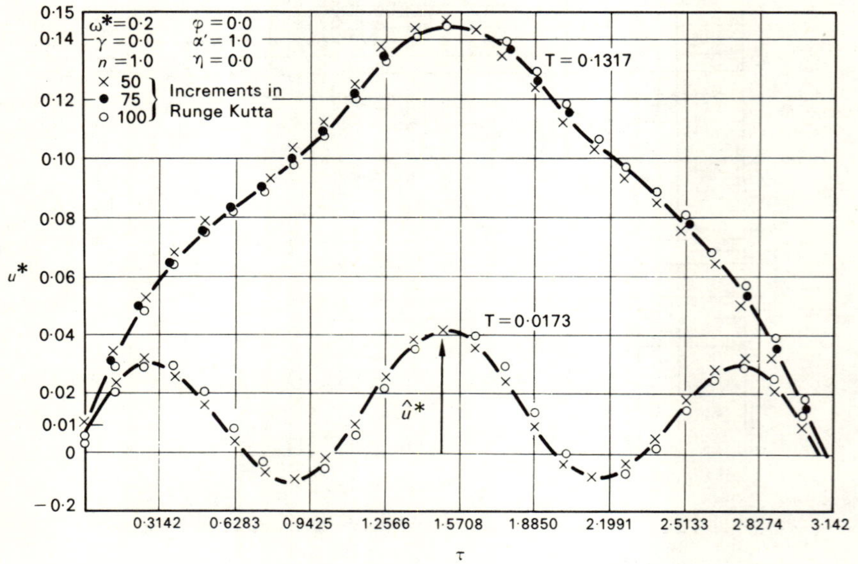

Fig. 12.29 – Effect of computational increment on computer results.

12.11 METHODS OF PERFORMANCE IMPROVEMENT

The loop gain of an electro-hydraulic servo can be conveniently divided into two groups:

The electrical gain, which includes the servo amplifier and position feedback characteristics.

The hydraulic gain, which includes the torque motor, valve and actuator characteristics.

If the servomotor has a high hydraulic gain, and an excess velocity capability over that necessary to achieve the frequency response bandwidth, then the electrical gain must be reduced in order to maintain the stability margin. The amplifier and torque motor are therefore less effective. The servovalve and actuator should be matched so that the saturation velocity is just sufficient to meet the response requirements. Amplifier, torque motor and velocity saturation points should be as near as possible. The loop gain can then be optimised in the electronic sections, which are more linear and more flexible,

12.11.1 Orifice bleeds and valve underlap

The instability and harmonic distortion which are due to the dynamic behaviour of the servo and its structure can be damped by an orifice bleed across the actuator or by underlapping the valve.

The orifice within the actuator is different from simple viscous damping in that it is a function of the pressure difference across the cylinder, and thus of output force. This device approximates to an acceleration feedback, but it is also operative when there is an ouput load at steady velocity. There is an optimum orifice size for maximum sensitivity, though this may not damp the harmonics sufficiently.

The difficulties with this method are [11, 16]:

It exaggerates the effects of piston seal friction; a dead zone is caused because the differential pressure necessary to overcome friction requires a flow through the orifice, with a corresponding control valve displacement and error signal.

For the same reason an error is inherent with orifice bleed stabilisation if the servo is required to hold an output load.

The pressure drop-flow characteristics is a square law, so that the damping virtually disappears at low velocities. This limits the gain of the system if a secondary instability is to be avoided, and this condition can be aggravated by the equivalent form of backlash, particualrly seal slip-stick effects.

The orifice bleed significantly reduces the bandwidth of the system.

However, the orifice is a cheap and easy feature to incorporate, and is sufficient in many cases. It has an advantage in that it combats cylinder pressure peaks, and indeed, capillary type bleeds are sometimes introduced with this as the primary objective.

12.11.2 Pressure feedbacks

It is generally necessary [13, 16] to introduce an acceleration term or an approximation to this. This may, of course, be for reasons other than threshold, for example, harmonic distortion due to peaking pressures. The general principle is to feed back the cylinder differential pressure. The use of transducers and their associated electronics is obvious, but not generally favoured because of complexity and cost. The hydraulic two-stage valve suggests an hydraulic solution, since the cylinder pressures are easily ducted to actuate the second stage spool. Simple pressure feedback of this type can result in reduced output system stiffness. Dynamic pressure feedback has been developed to overcome this problem. This method is effective but requires careful matching.

12.11.3 Electrical networks

Standard textbooks give details of many transducer circuit passive networks which have their use in particular circumstances. The following are briefly described because of their usefulness in a more general sense.

Velocity feedback

The usefulness of this form of damping is marred because its effect on the natural frequency harmonic of the load and actuator is limited. A signal which is strong enough to reduce the harmonics will give too much lag at lower frequencies. However, if another feature is introduced to deal with the harmonic, for example an orifice bleed, then velocity feedback will enable the gain to be increased, and a decrease in dead zone will be achieved.

Acceleration feedback

There is no doubt as to the effectiveness of an acceleration term in stabilising an hydraulic actuator. Accelerometers have been developed to provide a wide range of performance with high accuracy, but they are costly. Simple feedbacks can be achieved by using transient velocity feedback and shaping networks in conjunction with the servo amplifer. These are shown in Fig. 12.30 but must be 'tuned' to the load characteristics. This method has the advantage of being less sensitive to structural noise (which would be picked up by the accelerometer).

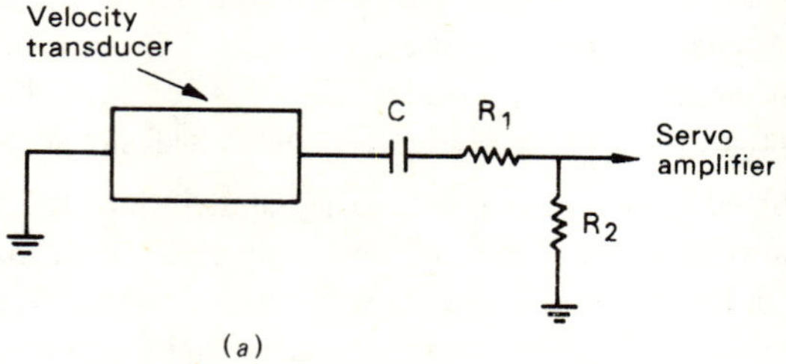

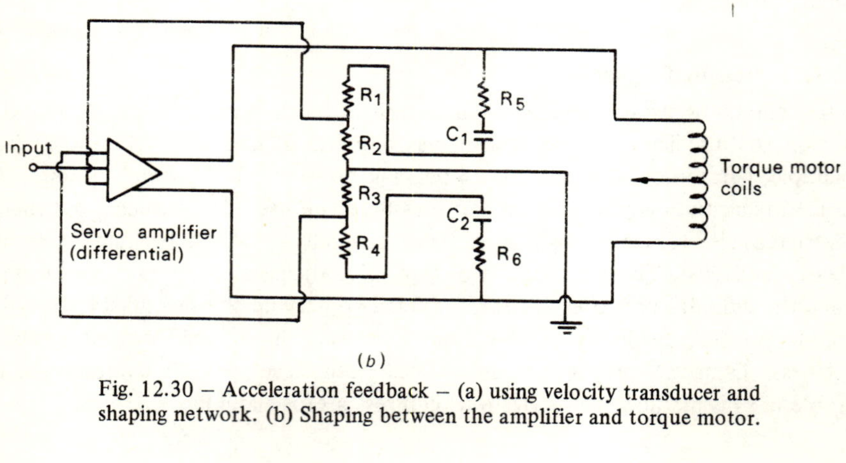

Fig. 12.30 – Acceleration feedback – (a) using velocity transducer and shaping network. (b) Shaping between the amplifier and torque motor.

Feedback shaping can be used to great effect, especially in closing the loop around the valve, torque motor and amplifier. This gives much greater control of the valve position which is effectively equivalent to output velocity.

12.12 EXAMPLES

1. An electro-hydraulic positional control system consists of an amplifier and a loaded servomotor having an inertia of $1.25 \times 10^{-6}\,\text{kg m}^2$. The speed ratio of motor to load shaft is 100, and the viscous friction of the system, measured at the motor, is $6.76 \times 10^{-5}\,\text{N m/rad/s}$. The feedback is via a gearing differential. The gearing differential produces 1 V/deg error, and the motor produces a torque of 0.027 N m per 100 V supplied by the amplifier. The system damping is 0.25 of the critical damping value. Determine:

   (a) The required amplifier gain.
   (b) The frequency of damped oscillations.
   (c) The steady state error when the input is driven at a constant 20 rev/min.

   (*Answers*: (a) 94; (b) 16.6 Hz; (c) 0.56 deg.)

2. An electro-hydraulic control system is used to position the cutting head of a machine tool. The loop consists of:

   (a) A linear transducer at the input which is capable of giving 5 V output for 2.54 cm movement at its input.
   (b) An electronic amplifer of gain 7.5 mA/V and incorporating a simple lag network in series, with a time constant of 0.75 s.
   (c) An electromagnetic torque motor of gain 0.38 cm/mA, the dynamic behaviour of which can be represented by a simple exponential lag. Experimental results show that 98 per cent of steady state is reached after 1.85 s.
   (d) A hydraulic spool valve and actuator, the output rate of which is 26.67 cm/s for a spool displacement of 1.27 cm. Experimental results show that the time constant is 0.5 s and that the damping ratio is 0.3.
   (e) An output transducer with a gain of 0.98 V/cm.

   Find the transfer function for the system and establish the system stability.

   (*Answer*: System is unstable.)

12.13 REFERENCES

[1] Royle, J. K. (1959), Inherent non-linear effects in hydraulic control systems with inertia loading. *Proc. Inst. Mech. Engrs.*, London, **173**, p. 257.

[2] Wang, P. K. C. and Ma, J. T. C. (1962), Cavitation in valve controlled hydraulic actuators. *J. Appl. Mech.*, **29**, *Trans. A.S.M.E.*, **84**, p. 435.

[3] McCloy, D. and Martin, H. R. (1963–64), Some effects of cavitation and flow forces in the electrohydraulic servo mechanism. *Proc. Inst. Mech. Engrs.*, London, **178**.

[4] McCloy, D. (1965), 'Some effects of oil compressibility and valve lap on the performance and stability of the hydraulic servo mechanism'. Ph.D. thesis, Queen's University, Belfast.

[5] Martin, H. R. (1968), 'A theoretical study into peaking pressure reduction and its effects on hydraulic servo performance'. Ph.D. thesis, Nottingham University.

[6] Martin, H. R. and Lichtarowicz, A. (1966–67), Theoretical investigations into the prevention of cavitation in hydraulic actuators. *Proc. Inst. Mech. Engrs.*, London, **181**, Pt. 1, No. 18.

[7] Williams, H. (1960), Effect of oil momentum forces on the performance of electrohydraulic servomechanisms. *Proc. Conf. Recent Mechanical Engineering Developments in Automatic Control*, **31**, Inst. Mech. Engrs., London.

[8] Eynon, G. T. (1960), Developments in high performance electromechanical servomechanisms at the Royal Aircraft Establishment, Farnborough, (ibid.).

[9] Carrington, J. E. and Martin, H. R. (1961), Some uses of electrohydraulic servo systems in connection with aircraft cockpit simulators. *Proc. Conf. on Oil and Hydraulic Power, Transmission and Control*, Inst. Mech. Engrs., London, November.

[10] Guillon, M. (1961), *Hydraulic Servo Systems*, Butterworths.

[11] Carrington, J. E. and Martin, H. R. (1960), Threshold problems in electro-hydraulic servomotors. *Proc. Inst. Mech. Engrs.*, London, **180**, Pt. 1, No. 37.

[12] Martin, H. R. and Lichtarowicz, A. (1969), Small amplitude response characteristics in hydraulic servomechanisms. *4th Congress of I.F.A.C.*, Warsaw, June.

[13] Bell, R., and de Pennington, A. (1969–70), Active compensation of lightly damped electro-hydraulic cylinder drives using derivative signals, *Proc. Inst. Mech. Engrs.*, **184**.

[14] McCloy, D. (1973), *Cavitation effects in on–off controlled hydraulic servos*, A.S.M.E. Paper 72-ACT-A.

[15] McCloy, D., Armstrong, P. J. (1973), Design of high speed hydraulic position servomechanisms using optimissation techniques, *Proc. Inst. Mech. Eng.*, **187**.

[16] McCloy, D., and Martin, H. A. (1978), Some effects of non-linear friction on the response of an electrohydraulic servomechanism, *8th Nat. Conf. Fluid Power*, Coimbatore, India.

CHAPTER 13

Pneumatic servomechanisms

13.1 INTRODUCTION

The analysis of the pneumatic servo has received relatively little attention in comparison to the hydraulic servo. This is particularly due to the fact that the equations for compressible flow are more difficult to handle and to the fact that this type of servo is not so popular. The reason for this will become apparent as comparisons are made with its hydraulic counterpart.

The fact that pneumatic servos use a very compressible fluid accounts for the two main differences in performance. If, for example, air is introduced into a cylinder when the valve is opened, then sufficient gas must flow into the cylinder volume to increase the pressure and build up enough force to overcome any spring pre-load or static friction. In hydraulic systems a very small valve displacement will cause immediate pressure rise in the cylinder. Hence hydraulics have a rapid initial response while pneumatics exhibit a time delay.

Because the fluid is compressible, the servo will exhibit a lack of stiffness, especially to external load disturbances, and in order to achieve the stiff characteristics of hydraulics, these systems must either be large enough to absorb any load variation with relatively small change in pressure, or be provided with a large servovalve capable of rapidly adding or removing gas.

On the other hand, a compressible fluid can be used to store energy as in the case of the hydro-pneumatic accumulator discussed in Chapter 3.

The efficiency of pneumatic systems is much lower than that of hydraulic systems. Two important factors causing loss are:

Cooling of the fluid, which dissipates energy in the form of heat.

Use of positive displacement actuators causes loss of energy by expansion.

In addition it is difficult to achieve satisfactory sealing, and several kilowatts may be used in making up leakage. This puts a limit on system pressure. The poor lubricating properties of gases compared to oil, also lead to energy losses. Table 13.1 highlights the main differences from the fluid mechanic viewpoint. One big advantage of the gas actuator is its ability to work above 500°C, where both electrical and hydraulic systems would fail.

The basic equations of pneumatic systems are non-linear and tend to be more complex than those of incompressible flow systems. This may explain why this field of analysis has received little attention. One of the first analyses was carried out by Shearer, who studied an open loop system consisting of a pneumatic servomotor driving an inertia load acted on by an external force. He used a linear approach, and this will be developed later in this chapter [1, 2].

Table 13.1.

	Orifice flow	Capillary flow
Gas	Flow α upstream pressure (choked)	Flow α (upstream pressure)^2
Hydraulics	Flow α $\sqrt{}$(pressure drop)	Flow α pressure drop

Over the last few years there has been a renewed interest in the analysis of pneumatic systems, mainly brought about by the easy access to computers and in particular the use of digital continuous system simulation languages [3, 4, 5, 6].

13.2 PNEUMATIC SERVOS

The pneumatic servo configuration is very similar to that considered in the chapter dealing with hydraulic systems. Most of the earlier pneumatic control systems were used in the process control industries, where low pressure air (7 bar) was easily obtainable and gave a sufficiently fast response. Nowadays, owing to the constantly increasing temperature requirements for missile and aircraft components, a renewed interest has developed for high speed pneumatic control devices using pressures in the region of 150 bar [7].

In the system considered, a five-port valve is used to control a double acting cylinder. A typical arrangement is shown in Fig. 13.1.

13.2.1 A model for a pneumatic valve and actuator

Construction of a mathematical model involves three major considerations:

(1) the determination of the mass flow rate through the valve,
(2) the determination of the pressure, volume and temperature of the air in cylinder,
(3) the determination of the dynamics of the load.

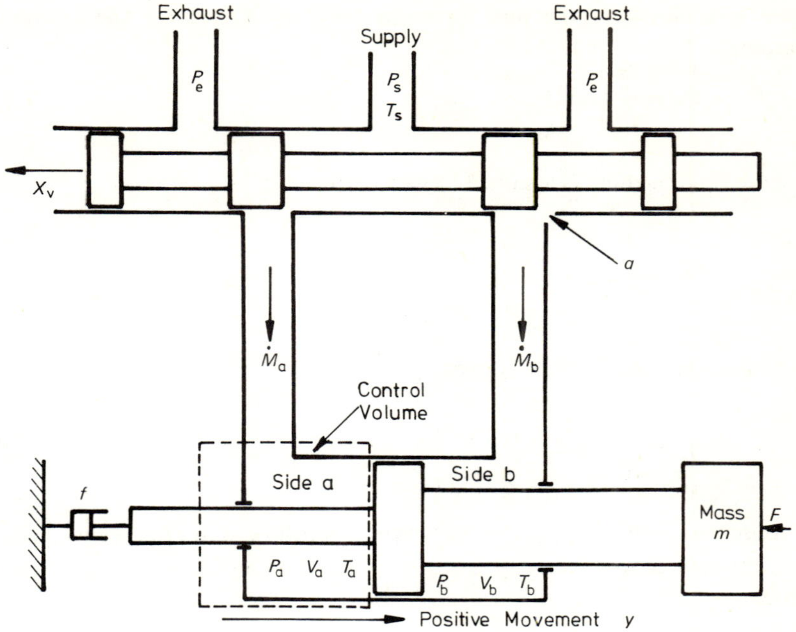

Fig. 13.1 – The pneumatic actuator.

13.2.2 The valve characteristics

The characteristics of pneumatic valves have been discussed in section 4.3. In this case the five-port valve is treated as equivalent to two three-port valves, one for each side of the cylinder. Thus the mass flow rates are given by:

$$\dot{M}_a = C_{qa} C_{ma} a_a P_s / \sqrt{T_s} \tag{13.1}$$

and

$$\dot{M}_b = -C_{qb} C_{mb} a_b P_b / \sqrt{T_b} \tag{13.2}$$

where C_{qa} and a_a are the flow coefficient and area of the supply port and C_{qb} and a_b the same for the exhaust port. $C_{ma} = f(P_a/P_s)$ and $C_{mb} = f(P_e/P_b)$ and can be determined from Fig. 4.5.

Equations (13.1) and (13.2) assume that there is only one controlling orifice on each side of the cylinder. A more detailed analysis, accounting for the presence of series restrictions in the valve, has been developed by Sanville [11].

13.2.3 The charging process

Considering the left-hand side of the cylinder, Fig. 13.1, the thermodynamic system is enclosed in the box, or control volume. Applying the continuity equation:

$$\dot{M}_a = \frac{d}{dt}(\rho_a V_a)\,. \tag{13.3}$$

Since the fluid medium is assumed to be a perfect gas:

$$\rho_a = \frac{P_a}{RT_a}$$

Substitution in equation (13.3) gives:

$$\dot{M}_a = \frac{1}{R}\frac{d}{dt}\left(\frac{P_a V_a}{T_a}\right). \tag{13.4}$$

The conservation of energy states that the energy inflow minus the energy outflow equals the rate at which energy is stored inside the control volume; this can be expressed:

$$\Sigma \dot{M}_{in} h'_{in} - \Sigma \dot{M}_{out} h'_{out} + \left\{\frac{dQ_h}{dt} - \frac{d\Phi}{dt}\right\} = \frac{dE}{dt} \tag{13.5}$$

where Q_h = heat flow to the control volume
Φ = work done by the system
h' = total energy per unit mass of fluid

$$h' = u + \frac{P}{\rho} + \frac{V^2}{2} + gz = h + \frac{V^2}{2} + gz$$

h = enthalpy of the fluid
u = internal energy
E = total internal energy inside the control volume
z = elevation
v = fluid velocity.

The system control volume gives:

$$C_p \dot{M}_a T_a - P_a \frac{dV_a}{dt} + \frac{dQ_h}{dt} = \frac{d}{dt}(C_v \rho_a V_a T_a) \tag{13.6}$$

since for a perfect gas, flowing between points 1 and 2:

$$C_p(T_1 - T_2) = \frac{V_2^2 - V_1^2}{2}.$$

If the cylinder is perfectly insulated (adiabatic conditions) and again assuming a perfect gas, (13.6) can be rewritten:

$$\dot{M}_a T_a - \frac{P_a}{C_p}\frac{\mathrm{d}V_a}{\mathrm{d}t} = \frac{1}{\gamma R}\frac{\mathrm{d}}{\mathrm{d}t}(P_a V_a)$$

or

$$\dot{M}_a = \frac{1}{RT_a}\left\{P_a \frac{\mathrm{d}V_a}{\mathrm{d}t} + \frac{V_a}{\gamma}\frac{\mathrm{d}P_a}{\mathrm{d}t}\right\}. \tag{13.7}$$

since

$$\gamma = \frac{C_p}{C_v} \quad \text{and} \quad \rho = \frac{P}{RT} \quad \text{and} \quad C_p = \frac{\gamma R}{\gamma - 1}.$$

By a similar argument:

$$-\dot{M}_b T_b - \frac{P_b}{C_p}\frac{\mathrm{d}V_b}{\mathrm{d}t} = \frac{1}{\gamma R}\frac{\mathrm{d}}{\mathrm{d}t}(P_b V_b)$$

and

$$-\dot{M}_b = \frac{1}{RT_b}\, P_b \frac{\mathrm{d}V_b}{\mathrm{d}t} + \frac{V_b}{\gamma}\frac{\mathrm{d}P_b}{\mathrm{d}t}\ . \tag{13.8}$$

13.2.4 The load dynamics

The load equation is:

$$P_a A_a - P_b A_b - P_{atm}(A_a - A_b) = m\frac{\mathrm{d}^2 y}{\mathrm{d}t^2} + f\frac{\mathrm{d}y}{\mathrm{d}t} + F + F_f\ . \tag{13.9}$$

where A_a and A_b are the effective piston areas in the left- and right-hand sides of the cylinder, m is an inertia load, f a viscous friction load, F a constant external force and F_f a friction force. The third term on the left-hand side of equation (13.9) is included to account for the fact that the pressures are absolute.

The friction force F_f can be highly non-linear, one formulation being:

$$F_f = F_c \operatorname{sgn}(\mathrm{d}y/\mathrm{d}t) + C_1(\mathrm{d}y/\mathrm{d}t) + C_2(P_a - P_b)\ . \tag{13.10}$$

The total friction is a combination of load friction and seal friction. It is difficult to predict with accuracy and is affected by such things as dwell time, ambient temperature, amount of lubrication present, dryness of the air etc.

13.3 SOLUTION OF THE MODEL EQUATIONS

There are five basic equations in the mathematical model, (13.1) and (13.2) for the mass flow rates, (13.7) and (13.8) for the relationship between mass flow rates and conditions in the cylinder, (13.9) for the relationship between the conditions in the cylinder and the motion of the load. There are several non-linearities present making analytical solution impossible. Several computer solutions have been derived, however [3, 4, 6].

Before discussing solutions it would be instructive to consider experimental measurements of the response of a pneumatic valve-actuator system. PERA, the Production Engineering Research Association, have carried out extensive tests, reported in [3]. Typical results for step responses of cylinders with equal size ports, restricted inlet ports and restricted outlet ports are shown in Fig. 13.2. These illustrate several important features:

(a) the effect of valve lag,
(b) the delay in the cylinder response caused by static friction forces,
(c) the oscillatory nature of the response where ports are restricted.

These effects will be discussed in more detail in the following chapter. The present chapter is more concerned with the use of the pnematic system as a servomechanism. Thus stability is of prime importance and a linearised analysis will be used for its investigation.

13.4 LINEARISED ANALYSIS OF THE PNEUMATIC SERVO

In the system considered, a five-port valve is used to control a double-acting through rod cylinder, so that $A_a = A_b = A$ in Fig. 13.1. The valve could be operated directly by a torque motor; is some cases a flapper-nozzle valve is favoured.

Linear analysis, first pursued by Shearer, requires the following assumptions to be made [1, 2]:

(1) Supply pressure P_s and temperature T_s are constant.
(2) No heat transfer between the fluid and its environment.
(3) The piston moves only small distances from the central postion.
(4) The cylinder pressures vary only by small amounts from an initial steady value P_i.

Other assumptions will be encountered in the analysis.

The first step is to linearise the valve characteristics given in equations (13.1) and (13.2). This has already been done in section 7.4. The linearised equations are:

$$\delta \dot{M}_a = k_{1a}\delta a_a + k_{2a}\delta P_s + k_{3a}\delta P_a + k_{4a}\delta T_a$$

$$\delta \dot{M}_b = k_{1b}\delta a_b + k_{2b}\delta P_b + k_{3b}\delta P_e + k_{4b}\delta T_b \ , \qquad (13.11)$$

where the k represents the relevant partial derivative evaluated at the initial condition. The partial derivatives may be evaluated experimentally as in Fig. 7.5 or determined from equation (7.14).

Now $\delta P_s = \delta P_e = 0$, and since the valve is symmetrical $\delta a_a = \delta a_b = \delta a$.

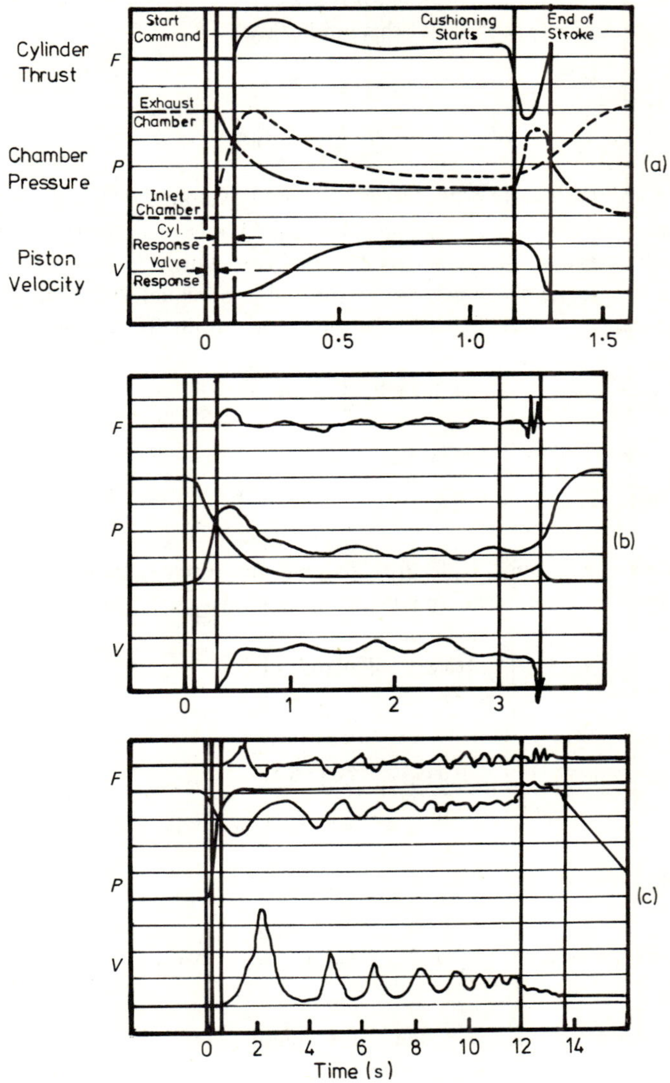

Fig. 13.2 – Experimental results from a pneumatic actuator (a) equal size ports (b) restricted inlet (c) restricted outlet (after [3]).

Also for ease of analysis it is assumed that $\delta T_a = \delta T_b = 0$. Equations (13.11) now become:

$$\delta \dot{M}_a = k_{1a}\delta a + k_{3a}\delta P_a$$

and

$$\delta \dot{M}_b = k_{1b}\delta a + k_{2b}\delta P_b \;. \tag{13.12}$$

The analysis can be further simplified if it is assumed that $k_{3a} = -k_{2b}$. Referring back to equation (7.14) shows that this implies that:

$$\left\{\frac{\partial \dot{M}_a}{\partial P_a}\right\}_i = -\zeta\frac{\dot{M}_i}{P_{ai}} = -\left\{\frac{\partial \dot{M}_b}{\partial P_b}\right\}_i = -\frac{\dot{M}_i}{P_{bi}}(1+\zeta)\;.$$

This will be approximately true, bearing in mind assumption (4) above, when ζ is very large or when the orifice pressure ratio tends to unity (Fig. 7.6). This implies small pressure drops across the orifices, which is consistent with a linearised analysis. These coefficients would also be equal, but both would be zero for the case of a zero lapped valve operating about its zero position (since $\dot{M}_i = 0$).

Thus equations (13.12) can now be written:

$$\delta \dot{M}_a = k_1\delta a + k_2\delta P_a$$

$$\delta \dot{M}_b = k_1\delta a - k_2\delta P_b \;. \tag{13.13}$$

The equations (13.7) and (13.8) for charging and discharging now need to be linearised. They become, neglecting products of small quantities:

$$\delta \dot{M}_a = \frac{1}{RT_s}\left\{P_i\frac{\mathrm{d}(\delta V_a)}{\mathrm{d}t} + \frac{V_i}{\gamma}\frac{\mathrm{d}(\delta P_a)}{\mathrm{d}t}\right\} \tag{13.14}$$

and

$$-\delta \dot{M}_b = \frac{1}{RT_s}\left\{P_i\frac{\mathrm{d}(\delta V_b)}{\mathrm{d}t} + \frac{V_i}{\gamma}\frac{\mathrm{d}(\delta P_b)}{\mathrm{d}t}\right\} \tag{13.15}$$

where the subscript i denotes initial conditions. Hence:

$$\delta \dot{M}_a + \delta \dot{M}_b = \frac{1}{RT_s}\left\{P_i(\dot{V}_a - \dot{V}_b) + \frac{V_i}{\gamma}(\dot{P}_a - \dot{P}_b)\right\}. \tag{13.16}$$

At any instant the volumes each side of the cylinder are:

$$V_a = V_i + A\delta y$$

$$V_b = V_i - A\delta y\;.$$

Hence $V_a - V_b = 2A\delta y$. (13.17)

Substituting in equation (13.16) gives:

$$\delta \dot{M}_a + \delta \dot{M}_b = \frac{1}{RT_s}\left\{2P_i A\dot{y} + \frac{V_i}{\gamma}(\dot{P}_a - \dot{P}_b)\right\}. \quad (13.18)$$

Using equation (13.13) gives:

$$2k_1\delta a + k_2(\delta P_a - \delta P_b) = \frac{1}{RT_s}\left\{2P_i A\dot{y} + \frac{V_i}{\gamma}(\dot{P}_a - \dot{P}_b)\right\}. \quad (13.19)$$

Introducing the Laplace operator and rearranging gives:

$$As(\delta y) = k_4(\delta a) - (k_5 + k_6 s)(\delta P_a - \delta P_b) \quad (13.20)$$

where $k_4 = k_1 RT_s/P_i$
$k_5 = -k_2 RT_s/2P_i$
$k_6 = V_i/2\gamma P_i$.

Finally it is necessary to linearise the load equation given by (13.9). For the case under consideration seal friction is neglected and this becomes:

$$A(\delta P_a - \delta P_b) = m\,\mathrm{d}^2y/\mathrm{d}t^2 + f\,\mathrm{d}y/\mathrm{d}t + F .$$

This is linearised to become:

$$A(\delta P_a - \delta P_b) = m\ddot{y} + f\dot{y} + \delta F . \quad (13.21)$$

Combining this with equation (13.20) gives the open loop transfer function for the valve-cylinder combination:

$$\left\{\frac{mk_6}{fk_5 + A^2}s^2 + \frac{mk_5 + fk_6}{fk_5 + A^2}s + 1\right\}(s\delta y)$$

$$= \frac{k_4 A(\delta a) - (k_5 + k_6 s)\delta F}{fk_5 + A^2} . \quad (13.22)$$

Thus the natural frequency ω_n is:

$$\omega_n = \sqrt{\frac{fk_5 + A^2}{mk_6}} \quad (13.23)$$

and the damping ratio c is:

$$c = \frac{mk_5 + fk_6}{2\sqrt{\{(fk_5 + A^2)m K_6\}}} . \qquad (13.24)$$

In a practical system, operating about valve zero, k_5 will be small. This combined with the fact that f is also likely to be small allows equation (13.23) to be written as:

$$\omega_n = \sqrt{(A^2/mk_6)} = \sqrt{(2\gamma P_i A^2/mV_i)} = \sqrt{(4\gamma P_i A^2/mV)}$$

where V = swept volume.

For the equivalent hydraulic system:

$$\omega_n = \sqrt{(4NA^2/mV)} .$$

The difference between these two formulae is the use of the pneumatic bulk modulus γP_i instead of the liquid one. The hydraulic natural frequency is clearly much greater than that for the pneumatic system. This tends to give a better dynamic response in the hydraulic case.

13.5 LOAD SENSITIVITY

Since the fluid is so compressible, the sensitivity to external loading can be high. From equation (13.22) it is seen that the steady state relationship for load velocity is:

$$(s\delta y)_{ss} = \frac{k_4 A(\delta a) - k_5 \delta F}{fk_5 + A^2} . \qquad (13.25)$$

Load sensitivity will be zero if k_5 is zero. Otherwise readjustments in δa will be necessary in order to hold a steady velocity against a changing load.

The sensitivity to load disturbances can also be found under dynamic conditions by considering equation (13.20):

$$As(\delta y) = k_4(\delta a) - (k_5 + k_6 s)(\delta P_a - \delta P_b) . \qquad (13.26)$$

If a position loop is arranged round the actuator, then:

$$\delta a = \delta a_i - n\delta y .$$

The external loading can be expressed by:

$$\delta F = -(\delta P_a - \delta P_b)A . \qquad (13.27)$$

Combining these three equations and imposing the condition that the input is held fixed at $\delta a_i = 0$:

$$As(\delta y) = -nk_4\delta y + (k_5 + k_6 s)\frac{\delta F}{A}$$

or

$$\frac{\delta F}{\delta y} = \frac{A(nk_4 + As)}{k_5 + k_6 s} \,. \tag{13.28}$$

Converting into the frequency domain and reducing the expression to a single complex number:

$$\frac{\delta F}{\delta y} = \frac{A[nk_4k_5 + Ak_5\omega^2 + (Ak_5 - nk_4k_6)\omega j]}{k_5^2 + k_6^2\,\omega^2} \,. \tag{13.29}$$

For small values of ω, this gives:

$$\frac{\delta F}{\delta y} = \frac{nAk_4}{k_5}$$

which is dependent on the valve characteristics. For large values of ω, equation (13.20) gives:

$$\frac{\delta F}{\delta y} = \frac{A^2}{k_6} = \frac{2\gamma P_i A^2}{V_i}$$

which is the stiffness of the fluid in the cylinder.

13.6 METHODS OF STABILISATION

Putting $\delta a = h\delta x_v$, the open loop equation for the system can be written ($\delta F =$ 0):

$$\frac{\delta y}{\delta x_v} = \frac{hk_4A/(k_5f + A^2)}{s\left\{\dfrac{1}{\omega_n^2}s^2 + \dfrac{2c}{\omega_n}s + 1\right\}} \,. \tag{13.30}$$

Practical experience shows that it is reasonable to assume that $k_5 = f = 0$ and therefore (13.30) reduces to:

$$\frac{\delta y}{\delta x_v} = \frac{hk_4\omega_n^2/A}{s(s^2 + \omega_n^2)} = KG(s) \,. \tag{13.31}$$

In order to improve the dynamic response, greater than 90° phase lead is needed in the feedback loop. This can be achieved by the addition of acceleration and velocity terms to the position feedback loop. From Fig. 13.3:

$$\frac{\delta y'}{\delta y} = (K_a s^2 + K_v s + K_y)\,. \tag{13.32}$$

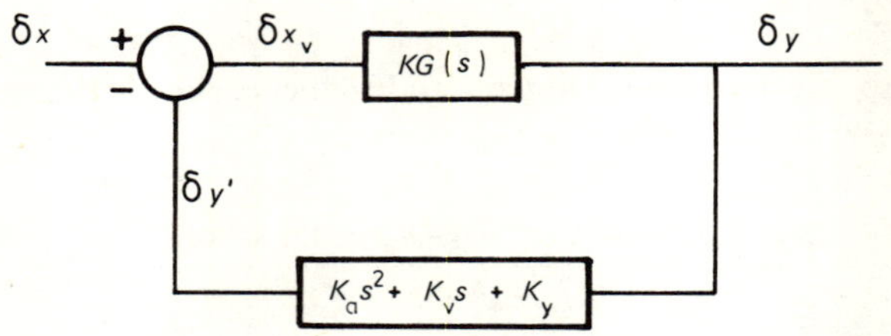

Fig. 13.3 – Block diagram of pneumatic servo with position, velocity and acceleration feedback.

Hence the loop transfer function is:

$$\frac{\delta y}{\delta x_v} = \frac{\dfrac{hk_4\omega_n^2}{A}\Big/ s(s^2+\omega_n^2)}{1 + h(K_a s^2 + K_v s + K_y)\dfrac{k_4\omega_n^2}{A}\Big/ s(s^2+\omega_n^2)} \tag{13.33}$$

A root locus plot can then be obtained from the characteristic equation:

$$h\frac{k_4}{A}\,\omega_n^2 K_a\,\frac{s^2 + (K_v/K_a)s + K_y/K_a}{s(s^2+\omega_n^2)} = -1\,. \tag{13.34}$$

The root locus plot for various pole-zero positions is shown in Fig. 13.4 [8]. It can be seen that Fig. 13.4(a) is completely unstable and Fig. 13.4(c) is conditionally unstable. This clearly confirms that the feedback loops need to be carefully tuned, otherwise any advantage is lost. The cases considered are obviously the worst condition since some inherent damping will be present in reality.

Pure acceleration feedback can be expensive and double differentiation of the position signal procedures a signal with a high noise content, but since the pressure in the cylinder determines the acceleration of the output load, it can be used as an acceptable alternative. However, there are certain problems associated with this approach, particularly if the output load contains significant damping and spring forces. Also if an external load is applied, the initial pressure feedback would have the opposite effect of acceleration feedback.

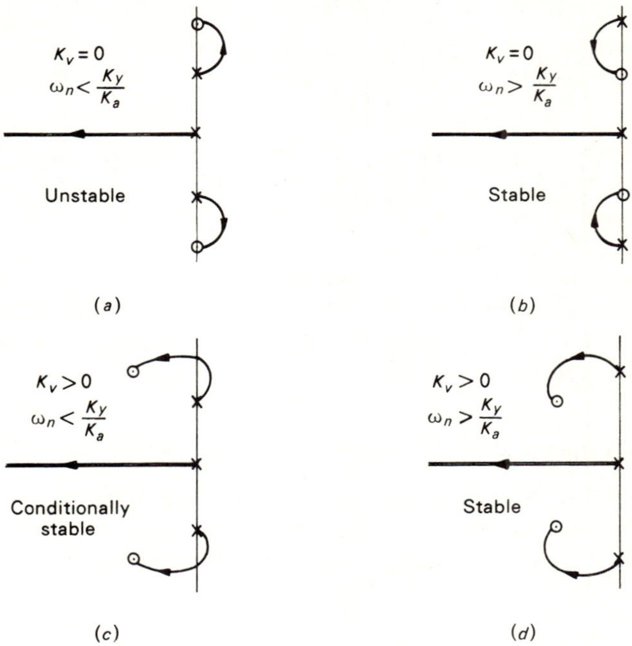

Fig. 13.4 – Effect of feedback combination on root locus (after [8]).

If the load consists only of mass:

$$A(\delta P) = m \frac{\mathrm{d}^2}{\mathrm{d}t^2}(\delta y) \tag{13.35}$$

and if the pressure feedback is $K_\mathrm{p}\delta p$, equation (13.30) can now be expressed:

$$h\frac{k_4}{A}\,\omega_\mathrm{n}^2\,\frac{m}{A}K_\mathrm{p}\,\frac{s^2 + (K_\mathrm{v}A/K_\mathrm{p}m)s + (K_\mathrm{y}\,A/K_\mathrm{p}\,m)}{s(s^2 + \omega_\mathrm{n}^2)} = -1 \tag{13.36}$$

which is similar to equation (13.34).

One disadvantage of pressure feedback in a real system is that for a small positional error the system must build up a pressure differential across the ram to move the piston and thence eliminate the error. Pressure feedback will tend to prevent this process. Acceleration feedback on the other hand will not affect the system until the piston is actually moving.

Transient pressure feedback is an alternative method, in which the pressure signal is fed through a high pass filter so as to be only effective at the higher frequencies.

Incorporating the high pass filter transfer function in equation (13.36) gives:

$$h\,\frac{k_4}{A}\,\omega_n^2\,\frac{K_p(m/A)s^2\,\tau s/(\tau s+1)+K_v s+K_y}{s(s^2+\omega_n^2)}=-1$$

or

$$h\,\frac{k_4}{A}\,\omega_n^2\,\frac{K_p(m/A)s^3+K_v s^2+(K_v/\tau+K_y)s+(K_y/\tau)}{s(s^2+\omega_n^2)(s+1/\tau)}=-1\,. \qquad (13.37)$$

In fact, this type of feedback is stiffer in relation to external load disturbances at the lower frequencies. Once more it is seen that careful selection of time constants is required to avoid instability. If τ is very big, then equations (13.36) and (13.37) become identical.

13.7 STABILISATION USING AUXILIARY TANKS

In his paper [1], Shearer shows that the same effect can be achieved using transient flow stabilisation, Fig. 13.5. Here flow from auxiliary tanks is diverted from or added to the valve flow during transient conditions. The transient flow rate $\dot{M}_t$ takes the form:

$$\frac{\tau s}{\tau s+1}(\delta P_a-\delta P_b)\,. \qquad (13.38)$$

This feedback is non-zero when $(\delta P_a-\delta P_b)$ is time variant, but for steady values of pressure drop across the piston $\dot{M}_t$ is zero.

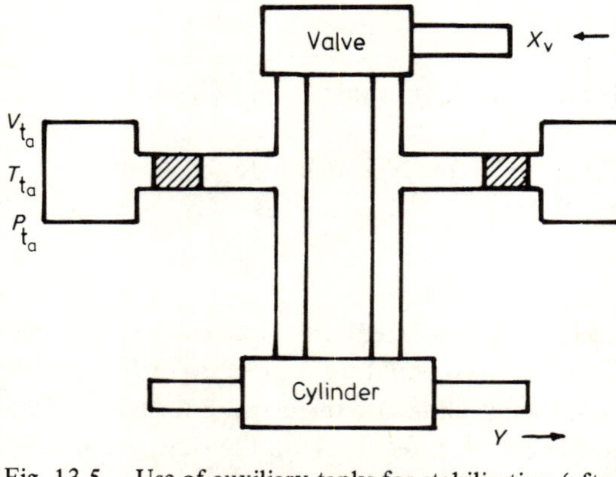

Fig. 13.5 – Use of auxiliary tanks for stabilisation (after [1, 2]).

Shearer used capillary flow restrictors because their characteristics are better than orifices. The equations for the flow in these passages are:

$$\dot{M}_{t_a} = \frac{ab^3}{24\mu RT_s l}(P_a^2 - P_{t_a}^2) \tag{13.39}$$

$$\dot{M}_{t_b} = \frac{ab^3}{24\mu RT_s l}(P_b^2 - P_{t_b}^2) \tag{13.40}$$

where a = capillary passage width
b = capillary passage thickness
l = capillary passage length.

Let:

$$C = \frac{ab^3}{12\mu l}$$

and assuming small changes about the initial steady conditions such that $P_a = P_b = P_{a_t} = P_{b_t} = P_i$, then:

$$\delta\dot{M}_{t_a} = \frac{CP_i}{RT_s}(\delta P_a - \delta P_{t_a}) \tag{13.41}$$

$$\delta\dot{M}_{t_b} = \frac{CP_i}{RT_s}(\delta P_b - \delta P_{t_b}) . \tag{13.42}$$

The energy equation applied to the identical tanks result in the equations:

$$\delta\dot{M}_{t_a} = \frac{V_t}{\gamma RT_s}\frac{d}{dt}(\delta P_{t_a}) \tag{13.43}$$

$$\delta\dot{M}_{t_b} = \frac{V_t}{\gamma RT_s}\frac{d}{dt}(\delta P_{t_b}) . \tag{13.44}$$

Combining equation (13.41) with (13.43) and (13.42) with (13.4) gives:

$$\frac{V_t}{\gamma CP_i}\frac{d}{dt}(\delta\dot{M}_{t_a}) + \delta\dot{M}_{t_a} = \frac{V_t}{\gamma RT_s}\frac{d}{dt}(\delta P_a) \tag{13.45}$$

$$\frac{V_t}{\gamma CP_i}\frac{d}{dt}(\delta\dot{M}_{t_b}) + \delta\dot{M}_{t_b} = \frac{V_t}{\gamma RT_s}\frac{d}{dt}(\delta P_b) . \tag{13.46}$$

The net flows to chambers a and b are:

$$(\delta\dot{M}_a - \delta\dot{M}_{t_a})$$

and

$$-(\delta\dot{M}_b + \delta\dot{M}_{t_b})\,.$$

The energy equations (13.7) can now be applied to the chambers on each side of the cylinder:

$$\delta\dot{M}_a - \delta\dot{M}_{t_a} - \frac{P_i A}{RT_s}\frac{d}{dt}(\delta y) = \frac{V_i}{\gamma RT_s}\frac{d}{dt}(\delta P_a) \tag{13.47}$$

$$-\delta\dot{M}_b - \delta\dot{M}_{t_b} + \frac{P_i A}{RT_s}\frac{d}{dt}(\delta y) = \frac{V_i}{\gamma RT_s}\frac{d}{dt}(\delta P_b)\,. \tag{13.48}$$

Combining (13.47) with (13.45) and (13.48) with (13.46) gives:

$$\delta\dot{M}_a - \frac{(V_t/\gamma RT_s)s(\delta P_a)}{(V_t/\gamma CP_i)s+1} - \frac{P_i A}{RT_s}s(\delta y) = \frac{V_i}{\gamma RT_s}s(\delta P_a) \tag{13.49}$$

$$-\delta\dot{M}_b - \frac{(V_t/\gamma RT_s)s(\delta P_b)}{(V_t/\gamma CP_i)s+1} + \frac{P_i A}{RT_s}s(\delta y) = \frac{V_i}{\gamma RT_s}s(\delta P_b)\,. \tag{13.50}$$

Subtract (13.50) from (13.49) giving:

$$(\delta\dot{M}_a + \delta\dot{M}_b) - \frac{(V_t/\gamma RT_s)s(\delta P_a - \delta P_b)}{(V_t/\gamma CP_i)s+1} - \frac{2P_i A}{RT_s}s(\delta y)$$

$$= \frac{V_i}{\gamma RT_s}s(\delta P_a - \delta P_b)\,. \tag{13.51}$$

But, from equation (13.13):

$$\delta\dot{M}_a + \delta\dot{M}_b = 2hk_1\,\delta x_v + k_2(\delta P_a - \delta P_b)$$

giving:

$$hk_1(\delta x_v) - k_2(\delta P_a - \delta P_b) - K\frac{\tau s(\delta P_a - \delta P_b)}{\tau s+1} - As(\delta y)$$

$$= k_6 s(\delta P_a - \delta P_b) \tag{13.52}$$

where $K = C/2$
$\tau = V_t/\gamma CP_i\,.$

Comparing this with equation (13.20) for the uncompensated system highlights the additional transient flow component (the third from the left in equation (13.52)).

13.8 HOT GAS SERVOMOTORS

Hot gas servomotors, which operate around 500°C and 70 bar, are particularly useful for space applications where propellant exhaust gases can be usefully recirculated. One particular difference between this type of system and the more conventional pneumatic actuator is the dirtiness of the fluid medium [8, 9, 10]. This aspect and the higher temperature create special problems in actuator design.

One possible configuration [10] is shown in Fig. 13.6. Several useful features of this assembly are:

1. The application of metal bellows in place of the more conventional cylinder eliminates sealing friction and lubrication problems. Their use will also reduce inertia. In addition, the thin-walled bellows with their large corrugated heat transfer area in contact with ambient air, will lower the temperature of the actuator chamber, thus reducing material creep and fatigue problems.
2. The use of a thin-walled metal bellows has advantages over the pipe of the jet pipe valve, since the reduced inertia will give a faster response.

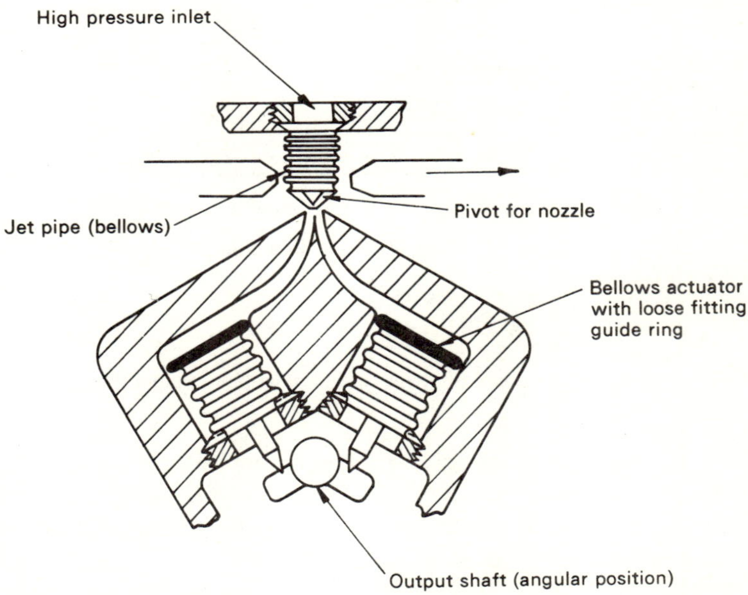

Fig. 13.6 – Single stage hot gas jet pipe valve with bellows actuator.

3. The use of pivoting push rods rather than the normal connecting rods, will help to eliminate binding and frictional effects.
4. Dual bellows actuators will eliminate backlash and frictional hysteresis. Metal bellows with a spring stiffness of 20 kN/m and operating pressures of 200 bar, capable of withstanding 4000 reversals at 500°C are available. It is also possible to spray the hot contact side with a ceramic coating, thus reducing the metal temperature.

Another possible arrangement, shown in Fig. 13.7, is suitable for long stroke actuators or large flows [10]. The valve has a jet pipe first stage followed by a poppet arrangement in the second stage. The poppet valve, which uses seating rather than shear action, has considerable advantage over the spool valve in high temperature, dirty fluid applications.

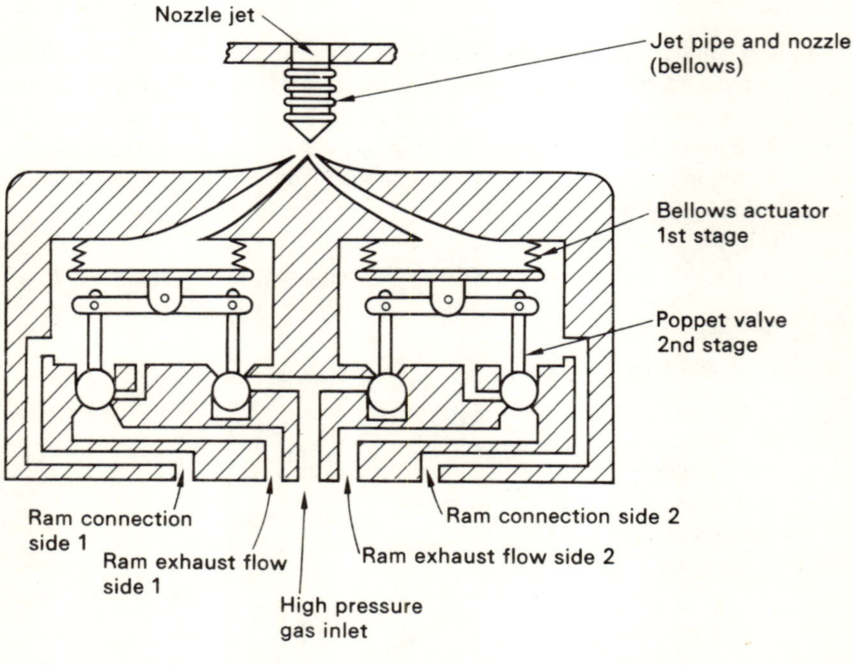

Fig. 13.7 – Two-stage hot gas control valve, first stage jet pipe, second stage poppet.

Since hot gas is very compressible, the natural frequency is low and control of the dynamic response is more difficult. Compressibility allows considerable excess accelerating forces at the instant of release of breakaway friction. The dynamic response can be sensitive to the amplitude of the command signal as well as the magnitude of the friction load. Dither is a possible alternative to reduce friction effects.

Gas systems are very expensive and their use can only be justified in aerospace applications, where minimum weight, reliability and high environmental temperatures are important. Cost of propellants are high, although in time the system may become feasible as an industrial application.

There are a number of suitable propellants, for example solid propellants for short duration applications and liquid propellants for systems requiring large quantities. Bipropellants, having liquid fuel and oxidisers, are normally not used because of the storage tank complication. Monopropellants such as hydrazine and hydrogen peroxide, are unstable compounds which decompose spontaneously at high temperature and pressure to produce hot gas around 500–1000°C. A typical early type gas generator in which a liquid propellant is converted to a gas is shown in Fig. 13.8.

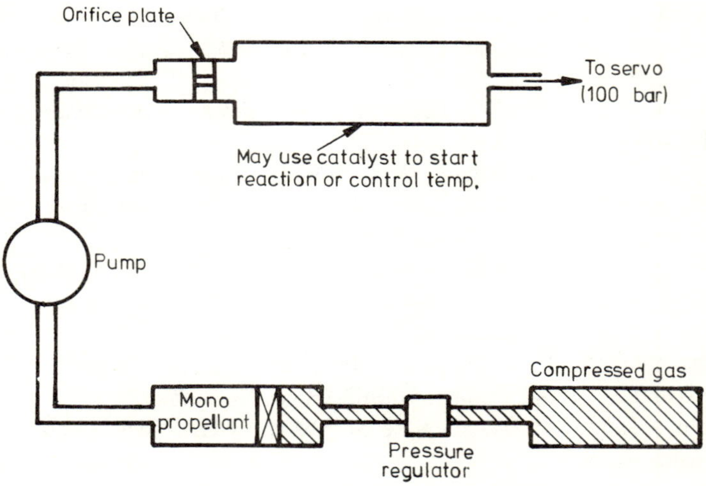

Fig. 13.8 – An early method of generating gas.

13.9 EXAMPLES

13.9.1

(a) A frictionless piston of mass M is at rest on top of a column of air of volume V_i in a cylindrical vessel of cross-sectional area A. Use a linearised analysis to determine the natural frequency of oscillation if the piston is disturbed.

(b) If there is leakage past the piston (and still assuming that friction is negligible) use the linear analysis to show that the leakage damps the oscillation.

(*Answer*: (a) $\omega_n^2 = \gamma A^2 (Mg/A + P_{at})/MV_i$.)

13.9.2

Air discharges from a vessel to atmosphere through an orifice. Develop the mathematical model for this process. Linearise the model and show how it can be used to determine how long it takes for the pressure to drop by a given amount.

13.10 REFERENCES

[1] Shearer, J. L. (1956), Study of pneumatic processes in the continuous control of motion with compressed air, Parts I and II, *Trans. A.S.M.E.*, **78**, 233.

[2] Shearer, J. L. (1957), Non-linear analogue study of a high pressure servomechanism, *Trans. A.S.M.E.*, **79**, 465.

[3] *Pneumatics – dynamic behaviour of double acting air cylinders* (1965), PERA rep. 135.

[4] Burrows, C. R. and Webb, C. R. (1967–68), Simulation of an ON-OFF pneumatic servo-mechamism, *Proc. Inst. Mech. Engrs.*, London, **182**, Pt. 1.

[5] Andersen, B. W. (1967), *The analysis and design of pneumatic systems*, John Wiley (New York).

[6] Bowns, D. E. and Ballard, R. L. (1972), Digital computation for the analysis of pneumatic actuator systems, *Proc. Inst. Mech. Engrs.*, London, **186**, 881–889.

[7] Botting, L. R., Eynon, G. T., and Foster, K. (1969–70), The response of a high-pressure pneumatic servomechanism to step and sine wave inputs, *Proc. Inst. Mech. Engrs.*, London, **184** (Pt. 1), 54.

[8] Halstenberg, R. V. (1959), Some linear and non-linear aspects of hot gas servo design, *National Control Conf.*, Dallas, November.

[9] Mann, J. R. (1961), *Hot gas bang-bang servo*, WADD Tech. Report.

[10] Davison, E. and McCloy, D. (1965), Recent developments in gas servos, *Hydraulic Pneumatic Power*, September.

[11] Sanville, F. E. (1971), A new method of specifying the flow capacity of pneumatic fluid power valves, *2nd International Fluid Power Symposium*, B.H.R.A., Cranfield, January.

CHAPTER 14

Dynamics of pressure and speed control systems

14.1 INTRODUCTION

In order to control the speed and applied force of output devices such as cylinders and motors, pressure and flow control valves are incorporated in the circuit [1]. Chapter 6 has dealt with the application of these and other valve combinations in circuit design. This chapter is concerned with some of the dynamic problems that are present in these types of components and circuits.

Pressure control valves act to regulate pressure in a circuit and may function as either relief or reducing valves [3]. In the relief valve mode of operation the valve is normally closed and opens at a predetermined pressure, by passing the excess flow. This mode of operation generates a major part of the heat in the circuit as well as radiating the most noise while relieving pressure. In the pressure reducing mode, the valve is normally open and closes to maintain a minimum pressure by restricting flow in the line. Hence, the relief valve is a parallel or bleed device and the reducing valve is a series device.

The simplest form of speed control for both pneumatic and hydraulic systems is a needle type flow restrictor of the type shown in Fig. 6.17. The orifice area is easily varied between zero and fully open by turning the adjusting screw.

However, a problem with this type of control of speed is that the flow rate and hence the speed are load dependent. To overcome this problem pressure compensation is introduced into the valve design, as shown in Fig. 14.1.

14.2 PRESSURE COMPENSATED FLOW CONTROL

The setting of the needle valve in Fig. 14.1 establishes the flow rate through the valve. If the outlet pressure varies the spool moves in a direction to compensate for the change in outlet pressure. For example, if pressure P falls, the spool moves downwards restricting the flow Q_1 and reducing the pressure P_1. By this means the pressure drop across the neddle valve is maintained constant irrespective of load changes.

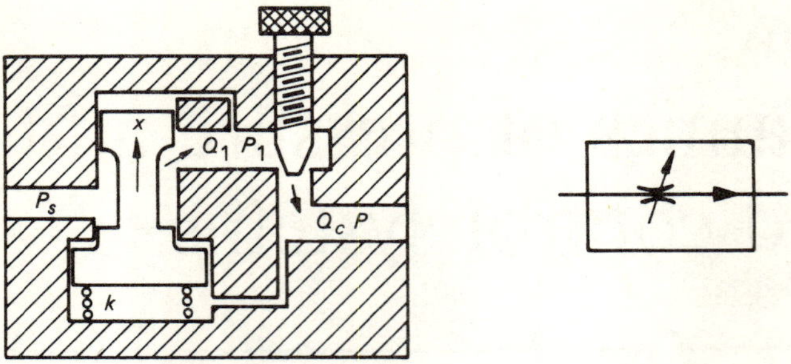

Fig. 14.1 – Schematic of a pressure compensated hydraulic flow control valve.

A static analysis [4] of this valve brings to light the important parameters.

$$\text{Flow rate } Q_1 = C_{q1} f(x) \sqrt{\left\{\frac{2}{\rho}(P_s - P_1)\right\}} \tag{14.1}$$

where the orifice area $a_1 = f(x)$ and x is measured from the closed position. The flow rate Q_c through the control orifice is:

$$Q_c = Q_1 = C_{qc} a_c \sqrt{\left\{\frac{2}{\rho}(P_1 - P)\right\}} \tag{14.2}$$

where a_c = area of the control orifice.

Consideration of the forces acting on the spool gives:

Pressure force + Bernoulli force = Spring force

$$A(P_1 - P) + \frac{\rho Q_1^2}{f(x)} \cos 69^\circ = k(l_o - x) \tag{14.3}$$

where l_o = initial spring compression when x is zero.

Equation (14.2) gives:

$$P_1 - P = \rho Q^2/(2C_{qc}^2 a_c^2)\,. \tag{14.4}$$

Substitution of equation (14.4) in equation (14.3) gives:

$$Q^2 = \frac{2C_{qc}^2 a_c^2 k(l_o - x)}{[\rho A + 2C_{qc}^2 a_c^2 \rho \cos 69^\circ / f(x)]}\,. \tag{14.5}$$

The dependency of Q on x can be eliminated if l_o is made much greater than the spool movement x and if $(C_{qc}a_c)^2$ is much less than $f(x)$, the spool orifice area. Equation (14.5) then reduces to:

$$Q^2 = 2(C_{qc}a_c)^2\,kl_o/\rho A \tag{14.6}$$

and the flow rate is independent of load. Chapter 7 showed that the flow coefficient for an orifice varies with flow number. If a flow control valve is operating at low flow numbers, where C_q varies rapidly with flow number ($\lambda < 300$), that is, laminar flow, then some means must be provided for offsetting temperature effects. (C_q increases with temperature if the flow is laminar.) Temperature compensation can be achieved by making the stem of the adjusting screw of a material with a high coefficient of thermal expansion. As temperature increases the increasing C_q will be offset by the reduction in orifice area caused by the spool expansion.

The dynamic performance of pressure compensated flow control valves is described in [4]. A typical commercial valve is there shown to have an oscillatory flow response to a step input in load pressure P. For example, a pressure step of 22 bar gave rise to a damped oscillating flow rate Q of initial amplitude 30 ml and natural frequency 25 Hz. The flow settled out after 0.12 s.

14.3 STABILITY OF PRESSURE CONTROL VALVES

The analysis of the relief valve is similar to that developed by Ma [6] in which the dynamic behaviour of the pressure reducing valve is studied. A typical arrangement is shown in Fig. 14.2 where V_c is the volume of the pressure control

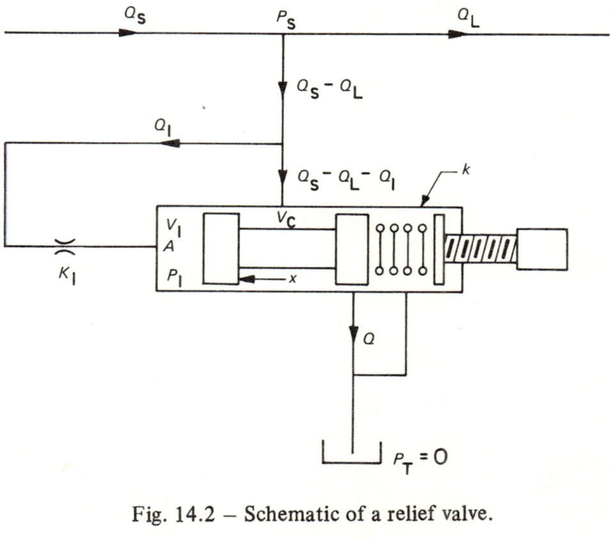

Fig. 14.2 – Schematic of a relief valve.

chamber and V_1 the volume of the pressure sensing chamber. The spring pre-load is set to determine the desired relief pressure, and this results in a force kx_0 acting on the spool. The difference between the pre-load force and the pressure in the pressure sensing chamber, will cause the spring to deflect an amount x, the movement of the spool. Under dynamic conditions force balance is:

$$kx_0 - AP_1 = M\frac{d^2x}{dt^2} + f\frac{dx}{dt} + kx + F_\beta \tag{14.7}$$

where M is the mass of the moving parts, x is the valve opening and F_β are the flow force effects which will contain steady state and transient force components. Only one set of orifices are involved, therefore:

$$F_\beta = 2C_qKx(P_s - P_T)\cos\phi - \rho L\frac{dQ}{dt}. \tag{14.8}$$

Hence, combining (14.7) and (14.8):

$$kx_0 + \rho L\frac{dQ}{dt} = M\frac{d^2x}{dt^2} + f\frac{dx}{dt} + kx + 2C_qKxP_s\cos\phi + P_1A. \tag{14.9}$$

The flow through the main orifice, Q, can be approximated by considering small flow increments as shown in Chapter 7.

$$Q = \frac{\partial Q}{\partial x}x - \frac{\partial Q}{\partial P}P_s$$

$$= C_x x + C_p P_s. \tag{14.10}$$

If, for a simple orifice:

$$Q = C_qKx\sqrt{(2/\rho)}\sqrt{(P_s - P_T)}$$

then by differentiation:

$$\frac{\partial Q}{\partial x} = C_x = \frac{C_qKx\sqrt{(2/\rho)}}{\sqrt{(P_s - P_T)}}$$

$$\frac{\partial P}{\partial x} = C_p = C_qK\sqrt{\left\{\frac{2}{\rho}(P_s - P_T)\right\}}.$$

Differentiation of (14.10) gives:

$$\frac{\partial Q}{\partial t} = C_x \frac{\partial x}{\partial t} + C_p \frac{\partial P_s}{\partial t} . \tag{14.11}$$

These results can now be substituted into (14.9) giving:

$$kx_0 + \rho L C_x \frac{dx}{dt} = P_1 A + \rho L C_p \frac{dP_s}{dt} + M \frac{d^2x}{dt^2} + f \frac{dx}{dt} + kx + K'P_s x$$

where $K' = 2C_q K \cos \phi$.

Now let

$(K'P_s + k) = k_e$ the equivalent spring stiffness

and

$(f - \rho L C_x) = f_e$ the equivalent friction coefficient.

Then

$$kx_0 - P_1 A + \rho L C_p \frac{dP_s}{dt^2} = M \frac{d^2x}{dt^2} + f_e \frac{dx}{dt} + k_e x . \tag{14.12}$$

The term $\rho L C_p (dP_s/dt)$ can be considered small. Then by letting $c = f_e/\sqrt{2Mk_e})$, (14.6) becomes, when transformed:

$$kx_0 - P_1 A = k_e \left\{ \frac{s^2}{\omega_n^2} + \frac{2c}{\omega_n} s + 1 \right\} \tag{14.13}$$

where $\omega_n = \sqrt{(k_e/M)}$ is the natural frequency of the spool and equivalent spring.

Flow into and out of the pressure sensing chamber can be defined by:

$$Q_1 = K_1 (P_s - P_1) = \frac{V_1}{\beta} \frac{dP_1}{dt} - A \frac{dx}{dt} . \tag{14.14}$$

This assumes the orifice to be small enough for laminar flow to exist and takes into account compressibility effects in this chamber. Equation (14.14) can be expressed in terms of P_1, the pressure in the sensing chamber.

$$P_1 = \frac{K_1 P_s + Asx}{K_1 (s/\omega_s + 1)} \tag{14.15}$$

where $\omega_s = \beta K_1 / V_1$ is the natural frequency of the pressure sensing chamber.

Flow through the pressure control chamber can be defined by:

$$(Q_s - Q_L) - Q_1 - K_1 P_s = \frac{V_c}{\beta}\frac{dP_s}{dt} - Q \qquad (14.16)$$

where K_l is the leakage coefficient. Again compressibility effects in this chamber are included.

Substitution from (14.11) and (14.14) into (14.16) gives:

$$(Q_s - Q_L) - K_1 P_s + K_1 \left\{\frac{K_1 P_s + Asx}{K_1 (s/\omega_s + 1)}\right\} - K_1 P_s - C_p P_s + C_x x$$

$$= \frac{V_c}{\beta}\frac{dP_s}{dt}$$

or

$$(Q_s - Q_L)\left(\frac{s}{\omega_s} + 1\right) + C_x \left[1 + \left(\frac{1}{\omega_s} + \frac{A}{C_x}\right) s\right] x$$

$$= \left[\left(\frac{s}{\omega_s} + 1\right)\left(1 + \frac{K_1}{K_s} + \frac{s}{\omega_p}\right) K_s - K_1\right] P_s \qquad (14.17)$$

where $\omega_p = \beta K_s / V_c$ is the natural frequency of the pressure control chamber and:

$$K_s = C_p + K_1 .$$

Expanding the right-hand side of (14.17) gives:

$$(Q_s - Q_L)\left(\frac{s}{\omega_s} + 1\right) + C_x \left[1 + \left(\frac{1}{\omega_s} + \frac{A}{C_x}\right) s\right] x$$

$$= K_s \left[\frac{s^2}{\omega_s \omega_p} + \frac{1}{\omega_p}\left(1 + \frac{\omega_p}{\omega_s} + \frac{V_1}{V_c}\right) s + 1\right] P_s . \qquad (14.18)$$

This equation is essentially the same as that derived by Merritt [12]. The right-hand side of (14.18) can be further simplified if it is assumed that $V_1/V_c < 1$, giving:

$$(Q_s - Q_L) \quad \frac{s}{\omega_s} + 1 \quad + C_x \quad 1 + \quad \frac{1}{\omega_s} + \frac{A}{C_x} \quad s \quad x$$

$$= K_s \quad 1 + \frac{s}{\omega_s} \quad 1 + \frac{s}{\omega_p} \; P_s . \qquad (14.19)$$

Equations (14.13) and (14.15) can be combined to give:

$$kx_0 - \frac{AP_s}{\left(\dfrac{s}{\omega_s}+1\right)} - \frac{A^2 sx}{K_1\left(\dfrac{s}{\omega_s}+1\right)} = K_e\left[\frac{s^2}{\omega_n^2} + \frac{2c}{\omega_n}s + 1\right]x \qquad (14.20)$$

and this allows the block diagram in Fig. 14.3 to be drawn [2].

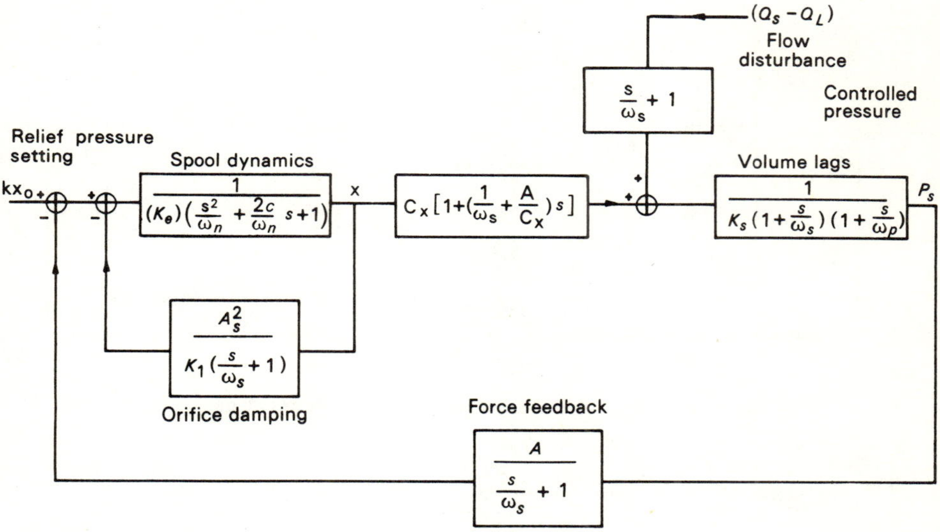

Fig. 14.3 – Block diagram of relief valve.

If the orifice controlling flow into the pressure sensing chamber were removed, then $K_1 = \infty$ and $s/\omega_s = 0$. Equations (14.19) and (14.20) then reduce to:

$$(Q_s - Q_L) + (C_x + As)x = K_s\left(1 + \frac{s}{\omega_p}\right)P_s \qquad (14.21)$$

and

$$kx_0 - AP_s = K_e\left\{\frac{s^2}{\omega_n^2} + \frac{2c}{\omega_n}s + 1\right\}x \qquad (14.22)$$

which gives the new block diagram of Fig. 14.4.

Examination shows that stability will be affected since a secondary damping loop around the spool has been removed. This system now relies on the inherent damping of the valve which is usually small. This type of instability results in valve chatter [5].

Consider the loop shown in Fig. 14.4, when a load disturbance has settled out so that $(Q_s - Q_L) = 0$. Then:

$$\frac{P_s}{kx_0} = \frac{(C_x + As)\Big/K_sK_e\left\{\dfrac{s^2}{\omega_n^2} + \dfrac{2c}{\omega_n}s + 1\right\}\left(1 + \dfrac{s}{\omega_p}\right)}{1 + A(C_x + As)\Big/K_sK_e\left\{\dfrac{s^2}{\omega_n^2} + \dfrac{2c}{\omega_n} + 1\right\}\left(1 + \dfrac{s}{\omega_p}\right)} \quad . \tag{14.23}$$

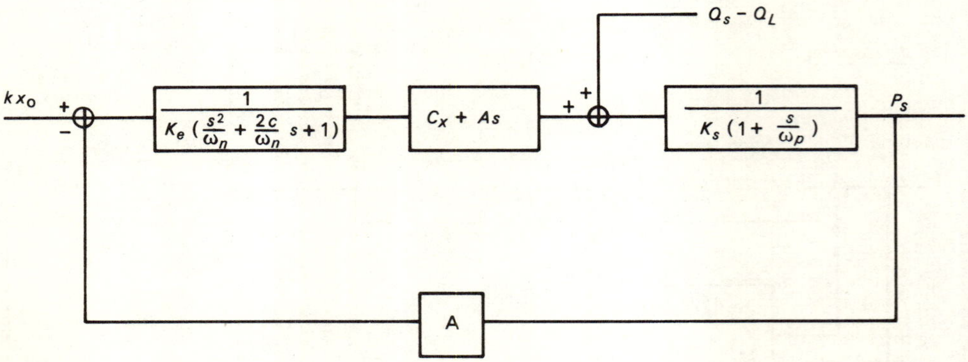

Fig. 14.4 – Block diagram of relief valve without orifice damping.

The characteristic equation of (14.23) is:

$$\frac{s^3}{\omega_n^2\omega_p} + \left(\frac{1}{\omega_n^2} + \frac{2c}{\omega_n^2\omega_p}\right)s^2 + \left(\frac{1}{\omega_p} + \frac{2c}{\omega_n} + \frac{A^2}{K_sK_e}\right)s + \left(1 + \frac{AC_x}{K_sK_e}\right) = 0\,.$$

The Routh–Hurwitz criterion requires that:

$$\frac{1}{\omega_p} + \frac{2c}{\omega_n} + \frac{A^2}{K_s K_e} - \frac{(1/\omega_n^2\omega_p)/(1 + AC_x/K_s K_e)}{1/\omega_n^2 + 2c/\omega_n\omega_p} \geqslant 0 \text{ for stability}$$

or

$$\frac{2c\omega_p}{\omega_n} + \frac{2c\omega_n}{\omega_p} + \frac{A^2}{K_sK_e}(\omega_p + 2c\omega_n) > \frac{AC_x}{K_sK_e}\,. \tag{14.24}$$

If the damping term $c = 0$ then for stability:

$$\omega_p \geqslant C_x/A\,.$$

A pressure relief valve will respond to load changes if no restrictor is present, but will have poor stability and pressure control accuracy. The poor stability characteristics make the valve noisy owing to valve chatter. Use of a restrictor will alleviate the situation but will give a slower response. Two-stage pressure control valves have much better stability characteristics.

14.4 SPEED CONTROL IN FLUID POWER SYSTEMS

The simplest adjustable restrictor shown in Fig. 14.5 can be used in several circuit configurations:

(a) Meter-in.
(b) Meter-out.
(c) Bleed-off.

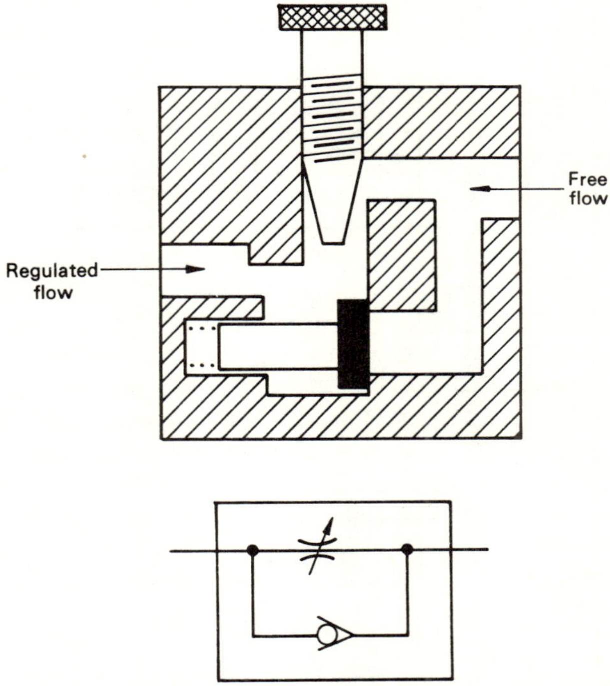

Fig. 14.5 – Variable restrictor and non-return valve as an integral unit.

When the restrictor is placed in the supply line upstream of the cylinder the system is said to be under **meter-in speed control**, Fig. 14.6(a). The rod speed is controlled when it is extending, but is free to return at maximum speed. This type of control is best suited for resistive loads, that is loads that the cylinder has to push against. The arrangement is not suitable for over-running loads. **Meter-out control** is shown in Fig. 14.6(b), where the restrictor is placed in the return line. The valve controls the rate at which fluid can leave the rod end of the cylinder, thus controlling speed. Over-running loads cannot pull the rod out of the cylinder at a faster rate than that set by the flow control valve.

In general, cylinders and motors can experience the following type of loading:

(a) **Resistive** in which the load reaction on the output drive opposes the motion of the actuator.

(b) **Over-running** in which the load reaction on the output device has the same direction as the motion of the actuator.

(c) **Inertial** in which the load is governed by Newton's second law of motion.

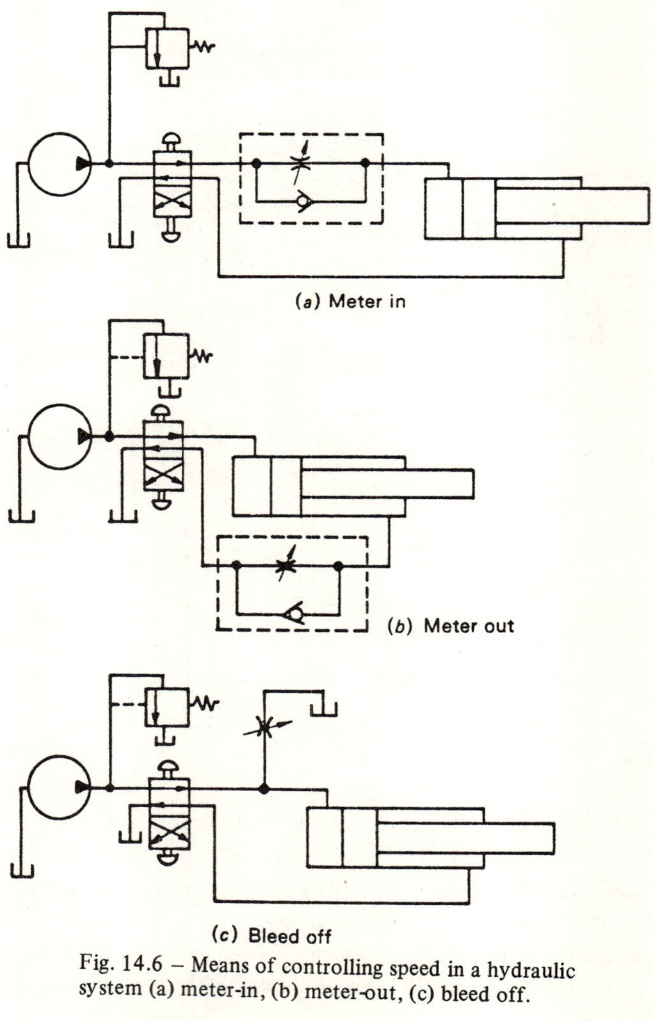

Fig. 14.6 – Means of controlling speed in a hydraulic system (a) meter-in, (b) meter-out, (c) bleed off.

In reality most situations are a combination of the above with one dominating.

In **bleed-off control**, Fig. 14.6(c), the flow control device is placed in parallel to the cylinder feed line. Hence a portion of the main flow is bypassed to the tank. This method can be used to adjust the cylinder speed over a range that is a smaller portion of the maximum speed. It has the advantage of needing a control valve large enough only to handle the maximum bleed flow and not the total flow. In addition it does not introduce a pressure drop into the main line of the circuit.

14.5 SPEED CONTROL IN PNEUMATIC SYSTEMS

In order to appreciate better the relative advantages and disadvantages of meter-in and meter-out controls, it is necessary to look at system dynamics. Fig. 14.7,

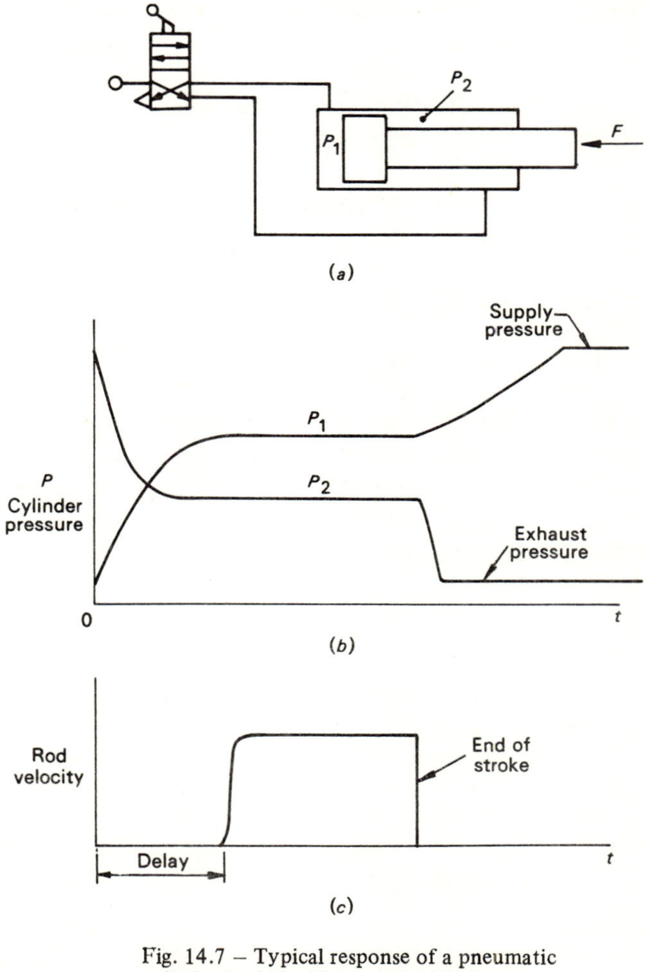

Fig. 14.7 – Typical response of a pneumatic cylinder loaded with a constant force.

for example, shows a typical response of a pneumatic cylinder loaded with a constant force F. At zero time the valve is actuated to make the rod extend. The blank end of the cylinder is connected to supply and the pressure P_1, which was initially at exhaust pressure, rises. At the same time the rod end is connected to exhaust and pressure P_2 which was initially at supply pressure, falls. The speed of the pressure changes depends on the compressibility of the air, the size of the valve ports, supply pressure and cylinder size. The cylinder rod will not move until the differential in the pressures P_1 and P_2 is sufficient to provide a force F to equal the load on the rod. The rod then extends at constant velocity, the pressures P_1 and P_2 remaining constant. When the rod stops at the end of its stroke the blank end pressure P_1 rises to supply pressure while the rod end pressure P_2 falls to atmosphere.

A comparison between meter-in and meter-out performance can best be illustrated by the following example. Consider a standard pneumatic cylinder with:

Bore diameter = 50.8 mm $A_B = 20.26\ \text{cm}^2$ = bore area.
Rod diameter = 12.7 mm $A_R = 1.27\ \text{cm}^2$ = rod area.

If the force F, Fig. 14.7, opposing the piston rod extension is constant, then the force balance gives:

$$(P_1 - P_a)A_1 - (P_2 - P_a)A_2 = F$$

or

$$(P_1 - P_2)A_B + (P_2 - P_a)A_R = F\,. \tag{14.25}$$

It is important to note that these are absolute pressures. P_a is atmospheric. In the meter-in configuration, Fig. 14.6(a), the pressure P_2 will be nearly atmospheric if the demanded speed is small. This is shown in Fig. 14.8 for two load conditions.

Using equation (14.25), the maximum force the cylinder can move at very slow speed is:

$$\hat{F} = (P_s - P_a)A_B$$

If the supply pressure is 7.9 bar then:

$$\hat{F} = \frac{6.9 \times 10^5 \times 20.26}{10^4}$$

$$= 1.397\ \text{kN}\,.$$

This represents a resistive load.

Again using equation (14.25) the maximum over-running load the cylinder can control is:

$$\check{F} = -P_a A_B = -202 \text{ N} .$$

At this force level the pressure P_1 is at absolute zero. Any larger negative direction force or pull would cause the cylinder rod to 'run away'.

The dynamic response of a meter-in control for two load conditions is shown in Fig. 14.8. For the smaller load, the steady state value of P_1 from equation (14.25) is P_{1x} where:

$$P_{1x} = P_a + F/A_B = 4.26 \text{ bar} . \tag{14.26}$$

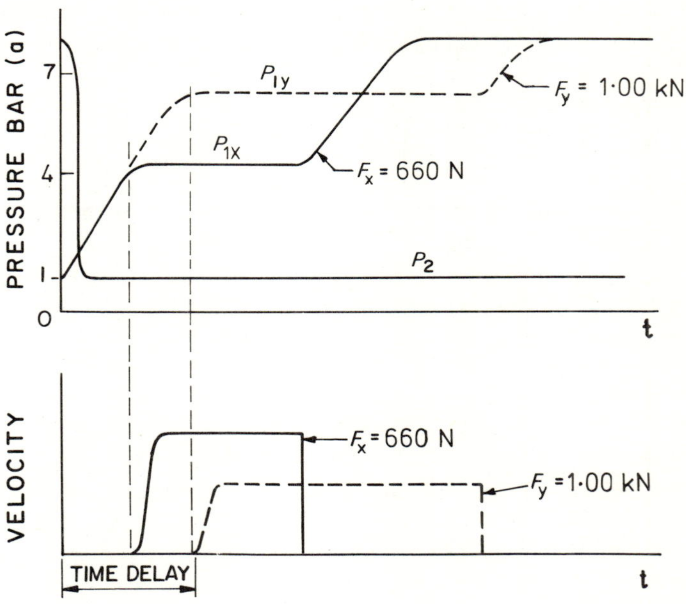

Fig. 14.8 – Effect of load change on speed of pneumatic cylinder with meter-in speed control.

No rod movement occurs until P_1 has risen from atmospheric pressure to this value. If a low rod speed is demanded the metering orifice will be adjusted to present a large resistance to inflow and under these conditions the delay required for the build up of P_1 can be considerable. This is another disadvantage of meter-in control in pneumatic circuits. If the load is increased to 1.00 kN, the steady state value of P_1 (now P_{1y}) is increased to 5.94 bar and the pressure build-up time is further increased. An important point to note is that the steady state velocities for the two loads are different. The speed is load dependent, and this is a serious disadvantage of a speed control system. The ratio of the two speeds can be calculated. The incoming mass flow rate $\dot{M}$ is:

$$\dot{M} = \dot{M}_r f(P_1/P_s)$$

where $\dot{M}_r$ is constant and given by section 4.3, using reference values. If the steady mass flow rates at $F_x = 660$ N and $F_y = 1.00$ kN are respectively $\dot{M}_x$ and $\dot{M}_y$, then:

$$\frac{\dot{M}_y}{\dot{M}_x} = \frac{f(P_1/P_s)y}{f(P_1/P_s)x} = \frac{f(0.75)}{f(0.54)} . \qquad (14.27)$$

From Fig. 4.5, a value of $(P_1/P_s) = 0.75$ gives a $C_m = 0.0360$ while a value of $(P_1/P_s) = 0.54$ gives a $C_m = 0.0405$ which is nearly a choked flow condition.

The ratio of mass flows, equation (14.27), now becomes:

$$\frac{\dot{M}_y}{\dot{M}_x} = \frac{0.0360}{0.0405} = 0.89 .$$

The mass flow rate into the cylinder has been reduced at the higher loading case because the pressure P_1 has risen.

It is the volumetric rate into the cylinder that dictates the rod speed and, unlike the hydraulic oil case, it is dependent on fluid density.

Mass flow rate $\dot{M}$ and volumetric flow rate Q are related by:

$$\dot{M} = \rho Q$$

therefore

$$\frac{Q_y}{Q_x} = \frac{\dot{M}_y \rho_x}{\dot{M}_x \rho_y} . \qquad (14.28a)$$

If the process is isothermal then:

$$\frac{\rho_x}{\rho_y} = \frac{P_{1x}}{P_{1y}}$$

therefore

$$\frac{Q_y}{Q_x} = \frac{\dot{M}_y}{\dot{M}_x} \frac{P_{1x}}{P_{1y}} \qquad (14.28b)$$

$$= 0.89 \times \frac{4.26}{5.94} = 0.64 .$$

Hence the speed is reduced under the heavier loading case.

If the meter-out control is now considered, Fig. 14.6(b), and the steady blank end pressure is assumed to equal the supply pressure of 7.9 bar(a), equation (14.25) can be rearranged to give the rod end pressure as:

$$P_2 = (P_s A_B - P_a A_R - F)/(A_B - A_R)\,. \qquad (14.29)$$

Hence the stall load, when $P_2 = P_a$, now becomes 1.40 kN.

A very important feature of meter-out control is that over-running loads can be handled. Equation (14.29) shows that P_2 becomes greater when F becomes more negative.

The dynamic response of a meter-out control is sketched in Fig. 14.9. On operation of the directional control valve, P_1 rises to near supply pressure. Pressure P_2 falls slowly as air, initially at supply pressure, escapes through the meter-out restriction. When the required pressure differential is reached the piston moves. Here again, as with meter-in, the delay before the cylinder rod moves increases as the load increases. In this case, however, the steady state rod velocity is the same under both loading conditions. This is because the flow through the restriction is:

$$\dot{M} = \dot{M}_r (P_2/P_r) f(P_e/P_2) \qquad (14.30)$$

where $\dot{M}_r$ and P_r are constants (see Chapter 4). Choked flow occurs when $P_e/P_2 <$ 0.528 for air and hence in this case the outflow is choked when $P_2 > 1.93$ bar.

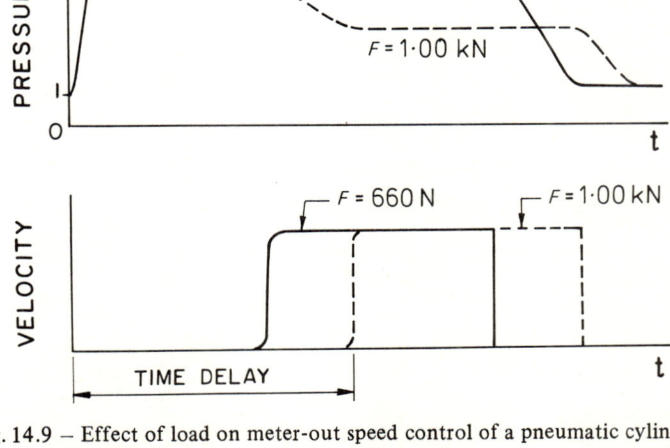

Fig. 14.9 – Effect of load on meter-out speed control of a pneumatic cylinder.

The load equation, equation (14.29), then indicates that choking occurs if the load is less than 1.22 kN (87 per cent of stall load). If again mass flow rates at loads of 660 N and 1.00 kN are $\dot{M}_x$ and $\dot{M}_y$, then since the flow is choked:

$$\frac{\dot{M}_x}{\dot{M}_y} = \frac{P_{2x}}{P_{2y}} . \tag{14.31}$$

Equation (14.28a) then gives:

$$\frac{Q_y}{Q_x} = \frac{P_{2y}\rho_x}{P_{2x}\rho_y} = 1 . \tag{14.32}$$

Hence, provided that the outflow is choked, the volumetric flow rates and hence the steady state velocities are independent of load. The meter-out control of the pneumatic cylinders has thus two decided advantages over meter-in:

1. Speed is less affected by load changes.
2. It will not run away under the influence of overhauling loads.

These factors make meter-out control by far the more popular for pneumatic circuits.

14.6 PNEUMATIC METER-OUT CONTROL OF MASS LOADS

The above discussion of speed control assumed the simplest of loads. In practice, however, inertia loads, friction and spring loads are encountered and it is sometimes necessary to predict their effects on system performance. The following analysis shows how inertia load effects can be determined. For the system shown in Fig. 14.10, the mass M_2 of air in the exhaust side of the cylinder is given by the equation of state as:

$$M_2 = P_2V_2/RT_2 . \tag{14.33}$$

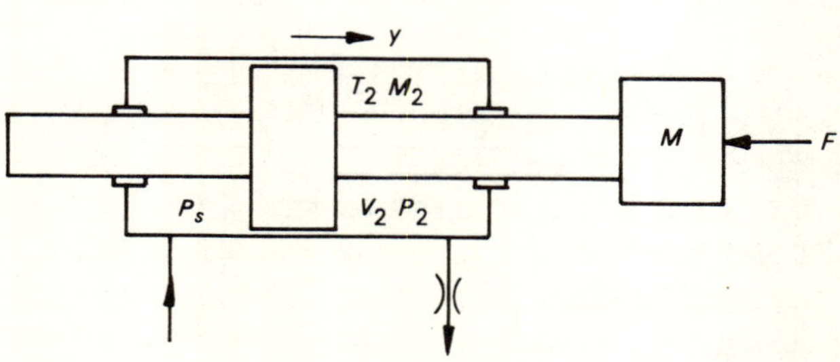

Fig. 14.10 – Schematic of pneumatic cylinder with meter-out control.

The rate of change of mass is, assuming for simplicity an isothermal process:

$$\dot{M}_2 = \left(\frac{P_2}{RT_2}\right)\dot{V}_2 + \left(\frac{V_2}{RT_2}\right)\dot{P}_2 \tag{14.34}$$

or

$$\frac{\dot{M}_2}{M_2} = \frac{\dot{V}_2}{V_2} + \frac{\dot{P}_2}{P_2} \,. \tag{14.35}$$

If the flow is choked, as it usually is in meter-out circuits:

$$\dot{M}_2 = -kP_2 \tag{14.36}$$

where $k = \dot{M}_r / P_r$.

The load equation is:

$$P_s - P_2 = (M/A)\ddot{y} + F/A \,. \tag{14.37}$$

Note that the pressure on the left-hand side of the cylinder is assumed to remain constant at P_s (infinite capacity). F is a step load applied at time $t = 0$.

Since

$$V_2 = V_2(0) - Ay \tag{14.38}$$

equation (14.37) becomes:

$$P_s - P_2 = F/A - (M/A^2)\ddot{V}_2 \,. \tag{14.39}$$

Equations (14.35), (14.36) and (14.39) combine to give:

$$\dddot{V}_2 + \frac{kP_2}{M_2}\ddot{V}_2 + \frac{P_2 A^2}{MV_2}\dot{V}_2 = \frac{kP_2 A^2}{MM_2}\left[\frac{F}{A} - P_s \,.\right] \tag{14.40}$$

The fact that the coefficients of this equation are functions of the dependent variables makes it difficult to solve. Note, however, that in the new steady state following the applications of the step load F, the velocity is $\dot{V}_{2s}$ given by:

$$\begin{aligned}\dot{V}_{2s} &= \frac{MV_{2s}}{P_{2s}A^2}\,\frac{kP_{2s}A}{MM_{2s}}\,[P_s - P_{2s} - P_s] \\ &= \frac{-kV_{2s}P_{2s}}{M_{2s}} = -kRT_{2s} \,.\end{aligned} \tag{14.41}$$

Initially equation (14.36) gives:

$$\dot{M}_2(0) = -kP_2(0)$$

and equation (14.35) gives:

$$\dot{V}_2(0) = [V_2(0)/M_2(0)]\dot{M}_2(0) = -[V_2(0)/M_2(0)]\,kP_2(0)$$
$$= -kRT(0)\,. \tag{14.42}$$

If the process is isothermal $T_s = T(0)$ and the steady state velocities are the same before and after the load application. This agrees with conclusions drawn earlier in this chapter.

Some indication of system behaviour can be obtained quickly if the variables are assumed to have only small changes about their initial values. In this case equation (14.40) can be approximated by:

$$\dddot{V}_2 + \frac{kP_2(0)}{M_2(0)}\ddot{V}_2 + \frac{P_2(0)A^2}{MV_2(0)}\dot{V}_2 = \frac{kP_2(0)A^2}{MM_2(0)}\left[\frac{F}{A} - P_s\right] \tag{14.43}$$

or

$$\dddot{V}_2 + 2c\omega_n\ddot{V}_2 + \omega_n^2\dot{V}_2 = \omega_n^2 V_2(0)\,. \tag{14.44}$$

The natural frequency ω_n is given by:

$$\omega_n^2 = P_2(0)A^2/(MV_2(0))\,. \tag{14.45}$$

Equation (14.45) shows also that the natural frequency tends to increase as the rod moves to the right reducing V_2.

The damping ratio c is:

$$c = -\frac{1}{2}\frac{\dot{V}_2(0)}{A}\sqrt{\frac{2M}{P_sV_2(0)}} = \frac{1}{2}\dot{y}(0)\sqrt{\frac{2M}{P_sV_2(0)}}\,. \tag{14.46}$$

The damping depends on the initial rod velocity or valve mass flow. The response of the system is now reasonably well defined. Equations (14.41) and (14.42) show that the steady state velocities are the same before and after load application. Equation (14.44) shows that there is a transient oscillation after the load application, and equations (14.45) and (14.46) enable the damping and frequency of the oscillation to be calculated.

The same results can be found by using the linearised orifice flow equation developed in Chapter 4. As an exercise in using valve coefficients it is worthwhile repeating the analysis by this method. The equation of state, equation (14.33), can be written:

$$\frac{\delta M_2}{M_2} = \frac{\delta V_2}{V_2} + \frac{\delta P_2}{P_2} - \frac{\delta T_2}{T_2} \tag{14.47}$$

where δ is a small increment. If a polytropic relationship exists between T_2 and P_2 then:

$$\frac{\delta T_2}{T_2} = \frac{n-1}{n}\frac{\delta P_2}{P_2} \tag{14.48}$$

where $n = 1$ for an isothermal process and $n = 1.4$ for air for an adiabatic process. Equation (14.47) and (14.48) give:

$$\frac{\delta M_2}{M_2} = \frac{\delta V_2}{V_2} + \frac{1}{n}\frac{\delta P_2}{P_2}\ . \tag{14.49}$$

Assume small perturbations about the initial conditions and put:

$$\begin{aligned} M_2 &= M_2(0) + \delta M_2 \\ P_2 &= P_2(0) + \delta P_2 \\ V_2 &= V_2(0) + \delta V_2\ . \end{aligned} \tag{14.50}$$

Equation (14.49) is then:

$$\frac{\delta M_2}{M_2(0)} = \frac{\delta V_2}{V_2(0)} + \frac{1}{n}\frac{\delta P_2}{P_2(0)}\ . \tag{14.51}$$

The linearised orifice flow equation is developed in section 7.4:

$$-\delta \dot{M}_2 = -\delta \dot{M}_2(0)(1+\zeta)\frac{\delta P_2}{P_2(0)} + \frac{1}{2}\dot{M}_2(0)\frac{\delta T_2}{T_2(0)} \tag{14.52}$$

and using equation (14.48):

$$\delta \dot{M}_2 = \dot{M}_2(0)\frac{\delta P_2}{P_2(0)}\left\{(1+\zeta) - \frac{1}{2}\frac{n-1}{n}\right\}. \tag{14.53}$$

Equation (14.51) also gives:

$$\frac{\dot{M}_2}{M_2(0)} = \frac{\dot{V}_2}{V_2(0)} + \frac{1}{n}\frac{\dot{P}_2}{P_2(0)} \tag{14.54}$$

or

$$\frac{\delta \dot{M}_2}{M_2(0)} = \frac{\delta \dot{V}_2}{V_2(0)} + \frac{1}{n}\frac{\delta \dot{P}_2}{P_2(0)} . \tag{14.55}$$

The load equation is:

$$\delta P_2 = -(M/A)\,\delta \ddot{y} - F/A \tag{14.56}$$

and

$$\delta V_2 = V_2(0) - A\delta y .$$

Equations (14.53), (14.55), (14.56) combine to give for $n = 1$;

$$\dddot{y} - \frac{\dot{M}_2(0)(1+\zeta)}{M_2(0)}\ddot{y} + \frac{P_2(0)A^2}{MV_2(0)}\dot{y} = \frac{-\dot{M}_2(0)(1+\zeta)F}{MM_2(0)} . \tag{14.57}$$

This equation is second order in y and has the same natural frequency and damping ratio as equation (14.44). Experimental results of pneumatic speed control are presented in [7].

14.7 HYDRAULIC METER-OUT CONTROL OF MASS LOADS

For simplicity's sake it is assumed that the flow control valve's response is much faster than that of the rest of the system so that its dynamics can be neglected. The continuity equation then gives, Fig. 14.11:

$$Q = Av(0) = Av - \frac{V_2}{\beta}\frac{\mathrm{d}P_2}{\mathrm{d}t} \tag{14.58}$$

where $v(0)$ is the demanded speed of the cylinder rod and Q is the constant

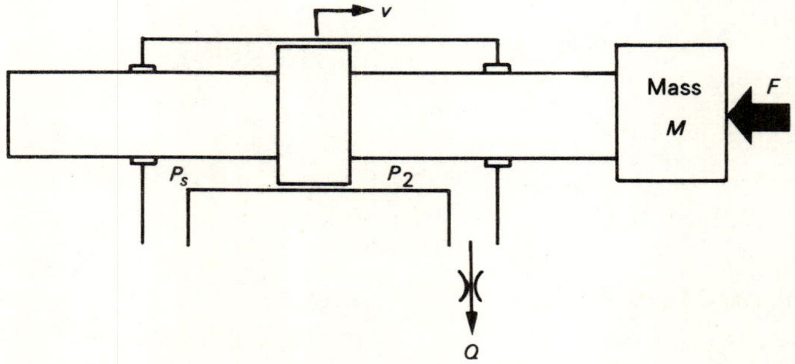

Fig. 14.11 – Schematic of hydraulic meter-out control.

flow rate through the flow control valve. The system is loaded with a constant force F and inertia M. Hence:

$$A(P_s - P_2) = AP_L = M\frac{dv}{dt} + F \,. \tag{14.59}$$

The pressure in the left-hand side of the cylinder is assumed to remain constant at P_s.

Equations (14.58) and (14.59) combine to give:

$$\frac{d^2P_L}{dt^2} + \omega_n^2 P_L = \omega_n^2 \frac{F}{A} \tag{14.60}$$

$$P_L = F/A + [P_L(0) - F/A] \cos \omega_n t \,. \tag{14.61}$$

This solution represents an undamped oscillation of frequency $\omega_n = (\beta A^2/MV_2)$.

System damping is zero because the flow control valve, by maintaining a constant flow rate, eliminates valve damping. In practice, seal friction and load friction will attenuate the oscillation.

Chapter 11 discussed the stability of the hydraulic servo and this was seen to depend on valve damping – the decrease of flow rate with increase of load pressure. The meter-in control of a pneumatic cylinder discussed earlier in this chapter also depended on valve damping for its stability.

Equation (14.51) can also be written as:

$$P_2^* = P_s^* - F/A - \{P_s^* - P_2^*(0) - F/A\} \cos \omega_n t \tag{14.62}$$

where, as before, the asterisk denotes that the pressure is measured relative to tank pressure P_e. Equations (14.59) and (14.62) give:

$$\Delta v = v - v(0) = \frac{V_2 \omega_n}{\beta A} \left\{ P_s^* - P_2^*(0) - F/A \right\} \sin \omega_n t \,. \tag{14.63}$$

The response of a cylinder controlled by meter-out is shown in Fig. 14.12. Here the initial load is zero before a step load $F = 0.4P_s^*$ is applied to the cylinder rod.

Generally speed control by hydraulics is much superior to that by pneumatics. Time delays are minimal in hydraulic systems where compressibility is low and the pressure compensated flow control valve makes possible load independent speed control. At very low demanded speeds stick slip oscillation [8] can also cause problems in hydraulic systems.

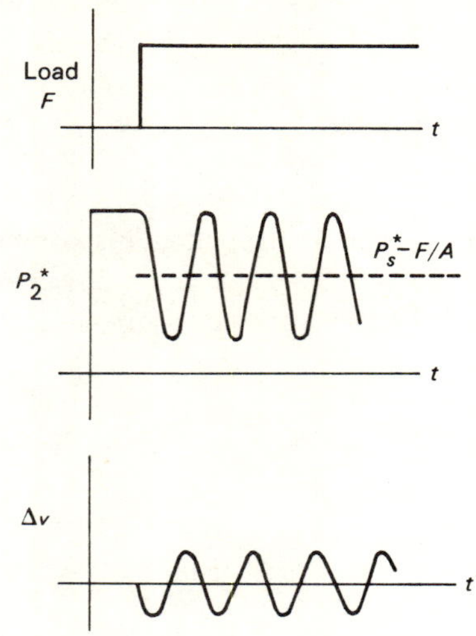

Fig. 14.12 – Response of a hydraulic cylinder with meter-out speed control.

14.8 EXAMPLE

1. A vertical cylinder is used to lift a 100 kg load. Air is supplied at the blank end at 7 bar abs. A metering orifice is inserted in the exhaust line from the rod end of the cylinder; so that the system is operating under meter-in control. The diameter of the cylinder bore is 60 mm and the rod diameter is 20 mm.

   (a) Assuming that the pressure in the blank end remains constant, determine the steady state pressure in the rod end.

   (b) Assuming that the metering orifice is a sharp edged circular hole of 2 mm diameter ($C_q = 0.6$), determine the steady state load velocity. You may assume isothermal conditions. The density and temperature of the ambient air are 1.2 kg/m^3 and 20°C, respectively.

   (*Answers*: (a) 3.846 bar abs.; (b) 148 mm/s.)

14.9 REFERENCES

[1] Henke, R. W. (1970), *Introduction to Fluid Power Circuits and Systems*, Addison-Wesley.

[2] Thoma, J. U. (1975), *Introduction to Bond Graphs and Their Applications*, Pergamon.

[3] Tsai, D. H. and Cassidy, E. C. (1961), Dynamic Analysis of a Simple Pneumatic Pressure Reducer, *J. Basic Eng.*, A.S.M.E., **83**, 253, June.

[4] Takenaka, T., and Urata, E. (1968), Static and Dynamic Characteristics of Oil Hydraulic Control Valves. *Fluid Power Int. Conf.*, London, September.

[5] Foster, K. (1961), Dynamic Analysis of a Two Stage Relief Valve, *Proc. Conf. on Oil Hydraulic Power Transmission and Control*, Inst. Mech. Engrs., London.

[6] Ma, C. Y. (1966), The Analysis and Design of Hydraulic Pressure Reducing Valves, *Trans. A.S.M.E.*, paper 66-WA/MD-4.

[7] *Pneumatics – Dynamic Behaviour of Double Acting Air Cylinders* (1965), P.E.R.A. Melton Mowbray, England, Report No. 135.

[8] Stringer, J. D. and Healey, A. J. (1968-69), Dynamic Characteristics of an Oil Hydraulic Constant Speed Drive, *Proc. Inst. Mech. Engrs.*, London, **183**, Pt. 1.

[9] Yeaple, F. D. (1966), *Hydraulic and Pneumatic Power and Control*, McGraw-Hill.

[10] Keller, G. R. (1969), *Hydraulic Systems Analysis*, Industrial Publishing Co.

[11] Ernst, W. (1960), *Oil Hydraulic Power and its Industrial Application*, McGraw-Hill.

[12] Merritt, H. E. (1967), *Hydraulic Control Systems*, Wiley.

CHAPTER 15

Boolean Algebra and Logic Gates

15.1 INTRODUCTION

The word **fluidic** was derived from the two words fluid logic-logic operations using fluid devices. In this context logic is the means of achieving the desired system behaviour by means of a set of unambiguous decisions.

15.2 BOOLEAN ALGEBRA

Boolean algebra is the algebra of logic, and although it was originally an abstract mathematical form, including a switching algebra of sets, and a propositional calculus, it can be applied with relative ease to the design of fluidic systems. George Boole, its inventor, published the important paper 'Mathematical analysis of logic' in 1847. It was not until 1938, however, that Shannon, an electrical engineer, showed how Boole's work could be used in the design of switching circuits in digital computers and telephone exchanges.

Modern computing and logic devices use binary arithmetic. With such a system any desired function can be generated using a simple two-state device, such as a switch which can be either on or off [1]. In addition, manipulation of binary numbers is simple. The three basic operations in Boolean algebra are NOT, AND, and OR.

15.2.1 The NOT operation

NOT implies inversion and is written as:

$$\bar{A} = \text{NOT } A. \tag{15.1}$$

Thus if at any particular time the variable A is 1, then $\bar{A} = 0$. The symbol for the NOT operation is shown in Fig. 15.1. This can be accomplished in practice by many fluid elements and the figure shows how the turbulence amplifier, the jet deflection element and the spool valve can act as invertors. (The use of conventional valves will be discussed in more detail later.)

In the turbulence amplifier a laminar jet is directed at a collector tube in which a substantial proportion of the jet's energy is recovered [2]. If the conditions are right, a control jet A placed at right angles to the main jet can change the laminar stream into a turbulent one, thereby causing a considerable reduction in the recovered energy.

The spring logic element of Fig. 15.1(d) uses a close coiled spring as a variable restrictor [3]. The signal A makes the diaphragm move, deflecting the spring and allowing air to escape through the open coils.

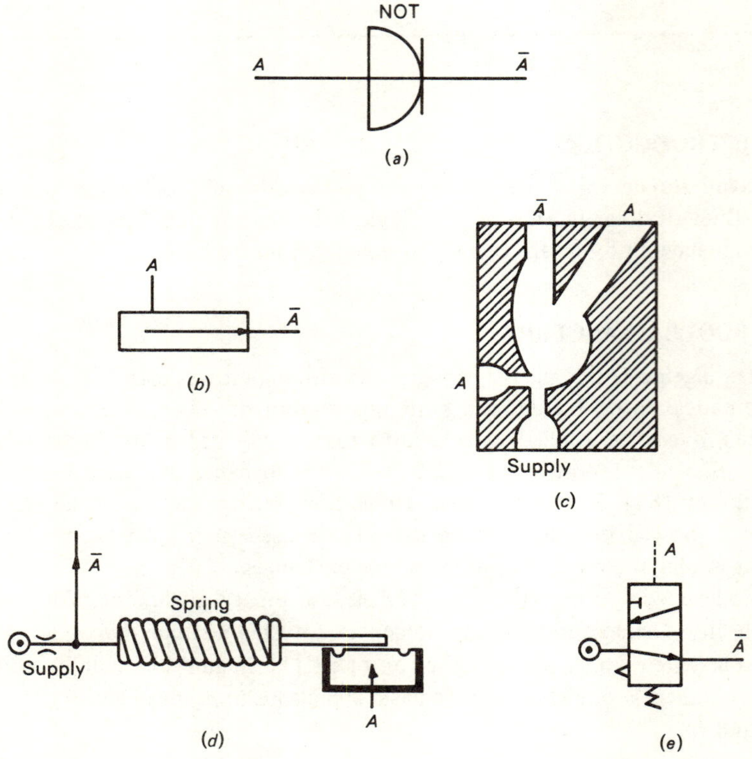

Fig. 15.1 – (a) Symbol for inversion, (b) symbol for a turbulence amplifier invertor, (c) an invertor using jet deflection, (d) inversion using a spring and diaphragm, (e) a spool valve invertor.

15.2.2 The AND operation

It is necessary at this stage to introduce the **logic gate**, which is a device having several inputs and designed so that there is an output when and only when a certain set of input conditions is met. The AND gate has unity output only when all of its inputs are unity. This is written as:

$$S = A \cdot B \cdot C \cdot D \tag{15.2}$$

or sometimes

$$S = ABCD$$

meaning that the output S is one, only if the inputs *A and B and C and D* are each one. Fig. 15.2 shows the symbol for the AND gate and a few devices that are capable of generating this function. The jet interaction device is passive, in that no power is supplied to it other than by the control signals. The fluid will only flow out of the output port if both controls *A and B* are one. The spool valve element can handle much more power but again only gives an output if both inputs are present. It is also a passive device.

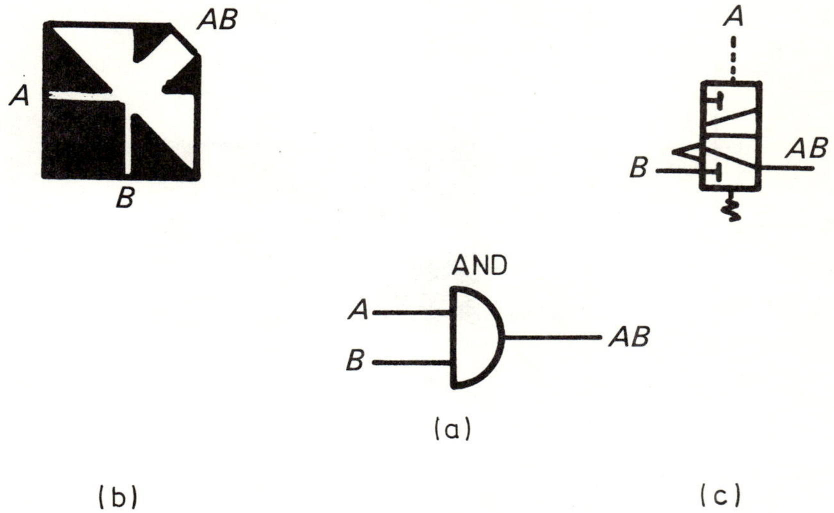

Fig. 15.2 – (a) AND gate symbol, (b) a passive jet interaction device, (c) a passive spool valve AND gate.

It is sometimes useful to construct a table showing the value of the output of a gate for all possible combinations of the inputs. This is known as a **truth table**, so called after its original connection with the truth and falsity of statements in symbolic logic. The truth table for an AND gate is:

A	B	S
0	0	0
0	1	0
1	1	1
1	0	0

Note that in going from one row to the next only one input variable is changed. The number of rows in the table is 2^n where n is the number of input variables.

15.2.3 The OR operation

The OR gate has unity output if any one of the inputs is unity. This is written:

$$S = A + B + C + D + \ldots \tag{15.3}$$

meaning that the output S is one only if the input A *or* B *or* C *or* D etc. is one. Two OR gates are shown in Fig. 15.3. In the wall attachment element the flow is normally attached to the upper wall. It is a monostable device. Controls A or B or C will cause the jet to switch to the lower wall.

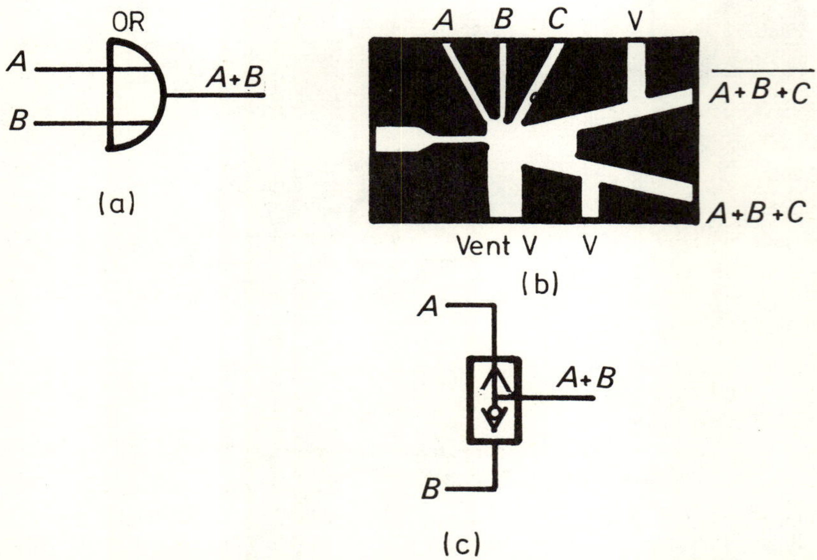

Fig. 15.3 – (a) OR gate symbol, (b) a wall attachment OR/NOR gate, (c) use of a shuttle valve.

The simple shuttle valve, introduced in section 6.2.8, is basically an OR gate and will be discussed in more detail later.

The truth table for a two input OR gate is:

A	B	S
0	0	0
0	1	1
1	1	1
1	0	1

15.3 SOME OTHER LOGIC GATES

There are other logic operations in addition to the above three, but it is important to note that any logic operation can be implemented using only these three basic gates NOT, AND, and OR [1, 4].

For example the NOR operation is the inversion of OR and is written:

$$S = \overline{A + B + C + \ldots} \,. \tag{15.4}$$

The output S is one, only if all inputs are zero.

The turbulence amplifier with two inputs is a very simple NOR gate, Fig. 15.4(b). The Stanford ball element [5] which is capable of working at high pressures with either liquids or gases, is illustrated in Fig. 15.4(c). Imbalance of forces ensures that the small ball will be moved to the left, closing off the supply, when either input A or B is activated. Operation of the spool valve element of Fig. 15.4(d) is self evident. Reference again to Fig. 15.3(b) will now show that the wall attachment OR element is in fact an OR-NOR gate, depending which leg output is used. NOR can also be obtained by cascading NOT and OR gates.

The NAND operation is the inverse of AND and is commonly used in logic circuits. It is described by the equation:

$$S = \overline{ABCD\ldots} \tag{15.5}$$

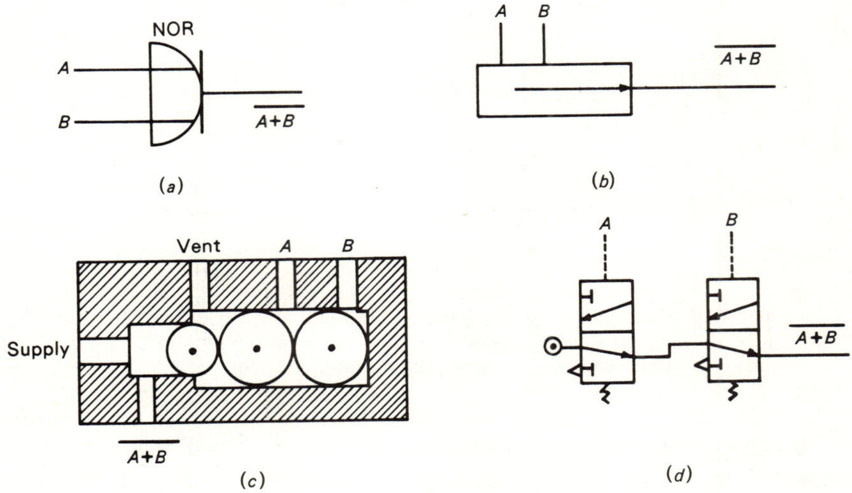

Fig. 15.4 – (a) NOR gate symbol, (b) a turbulence amplifier used as a NOR, (c) the Stanford ball element, (d) use of spool valves in series.

The **flip-flop**, or **bistable**, is another important device. Its symbol is shown in Fig. 15.5(a). If the last input were A, output S_2 would be on. If it were B, then output S_1 would be on. The memory inherent in these devices results in the state of the output being affected by not only its present inputs but also its past inputs. The bistable amplifier, Fig. 15.5(b), relying on reattachment for its operation, is perhaps the simplest fluidic bistable. Having been switched to one side by a signal A it will remain there even when signal A is removed. Further signals A will not change the state of the output. This can only be changed by a signal B.

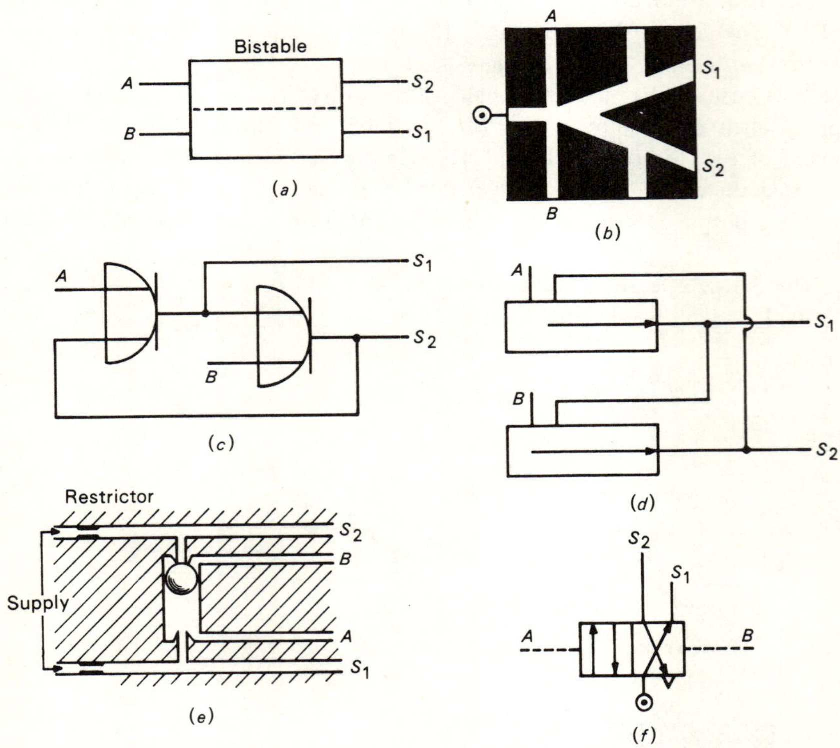

Fig. 15.5 – (a) Symbol for a bi-stable, (b) reattachment device, (c) two NOR gates combined to make a bi-stable, (d) use of turbulence amplifiers, (e) the Kearfott ball element, (f) a double pilot operated four-port valve.

Bistable action can also be achieved by the use of two NOR gates interconnected as in Fig. 15.5(c). Turbulence amplifier NOR gates can be used as in Fig. 15.5(d). If an input pulse is supplied at B, the lower amplifier is turned off, no signal exists at S_2 and an output S_1 is obtained. This output supplies a signal which continues to keep the lower amplifier off even when the pulse B

is removed. This is the memory. A pulse at A turns off the top amplifier and the lower amplifier comes on giving an output S_2.

The element in Fig. 15.5(e), the Kearfott ball element, is a moving part bistable [6]. As shown, there is no input and the ball is held in the upper position by a force imbalance achieved by a careful sizing of the orifices. Output S_2 is then pressurised and S_1 sits at a much lower pressure level. A pressure pulse at B will disturb the equilibrium and the ball will fall to its lower position, in which output S_1 is pressurised and S_2 relieved. The ball will remain there after the removal of signal $\bar{B}$.

In conventional pneumatic circuits the pilot operated, two-position, four-port valve is commonly used as a memory, Fig. 15.5(f). Here, friction between spool and valve body provides the memory mechanism. This, of course, is unreliable if used under conditions of severe vibration or impact where the spool may move without the application of a pneumatic signal.

15.4 MANIPULATION OF LOGIC EXPRESSIONS

There are several rules and theorems involved with Boolean algebra, and these must be appreciated before an attempt can be made to design logic circuits with the basic fluidic elements [1].

1. **Identities**

$$A + 0 = A$$

$$A \cdot 1 = A\,.$$

In the first case the OR function is solely dependent on the value of A. In the second, since one input to the AND always exists, then the output depends only on the value of A.

2. The **idempotent laws**

$$A + A = A$$

$$AA = A\,.$$

3. The **commutative laws**

$$A + B = B + A$$

$$AB = BA\,.$$

These indicate that the order in which signals are consided is of no significance.

4. The **associative laws**

$$(A + B) + C = A + (B + C)$$

$$(AB)C = A(BC)\,.$$

These show that the way in which functions are grouped is of no significance.

5. The **distributive law**

$$A(B+C) = AB + AC\,.$$

The reader should verify this from a truth table.

6. $A + AB = A\,.$

The truth of this equality can be checked by means of a truth table, but it can also be demonstrated by deduction. It states the left-hand side is independent of B. If A is one, then regardless of the value of B, the OR function will have a value of one. If A is zero then AB is zero and the OR function also has zero value.

7. $A + \bar{A} = 1\,.$

8. $A\bar{A} = 0\,.$

9. $A0 = \bar{A}\,.$

10. $A + 1 = 1\,.$

11. **De Morgan's theorems**

$$\overline{ABC} = \bar{A} + \bar{B} + \bar{C}$$

$$\overline{A+B+C} = \bar{A}\bar{B}\bar{C}\,.$$

These show that the inverse of a function can be obtained by inverting all the variables and then changing the AND to OR or the OR to AND.

15.4.1 Example of reduction of a logic function

Consider the expression:

$$(XY + ABC)(XY + \bar{A} + \bar{B} + \bar{C})\,.$$

This expands to:

$$XYXY + \bar{A}XY + \bar{B}XY + \bar{C}XY + ABCXY + A\bar{A}BC + AB\bar{B}C + ABC\bar{C}$$

Now from theorem 2:

$$XYXY = XY$$

from theorems 8 and 9:

$$A\bar{A}BC = 0BC = 0$$

$$AB\bar{B}C = 0AC = 0$$

$$ABC\bar{C} = AB0 = 0\,.$$

The expression then reduces to:

$$XY + \bar{A}XY + \bar{B}XY + \bar{C}XY + ABCXY$$

$$= XY(1 + \bar{A} + \bar{B} + \bar{C} + ABC)\,.$$

But from theorem 10:

$$1 + \bar{A} + \bar{B} + \bar{C} + ABC = 1\,.$$

Hence the original expression reduces to $XY1 = XY$.

15.4.2 Example of the use of De Morgan's theorems

De Morgan's theorems are extremely useful for determining how one logic function can be derived from the others; for example how can AND logic be obtained using NOR logic? This would be very important if the basic elements at hand were NOR units and AND was desired.

Consider:

$$S = AB\,.$$

De Morgan's theorem gives:

$$\bar{S} = \bar{A} + \bar{B}\,.$$

To put this in NOR form invert again to give:

$$S = \bar{\bar{S}} = \overline{\bar{A} + \bar{B}}\,.$$

This is illustrated in Fig. 15.6 in which the turbulence amplifier, a basic NOR element, has been used. Note that the number of bars in the logic expression determines the number of turbulence amplifiers required – three in this case.

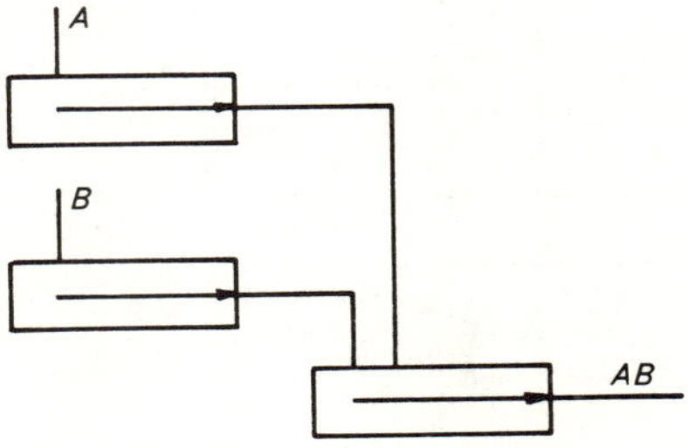

Fig. 15.6 – AND logic using NOR elements.

15.5 A CIRCUIT DESIGN

Consider the following problem. Design the logic machine for controlling an automatic drill operation. The drill should only operate if the start button is pressed, if the workpiece is in position, and if the clamp is operated. The drill should not operate if the tool is broken or if the number of operations with the tool has exceeded a certain limit. In addition the operator should be able to start the operation from a remote position.

Firstly, it is assumed that transducers are available to provide signals indicating the state of the machine:

Signal A = start button pressed
B = workpiece in position
C = guard in position
D = tool is broken
E = tool life is exceeded
F = start from remote position .

If S is the drill actuating signal then:

$$S = (A + F)BC(\overline{D + E}) .$$

The logic circuit is shown in Fig. 15.7, and this can be implemented in practice by using the OR, NOR and AND gates described earlier. Sometimes, however, the number of available different gates is limited and the logic circuit must be changed. If, for example, only turbulence amplifiers are available, the NOR logic must be used. Inversion gives:

$$\bar{S} = (\overline{A + F}) + \bar{B} + \bar{C} + (D + E)$$

and a further inversion gives:

$$S = \bar{\bar{S}} + \overline{(A + F) + B + C + D + E} .$$

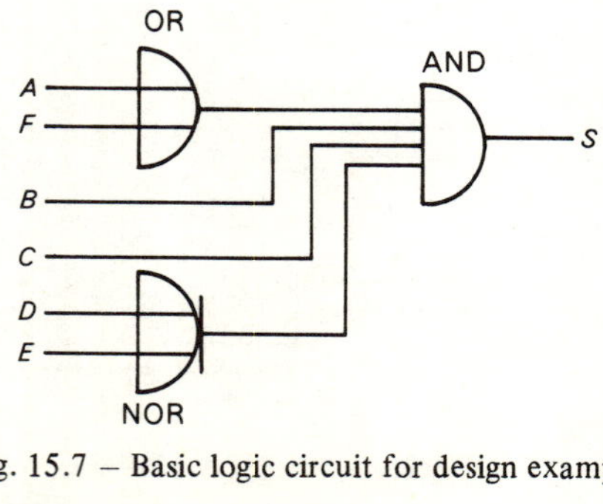

Fig. 15.7 – Basic logic circuit for design example.

Four turbulence amplifiers (the number of bars in the equation) are necessary. The circuit is shown in Fig. 15.8. Note that the NOR gate has five inputs, and this may be difficult to realise in practice. If only four inputs can be handled then the logic expression can be rewritten as:

$$S = \overline{\overline{A+F}+\bar{B}+\bar{C}+\overline{\overline{D+E}}}\,.$$

Six turbulence amplifiers are now necessary.

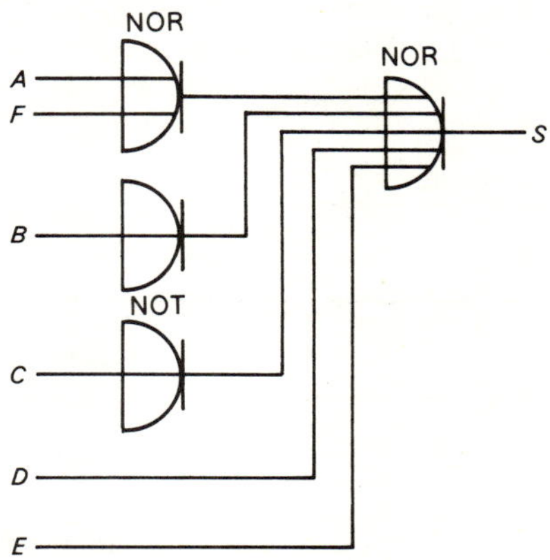

Fig. 15.8 – Logic circuit if NOR gates only are available.

15.6 LOGIC WITH CONVENTIONAL VALVES

Conventional valves can be used as logic gates but it is only since the advent of the relatively new technology of fluidics that engineers have come to appreciate or to take advantage of this fact [7, 8, 9, 10, 11]. Check valves, shuttle valves and the directional control valves can be combined to perform a wide variety of logic functions, and as a result circuits are now being designed by logic methods rather than by the very commonly used ‘cut and try’ or ‘rule of thumb’ methods [4]. This leads to a more economical use of valves.

The conventional valve can thus be used for information transfer as well as for power transfer, and this eliminates the need for step-up relays and the fine filtration required in logic circuits using dynamic elements. Unfortunately many valves, especially the hydraulic ones, were too large and expensive to allow their use in any but the simplest of logic circuits. However, miniaturisation is now an important fluid power development area, and five-port valves are available with a volume less than $20\,\text{cm}^3$. Many of these valves are designed on a modular basis, allowing each valve to be conveniently plugged into the manifold block which includes supply, exhaust and control connections. This eliminates the bulky interconnections associated with conventional pneumatics or hydraulics and gives a system which takes up a minimum of space and is easily interchanged.

15.6.1 Use of the check valve

The check valve or non-return valve (section 6.2.8) allows flow in only one direction. It can be used in the OR gate, Fig. 15.9, where its unidirectional flow characteristic allows each input to the gate to be isolated from the others. Note that the inputs A, B and C must be fluid signals. If mechanical inputs are to be used, these must first be transformed to fluid signals before they can be processed by this gate. The OR gate requires a bleed in the output line in order to allow the outlet to assume the zero condition.

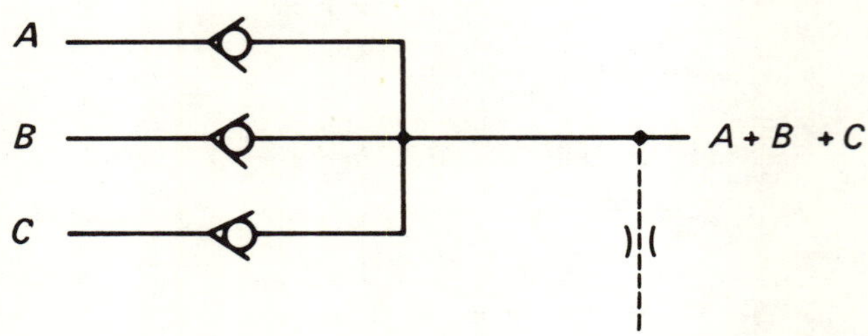

Fig. 15.9 – OR gate using non-return valves.

15.6.2 Use of the two-port valves

A pilot operated, spring return two-port valve is shown in Fig. 15.10. The output is either connected to the input γ or is blocked off depending on the value of the other input α. The combined effect of both inputs gives AND logic:

$$S = \alpha\gamma\,. \tag{15.6}$$

The input α can be either fluid or mechanical, but γ clearly must be a fluid signal. Since there is no direct provision for exhausting the output other than through the load, the device requires a bleed as shown. An OR gate with mechanical inputs A, B and C is shown in Fig. 15.10(b). An AND gate with fluid signal inputs is shown in Fig. 15.10(c). In this case if the interconnecting lines are long it may be necessary to install bleeds between each valve as well as in the output line. False signals may arise if this is not done. For example if both inputs A and B are on, the interconnecting lines are pressurised. If now A and B are removed and C is applied, the capacitance of the line may be sufficient to provide an output signal pulse, and if the following element were a bi-stable this could well be troublesome.

15.6.3 Use of the shuttle valve

This valve is an extremely simple and inexpensive OR gate. It is a natural development of the two input check valve OR gate, Fig. 15.11(a). Examination of this figure reveals that there is actually no need for both moving elements – one is sufficient for isolating the inputs, Fig. 15.11(b). This is the shuttle valve. Removal of the extra element allows the output to exhaust back through the input connection so that a bleed in the load line is not required. Each valve is limited to two inputs, but several can be connected as in Fig. 15.11(c) to accept more inputs.

15.6.4 Use of three-port valves

It was seen earlier that the two-port valve required a bleed in the load line. With the three-port valve the extra port allows the output to assume the 0 state

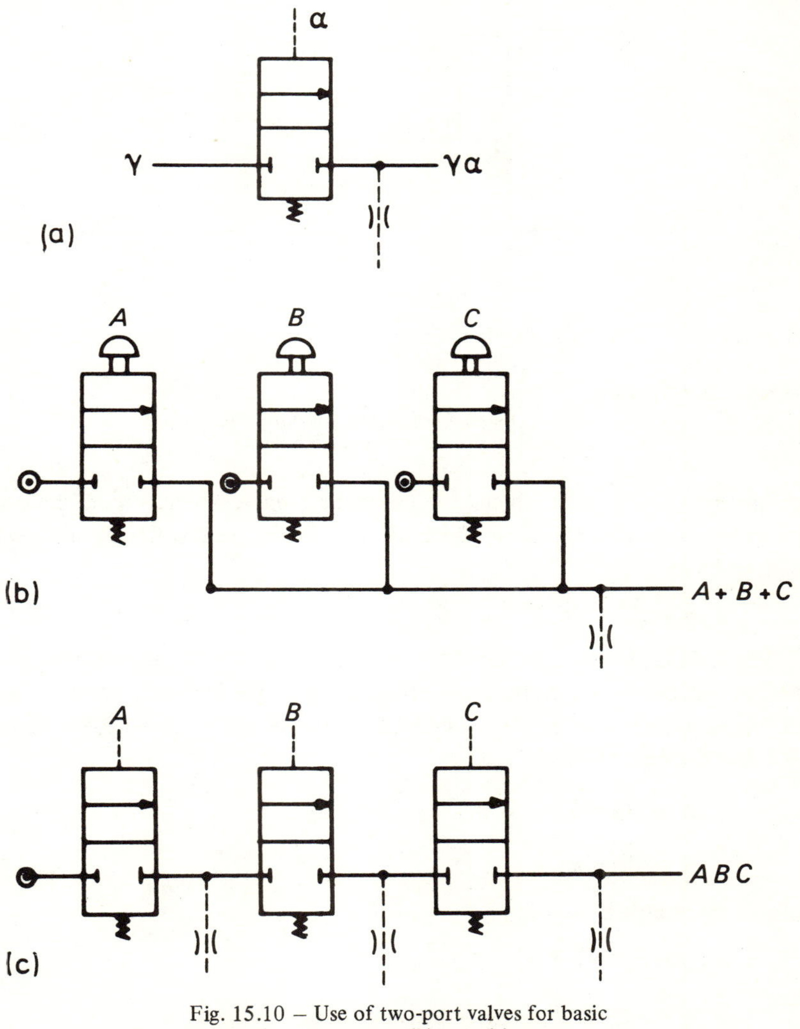

Fig. 15.10 – Use of two-port valves for basic logic, (a) passive AND, (b) OR, (c) AND.

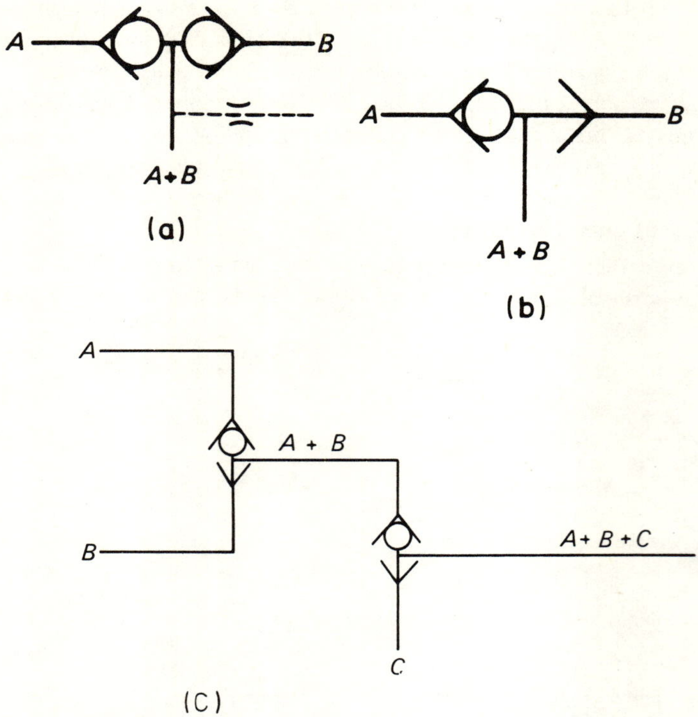

Fig. 15.11 – OR logic using, (a) two non-return valves, (b) one shuttle valve, (c) two shuttle valves.

without the use of such a bleed. A pilot operated spring returned three-port valve is shown in Fig. 15.12(a). There are three inputs, α, β and γ and the output S is given by:

$$S = \beta\bar{\alpha} + \gamma\alpha\,. \tag{15.7}$$

Fig. 15.12(b) shows an identity element, an AND gate, an OR gate and an invertor. All of these except the AND gate are active elements in which either β or γ is equal to 1, that is, it is connected to supply. In the OR gate, for example, $\gamma = 1$ and the output is:

$$S = \beta\bar{\alpha} + 1.\alpha$$

or

$$S = \beta\bar{\alpha} + \alpha\,.$$

Using De Morgan's Rule:

$$\bar{S} = (\overline{\beta\bar{\alpha}})\,\bar{\alpha} = (\bar{\beta} + \alpha)\bar{\alpha} = \bar{\beta}\bar{\alpha}\,.$$

Hence

$$S = \beta + \alpha\,.$$

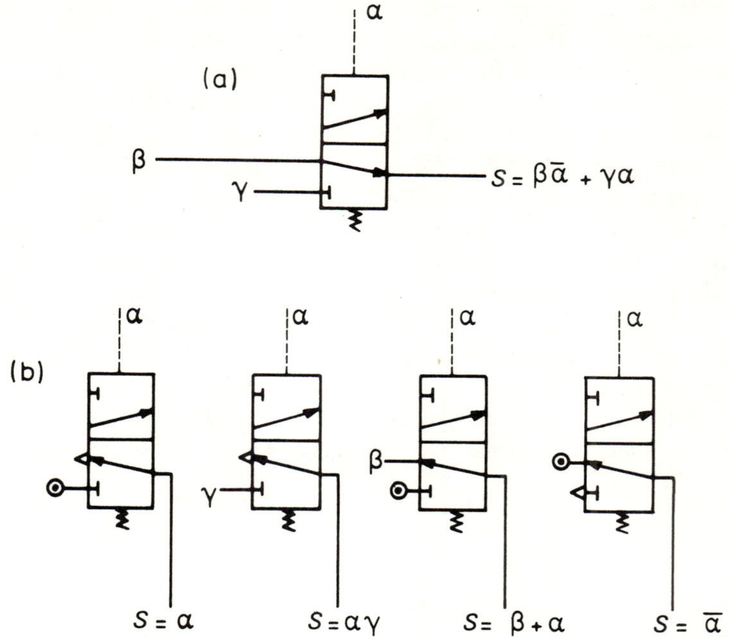

Fig. 15.12 – (a) The three-port valves as a logic element
(b) identity, AND, OR, NOT logic.

If mechanical inputs are needed, then the OR and AND gates of Fig. 15.13 can be used. These are both active and are often employed in machine control

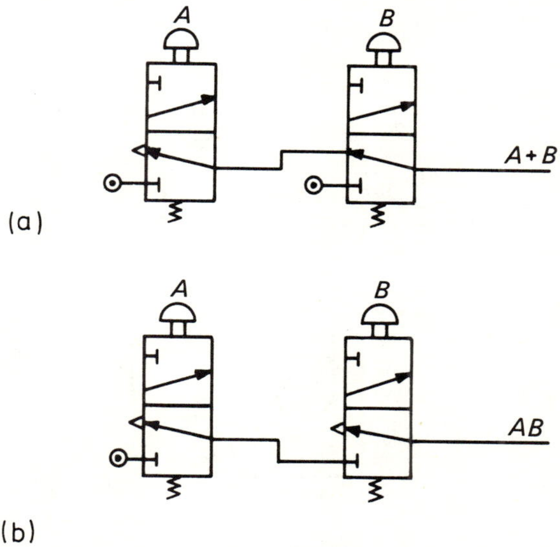

Fig. 15.13 – Logic gates with mechanical/manual inputs.

circuits. The OR gives an operator the facility for starting a machine from two remote locations, and the AND gate is used in two hand safety circuits which protect the operator's hands during a machining operation.

In most cases the logic function is inverted if supply and exhaust ports are interchanged. If this is done in Fig. 15.13, NOR and NAND logic circuits ensue.

The combination of shuttle valve and three-port valve can be useful in some circumstances. For example if fluid inputs *A* and *B* are available, a shuttle valve followed by a three-port valve invertor, Fig. 15.14(a), gives NOR logic.

Again, the OR gate of Fig. 15.13(a) may not function reliably if the line joining valves *A* and *B* is excessively long. This can be halved in length if the gate of Fig. 15.14(b) is used. Here two identity elements supply inputs to a shuttle valve which can be placed half way between the valves.

These examples show that the basic logic functions can be derived quite economically with the conventional three-port valves. This economy is further illustrated by reference to some more complicated functions. One such is the

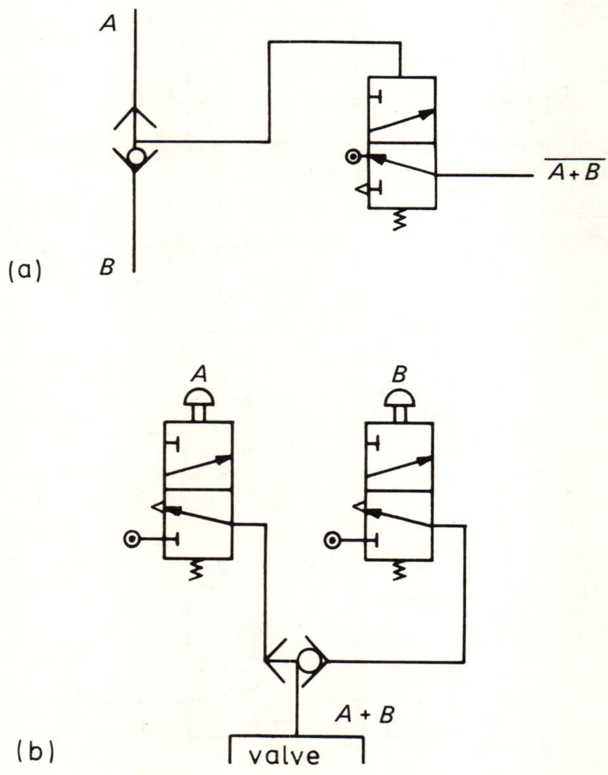

Fig. 15.14 – Combinations of shuttle and three-port valves, (a) for NOR logic, (b) for OR logic.

exclusive OR function. In this case, if the inputs are X and Y, there is an output when X or Y is present, but not when both are present. In logic symbols:

$$S = X \oplus Y = Y\bar{X} + X\bar{Y} .$$

The output of a three-port valve is:

$$S = \beta\bar{\alpha} + \gamma\alpha$$

and hence the exclusive OR is derived if:

$$\alpha = X, \quad Y = \beta, \quad \gamma = \bar{Y} .$$

Two valves are needed, Fig. 15.15, one being an invertor.

Another example is the modulo n gate which gives an output only if there is an odd number of inputs. In fact the exclusive OR is an elementary 2 input form of this gate. For a 3 input gate:

$$S = A\bar{B}\bar{C} + \bar{A}B\bar{C} + \bar{A}\bar{B}C + ABC .$$

This has to be written in a form suitable for construction with three-port valves, and reference to equation (15.7) shows that this is:

$$S = \bar{C}(A\bar{B} + \bar{A}B) + C(\bar{A}\bar{B} + AB)$$

or

$$S = \bar{C}S_1 + CS_2 .$$

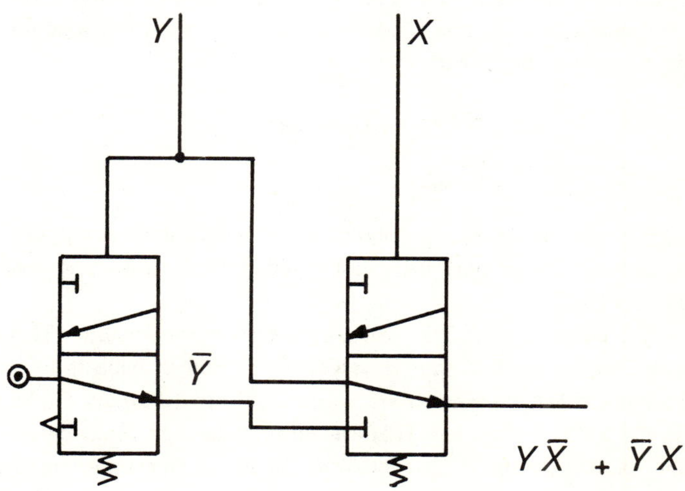

Fig. 15.15 – Exclusive OR.

The function S_1 is provided by the exclusive OR circuit of Fig. 15.15 and requires valves 1 and 2 in Fig. 15.16. The function S_2 is derived in a similar manner except that the A and $\bar{A}$ ports are interchanged. A total of four valves is required.

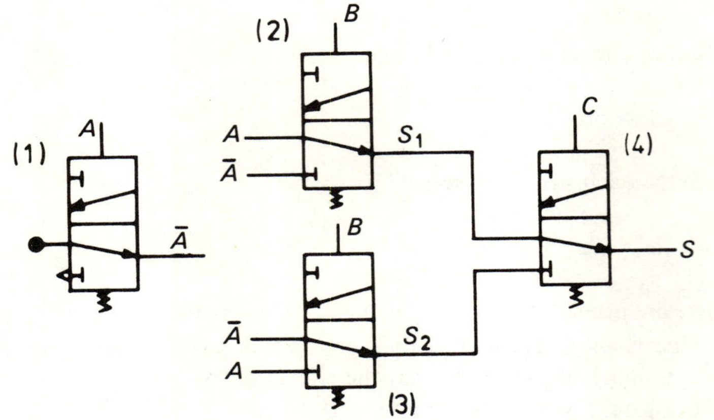

Fig. 15.16 – Circuit for a three input modulo n gate.

15.6.5 Use of four and five-port valves

The four-port valve is equivalent to two three-port valves on a common stem. The two outputs or cylinder connections are $S_1 = \beta\bar{\alpha} + \gamma\alpha$, and $S_2 = \beta\alpha + \gamma\bar{\alpha}$. Fig. 15.17(a) shows that S_1 corresponds to the output of a three-port valve with inputs α, β and γ whilst S_2 is the output of a three-port valve with inputs $\bar{\alpha}$, β and γ. The student should examine Figs. 6.3 and 6.7 and ensure that the equivalence of a four-port and two three-port valves is clear. Hence the four-port valve has twice the logic power of the three-port valve. For example, consider again the modulo n circuit with logic:

$$
\begin{aligned}
S &= \bar{C}(A\bar{B} + \bar{A}B) + C(\bar{A}\bar{B} + AB) \\
&= \bar{C}S_1 + CS_2 \ .
\end{aligned}
$$

In Fig. 15.16 two three-port valves were needed to generate signals S_1 and S_2 but now only one four-port valve is needed for this task. The final circuit is shown in Fig. 15.17(b).

The five-port valve of Fig. 15.18(a) is even more versatile. This allows for four inputs, and indeed if shuttle valves are placed at each input port in Fig. 15.8(b) an additional two inputs are available. Up to 10 inputs can be accommodated. In the valve in Fig. 15.18(b), the spring has been removed and the valve is double pilot operated. This provides bi-stable operation or the memory function. As shown, output S_1 is on and S_2 is off. If a pulse is applied at A or B, the valve will switch, turning S_1 off and S_2 on. Friction between valve spool and sleeve, or in same cases a mechanical detent, will maintain the valves is this state until a pulse is applied at *C* or *D*. A later chapter on sequencing will show that such a memory element is invaluable for cylinder control.

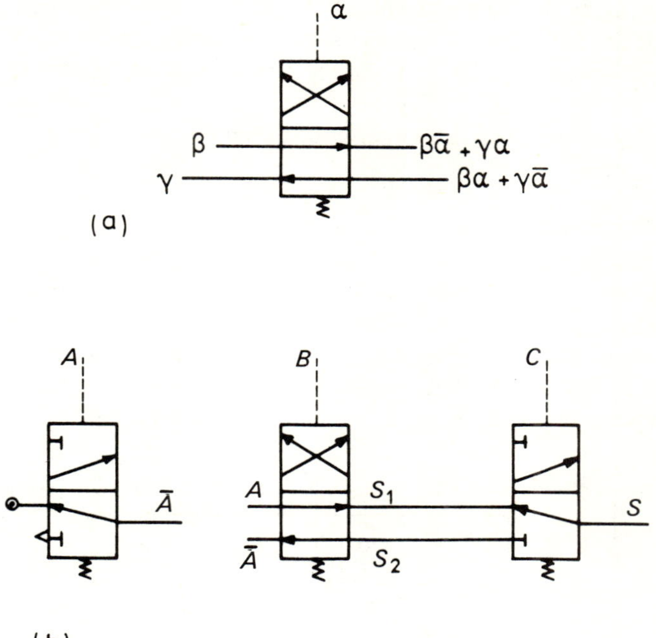

Fig. 15.17 – (a) The logic of the four-port valve, (b) circuit for a 3 input modulo *n* gate.

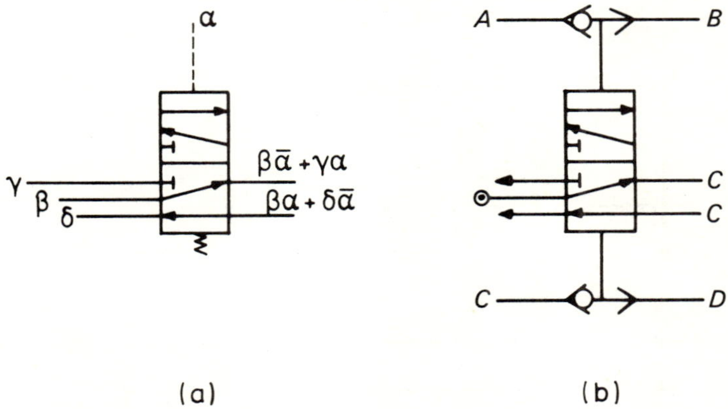

Fig. 15.18 – (a) The logic of the five-port valve, (b) increasing the number of inputs using shuttle valves.

15.7 COUNTING WITH VALVES

Many industrial processes use counting circuits. In packaging, for example, it is necessary to ensure that the correct number of elements is in position before the stack is placed in the package. Again in some sequences a pneumatic clock is used for timing the start and stop of various actions within a machine cycle. Such a clock is basically an oscillator which produces pulses at a constant rate. Since the pulses are of constant duration, time can be determined by counting them.

Counting circuits may be either coded or non-coded. The former will be discussed first.

15.7.1 Coded circuits

In coded circuits it is usual to employ the binary code so that the decimal 5 for example is coded as the binary 101. When a code is used it is usually necessary to provide some means for decoding again, and this will be demonstrated later in this chapter. The basic element used in binary counters is the binary stage which changes its state and retains its output for each input pulse. One such stage, well proven in practice, is shown in Fig. 15.19 [4, 12, 13, 14]. The function

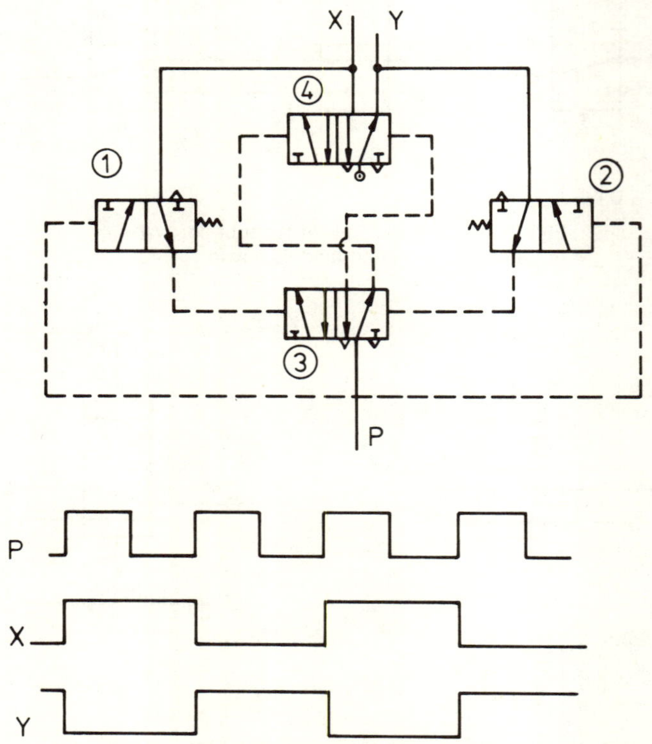

Fig. 15.19 – A binary stage with input P and outputs X and Y.

diagram shows that the mode of operation is basically that of dividing the input (P) frequency by two. As shown, output Y is on and X is off. There is no input P and hence the feedback from X and Y to valve 3 holds it in the position shown. An input pulse will now pass through valve 3, switching valve 4 which in turn switches X on and Y off. During the pulse application, valves 1 and 2 are actuated, blocking the feedback lines to valve 3 so that it does not move. Valve 3 moves after the termination of the pulse. Outputs X and Y remain on and off respectively until the arrival of the next pulse.

A useful circuit is provided if the outputs X and Y are connected to the ports of a double acting cylinder. If pulses are supplied by manual operation of a button-operated spring return valve, the cylinder rod will be extended and retracted by successive pulses of the *same* button. This circuit could be developed to allow a door to be opened or closed from either side. In this case a button-operated valve is needed on each side of the door, and their two outputs are connected through a shuttle valve to the binary counter stage.

Binary counting requires a series connection of binary stages, Fig. 15.20. The first pulse P turns X_1 on and Y_1 off. The second pulse returns the first stage to its initial state, hence sending an input Y_1 to the second stage, etc. The function diagram illustrates the division by two at each stage. The number X_3 X_2 X_1 represents in binary code the number of pulses P. For example $X_1 = 1$, $X_2 = 1$, $X_3 = 1$ gives the binary number 111, which is 7 in decimal.

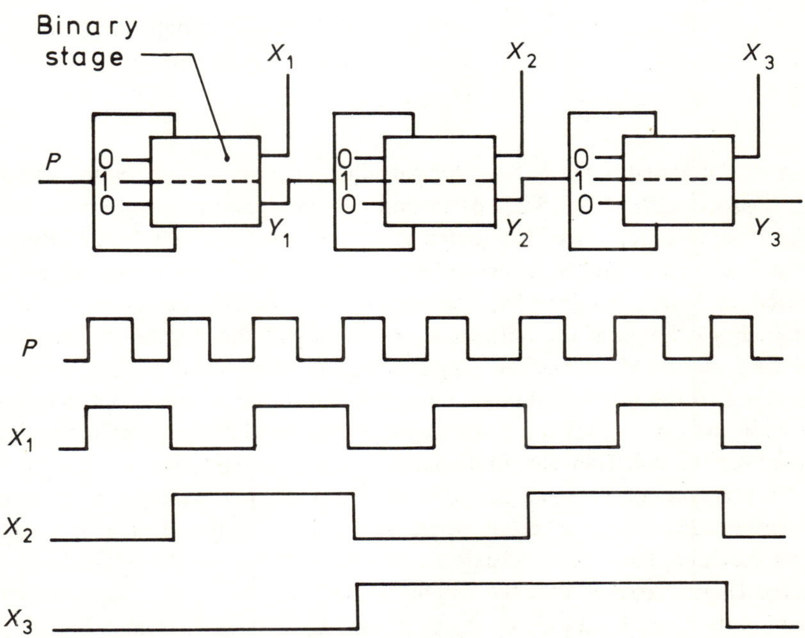

Fig. 15.20 – A three stage binary counter.

It was mentioned earlier that coded circuits usually need some form of decoding before they can be used practically. This is demonstrated in the following example [12]. Assume that in a packaging operation, articles are to be packed in twos or in sixes, and the actual number should be easily controlled by the operator. A three stage binary counter is necessary for this job. As in Fig. 15.20 there are three outputs from the counter, X_1, X_2 and X_3. If each article in turn trips a valve, sending a pulse to the counter, then when two have passed the code will read 010 or $\bar{X}_3\ X_2\ \bar{X}_1$. When six have passed the code will read 110 or $X_3\ X_2\ \bar{X}_1$. Decoding consists of taking action when these particular combinations of X_3, X_2 and X_1 occur. Fig. 15.21 shows that three valves are needed for decoding. The manually operated valve allows selection of either 2 or 6 articles. As shown it is in the up position so that when 2 articles are counted, $\bar{X}_3\ X_2\ \bar{X}_1 = 1$ and there is an output S which can be used to drive the packaging cylinder.

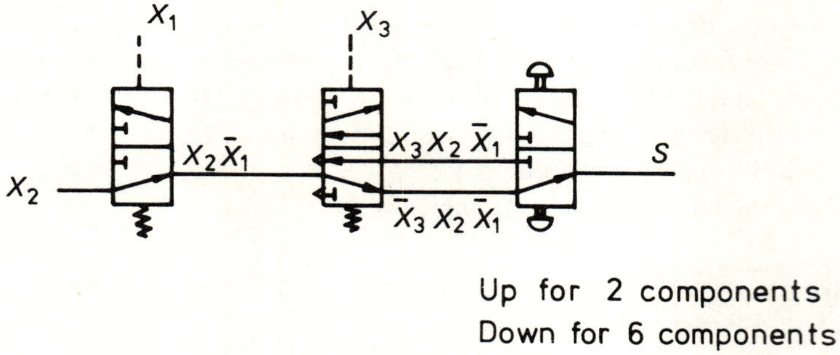

Fig. 15.21 – A decoder.

A digital actuator can be used for decoding purposes in positioning systems, [15]. This consists, Fig. 15.22, of several cylinders connected in series. Each cylinder can be independently actuated so that if there are n cylinders, 2^n discrete output positions y can be achieved. In addition if the cylinder strokes can be arranged in binary progression, the cylinder will decode binary input information. For example if the outputs $X_3\ X_2$ and X_1 of the counter in Fig. 15.20 were used to actuate the cylinder as shown, then the output position y would correspond to the number of input pulses applied to the counter. Such actuators have been used in aircraft and aerospace applications where on-board computers are used for control. They also find application in machine tools.

Very simple applications can also be envisaged. For example, if in a pipe mill one needed to cut a certain series of standard lengths of pipe on a single sawing machine, the digital actuator could be used to reset the stops for each required length. Assume that the lengths were 7, 5, 3 and 1 units, in which case coded inputs $X_3\ X_2\ \bar{X}_1, X_3\ \bar{X}_2\ X_1, \bar{X}_3\ X_2\ X_1$ and $\bar{X}_3\ \bar{X}_2\ X_1$ are required. In a simple case like this it is unlikely that a counter would be used. The operator could easily use button-operated three-port valves to provide the inputs X_3, X_2 and X_1 as required. In this case he would be acting as the coding machine.

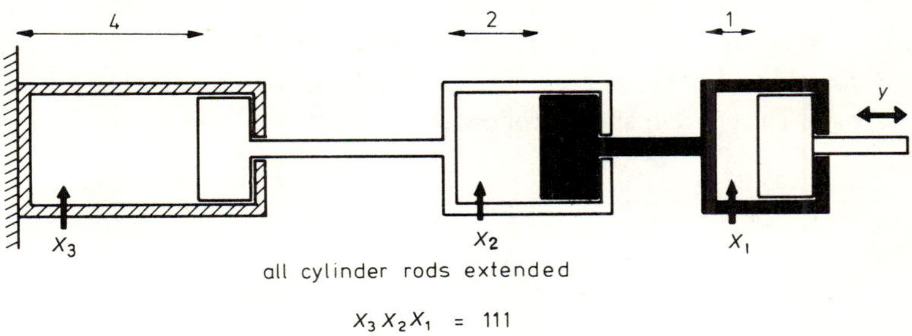

Fig. 15.22 – A digital actuator using series cylinders.

15.7.2 Non-coded counting circuits

In a ring counter, information is transferred from one stage to the next when a command or shift signal is applied. Thus, unlike the binary counter, only one output is on at a given time. A well proven pneumatic counter is shown in Fig. 15.23. This uses double pilot operated three-port valves as memory elements. The complementary inputs R (reset) and S (set) could be obtained as shown in Fig. 15.24. Use of two diaphragm-operated spring return valves ensures that the reset line is exhausted before the set line is pressurised, and vice versa. This is essential for reliable operation of the counter.

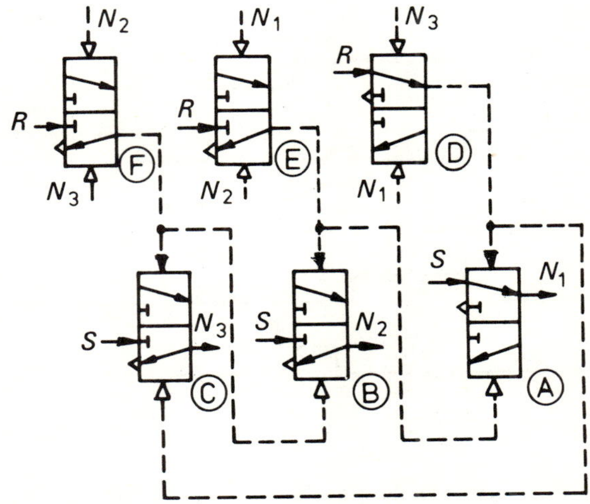

Fig. 15.23 – A ring counter using pneumatic valves.

The counter outputs are designated N_1, N_2 and N_3. As shown in the function diagram of Fig. 15.25 all outputs are initially off. The set signal S is off and R is on as in Fig. 15.24. The first pulse turns S on and R off (for the pulse duration). Output N_1 of valve A is then on and this switches valves D and E. When the pulse terminates, R is on and S is off. The output of valve E then switches valves A and B. Reception of the next pulse brings S on again and there is then an output N_2 at valve B. The student should follow the remainder of the count.

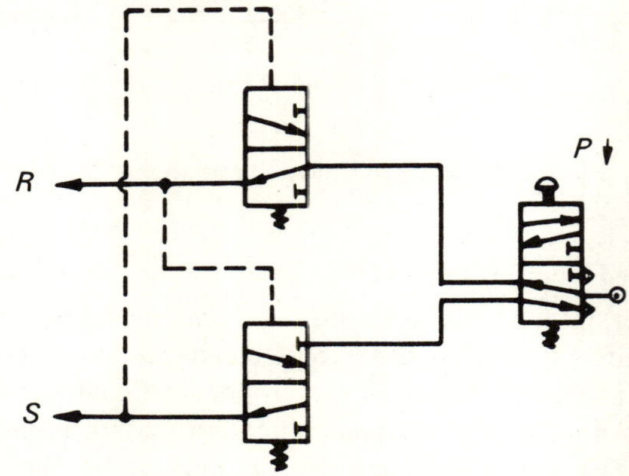

Fig. 15.24 – A means of generating the complementary inputs R and S for the ring counter of Fig. 15.23.

Fig. 15.25 – Inputs and outputs of the ring counter of Fig. 15.23.

This counter is useful in circuits which require the repetitive movement of a cylinder rod. For example, if a rod is to reciprocate three times within a machining operation, pulses can be generated by a limit valve at the end of the rod stroke. These can be passed to a three stage ring counter, like that of Fig. 15.23, and the output N_3 can be used to stop the oscillation.

Unfortunately, even with miniaturised components, pneumatic counting circuit can be bulky and expensive if the count is large – say greater than 16. Their use is likely to be restricted to small counts in which case the advantages of high pressure operation still predominate. For large counts fluidic and electronic counter are more economical.

15.8 EXAMPLES

15.8.1

Show, using section 15.4, that:

$$(A + B + C)(A + \bar{B} + \bar{C})(A + B + \bar{C}) = A + B\bar{C}.$$

15.8.2

Refer back to the design problem of section 15.5 and, using De Morgan's laws, design the logic circuit assuming that only NOT and AND elements are available.

(*Answers*: Five NOT elements, one 2 input AND and one 5 input AND are required.)

15.8.3

In the packaging operation referred to in section 15.7.1 the articles are to be packed in threes or in sixes. Assuming that a three stage binary counter is available design the necessary decoding circuit using conventional valves.

15.9 REFERENCES

[1] Hoernes, G. E. and Heilweil, M. R. (1964), *Introduction to Boolean Algebra and Logic Design*, McGraw-Hill.

[2] Auger, R. N. (1962), Turbulence amplifier design and application, *Fluid Amplification Symposium Proc. D.O.F.L.*, October.

[3] de Bruyne, W. A. (1965), Pneumatic NOR blocks, *1st Cranfield Fluidics Conf.*

[4] Foster, K. and Parker, G. A. (1970), *Fluidics – Components and Circuits*, Wiley Interscience.

[5] Eige, J. J. (1961), *Multiple ball pneumatic amplifiers*, Harry Diamond Laboratories (1) 1.

[6] Riordan, H. E. (1966), *High speed pneumatic digital operations with moving elements*, ibid.

[7] Prokes, J. (1970), Hydraulic logic and its application, *Proc. 1970 Fluid Power International Conf.*

[8] *Pneumatic Handbook* (1960), Trade and Technical Press.

[9] Glaettli, H. H. (1961), Hydraulic logic – what is its potential, *Control Eng.*, **8**, 83-86.

[10] Zahor, J. (1975), Some features of the mini hydraulic system of Czechoslovak design, *4th International Fluid Power Symp. B.H.R.A.*

[11] Stoneman, H.M. (1978), Pneumatic logic using miniature valves, *Chartered Mech. Eng.* (Inst. Mech. Engrs.), March.

[12] *A Course in Applied Pneumatics* (1966), Martonair.

[13] Henzgen, D. (1968), New developments in pneumatic control technique, *Fluid Power International Conf.*

[14] Smith, B.W. (1970), A simple method for the design of pneumatic circuits, *1970 Fluid Power International Conf.*

[15] Miles, K. E. (1970), A new type incremental piston, *Fluid Power International*, April.

CHAPTER 16

Control of sequences

16.1 INTRODUCTION

In many applications several actuators are needed with movements following a fixed pattern or sequence. For example, an automatic drilling machine may require two cylinders; one to deliver and to withdraw components from the machine and one to power the drill movement. If the cylinder handling the component is labelled A and the drill feed cylinder is B, the sequence can be described:

1. Cylinder rod A feeds component.
2. Cylinder rod B feeds drill.
3. Cylinder rod B withdraws drill.
4. Cylinder rod A withdraws component.

If the extended position of rod A is called the A_e position and the retracted position is the A_r position, the sequence can be described as $A_e\,B_e\,B_r\,A_r$.

Each part of the sequence will of course require a time for completion. This may be dictated by the machine characterisitics (for example, the time taken for a cylinder rod to extend) and hence there is little control over it. On the other hand operating times may be dictated by the designer by the deliberated introduction of time delays. The sequence of events and the times required by each event constitute the programme.

When one event is completed the control logic must initiate the next event. Sequence shifting can be accomplished using [1]:

(a) time based sequencing in which shifting is only dependent on time,
(b) step completion sequencing in which the completion of one event generates a signal to start the next,
(c) manual shifting.

16.2 FIXED PROGRAMMES, TIME BASED

16.2.1 The cam-shaft programmer

A cam-shaft can be used to sequence any number of actuators, whether hydraulic or pneumatic. The cams may actuate fluid power valves directly or they may interrupt air jets. The sequence of operations is determined by the shape of the cams and by their relative angular positions on the cam shaft, Fig. 16.1. The cam

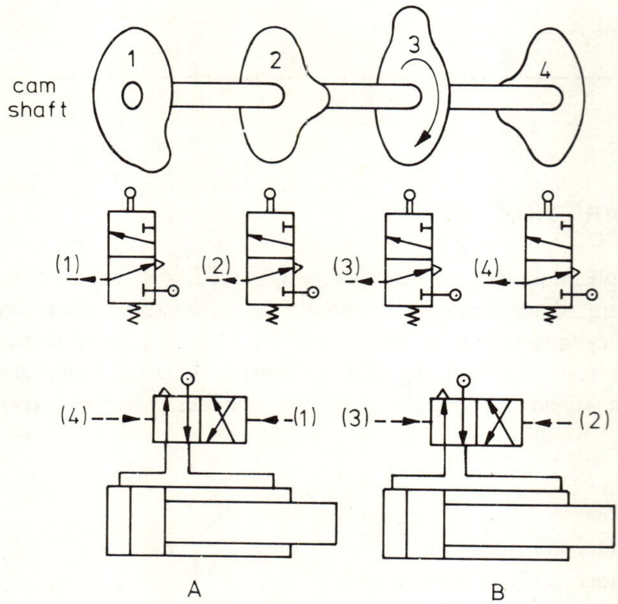

Fig. 16.1 – Time based sequence using a cam shaft programme.

often consists of several thin cams lying on top of each other, so that the cam profile can be changed by rotating the basic laminae relative to each other. For example, Fig. 16.2 shows how two laminae can be combined to form a variable shape cam.

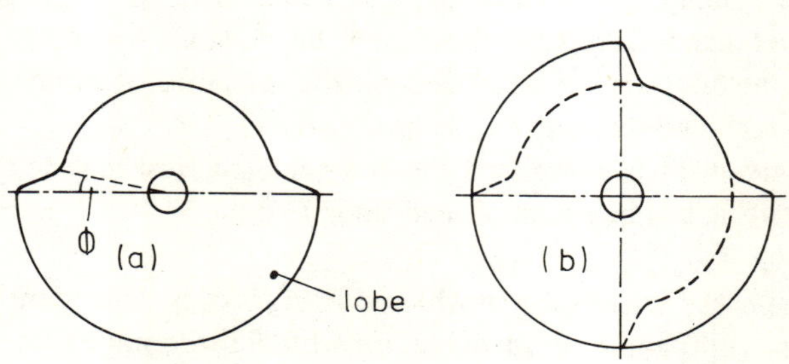

Fig. 16.2 – A cam with variable shape.

A circuit for the sequence A_e B_e B_r A_r is shown in Fig. 16.1. In this case the cam lobes operate pilot valves which in turn switch the double pilot operated valves or bi-stables. A pulse from a pilot valve is sufficient to switch a bi-stable so that short cam lobes are satisfactory for this application. A much simpler system results if single acting cylinders can be used. Fig. 16.3 shows that only two cams are necessary, their shapes being determined by the required timing of the sequence. For example, the angle α in Fig. 16.3(a) fixes the time between the extension of rod A and the extension of rod B. The actual time is α/ω, where ω is the angular velocity of the cam shaft. Both rods are retracted for a time β/ω.

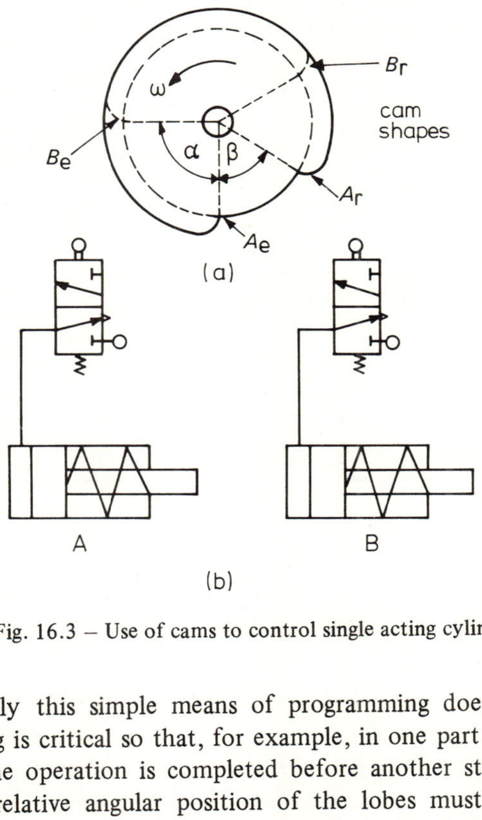

Fig. 16.3 – Use of cams to control single acting cylinder.

Unfortunately this simple means of programming does have some limitations. If timing is critical so that, for example, in one part of a sequence it is essential that one operation is completed before another starts, then the cam speed and the relative angular position of the lobes must be very carefully adjusted. The minimum value of the angle ϕ of the cam lobe, Fig. 16.2(a), and the shaft speed determine the smallest time interval that can be achieved with a cam shaft programmer. If, for example, ϕ is 10°, then the minimum time interval is 1/36 of the time for one revolution; this may not be sufficiently precise for many applications. It is possible, however, to obtain a resolution of as little as one second in a total time of 23 days, but this requires use of a complicated and expensive bank of shafts.

Problems also arise when cam shafts are driven at high speeds. Undesirable phenomena such as cam bounce begin to show themselves so that, on the whole, cam shaft programmers are best suited to low speed sequencing.

16.2.2 A counter controlled programmer

Control of fast sequences can be achieved by means of a clock driven binary counter, whose decoded outputs generate the required output movements at the desired times [2, 3, 4]. Consider for example the sequence timing diagram of Fig. 16.4; this is a typical 'feed, clamp and work' circuit. Note that overlap of signals has been arranged so that for example the feed cylinder A is not retracted until the component is clamped. Similarly the clamp is not removed until the work operation is completed. The timing diagram shows that this sequence is described by $A_e\, B_e\, A_r\, C_e\, C_r\, B_r$.

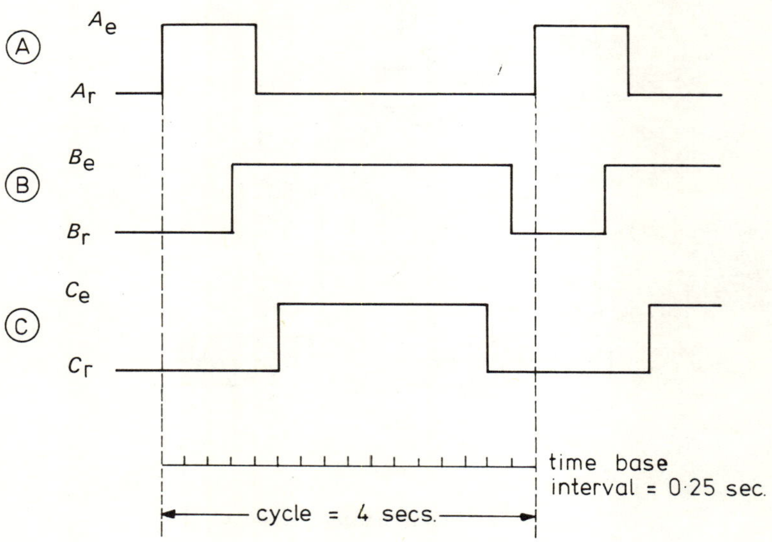

Fig. 16.4 – Timing diagram for a three cylinder sequence.

A block diagram of the control circuit is shown in Fig. 16.5. The clock, which could be a fluidic oscillator, feeds pulses to a binary counter. The outputs from the counter are fed to the decoder, its function being to activate the appropriate output when the correct count has been reached, (that is, when the clock has reached the right time). The bi-stables with inputs A_e^*, A_r^* etc. in turn actuate the cylinders, causing output motions A_e and A_r respectively. A pulse A_e^* will for example switch bi-stable A, causing rod A to extend.

The first step in designing the fluidic control for this system is to determine the minimum frequency for the clock. This is the reciprocal of the largest time interval, which when totalised, satisfies the time requirement of every function [12]. In this case, the feed A_e is activated for 1.0 sec, A_r for 3.0 sec, B_e for 1.5 sec, B_r for 1.0 sec, C_e for 2.25 sec and C_r for 1.75 sec. Hence 0.25 sec is the required clock resolution and its minimum frequency is then 4 Hz. The maximum count in a cycle is then 16, so that a four-stage binary counter is required.

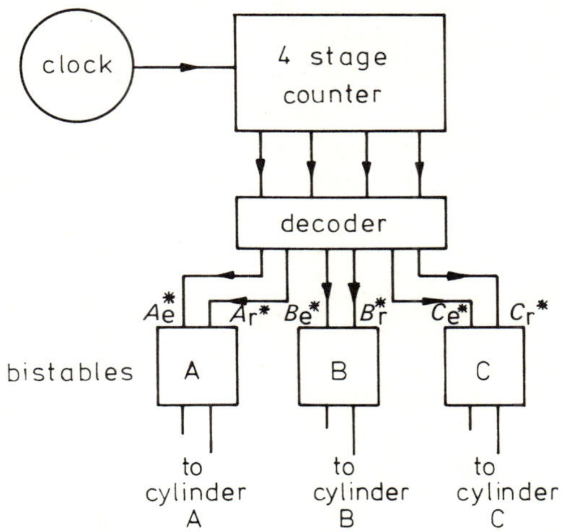

Fig. 16.5 – Block diagram of a counter controlled programmer for a three cylinder sequence.

The action of the system is summarised in the truth table shown in Fig. 16.6. At time zero, the start of the cycle, the feed cylinder should be extended; this

	Input				time	Output					
count	W	X	Y	Z	sec	A_e*	A_r*	B_e*	B_r*	C_e*	C_r*
0	0	0	0	0	0	1	0	0	d	0	d
1	0	0	0	1	0·25	d	0	0	d	0	d
2	0	0	1	0	0·50	d	0	0	d	0	d
3	0	0	1	1	0·75	d	0	1	0	0	d
4	0	1	0	0	1·00	0	1	d	0	0	d
5	0	1	0	1	1·25	0	d	d	0	1	0
6	0	1	1	0	1·50	0	d	d	0	d	0
7	0	1	1	1	1·75	0	d	d	0	d	0
8	1	0	0	0	2·00	0	d	d	0	d	0
9	1	0	0	1	2·25	0	d	d	0	d	0
10	1	0	1	0	2·50	0	d	d	0	d	0
11	1	0	1	1	2·75	0	d	d	0	d	0
12	1	1	0	0	3·00	0	d	d	0	d	0
13	1	1	0	1	3·25	0	d	d	0	d	0
14	1	1	1	0	3·50	0	d	d	0	0	1
15	1	1	1	1	3·75	0	d	0	1	0	d
16	0	0	0	0	4·00	1	0	0	d	0	d

Fig. 16.6 – Truth table for the timing diagram of Fig. 16.4.

is indicated by the 1 in the A_e column at time zero. The counter outputs are W, X, Y and Z. After an interval of 0.75 sec, the clamp is applied and this is indicated by the 1 in the B_e column when the counter output is 0011. The rest of the table is filled in in a similar fashion.

The 'd' stands for 'don't care'. Since bi-stables drive the cylinders it is only necessary to provide short pulse ouputs from the decoder. However it can sometimes be of advantage to prolong the signal to a bi-stable – it can simplify the logic. For example, rod A must extend at zero time, hence the bi-stable input A_e^* must be 1 at zero time. But it does not matter whether the input A_e^* is 1 or 0 at times 0.25, 0.5 and 0.75 sec, since the bi-stable memorises the input at zero time. A_e^* must become 0 at time 1.00 sec so that the rod can be withdrawn. Hence at $t = 0.25$, 0.5 and 0.75 sec it does not matter what value the input A_e^* assumes.

The function of the decoder is to look for specific combinations of the counting stage outputs, and having found them to activate the appropriate bi-stable input. For example when $WXYZ = 1111$, the decoder should activate B_r^*. The decoder logic can be written by inspection of the truth table. If the 'don't care' conditions are neglected the logic is:

$$\begin{aligned} A_e^* &= \bar{W}\bar{X}\bar{Y}\bar{Z} \\ A_r^* &= \bar{W}X\bar{Y}\bar{Z} \\ B_e^* &= \bar{W}\bar{X}YZ \\ B_r^* &= WXYZ \\ C_e^* &= \bar{W}X\bar{Y}Z \\ C_r^* &= WXY\bar{Z}\,. \end{aligned} \tag{16.1}$$

Implementation of this logic would require a considerable number of components, but fortunately consideration of the 'don't care' conditions can reduce these considerably. The logic is shown on the Karnaugh maps of Fig. 16.7. Six maps are required, one for each output.

The first step in the logic reduction is to group the 1s into the largest possible sub cube. The number of cells in a sub cube must be a power of 2, and the constituent cells must be adjacent. Such sub cubes are illustrated in Fig. 16.7, and in each case the 'd' is interpreted as 1. Note that in going from one cell to an adjacent cell, only one input changes value. Note also that adjacent cells are not necessarily physically adjacent on the map; for example cells 1000 and 0000 are adjacent as are 1100 and 1110.

Consider the sub cube for B_e^*. Interpreting the 'd' as 1 results in the logic expression:

$$B_e^* = \bar{W}\bar{X}YZ + \bar{W}XYZ \tag{16.2}$$

or

$$B_e^* = \bar{W}YZ\,(\bar{X} + X)$$

or

$$B_e^* = \bar{W} Y Z \,. \tag{16.3}$$

The reduction from (16.2) to (16.3) can be quickly and easily deduced from the map. (a) A sub cube is examined to determine which inputs do not change within it. In the case of B_e^*, W remains equal to 0 within the cube, Y remains at 1, Z remains at 1 – but X has two values 0 and 1. (b) The logic expression only contains the fixed input values, for example $B_e^* = \bar{W} Y Z$. Again consider the map for A_e^*. The sub cube in this case consists of four cells. Only two inputs remain constant within the cube – $W = 0$, $X = 0$. The logic is then $A_e^* = \bar{W}\bar{X}$. The other maps are treated in the same way to give:

$$\begin{aligned} A_e^* &= \bar{W}\bar{X} \\ A_r^* &= X \\ B_e^* &= \bar{W} Y Z \\ B_r^* &= W X Y Z \\ C_e^* &= \bar{W} X Y \\ C_r^* &= W X Y . \end{aligned} \tag{16.3}$$

Comparison of equations (16.3) and (16.2) shows that the decoder logic has been considerably simplified.

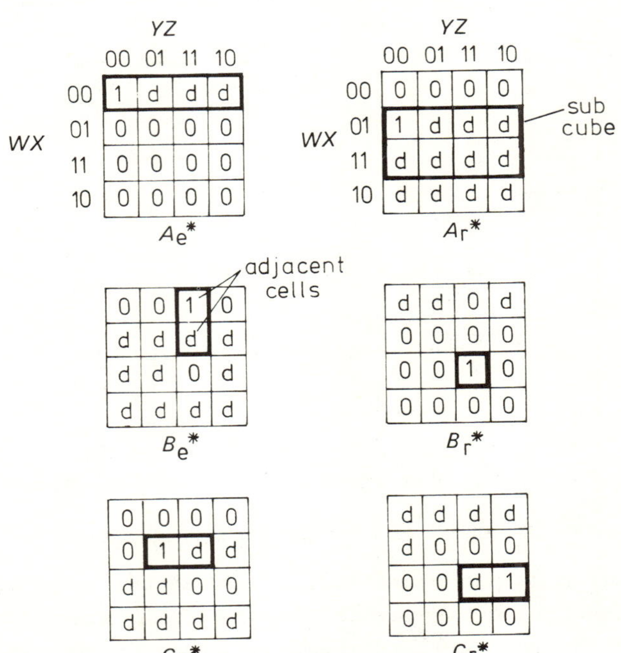

Fig. 16.7 – Use of Karnaugh maps to simplify the logic.

16.3 FIXED PROGRAMMES, STEP COMPLETION BASED

Time based sequences have a major weakness. If one part of a sequence is not satisfactorily completed, perhaps because of a jammed cylinder, the cam shaft or counter will move relentlessly onward to initiate the next movement. This may be a safety hazard in some applications and may require some form of signal to indicate that each part of the cycle is completed. This is referred to as **interlocking**.

16.3.1 Use of sequence valves

A sequence valve gives an indirect indication that a cylinder rod has completed its stroke. It allows fluid to pass only when its inlet pressure exceeds a predetermined value and can be used in both hydraulic and pneumatic circuits [5]. A typical 'clamp and drill' circuit A_e B_e B_r A_r is shown in Fig, 16.8. Cylinder A is the clamp and B operates the drill. When the four-port directional control valve E is operated manually, pressure is applied to the blank end of rod A which extends clamping the workpiece. Pressure build up in the blank end of A then causes the sequence valve C to operate, connecting the supply to the blank end of cylinder B and making its rod extend. The sequence A_e B_e is now completed. If the valve E is reset, rod B will retract, sequence valve D will operate and rod A will retract. The full sequence A_e B_e B_r A_r is complete. Not that manual sequence shifting has been included in this example.

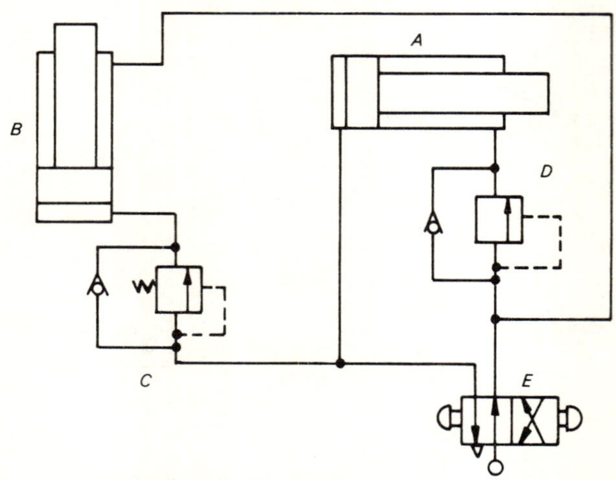

Fig. 16.8 – Sequence valves in a step completion based sequence.

Although this is a simple circuit it has several disadvantages:

(a) There is no positive interlock. If, for example, some 'stickiness' in cylinder A causes premature pressure build up, sequence valve C may operate, moving rod B before A has completed its stroke – the drill would operate before the component was clamped.

(b) Flow control valves cannot be used to control rod speeds. If meter-out speed control were attempted for cylinder B, the pressure in the rod end of B may build up and cause premature operation of sequence valve D.
(c) In hydraulic systems, care must be taken to ensure that premature pressure surges do not occur when the main directional control valve is switched.

On the whole, these valves must be used with extreme caution and should be reserved for the simplest of applications, such as that described. They are commonly used in the woodworking industry for clamping operations using pneumatics.

16.3.2 Use of limit valves

A mechanically operated limit valve, an interruptible jet or a back pressure sensor can be used to detect rod position. They give a direct indication of position and hence do not suffer from the disadvantages of the sequence valve.

An example of the use of roller operated three-port pilot valves is shown in Fig. 16.9. The sequence is A_e B_e A_r B_r. Provision is made for either single shot or continuous operation [6]. Note that the two input valves must be isolated from each other by means of a shuttle valve. If this were not done, the pressurised air obtained by actuating one input would escape to exhaust through the other input valve. When either input is actuated, the resultant signal switches the main valve (MVA) driving cylinder A. Fig. 16.9 shows the state of the circuit just after

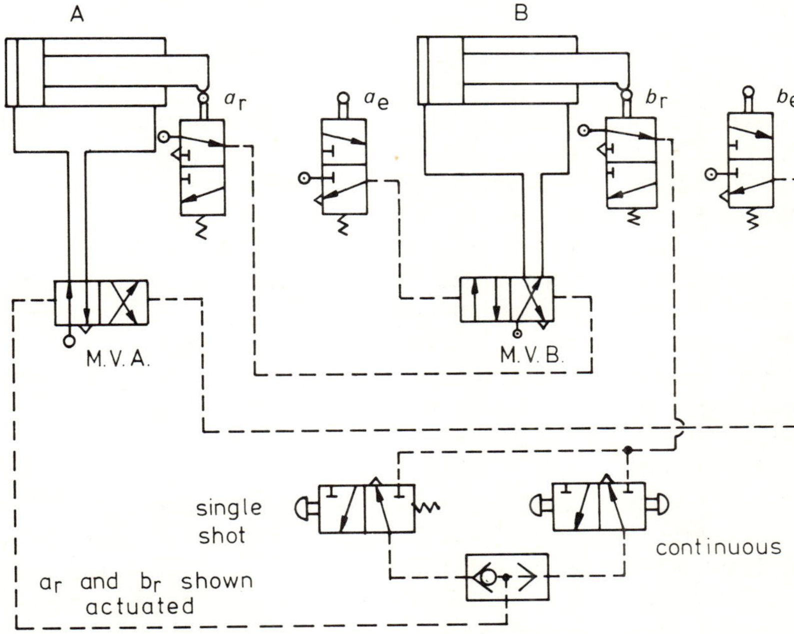

Fig. 16.9 – Use of limit valves in a step completion based sequence.

the input pulse has switched MVA. Rod A now extends, contacts pilot valve a_e which makes MVA switch. Rod B now extends, contacts pilot valve b_e which resets MVA – and so on.

This circuit is easy to construct. In fact all sequences in which retractions take place in the same order as extensions are easy to contruct, for example $A_e\ B_e\ C_e\ A_r\ B_r\ C_r$. If the order is different, problems of 'locked-on' signals arise [7]. Consider for example the sequence $A_e\ B_e\ B_r\ A_r$. Fig. 16.10 shows the state of

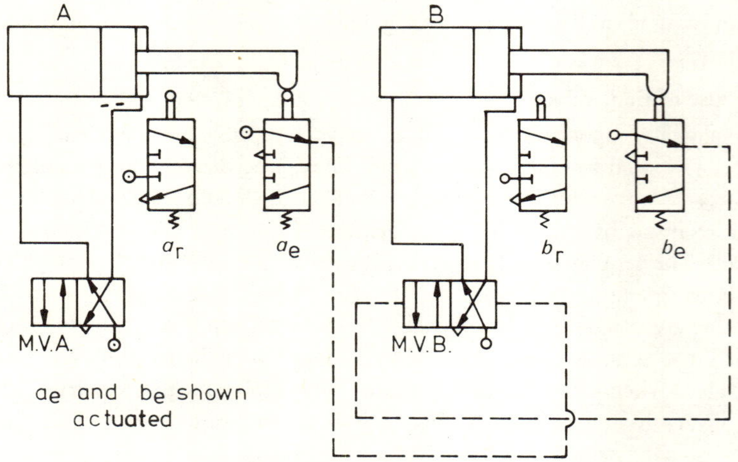

Fig. 16.10 – Part of the circuit for $A_eB_eB_rA_r$, illustrating clash of signals on MVB after rod B has extended.

the circuit after rod B has extended. Note that the signal from b_e, which should now reset MVB, cannot do so since it is opposed by the signal from a_e. This is called a **locked-on** signal condition. A similar clash arises between the signals to MVA from pilot valves b_r and a_r. The problem is solved by the introduction of a four or five-port selector valve which controls the supply to the limit valves and is so arranged that the limit valves are supplied only when necessary. Fig. 16.11 shows that the signal from b_e is now able to reset MVB because, although pilot a_e is actuated and is connected to the other side of MVB, the selector has removed its supply and hence there is no opposing signal on MVB. The complete circuit is shown in Fig, 16.12.

16.3.3 Use of electrics

The above mentioned techniques for avoiding locked-on signals are popular in pneumatic circuits but, in hydraulics, the use of electric limit switches and solenoid operated valves is common practice. In Fig. 16.13 the electrical circuitry for the sequence $A_e\ B_e\ B_r\ A_r$, is shown [6]. There are four main types of component: a start button, solenoids A_1, A_2, B_1 and B_2, microswitches a_e, b_e, and b_r, and a latching relay with four contacts R_1, R_2, R_3 and R_4. When the start button is depressed solenoid A_1 is actuated causing rod A to extend. Microswitch a_e then activates solenoid B, through contacts R_1 of the relay R. The rod B extends and b_e activates the relay, breaking contacts R_1 and making contacts R_2, R_3 and R_4. Solenoid B_2 is now activated through R_4 and the rod B retracts. The microswitch b_r then closes and activates solenoid A_2 through contacts R_3.

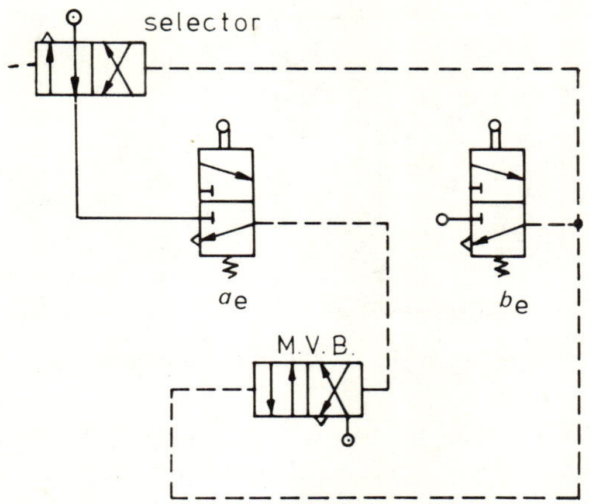

Fig. 16.11 – Use of a selector valve to remove the locked signal.

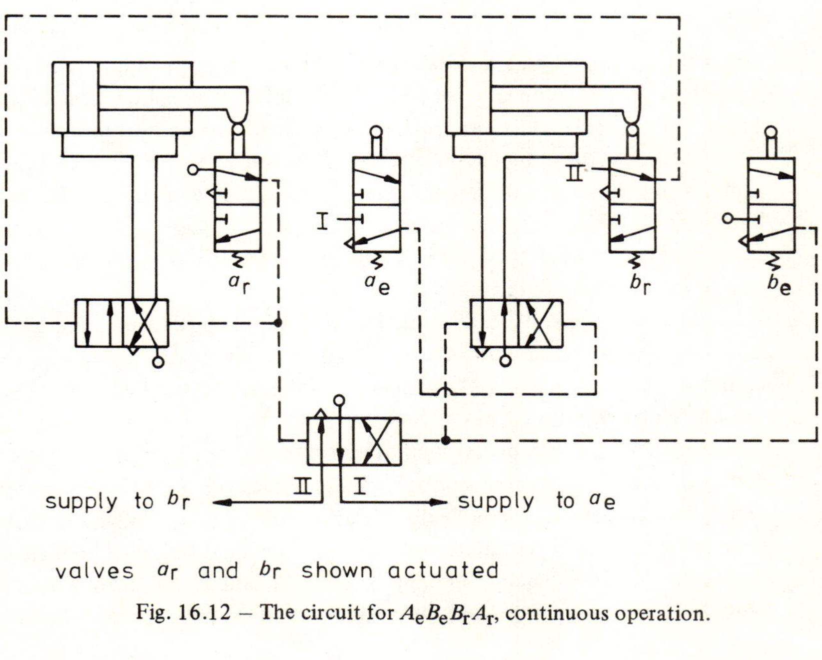

Fig. 16.12 – The circuit for $A_eB_eB_rA_r$, continuous operation.

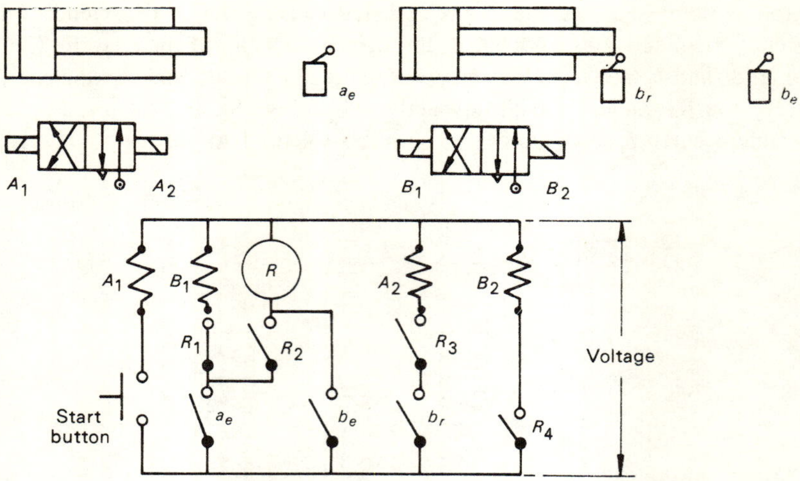

Fig. 16.13 – The circuit for $A_eB_eB_rA_r$ using electrical controls.

Rod A then retracts and as microswitch a_e breaks, the current to relay R is interrupted and its contacts revert to their original position. The relay fulfils the same function as the selector valve of the earlier example.

16.4 A SIMPLE DESIGN TECHNIQUE FOR SEQUENCE CIRCUITS [5]

This technique will be demonstrated by applying it to the sequence $A_e\, B_e\, B_r\, A_r$.

1. Draw the cylinders, main valves and pilot valves. Note, Fig. 16.14, the simplified symbol for a pilot valve. The left-hand connection is the supply to the valve and the right-hand connection is the output.
2. Determine the pilot signals acting on the main valves. In this case, the signal a_r must initiate the movement A_e and hence must act on the left-hand side of MVA. After A_e is completed, the signal from a_e must initiate B_e and hence must act on the left-hand side of MVB. The positions of b_r and b_e are similarly determined.
3. Draw the cylinder rods in their different sequential positions and hence determine if there are locked-on signals on the main valves. In this example, pilot signals a_r and b_r must not exist simultaneously since they act on opposite sides of MVA. Similarly, the simultaneously existence of signals a_e and b_e must be avoided. The first step of the chosen sequence shows a_r and b_r actuated. If both these valves are powered from the supply then there is a locked-on signal on MVA. Similarly, in the third step a_e and b_e are actuated and there is the possibility of locked-on signals on MVB. Consider the step 1 where a_r and b_r are both actuated. One of these signals must be eliminated – which one? Signal a_r initiates

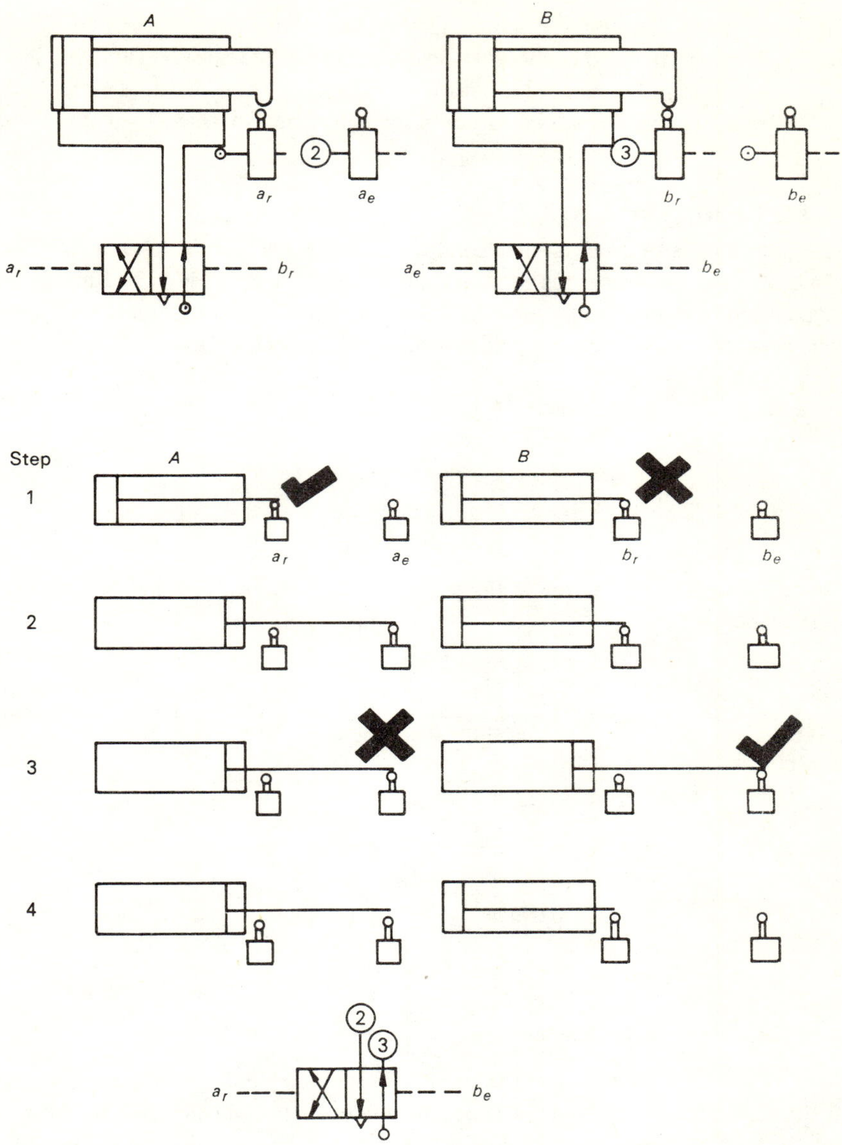

Fig. 16.14 – A breakdown of the design technique for the sequence $A_e B_e B_r A_r$.

the next step in the sequence to A_e and so must be retained; b_r must then be eliminated at this stage. This is indicated on the figure by putting an X at the b_r signal and a $\surd$ at the a_r signal. Similarly at step 3, a_e must be removed and b_e must be retained to initiate the next part of the sequence to B_r. A selector valve is needed. At the end of step 3, b_e

switches the selector removing the supply from a_e. At the end of step 1, a_r switches the selector valve to remove the supply from b_r. Since valves a_r and b_e control the selector valve, they must be supplied from the main supply. The two other pilot valves are supplied by the selector valve.

16.4.1 The sequence $A_e A_r B_e B_r$

This sequence is unusual in that conditions at the start of step 1 and step 3 are identical, that is, both rods retracted. The design method described above will be adopted, but in future problems the method will be reduced to tabular form. This particular sequence can be described in the following way:

Table 16.1.

Input	b_r	a_e	a_r	b_e
Output	A_e	A_r	B_e	B_r

The input on the upper row is the signal required to produce the movement on the row below.

Table 16.2.

Step	Output required	Input state			
		a_r	a_e	b_r	b_e
1	A_e	1	0	1	0
2	A_r	0	1	1	0
3	B_e	1	0	1	0
4	B_r	1	0	0	1

Use of Table 16.2 avoids the need to draw the cylinder rods in their various positions throughout the sequence. The table relates the inputs a_r a_e b_r b_e at the start of any step in the dequence to the movement required in that step. 1 indicates that a limit valve is actuated and a 0 indicates that the valve is not actuated, for example at the start of step 1 both rods are retracted, both valves a_r and b_r are actuated giving $a_r = b_r = 1$. The other two valves a_e and b_e are not operated. Hence when $a_r = b_r = 1$ and $a_e = b_e = 0$, the sequence demands that rod A extends to A_e.

Table 16.1 shows that MVA is operated in different directions by a_e and b_r. Similarly, main valve B is operated by inputs a_r and b_e. Table 16.2 is now inspected to see if locked-on signals occur. At the start of step 2 there is a possibility of a locked-on condition on main valve A. Input a_e initiates the next part of the sequence and must be powered, but b_r can be eliminated by removing its supply. Again in step 4 the supply must be removed from limit a_r to avoid a locked signal on MVB. The circuit is drawn in Fig. 16.15.

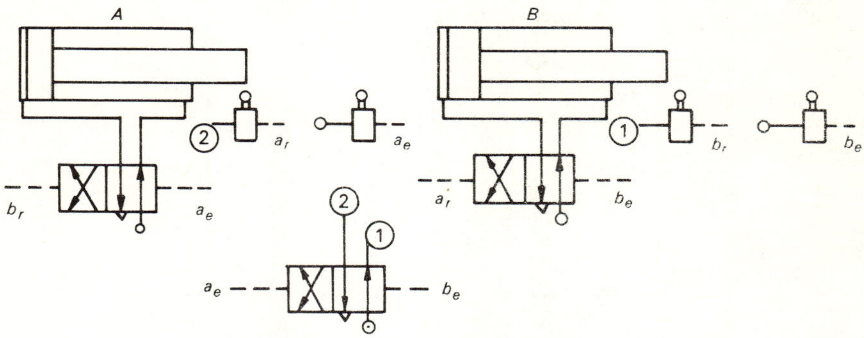

Fig. 16.15 – The circuit for the sequence $A_eA_rB_eB_r$.

The design method does not always produce the circuit with the minimum components. Here, four pilot valves and three four-port valves are needed. Fig. 16.16 shows a single cycle circuit which will give the same sequence of operations without the use of a selector valve. This circuit distinguishes between the similar conditions at the start of steps 1 and 3 by using the A_e movement to preset an air return, roller operated pilot valve a_e. This is reset later in the sequence. The valve a_r acts as a memory: it remembers that the last movement was A_e and only allows B_e while it is set.

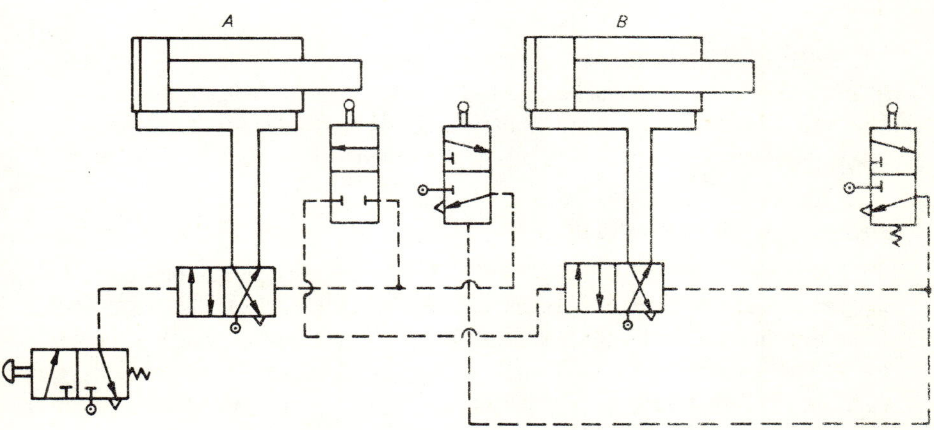

Fig. 16.16 – Circuit for the sequence $A_eA_rB_eB_r$, single shot, using an air return pilot valve.

16.4.2 The sequence $A_e\, B_e\, B_r\, C_e\, C_r\, A_r$

This design method can be extended to any number of cylinders. This example shows how it can be applied to a three cylinder sequence of the 'clamp and operate' type. In practice, cylinder A might clamp while cylinders B and C fed drills and taps respectively. Again the input-output relationships are presented in tabular form (Table 16.3).

Table 16.3.

Input	a_r	a_e	b_e	b_r	c_e	c_r
Output	A_e	B_e	B_r	C_e	C_r	A_r

Table 16.4.

Step	Output required	Input state					
		a_r	a_e	b_r	b_e	c_r	c_e
1	A_e	1	0	1	0	1	0
2	B_e	0	1	1	0	1	0
3	B_r	0	1	0	1	1	0
4	C_e	0	1	1	0	1	0
5	C_r	0	1	1	0	0	1
6	A_r	0	1	1	0	1	0

Table 16.4 shows that simultaneous existence of signals a_r and c_r, a_e and b_e, b_r and c_e must be avoided. There is a possibility of locked-on signals at steps 1, 3 and 5, and at these steps limits c_r, a_e and b_r respectively must have their supply removed. This is achieved by the use of two selector valves, Fig. 16.17.

16.5 THE CASCADE METHOD

This provides a straightforward method of designing sequential pneumatic circuits and always gives a workable circuit [7]. A clear set of rules has been laid down for this method with the result that a circuit can be designed on a step by step basis without a lot of mental effort. Consider again the sequence $A_e\, B_e\, B_r\, C_e\, C_r\, A_r$. The first task is to split this sequence into groups so that no letter appears twice in any group. In this case:

Group I is $A_e\, B_e$

Group II is $B_r\, C_e$
Group III is $C_r\, A_r$.

The importance of correct sequence splitting cannot be overemphasised, since it determines the number if components in the circuit. Splitting should be applied both in the forward and reverse directions as a check. For example,

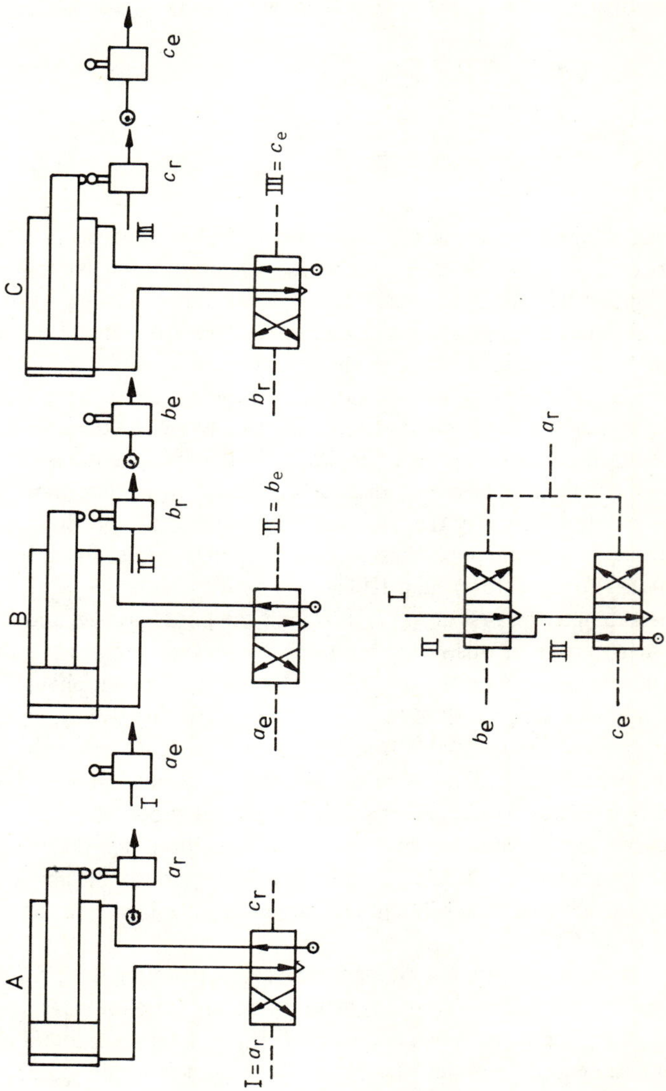

Fig.16.17 – Circuit for the sequence $A_eB_eB_rC_eC_rA_r$.

consider the sequence $A_e\ B_e\ A_r\ C_e\ C_r\ D_e\ D_r\ C_e\ C_r\ B_r$. Forward splitting gives apparently five groups:

$$\underbrace{A_e B_e}_{1} / \underbrace{A_r C_e}_{2} / \underbrace{C_r D_e}_{3} / \underbrace{D_r C_e}_{4} / \underbrace{C_r B_r}_{5}$$

Backward splitting makes it clearer that there are only four groups:

$$\underbrace{A_e / \underbrace{B_e A_r C_e}_{2} / \underbrace{C_r D_e}_{3} / \underbrace{D_r C_e}_{4} / C_r B_r}_{1}$$

This grouping simplifies the task of determining which pilot valves are likely to generate locked-on signals. The rule is simple – the last pilot valve to be operated in any group can be supplied directly from the main, but all other pilot valves in that group should be supplied by the appropriate selector valve. For example in the three cylinder circuit referred to above, valves b_e, c_e and a_r can be supplied from the main, whilst a_e, b_r and c_r should be supplied by selector valves. Pilot valves b_e, c_e and a_r, the last in their groups, are used to switch the selector valves.

The resultant circuit, Fig. 16.17, is similar to that designed in section 16.4.2. There is one difference, however, in that the signal to initiate movement in a group does not come directly from the last pilot valve in the previous group. The cascade method recommends that initiating signals come from the selector valves, that is, I for a_r, II for b_e and III for c_e.

In this example the first group is $A_e\ B_e$ and the pilot valves a_e and b_e must be supplied. At the start of this group both selector valves are moved to the left by signal a_r, the last signal from the preceding group $C_r\ A_r$. This supplies air to line I, switches MVA and sends rod A to A_e. The signal from valve a_e, which is also supplied from line I, switches MVB and rod B goes to B_e. It trips valve b_e, which is supplied from the main. This is the last signal from group I. It is used to remove all supplies from group I pilot valves and to supply the group II valves.

Sequences can be stopped at intermediate positions by placing manually operated three-port valves in the relevant lines. For example the above sequence can be stopped with all rods retracted if the line from a_r is interrupted by such a valve (see Fig. 16.9).

It is easy to work out the number of components required for a cascade circuit. If there are n cylinders, then n main valves and $2n$ pilot valves are needed. In addition $m - 1$ selector valves are needed, where m is the number of groups. The student is advised to design the circuit of the last section, $A_e\ B_e\ B_r\ A_r$, using the cascade method.

The main disadvantage of the method is that it does not always result in the circuit with the minimum number of components. As seen earlier (Fig. 16.9) the repetitive sequence $A_e\ B_e\ A_r\ B_r$ in which extensions and retractions are in the same order, does not need a selector valve. The cascade method would include one. In addition oscillating sequences such as $A_e\ A_r\ B_e\ B_r$ can be built more cheaply in other ways.

The procedures described above can be applied to any sequence, no matter how many cylinders are involved, provided that no movement occurs more than once. Examples of repetitive movements are:

1. $A_e\ B_e\ B_r\ B_e\ B_r\ A_r$ in which cylinder B oscillates, and
2. $A_e\ B_e\ B_r\ C_e\ B_e\ B_r\ C_r\ A_r$ in which cylinder B repeats its movements but there are other cylinder movements interspaced.

There are several ways of designing such circuits, the main problem being the need to distinguish between the repeated movements of a given cylinder. This can be done by allocating as many sets of limit valves to a cylinder as there are repetitive movements. Pneumatic counters, such as those described in section 16.7, can also be used to differentiate between movements. The reader interested in advanced circuit construction is recommended to consult the quoted references.

16.6 OTHER MEANS OF ELIMINATING LOCKED-ON SIGNALS

Locked-on signals can be avoided if the pilot valves produce short pulses instead of continuous signals [8, 9]. There are several ways of generating pulses, one being shown in Fig. 16.18. The three-port valve, a_e in Fig. 16.12, has its output port connected to a small reservoir, connected to the supply in the valve's normal position. When the valve is actuated the supply is cut off and the air reservoir is connected to the main valve. The signal operates MVB and then disperses through the bleed R, the size of which affects the period of application of the signal from a_e. When the signal has dispersed there is no pressure to oppose the signal from b_r, making the rod B retract.

The duration of the pulse is critically affected by the capacitance of the reservoir and lines and the impedance of the bleed orifice. Further developments have produced impulsing micro-valves which are basically limit valves that remove their output signals on sensing that the signal has completed its function, that is, operated a main valve. They do not rely on charge reservoirs.

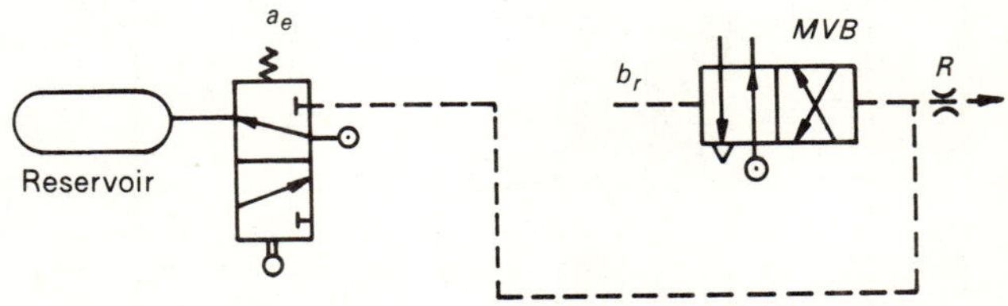

Fig. 16.18 – Another means of eliminating a locked signal – by limiting its duration.

Trip valves can also be used to generate pulses, Fig. 16.19. The trip is so designed that the valve is only actuated as the rod passes in one direction. Since the rod striker must pass over the trip in order to generate a pulse, it is necessary to place the valve a short distance from the end of the stroke.

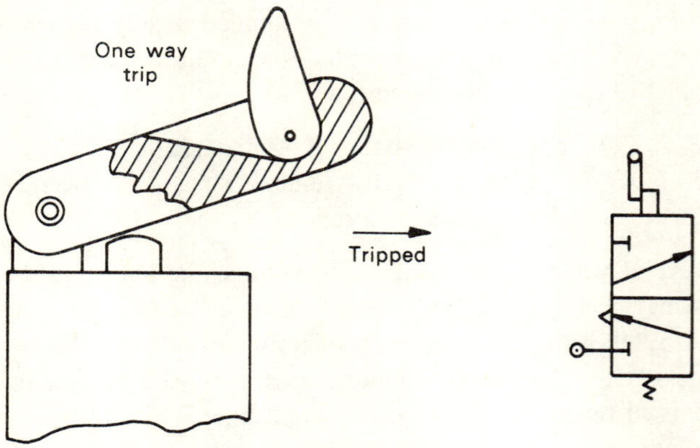

Fig. 16.19 – The trip valve provides a limited duration signal.

Use of impulse generators can reduce the number of components in a circuit. For example selector valves can be dispensed with. The design methods discussed earlier can be used to determine the number and location of the impulse generators. For the sequence $A_e\,B_e\,B_r\,A_r$ they would for example replace limit valves a_e and b_r.

A disadvantage of limited period signals is that on some occasions the pulse may not be sufficient to completely switch a main valve.

16.7 THE LOGIC OF THE CASCADE METHOD

The locked-on signal hazard can be explained in logic terms with the aid of the truth table for the sequence $A_e\,B_e\,B_r\,A_r$, Fig. 16.20. The inputs a_r, a_e etc. are the

Step	Inputs				Selector		Outputs				Group
	a_r	a_e	b_r	b_e	I	II	A_e^*	A_r^*	B_e^*	B_r^*	
1	1	0	1	0	1	0	1	0	0	d	I
2	0	1	1	0	1	0	d	0	1	0	
3	0	1	0	1	0	1	d	0	0	1	II
4	0	1	1	0	0	1	0	1	0	d	

Fig. 16.20 – Truth table for the sequence $A_eB_eB_rA_r$.

signals generated by the limit valves. The outputs A_e^*, A_r^* etc. are the signals driving the bi-stables as in Fig. 16.5. Note that the rows of the truth table follow in the correct sequence order. There are several 'don't care' conditions; for example once bi-stable A or MVA has been switched to make rod A extend it does not matter whether signal A_e^* is maintained or not, until the command to retract A is received at step 4.

Ignoring for the moment the column labelled 'selector', it can be seen that the same combination occurs at steps 2 and 4, although different outputs are required at these times. Some means must be devised to enable the logic to distinguish between these two states, and the easiest way to do this is to use a bi-stable so arranged that it is in different states when B_e^* and A_r^* should be 1. The selector valve fulfils this function. Its state is changed by the last signal of a group; in this case when b_e changes from 0 to 1, and when a_r changes from 0 to 1.

Inspection of the tables leads to the logic:

$$A_e^* = a_r \bar{a}_e b_r \bar{b}_e \mathrm{I}\, \overline{\mathrm{II}} .$$

This can be simplified. The physical arrangement of the system, Fig. 16.11, ensures that $a_e = 0$ when $a_r = 1$, that $b_e = 0$ when $b_r = 1$, and that II = 0 when I = 1. Hence the logic reduces to:

$$A_e^* = a_r b_r \mathrm{I} .$$

But the selector is switched to output I by the input a_r so that the expression further reduces to:

$$A_e^* = a_r b_r .$$

The output A_e^* is demanded at step 1 which, since the process is continuous, is preceded by step 4. At step 4, b_r is unity and it remains at unity in step 1. The logic can then be written:

$$A_e^* = a_r \cdot 1 = a_r .$$

This has already been demonstrated using the cascade method. It can also be done by mapping, but space does not permit a properly detailed account of it here. Similar arguments as the above can be used to show that:

$$B_e^* = \bar{a}_r a_e b_r \bar{b}_e \mathrm{I}\, \overline{\mathrm{II}} = a_e \mathrm{I}$$
$$B_r^* = \bar{a}_r a_e \bar{b}_r b_e \bar{\mathrm{I}}\, \mathrm{II} = b_e$$
$$A_r^* = \bar{a}_r a_e b_r \bar{b}_e \bar{\mathrm{I}}\, \mathrm{II} = b_r \mathrm{II} .$$

The resultant logic circuit is shown in Fig. 16.21. The combination of bi-stable and AND gates has been referred to as a double sequential AND [1]. In the practical circuit, Fig. 16.12, the limit valves provide the AND functions.

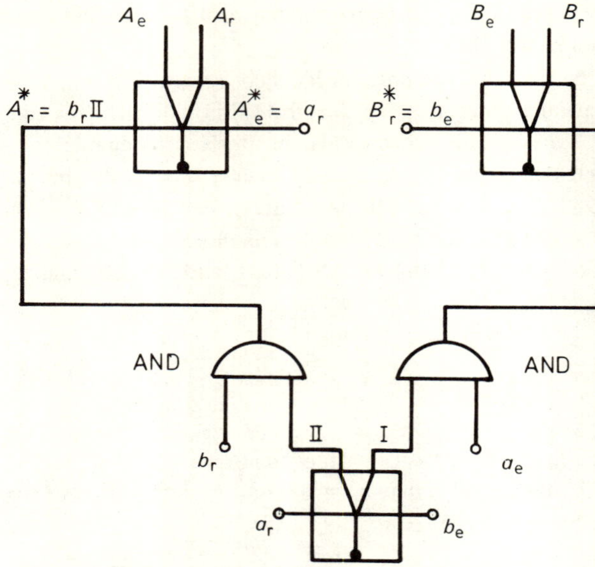

Fig. 16.21 – Logic circuit for the sequence $A_eB_eB_rA_r$.

In some applications non-contact sensing may be preferred [10]. Fig. 16.22 shows how interruptible jets can be used to sense rod position. In this case

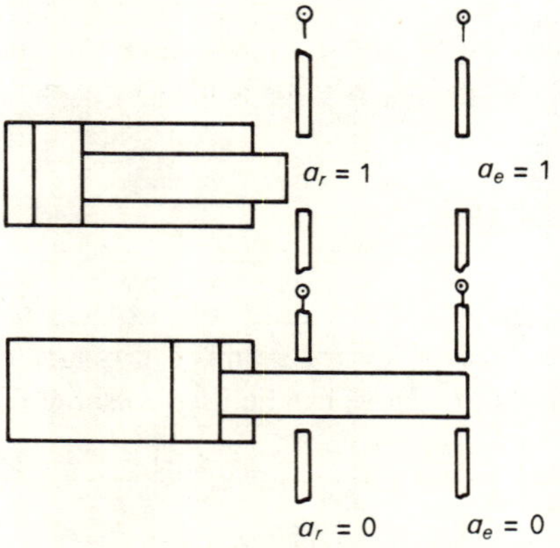

Fig. 16.22 – Use of interruptible jets as limit sensors.

the retracted position is indicated by $a_r = 1$ and the extended position by $a_e = 0$. This differs from the conventional valve circuit in which the extended position is indicated by $a_e = 1$. As far as the logic is concerned this means that the signals a_e and b_e in Fig. 16.21, should be replaced by their inverses. In practice this requires the use of an additional two NOT gates, the use of jets thus requiring a greater number of components.

16.8 VARIABLE PROGRAMMING

In the step completion sequences discussed above, the order of the sequence is built into the system, and if the order is to be changed, then the circuit must be redesigned. Thus step completion sequencing gives no flexibility, so far as sequence order is concerned. However in some variable programming systems, the time based functions are the only variable ones, and these requirements can be met by a step completion sequence working in conjunction with timers. Consider for example the sequence $A_e\ B_e\ A_r\ C_e\ C_r\ B_r$. The timing diagram is shown in Fig. 16.23, in which the full lines indicate the states of the output logic and the broken lines indicate the actual rod positions. The total cycle time consists of two parts:

1. inherent delays such as the time required for a cylinder rod to extend, and,
2. control delays introduced deliberately by the designer.

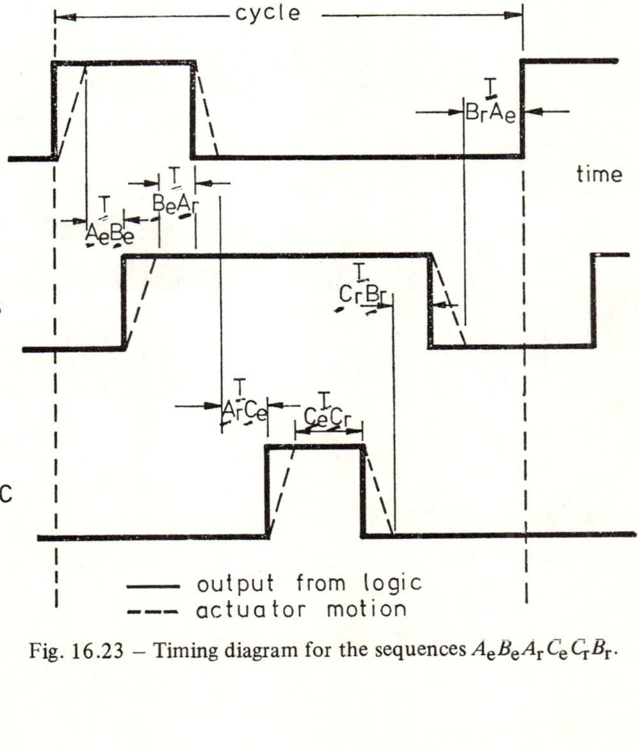

Fig. 16.23 – Timing diagram for the sequences $A_eB_eA_rC_eC_rB_r$.

Inherent delays can be increased by the use of speed controls. Their lower limit is determined by the cylinder and valve dimensions, loads and supply pressure. There are six control delays in the circuit of Fig. 16.23. For example $T(A_e\ B_e)$ is the time between the operation of limit valve a_e and the start of the extension of rod B. The six control delays require the use of six timers, one being placed in the output line of each of the appropriate limit sensors, timer $T(A_e\ B_e)$ being for example in the output line from sensor a_e.

Facility to vary sequence order as well as timing is essential in many systems, particularly machine tools such as lathes. The cam shaft programmer provides one possible solution, since the cam shapes, Fig. 16.3, can be readjusted to change the programme. This, however, can take some time and requires considerable skill. On the other hand it is now possible to have a removable cam shaft which can be quickly interchanged for another.

Punched cards and paper tape provide an extremely convenient means of varying a programme. Fig. 16.24 shows how a reading head, consisting of a bank of interruptible jets, can be used to convert the card information into a pneumatic output. The columns on the card represent functions and the rows determine the order of the sequence. As shown, the card is programmed for the sequence $A_e\ A_r\ B_e\ B_r$; the timing diagram is given in Fig. 16.25. The card is indexed past the reading head, sequence shifting in this case being partly time based and partly step completion based. Fig. 16.25 shows:

1. that A should extend when B has retracted,
2. there is a time delay ΔT_1 between the A_e and the A_r signals,
3. rod B extends when A has retracted,
4. there is a time delay ΔT_2 between the B_e and the B_r signals.

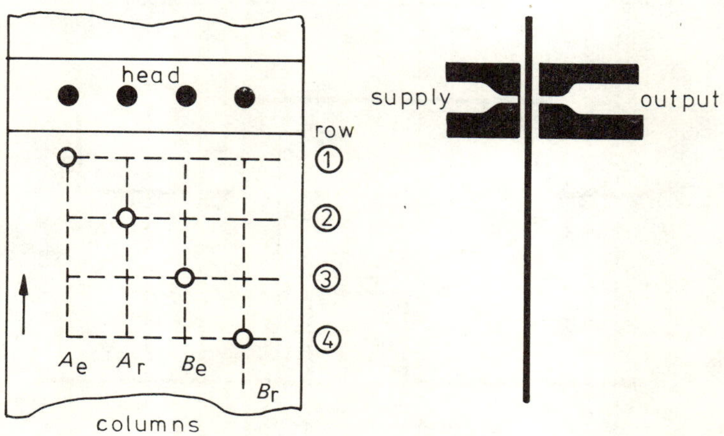

Fig. 16.24 – A pneumatic reading head for use in a variable programmer.

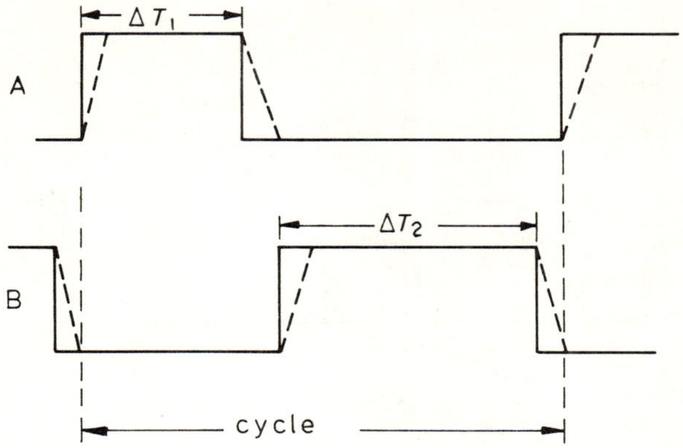

Fig. 16.25 – Timing diagrams for $A_e A_r B_e B_r$.

The shift signals are generated by the circuit shown in Fig. 16.26.

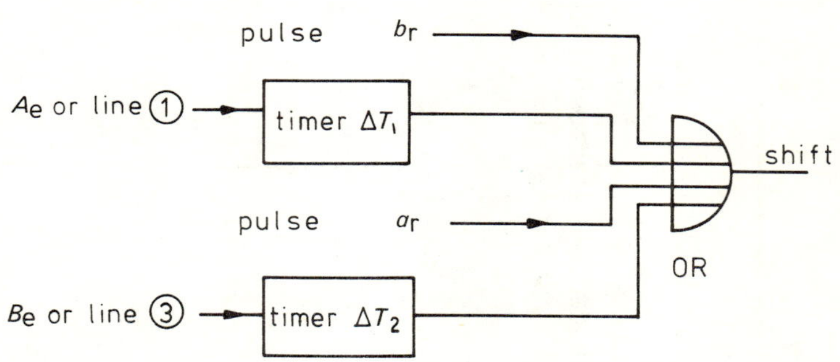

Fig. 16.26 – Means of generating shift signals for the sequence shown in Fig. 16.25.

The sequence advance can also be accomplished by means of a shift register or ring counter. This is illustrated in Fig. 16.27 which is again designed for the sequence $A_e\, A_r\, B_e\, B_r$ of Fig. 16.20. Information in the shift register is transferred from one stage to the next on reception of a shift pulse, which is again generated by the circuit of Fig. 16.25. In this case the last stage of the shift register is connected to the first stage, forming a ring counter. The programme is set up on a fluidic plug board, rather than on a card or tape. The plugboard is matrix type manifold block fitted with coupling valves and check valves or diodes. The arrangement of the closed coupling valves, Fig. 16.27, fixes the sequence of events. The diodes prevent cross-talk between the lines 1-4. As shown, the first pulse (generated from a start button or from b_r) energises line 1 making rod A extend. The timer ΔT_1 is energised and generates a new shift pulse after ΔT_1. The information is then transferred to line 2 and so on.

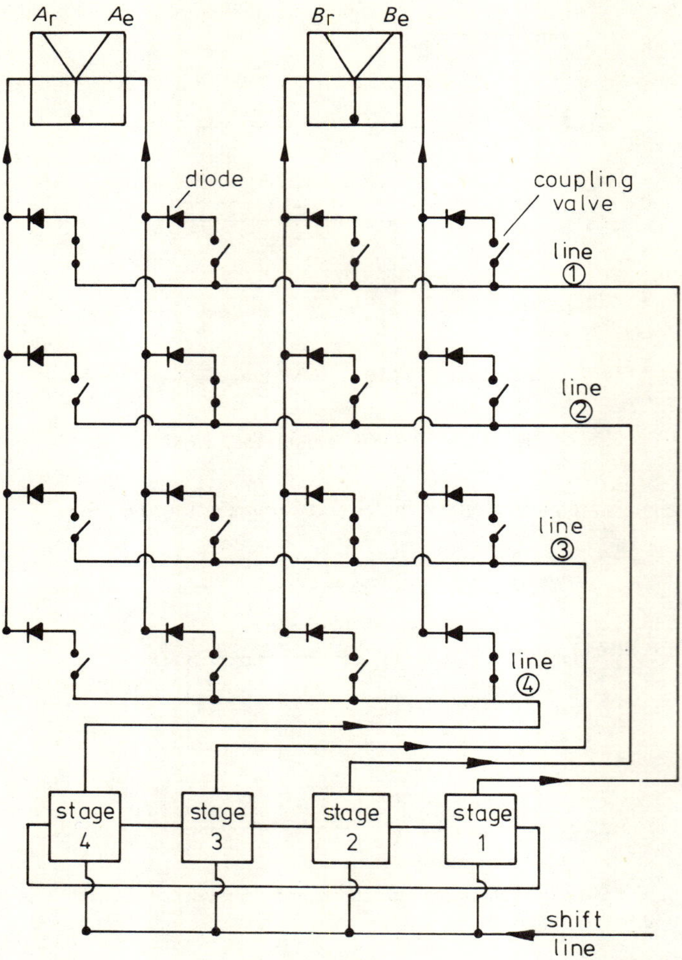

Fig. 16.27 – Use of a ring counter to generate the sequences $A_e A_r B_e B_r$.

16.9 EXAMPLES

16.9.1

Design circuits for the sequence A_e B_e C_e B_r A_r C_r using; (a) the cascade method, (b) impulse valves.

16.9.2

Determine, using logic gates, a circuit for the sequence A_e A_r B_e B_r.

16.10 REFERENCES

[1] Foster, K. and Parker, G. A. (1970), *Fluidics – Components and Circuits*, Wiley Interscience.

[2] Higgins, P. T. (1970), *Fluidic control for high speed machinery*, A.S.T.M.E. Tech. paper No. MS67-315.

[3] Standen, G. W. (1968), *Handbook for seminar on the advanced design of binary circuits for industrial machine controls*, Aviation Electric Ltd., Montreal.

[4] Standen, G. W. (1969), Industrial sequencing with fluidics, *1st Fluid Logic Conf.*, Milwaukee, Wisconsin.

[5] McCloy, D. and Martin, H. R. (1973), *The Control of Fluid Power*, Longman.

[6] Fawcett, J. R. (1968), *Pneumatic circuits and low cost automation*, Trade and Technical Press.

[7] *The Martonair cascade system of circuit analysis* (1967), Martonair rept. F6.

[8] Smith, B. W. (1970), *A simple method for the design of pneumatic circuits*, Fluid Power International Conf. Enots. Report.

[9] Davis, J. D. (1972), *Improving micro-valve helps simplify pneumatic systems*, Davis Pneumatic Systems Ltd.

[10] Cheng, R. M. H. and Foster, K. (1972), Systematic method of designing fluidic-pneumatic control circuits, *Proc. Inst. Mech. Engrs.*, **186**, 29/72.

CHAPTER 17

Discontinuously controlled and digital servos

17.1 INTRODUCTION

The familiar electro-hydraulic servo is an analogue device, and much effort has been directed at determining its stability and performance on a theoretical and practical basis. Nowadays, the digital computer and the microprocessor are commonly used in systems, either for presenting input data to the control system or as part of the control loop and, as a result, fluid power engineers have been trying to devise means of using conventional fluid power systems in conjunction with digital data. Digital systems have several advantages over their analogue counterparts.

1. Digital systems can be very accurate. The main limit on accuracy is the amount of data or bits of information that have to be handled.
2. Digital data is not impaired by manipulation, two-state devices being relatively insensitive to noise.
3. Systems are generally more reliable because of the simplicity of the basic two state element.

Their main disadvantages are:

1. Complexity of controls.
2. The additional cost of digital to analogue conversion, where necessary.

This chapter presents an introduction to methods of digital control, on-off control and to pulse-width modulation. It includes an anlaysis of cavity formation in on-off controlled hydraulic servos.

17.2 DIGITAL TO ANALOGUE CONVERTERS

Since the familiar analogue hydraulic servo is well established and is used on many machine tools, the most direct approach to making hydraulics and digital data compatible is to use digital to analogue conversion. In some systems the converter is placed between the digital data source and the input to the control system. In others, higher accuracy is achieved by placing two D/A converters in the loop, one for the error signal and one for the feedback signal. This necessitates the use of a digital comparator.

The fluid power engineer is well aware that bi-stable fluidic elements are ideal for digital data handling. Interface problems between the fluid power system and the data processor are reduced and in some cases eliminated. In addition, the added advantages of environmental tolerance and reliability make their use attractive. An example, proposed by Parker [1], of a pneumatic digital to analogue converter is shown in Fig. 17.1. This could be interposed between a

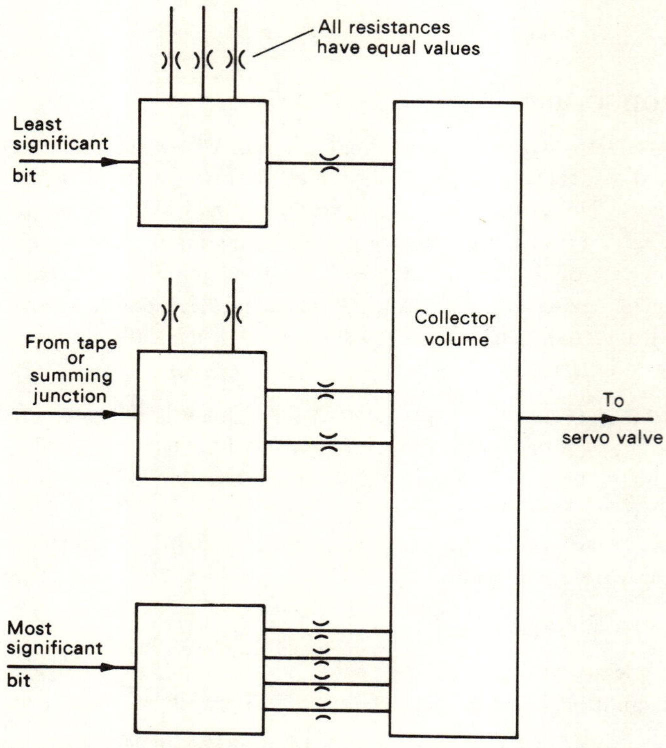

Fig. 17.1 – Digital to analogue converter (after Parker [1]).

fluidic binary counter and an air operated hydraulic servovalve. Basically, information (pneumatic pulses) is fed from fluidic elements to receivers and hence to a collector volume. Many fluidic elements are sensitive to load impedance changes and so the receivers in this case are designed to present a constant load impedance. This is achieved by ensuring that each input 'sees' the same orifice impedance – four similar orifices. The most significant bit will have the most significant effect on the pressure rise in the collector volume, since it supplies the volume through four orifices.

Another more direct approach is to use the valve itself as the converter so that the valve can accept digital signals directly without the use of prior decoding devices. Two possible arrangements are shown in Fig. 17.2. In the first [2], the first stage of the valve consists of four arms of a hydraulic bridge, in which the controlling resistances are two sets of normally closed flapper valves in parallel.

The orifices are sized in binary progression so that opening the first gives a

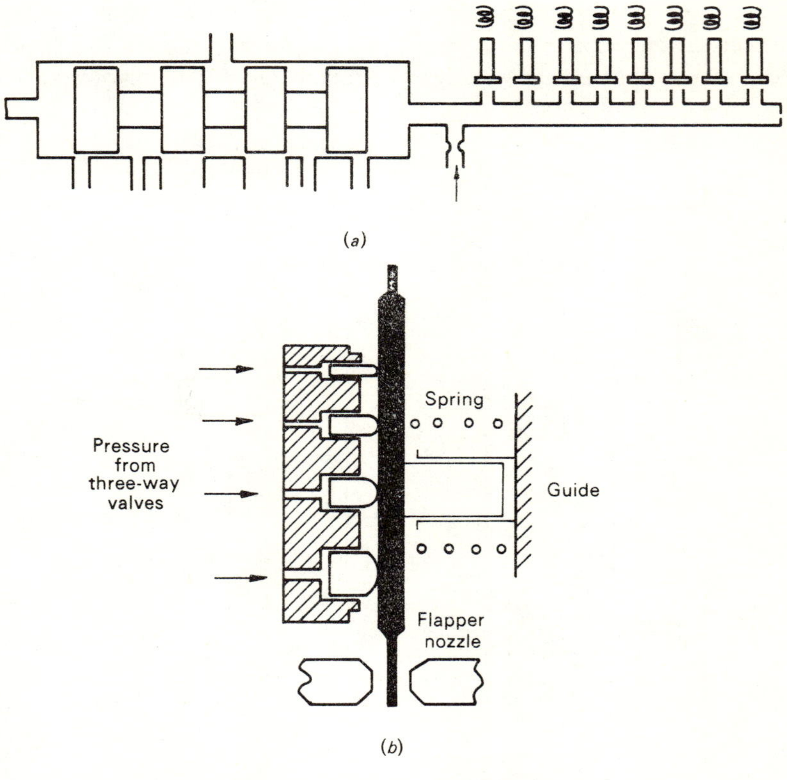

Fig. 17.2 – (a) Digital to analogue valve using binary orifices (after Wells [2]). (b) Digital to analogue valve using binary cylinders (after Himmler [3]).

pressure drop of n bar across the second stage spool and opening the second gives $2n$ bar and so on. The solenoids are driven by a parallel digital system. The second example uses binary forces [3], acting on a spring loaded flapper first stage. Both systems can accept a tape input, the magnitude of the signal not being critical, so that in the first case voltage stabilisation is unnecessary. The main disadvantage is that much data must be handled if high accuracy is desired.

17.3 PULSE CONTROLS

In the systems described above, parallel digital information is used, that is, all bits of information can be supplied simultaneously to the actuator. Another approach is to use sequential control, in which the number of control pulses arriving in sequence determines the output of the device. This method has been used successfully with fluid systems. The main element is the stepping motor, which rotates through a fixed movement for each applied pulse. Electrical stepping motors can operate at 1000 pulses/s, each of 1.5°. Fig. 17.3 shows the use of such a pulse system as part of a machine tool control system [4]. Each electrical pulse causes the stepping motor to rotate through a fixed amount and causes the hydraulic motor shaft to follow. Ott [5] has shown that the

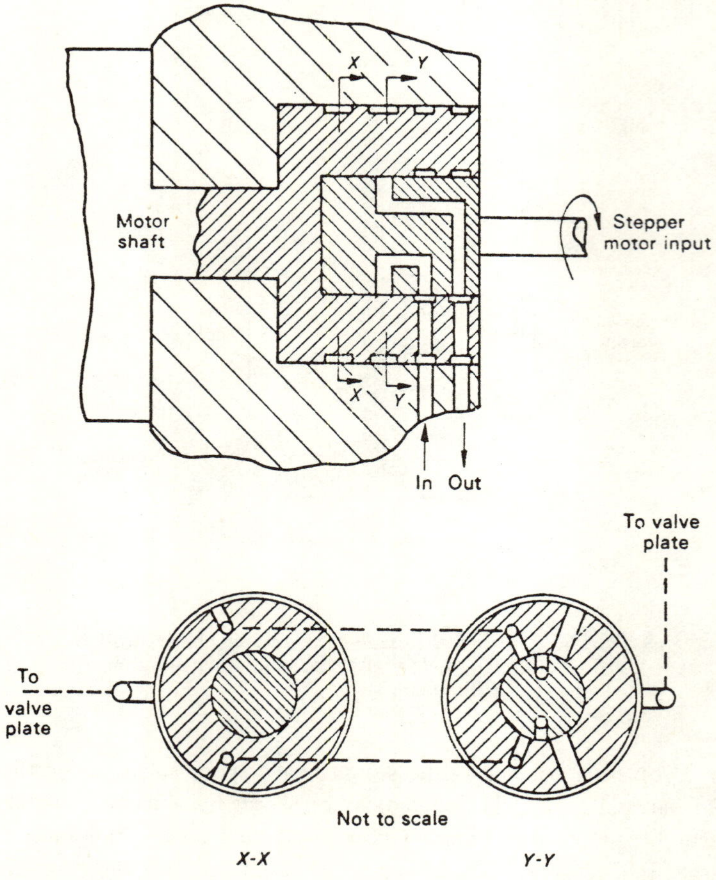

Fig. 17.3 – Hydraulic motor driven by stepper motor (after Kaiwa *et al.* [4]).

same thing can be achieved using a hydraulic impulse valve and a linear stepping motor. The impulse valve accepts low power positive and negative pulses (from perhaps a diaphragm driven by a bi-stable fluidic element) and sends high power pulses to a stepping motor. Fig. 17.4 shows that, for a rightward movement of the direction control valve, the input pulses $\pm\Delta$ will pressurise alternatively ports 1 and 2, while simultaneously exhausting ports 3 and 4. These ports 1–4 lead to the inlet and outlet passages of the fixed central element of the stepping motor. As shown, a pulse at 2 will send fluid to the right-hand side of the main piston, causing it to move to the left. As it moves it closes off the connection to passage 2 and when it has moved one port width it will stop. Other ports allow the fluid on the left-hand side of the piston to exhaust.

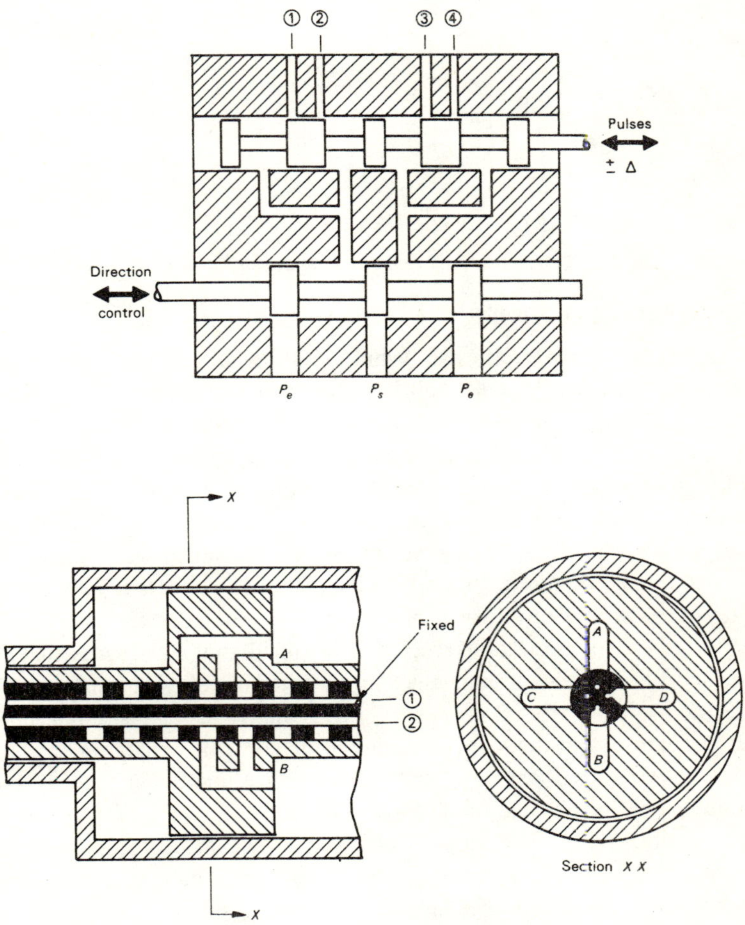

Fig. 17.4 – (a) Hydraulic impulse valve. (b) Linear stepping motor driven by the impulse valve (after Ott [5]).

Another example, Fig. 17.5, is the nutating gear stepping motor [6], which has been designed for use with fluidic circuits. There are only two moving parts – a driving gear which is free to wobble but not to rotate, and an output gear which can rotate but cannot nutate. Unequal pressurisation of the driving bellows causes the driving gear to tilt and contact the output gear. Since the output gear has 180 teeth and the input gear 181 teeth, every complete revolution of the point of contact will make the output gear advance by 1 tooth or rotate 2°. Fluid logic circuitry has been developed to provide the correct sequence of bellow pressurisation.

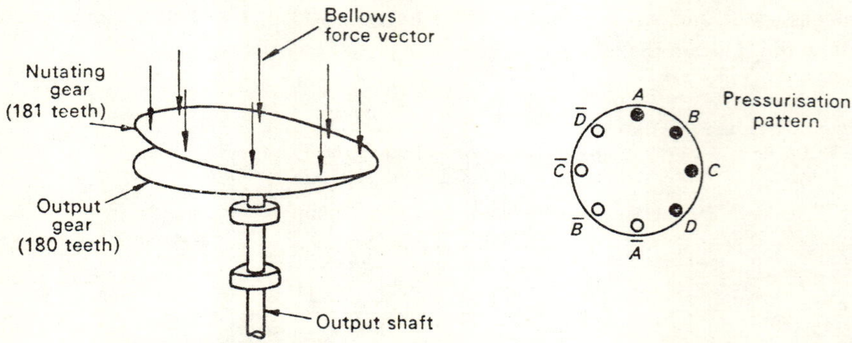

Fig. 17.5 – The nutating gear stepping motor (after Griffin [6]).

17.4 DIGITAL ACTUATORS

Nearly all of the above mentioned techniques have used, in one form or another, the analogue hydraulic servo. Another approach eliminates all analogue components, and hence the need for digital to analogue conversion, by relying on digital signals throughout; this uses a digital actuator which is capable only of positioning at one or more discrete positions. The simplest digital actuator is the conventional cylinder, positioned by a two-position valve. The cylinder rod can be considered as having only two discrete positions if such a valve is used, the positions being fully extended or fully retracted; and clearly the accuracy of such a positioning device depends on the accuracy of manufacture of the cylinder. An important feature with this digital actuator is that it is an open-loop system – positioning the rod at either of its two positions does not require the use of feedback and, as a result, stability problems should not exist. Digital actuators are used for machine tool control, programmed handling, indexing systems, gear changing and other applications where precise location at discrete positions is required. Fig. 17.6 shows a means of providing more discrete positions [7]. The stops are driven either mechanically or by fluid power, and setting the actuator in the desired position requires the movement of the appropriate stop and the pressurisation of the cylinder.

Another form of digital positioner uses cylinders connected in series, Fig. 17.7. With this device 2^n discrete positions are available if n cylinders are used.

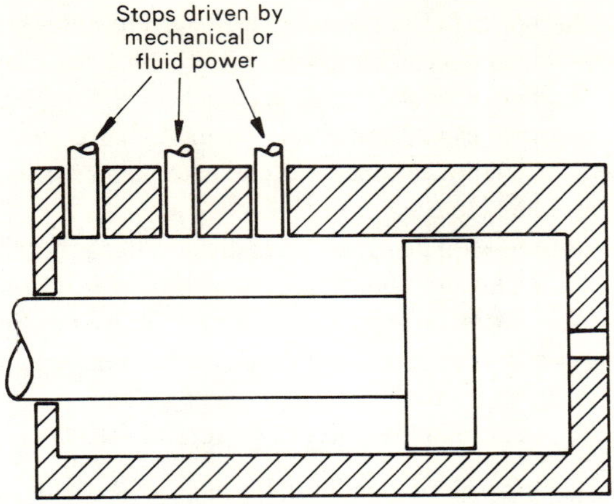

Fig. 17.6 – A discrete position cylinder using mechanical stops (after Miles [7]).

In addition, the cylinder strokes can be arranged in binary progression so that parallel binary information can be used to operate directly valves controlling each cylinder. One disadvantage of this simple system is its bulk, and Fig. 17.8

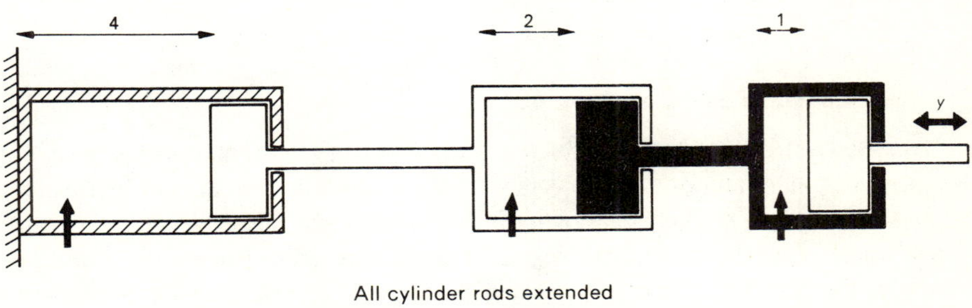

Fig. 17.7 – A digital positioner using series cylinders.

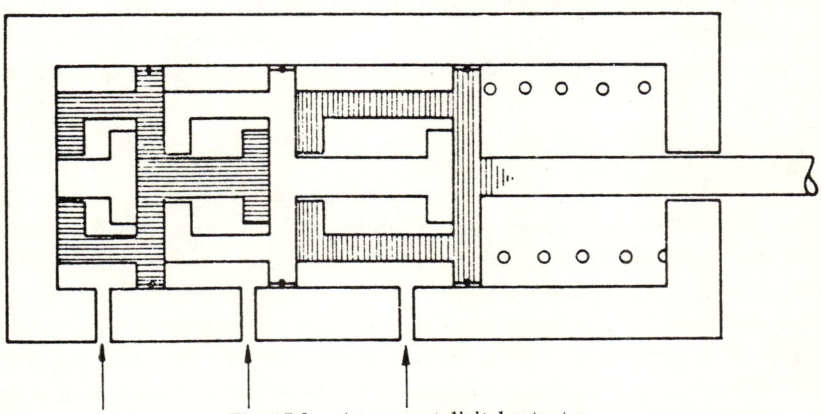

Fig. 17.8 – A compact digital actuator.

shows a more compact design [8]. The Production Engineering Research Association, England, have developed such a digital actuator with a stroke of approximately 0.8 m, smallest increment of 4 mm, an accuracy of 0.4 mm, and a speed of 15 m/min. Others of shorter stroke and greater accuracy have been developed [9] for aircraft and aerospace applications where existing on-board digital computers are used to operate the actuators.

Binary information can be supplied from a fluidic counter using for example the binary flip-flop which changes its output condition on the reception of each pulse. The outputs from the counter can be amplified by step-up relays to drive the appropriate cylinders. In some applications, conventional pneumatic valves are sufficiently fast in operation.

The main advantages of digital actuation can be summarised as:

1. They are basically open-loop systems. Compressible fluids, which can cause stability problems in analogue systems, give no trouble.
2. The two-position valves associated with the device are cheap, reliable and are not prone to contamination. A reported application uses hot gas at 660 K and 200 bar.
3. Lack of stability problems allows tighter accuracy and faster response than analogue systems.
4. Digital to analogue conversion is unnecessary.
5. They can be operated directly from tape parallel information or from sequential data from counters.

One final example, Fig. 17.9, of a discrete positioning actuator uses the calibrated volume method [10]. Here two valves are used, a directional control and an actuating valve. The directional control determines which way the output will move. Each time the actuating valve changes position (up to down or vice versa), the piston displaces a calibrated volume into the actuating cylinder and the output steps a fixed increment. Leakage from the calibrated volume can cause problems. This is an incremental digital actuator and can only be commanded to a position adjacent to its current position. Full value digital actuators, such as that shown in Fig. 17.8, can be commanded to any position at any time.

17.5 ON–OFF CONTROLLED SYSTEMS

Comment has been passed earlier on some advantages of two-state devices. The simple two- and three-position control valves also have several advantages over the sophisticated servovalve. They are much cheaper. They require less elaborate driving equipment; for example the torque motor is designed for maximum torque and not for maximum linearity. They are less sensitive to the environment, and dirty oil or hot gas can be handled readily. The price to be paid for these advantages is the reduced accuracy of a simple on–off controlled servo. The major problem of on–off control is the limit cycle [11], and attempts to remove this by using relay dead zone or valve overlap tend to aggravate the problem of inaccuracy. Nevertheless, there are applications, such as rudder control [12], where slight inaccuracy can be tolerated. In on-off controlled hydraulic servos, cavitation is likely to occur, and analysis at the end of this chapter shows how this affects response.

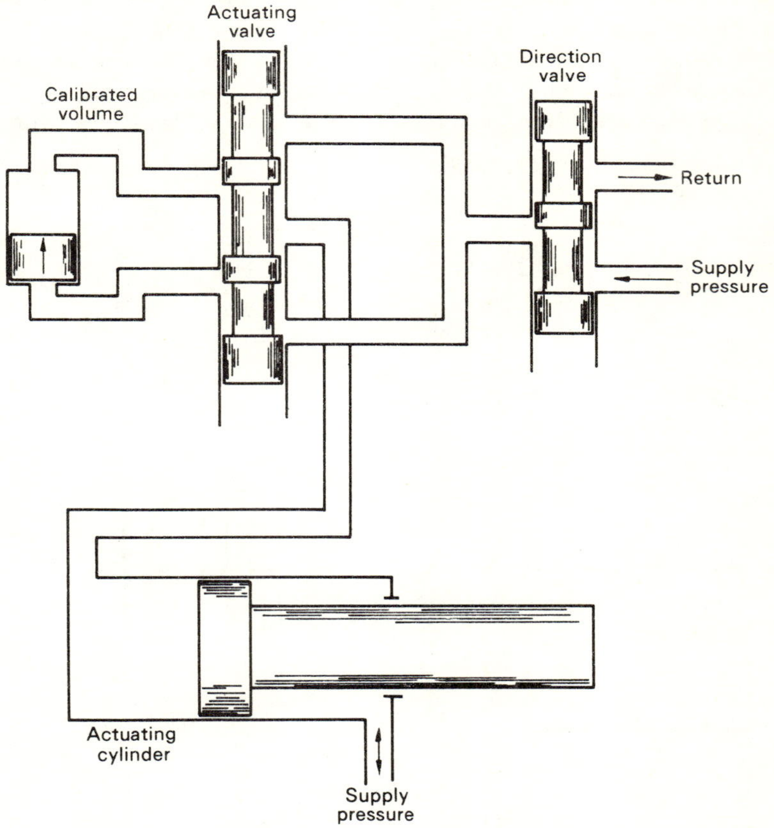

Fig. 17.9 – A discrete positioning actuator using a calibrated volume (after Raider and Scott [10]).

17.6 OSCILLATING CONTROL SYSTEMS – PULSE-WIDTH MODULATION

In order to retain the advantages of the two-position valve and to obtain better performance, the pulse-width modulated valve was developed. The feature which distinguishes this form of control from the conventional proportional control is that, instead of generating a signal proportional to the error, the amplifier generates a series of pulses which alternatively drive the valve to its two extreme positions. Pulse-width modulation of an input signal θ_1 can be achieved by adding a dither signal $D \sin \omega_d t$ to the input and passing the combined signal through a bi-stable element. Fig. 17.10 shows the bi-stable output for various inputs. A step input, for example, makes the valve dwell longer on one side than the other. In the acceleration switching valve [13], the output from the bi-stable drives a flapper-nozzle valve, whose output is amplified by a second stage spool. Because of the double integration inherent in this two-stage valve, the no-load cylinder rod acceleration is proportional to flapper position – hence the name 'acceleration switching valve'.

The addition of the dither signal to the input signal and the use of the bi-stable provides another means of processing analogue data in digital fashion. As discussed earlier, the use of the bi-stable element in a hydraulic

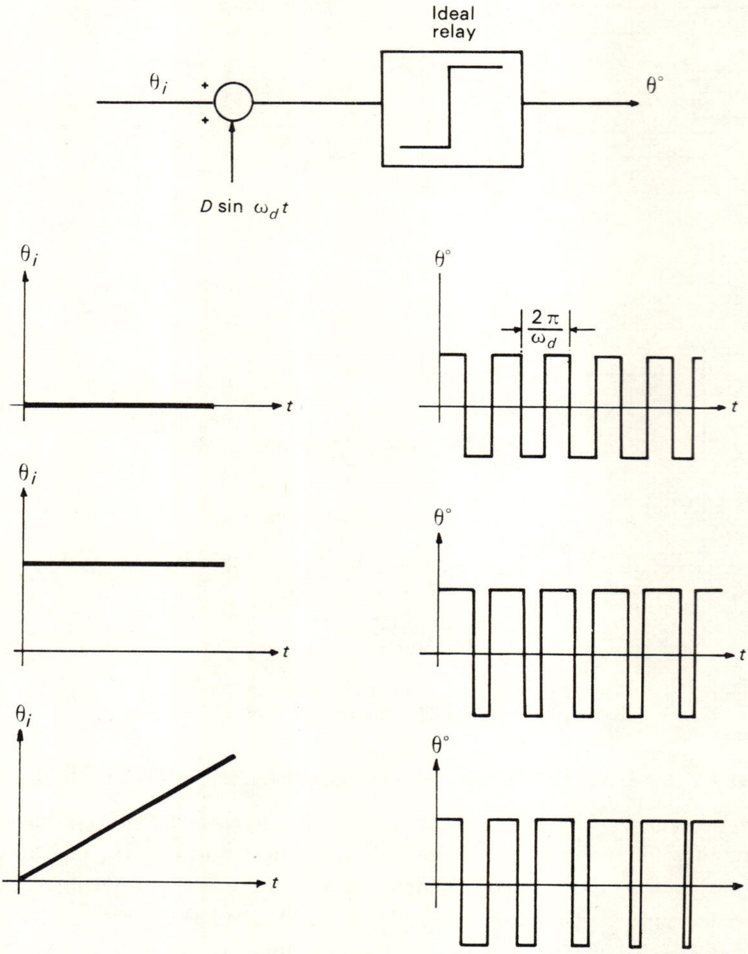

Fig. 17.10 – Pulse width modulation using a bi-stable, and response to different inputs.

servo can lead to stability problems but, fortunately, this is not so in this case. The dither signal effectively linearises the characteristics of the non-linear bi-stable, the modified characteristic being referred to by Taft and Nawaz [14] as the altered characteristic of the element. This is the plot of the average value of the output of the bi-stable over one cycle of the dither versus the input signal magnitude. For example, for sinusoidal dither, input to relay $= \theta_i + D \sin \omega_d t$. Then the average value $\bar{\theta}_o$ of output over one cycle can be shown to be:

$$\bar{\theta}_o = \frac{2}{\pi} \sin^{-1} \frac{\theta_i}{D} \,. \tag{17.1}$$

Fig. 17.11 shows the effect of dither amplitude on the altered characteristic. As this is reduced the effective gain is increased, but the linear range is reduced. This provides a means of varying system gain. Note that the device is saturated when the input signal magnitude exceeds the dither amplitude. The dither frequency should of course be high enough to avoid undesirable output oscillation. For a second-order system, a dither frequency five or more times the system natural frequency gives satisfactory results.

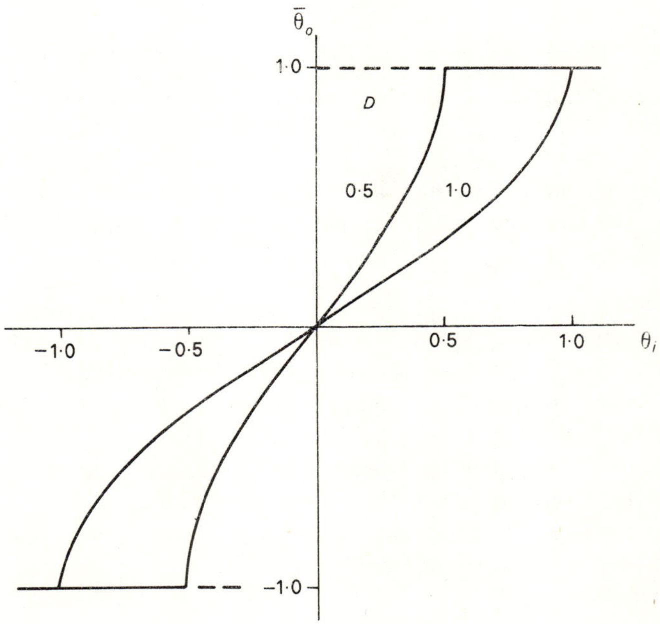

Fig. 17.11 – The altered characteristic of a bi-stable using a sinusoidal modulator (after Taft and Nawaz [14]).

Another method for obtaining pulse-width modulation in practice uses fluidics [15]. Fig. 17.12 shows that the dither is developed by a fluidic oscillator consisting of a bi-stable amplifier with positive feedback loops. The outputs of this device could be applied to the ends of the spool in a servovalve, so providing a means if driving and actuator.

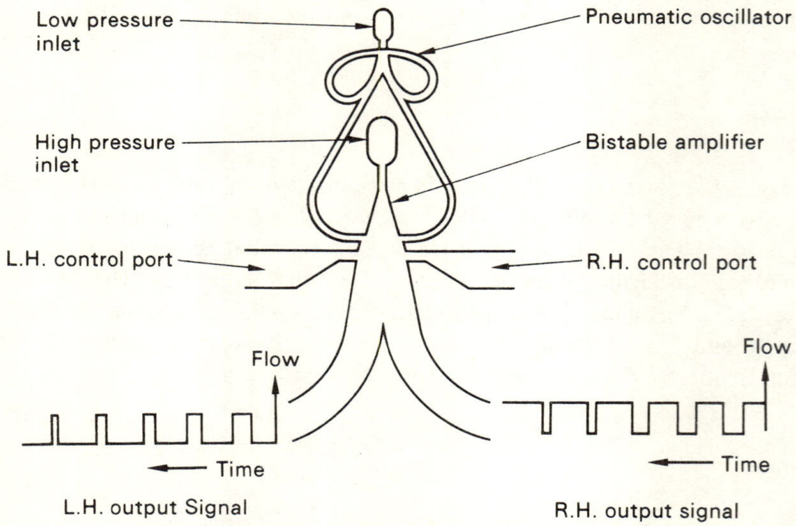

Fig. 17.12 – Fluid logic circuitry to produce a pulse width modulated output proportional to a differential input.

Besides the usual advantages of digital systems, pulse width modulated control of the servovalve is claimed to be superior to proportional control in several areas [13]:

1. Valve gain is essentially constant for practically all inputs less than the saturation value.
2. There is little centre shift of gain change due to temperature effects.
3. Excellent resolution is obtained because of the dither effect of the carrier frequency.
4. Increased reliability, since the torque motor has a large force output and foreign bodies are thus unlikely to obstruct motion.
5. Pressure gain is much larger than proportional valves, hence system stiffness is improved and the steady state errors due to actuator loading are reduced.

There is one significant disadvantage of this mode of operation; that is, the quiescent power drain when there is no input signal. Under these conditions, each side of the actuator is alternatively pressurised and exhausted and this can lead, particularly in pneumatic systems, to substantial energy wastage. This can be overcome by using the technique of differential pulse-width modulation which ensures that when there is no input signal, the control valve is closed. It has been shown [15] that this can be achieved by subtracting the carrier waveform from the modulated waveform, Fig. 17.13. A three-position

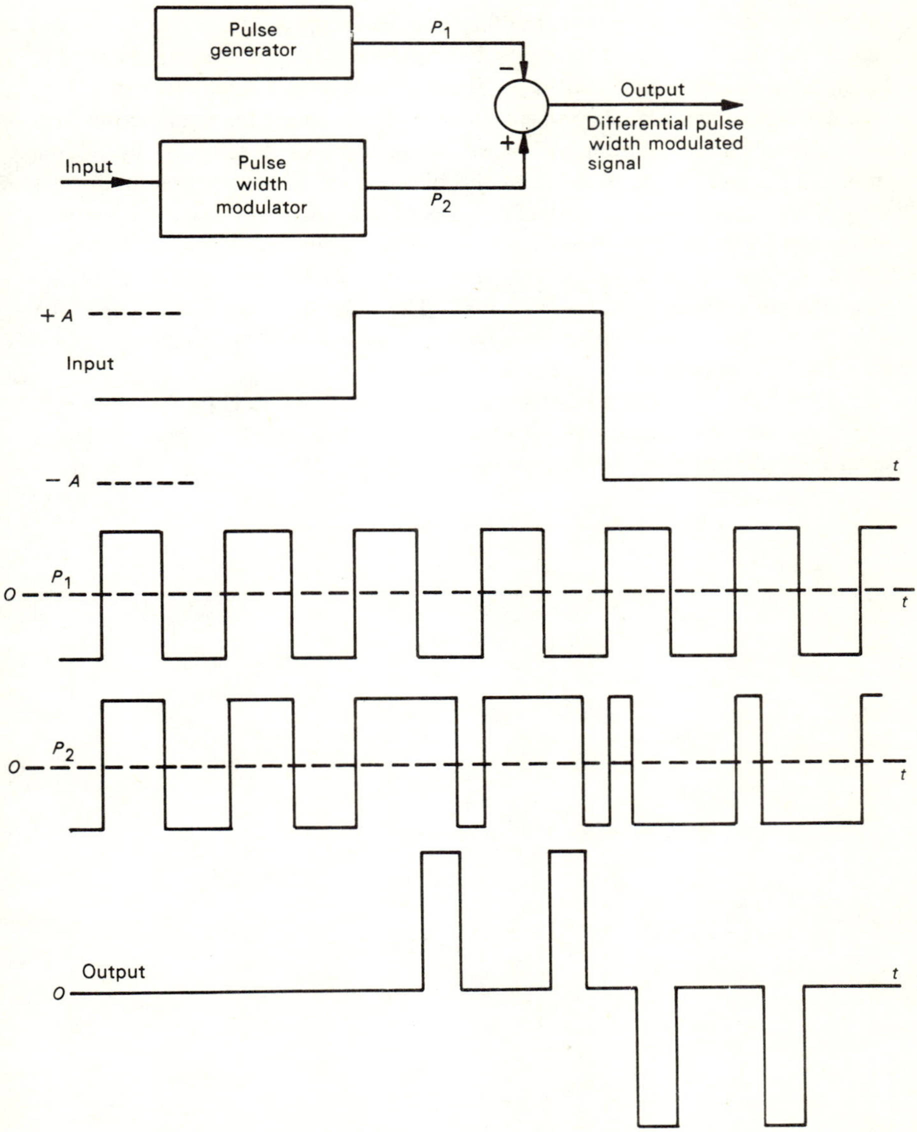

Fig. 17.13 – Differential pulse width modulation.

valve, with closed centre position, can be used in this case. Reference [16] discusses a floating flapper disk which can be used in the implementation of pulse length modulation. It combines the simplicity of pure fluid elements with the shut off capability of the mechanical valve.

17.7 RESPONSE IMPROVEMENTS USING DIGITAL TECHNIQUES

In analogue fluid power servos, stability compensation is achieved by using well established signal shaping techniques, but unfortunately, these cannot be used in digital systems unless digital to analogue conversion is used. Taft and Nawaz [14] have developed pure digital compensators which can be used in conjunction with the pulse-width modulation technique. Compensation is achieved by adding pulses to or subtracting pulse from the bi-stable output every time it changes state. The so-called posicast controller is a particular form of this compensator in which a pulse is subtracted from the output of the bi-stable. Fig. 17.14 shows how a posicast controller [16] effectively improves the damping of a second-order system with damping ratio $c = 0.2$. The pulse duration is one-half of the system's periodic time. This method of compensation has been implemented by using fluidic elements and has operated satisfactorily.

One of the principal advantages of the hydraulic servo is its fast response, and in view of this it is clearly important that more effort be applied to design based on the theory of the time optimal controller. When moving a load from

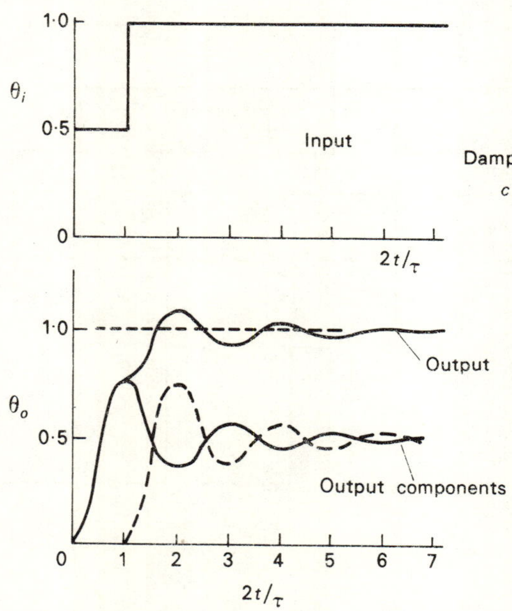

Fig. 17.14 – Posicast control of a second-order system with damping ratio $c = 0.2$ and period T.

one position to another, on-off control can give faster response than proportional control. For a second-order system, the valve is opened fully one way and the load moves with maximum acceleration. After an interval of time the valve is switched to its opposite extreme so that the load decelerates at maximum rate. The switching point is chosen so that the load stops at the desired position. The number of switching points increases with the order of the system and, if valve dynamics and oil compressibility are all taken into account, time optimal control of a hydraulic servo would require a very complicated controller, and the valve would need to switch position several times whilst driving the load from A to B in the shortest possible time. This is not feasible in practice. However, sub-optimal controllers can be designed [17] using a reduced mathematical model, and these, although being slower than the optimal, can still be appreciably faster than conventional controllers. For example, the system order can be reduced to two if valve dynamics and oil compressibility are ignored, and a few controllers based on this model are shown in Fig. 17.15. The simplest controller closes the valve when the output has reached the desired position and if, as is assumed, the fluid is incompressible the output will remain fixed. In practice, of course, the hydraulic spring will cause the output to oscillate about the desired position. The magnitude of this terminal oscillation can be reduced if the load velocity is arranged to be zero when the valve is closed. This is the second controller illustrated, and it can be seen that it is necessary to switch the valve twice so that the load reaches the desired position with zero velocity. However, although the terminal oscillation has been reduced in amplitude, it still exists because on valve closure there is still a pressure differential across the actuator piston. If cavitation occurs, the amplitude of the oscillation may still be large, and if this is intolerable then a controlled valve closure must be used, so arranged that the load reaches the desired position both with zero velocity and zero load pressure. Although this is possible [17], the advantages of on-off control are lost.

17.8 SOME PHENOMENA ASSOCIATED WITH ON-OFF CONTROLLED HYDRAULIC SERVOS

Discontinuous movements of the servovalve spool can lead to the occurrence of pressure peaks in the cylinder. In addition, cavitation is very likely to occur under these conditions, and it is therefore of interest to examine the step response of the valve cylinder combination [18, 19]. Such discontinuous valve movements are encountered in on-off servos and in pulsed servos where the valve spool driving takes the form of a sequence of rectangular pulses generated by sampler.

The valve actuator system is shown in Fig. 17.16. The system equations are repeated for the reader's sake.

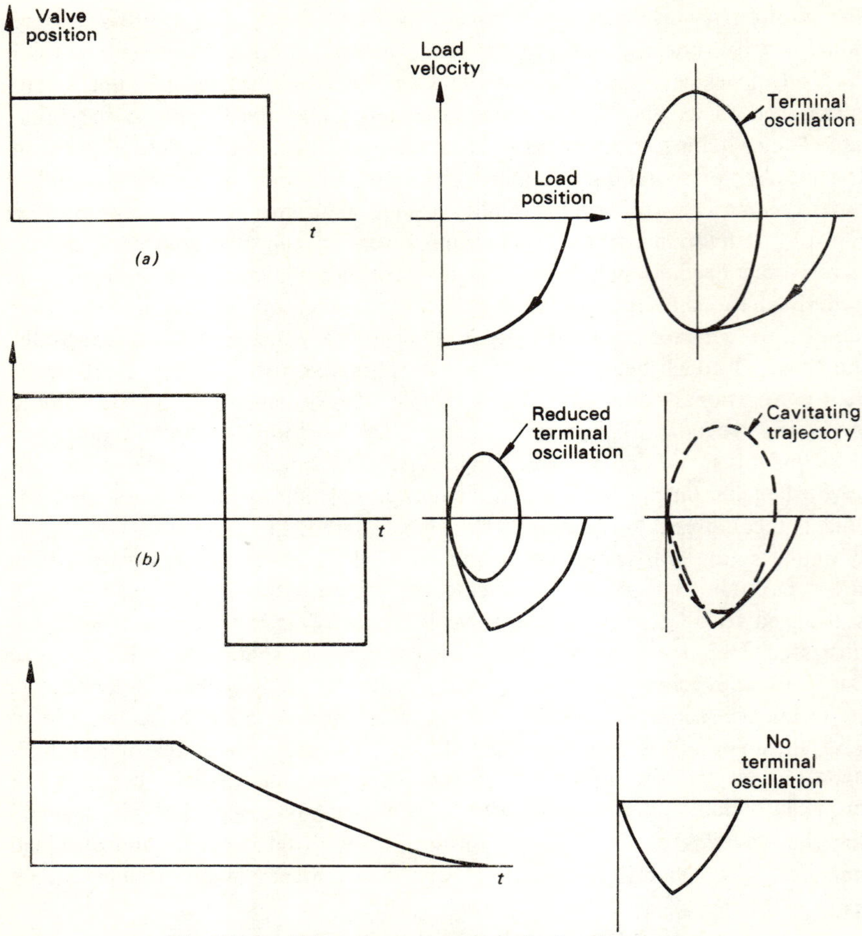

Fig. 17.15 – Sub-optimal control of a hydraulic actuator.

For spool displacement $x_v > 0$:

$$\text{Inflow:} \quad Q_1 = Kx_v[\operatorname{sgn}(P_s - P_1)]\sqrt{(|P_s - P_1|)} \tag{17.2}$$

$$\text{Outflow:} \quad Q_2 = Kx_v[\operatorname{sgn}(P_2 - P_e)]\sqrt{(|P_2 - P_e|)}\,. \tag{17.3}$$

For $x_v < 0$:

$$Q_1 = -K|x_v|[\operatorname{sgn}(P_1 - P_e)]\sqrt{(|P_1 - P_e|)} \tag{17.4}$$

$$Q_2 = -K|x_v|[\operatorname{sgn}(P_s - P_2)]\sqrt{(|P_s - P_2|)}\,. \tag{17.5}$$

Cavitation is at its worst when the load is predominantly inertia, and so for ease

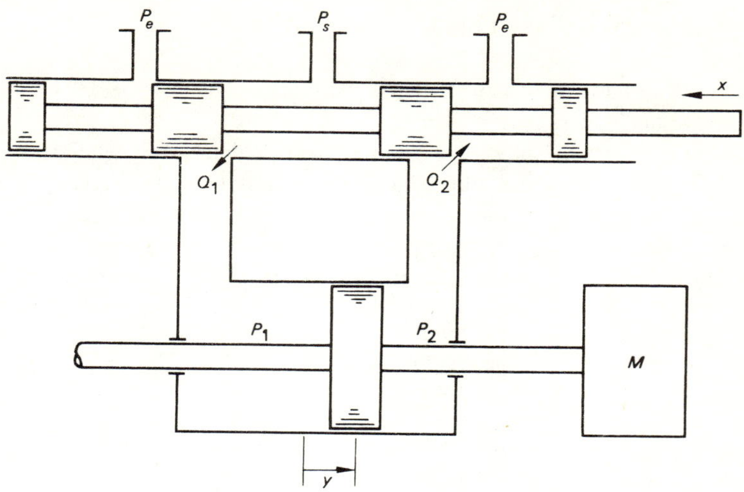

Fig. 17.16 – The hydraulic valve and actuator.

of analysis only this form of loading is considered. The load equation is then:

$$(P_1 - P_2)A = M\mathrm{d}^2y/\mathrm{d}t^2 \ . \qquad (17.6)$$

Fluid compressibility effects will also be ignored at this stage. If there are no cavities, the continuity equations are then:

$$Q_1 = Q_2 = Au \ . \qquad \text{where } u = \mathrm{d}y/\mathrm{d}t \qquad (17.7)$$

If a cavity exists then $Q = Au$ where Q is the flow rate into or out of the full cylinder side.

The response to a step input in spool position from $x_v = 0$ to $x_v = x_r$ will be considered first. The flow equations become:

$$Q_1 = Kx_r\sqrt{(P_s - P_1)} = Au \qquad (17.8)$$

$$Q_2 = Kx_r\sqrt{(P_2 - P_e)} = Au \ . \qquad (17.9)$$

These equations are made non-dimensional using the method described in Chapter 11. They reduce to:

$$\text{velocity } \bar{u} = \sqrt{(1 - \mathrm{d}\bar{u}/\mathrm{d}\bar{t})} \qquad (17.10)$$

with load pressure:

$$p_e = \mathrm{d}\bar{u}/\mathrm{d}\bar{t} \ . \qquad (17.11)$$

Equation (17.10) is of Riccati form and has solution:

$$\bar{u} = \tanh\{\bar{t} + \tanh^{-1}\bar{u}(0)\} \tag{17.12}$$

where $\bar{u}(0)$ is the initial velocity. Fig. 17.17 shows the response for $\bar{u}(0) = 0$ and for $P_e = 0$. When the spool step movement is initiated, the pressure P_1 jumps immediately to supply pressure, and P_2 falls to tank pressure. Ultimately both pressures settle at $\frac{1}{2}P_s$ and the load pressure is then zero in the new steady state. No cavities form as a result of this spool movement.

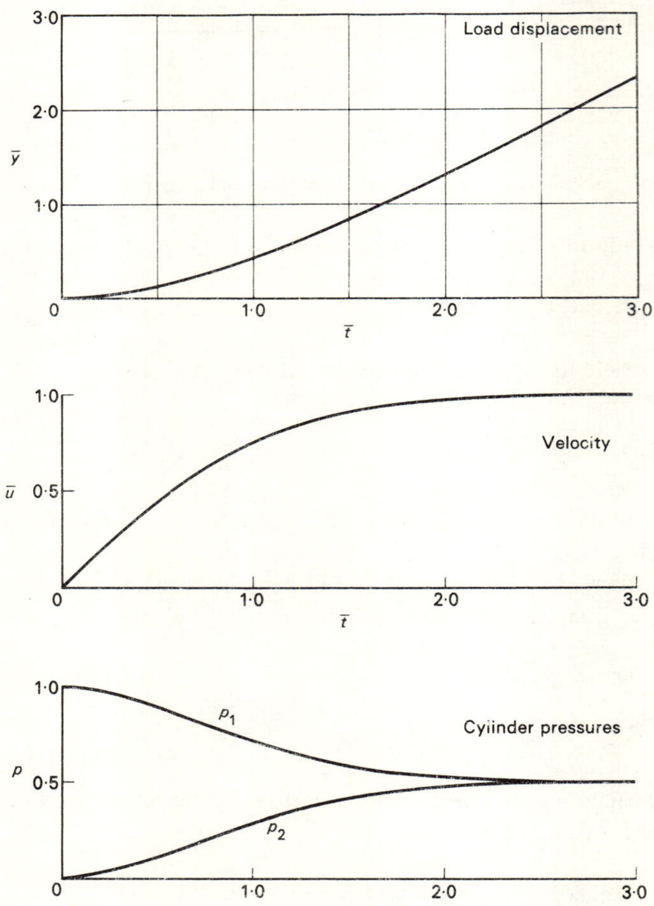

Fig. 17.17 – The response to a step valve opening.

The effects of compressibility have been determined from a full-scale digital computer simulation of the system, and the results are shown in Fig. 17.18. The inertia factor α is an inverse measure of the degree of compressibility:

$$\alpha = \frac{Kx_r}{A}\sqrt{\frac{2\beta M}{VP_s}}$$

where β = oil bulk modulus
V = total volume of oil in acuator.

For a high performance on-off controlled servo, α will usually exceed 2, and the results show that over this range incompressible analysis yields useful results.

In an on-off controlled servo, it is also possible to have step inputs in valve position from one extreme to another. Under these conditions cavities can form. Assume that the load is moving with steady velocity $+u(0)$ corresponding to spool displacement $+x_r$. There is no driving pressure required under these conditions, so that $P_1 = P_2 = \frac{1}{2}P_s$. The spool is then moved instantaneously to $-x_r$, a step input of magnitude $-2x_r$. Since the velocity is initially positive and the fluid is assumed incompressible, then P_2 must exceed P_s for $t = 0+$. Q_2 and u are then positive, oil flowing from cylinder to supply. Side 1 is connected to tank and pressure P_1 falls instantaneously to oil vapour pressure (assumed zero) in an attempt to make Q_1 fill side 1 of the cylinder. A cavity forms on side 1.

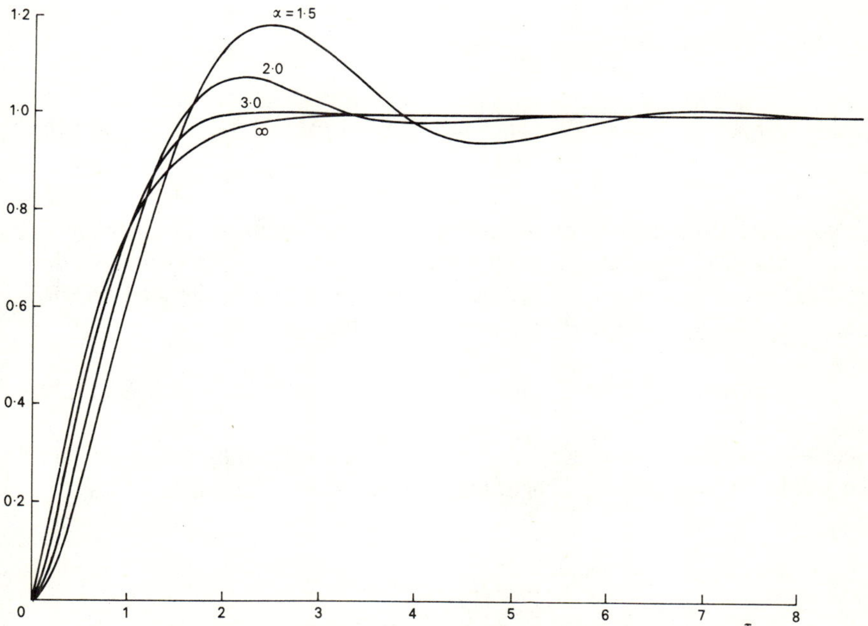

Fig. 17.18 – Effect of oil compressibility on step response.

The equations become:

$$Q_1 = Kx_r\sqrt{P_2} \tag{17.13}$$

$$Q_2 = Kx_r\sqrt{(P_2 - P_s)} = Au \tag{17.14}$$

$$P_2A = -M\mathrm{d}u/\mathrm{d}t\,. \tag{17.15}$$

These equations reduce to:

$$\bar{u} = \sqrt{\{-2(1 + \mathrm{d}\bar{u}/\mathrm{d}\bar{t})\}} \tag{17.16}$$

or, in Riccati form:

$$\mathrm{d}\bar{u}/\mathrm{d}\bar{t} = -(1 + \bar{u}^2/2)\,. \tag{17.17}$$

The solution is:

$$0 < \bar{t} < \bar{t}_s$$

$$\bar{u} = \frac{\sqrt{2}(a\cos\sqrt{2}\bar{t} - b\sin\sqrt{2}\bar{t})}{1 + a\sin\sqrt{2}\bar{t} + b\cos\sqrt{2}\bar{t}} \tag{17.18}$$

$$a = \frac{2\sqrt{2}\bar{u}(0)}{2 + \bar{u}^2(0)}\,; \qquad b = \frac{2 - \bar{u}^2(0)}{2 + \bar{u}^2(0)}\,.$$

$$\bar{y} = \ln\left\{\frac{1 + a\sin\sqrt{2}\bar{t} + b\cos\sqrt{2}\bar{t}}{1 + b}\right\} + \bar{y}(0)\,. \tag{17.19}$$

Note that at the instant of switching, pressure P_2 jumps to 1.5 times supply pressure. The load stops at $\bar{t}_s$ when pressure P_2 falls to supply pressure. At this point the load displacement is maximum positive. For $\bar{t} > \bar{t}_s$, the load velocity is negative, P_2 being less than P_s. Eqaution (17.17) then becomes:

$$\mathrm{d}\bar{u}/\mathrm{d}\bar{t} = \bar{u}^2/2 - 1 \tag{17.20}$$

which describes the motion until the cavity on side 1 is filled by piston displacement and backflow from tank. The solution of equation (17.20) is $\bar{t}_s < \bar{t} < \bar{t}_c$.

$$\bar{u} = \sqrt{2} - \frac{2\sqrt{2}}{1 + \mathrm{e}^{-\sqrt{2}(\bar{t} - \bar{t}_s)}} \tag{17.21}$$

$$\bar{y} = \bar{y}(\bar{t}_s) + \sqrt{2}(\bar{t}_s - \bar{t}) - 2\ln\left[\tfrac{1}{2}(1 + \mathrm{e}^{-\sqrt{2}(\bar{t} - \bar{t}_s)})\right] \tag{17.22}$$

The cavity size H is:

$$H = \int_0^t Au\,\mathrm{d}t - \int_0^t Kx_r\sqrt{P_e}\,\mathrm{d}t \tag{17.23}$$

or:

$$h = H/Ay_r = \bar{y} - \sqrt{2p_e}\,\bar{t}\,. \tag{17.24}$$

The cavity is filled at $\bar{t} = \bar{t}_c = \bar{y}/2p_e$. This is plotted in Fig. 17.19 which shows that the cavitation duration is strongly affected by backflow from tank. However, for the purpose of the present analysis, it is sufficient to assume that tank pressure is negligible compared to supply pressure, that is, $P_e = 0$. In this case, $\bar{t}_c = 1.8$. For $\bar{t} > \bar{t}_c$, pressure P_1 can rise and the system equation is:

$$\bar{u} = -\sqrt{(1 + \mathrm{d}\bar{u}/\mathrm{d}\bar{t})}\,. \tag{17.25}$$

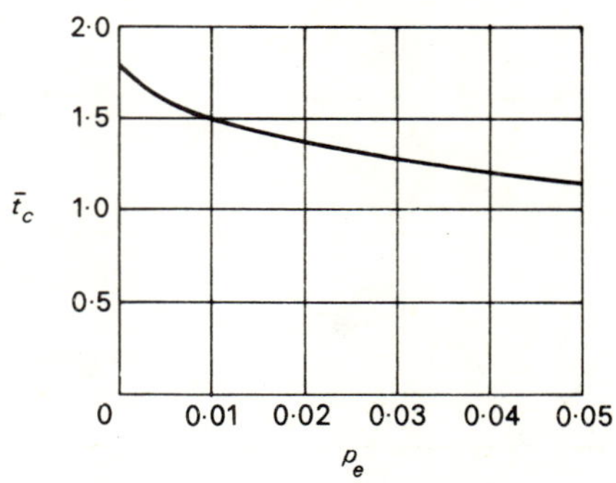

Fig. 17.19 – Effect of tank pressure on cavity filling time.

The solution of equation (17.25) is:

$$\bar{t}_c < \bar{t} < \infty$$

$$\bar{u} = 1 + \frac{2}{Be^{2(\bar{t}_c - \bar{t})} - 1} \tag{17.26}$$

$$B = \frac{\bar{u}(\bar{t}_c) + 1}{\bar{u}(\bar{t}_c) - 1}$$

$$\bar{y} = \bar{y}(0) - (\bar{t} - \bar{t}_c) - \ln \frac{e^{2\bar{t}_c} - Be^{2(\bar{t}_c - \bar{t})}}{e^{2\bar{t}_c} - B}\,. \tag{17.27}$$

Fig. 17.20 shows that the new steady state is reached at approximately $\bar{t} = 3.0$.

If the valve were now switched back to its original position, $+x_r$, the response could also be deduced from Fig. 17.20 and the cylinder pressures would be interchanged.

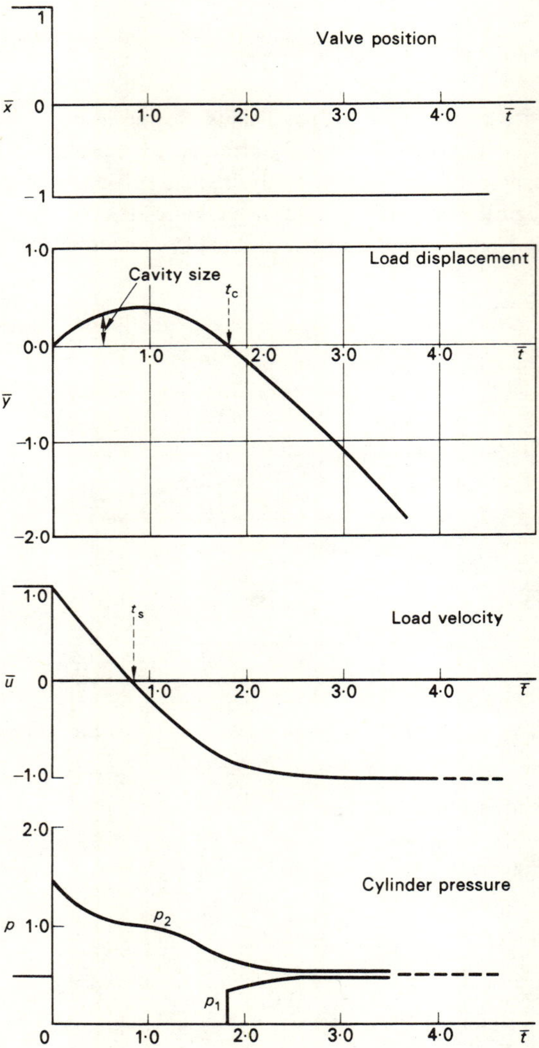

Fig. 17.20 – Response to a step in valve postion from $+x_r$ to $-x_r$.

17.9 REFERENCES

[1] Parker, G. A. (1968), Fluidic Digital to Analogue Converters, Paper K1, *3rd Cranfield Fluidics Conf.*, Turin.

[2] Wells, J. C. (1966), Digital techniques in fluid power engineering, *Hydraulics and Pneumatics*, September, 108-112.

[3] Himmler, C. R. (1968), Comparative performance of servo valves with fluidic and electrical inputs, *Proc. Fluid Power Int. Conf.*, London, September.

[4] Kaiwa, T. and Inaba, S. (1962), Electrohydraulic pulse motor, *Control Engineering*, January, 73-75.

[5] Ott, M. (1967), Digital control equipment for impulse controlled linear and rotary motors, *Hydraulic Pneumatic Power*, November, 635-640.

[6] Griffin, W. S. (1967), Development of high speed flueric logic circuitry for a novel pneumatic stepping motor. *Proc. A.S.M.E., Advances in Fluidics*, 402-414.

[7] Miles, K. E. (1970), A new type incremental piston, *Fluid Power International*, April, 33-34.

[8] Delmage, A. H. and Tremblay, J. J. (1965), Hydraulic digital actuator converts binary inputs to linear motor, *Control Eng.*, February, 69-70.

[9] Bidlack, J. D. (1963), Digital servo actuators work with compressible fluids, *Space and Aeronautics*, August, 118-121.

[10] Raider, J. W. and Scott, G. W. (1966), A digital hydraulic actuator, *Proc. Proc. 3rd Congress I.F.A.C.*, paper 25a, June.

[11] McCloy, D. and Martin, H. A. (1971), Some aspects of the response of bi-stable hydraulic servos, *2nd Int. Fluid Power Symposium*, B.H.R.A., January.

[12] Baeck, H. A. (1967), Hydraulic on-off servo: simple, rugged positioner. *Control Engrg.*, December, 79-82.

[13] Murtaugh, S. A. (1959), An introduction to the time modulated acceleration switching electro-hydraulic servomechanism, *A.S.M.E., J. Basic Engrg.*, June, 263-271.

[14] Taft, C. K. and Nawaz, S. (1970), Fluidic pneumatic control system compensator, *4th Cranfield Fluidics Conf.*, Coventry, paper E2.

[15] *Basic Applied Research in Fluid Power Control* (1964), Reports No. 8993-5-6, Dept. of Mathematical Engineering, Project Lab. M.I.T., Massachusetts, February.

[16] Gorbatenko, G. G. (1965), Posicast control by delayed gain, *Control Engrg.*, February, 74-77.

[17] Armstrong, P. J. and McCloy, D. (1970), *Sub-optimal solutions for time optimal control of a hydraulic servomechanism*, J.A.C.C. Paper 6B2, Atlanta.

[18] McCloy, D. (1969-70), Cavity formation in valve controlled hydraulic cylinders, *Proc. Inst. Mech. Engrs.*, London, **184**, pt. 1.

[19] McCloy, D. (1970), *Some phenomena associated with bang-bang controlled hydraulic servos*, J.A.C.C. Paper 6B1, Atlanta.

CHAPTER 18

Noise in fluid power systems

18.1 INTRODUCTION

Noise levels generated by industrial machinery are steadily increasing, owing to the introduction of more mechanisation and to higher power level usage. Awareness of this trend has resulted in recommended codes of practice and regulations concerning noise in the community and in the workplace.

Machinery noise arises from the generation of pressure fluctuations in air which are the result of either:

(a) movement of solid surfaces,
(b) action of turbulent discharge.

Examples of the processes described in (a) would be the noise from a hydraulic pump casing and the noise radiated by a cooling fan, while an example of process (b) would be an exhausting air jet from a pneumatic circuit.

In practice, the noise radiated from machinery can be made up of any combination of these processes, and the time history can be continuous or transient. In addition, many regions of a structure will vibrate at the same time. Thus it is seldom possible to establish theoretical methods of noise and vibration prediction in the real-life situation, except for establishing dominant frequencies. In any sort of noise control programme it is therefore essential to have both the equipment and the techniques of identifying noise sources, if any measure of progress in noise reduction is to be achieved.

On the other hand it is also possible to use this type of information to monitor the condition of machinery. Economic necessity, safety requirements, and in some cases technical complexity are making it attractive for engineers to consider on-line monitoring of critical areas of industrial machinery so that potential faults may be detected before a critical situation can arise.

Although noise is generally thought of as undesirable, 'peculiar' noises have been used for many years as a diagnostic aid, particularly in the field of medicine. The fact that an unusual noise may signal impending machine failure, implies that use can be made of this information for preventive maintenance.

For example the noises generated by a faulty bearing, a cracked gear, or escaping gas, produce characteristic spectral signatures.

Where it is not possible to reduce noise to an acceptable level, hearing protection can be achieved in several ways:

1. Acoustical filters placed in or around the ear.
2. Use of sound absorption, damping and barrier materials.
3. Remote control of the process.

The engineer must become concerned with the noise levels generated by his systems. Clearly an early introduction to this area will help to pin-point and solve the problems of noise pollution. Emphasis must be directed towards component design, rather than added features after the system is completed.

18.2 FUNDAMENTALS OF ACOUSTICS

It is impracticable to provide extensive details of acoustics in this text, and the reader is recommended to read references such as [1, 2] for background material. However, it is useful to provide a few of the fundamentals of acoustics to assist in the interpretation of material presented later.

Sound is the propagation of energy through an elastic medium and is characterised by dynamic fluctuations. In industrial equipment the sound can be radiated from a vibrating surface in which mechanical vibrations can be converted into acoustical energy, or it can be the result of eddy formation in a fluid. The important point to note is that sound is the result of a dynamic process and not a static one. Physically, sound is generated by moving or vibrating a body which results in the creation of air pressure fluctuations about the mean atmospheric pressure. This is the fundamental requirement – that a body must move in a gas to generate sound. Fig. 18.1 highlights the physical situation, where a continuous exchange between stored elastic energy (potential energy)

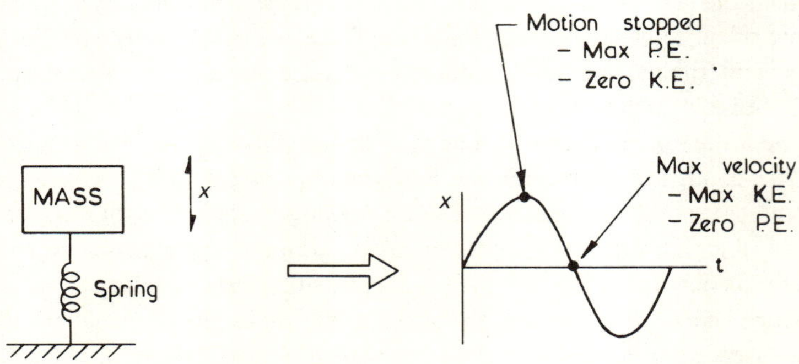

Fig. 18.1 – Sound generation process.

and stored kinetic energy (in the mass) occurs. The rate at which the exchange occurs (frequency of vibration) will determine the frequency of the sound level. Hence any device that rotates, oscillates, or reciprocates with frequencies in the range 50–15 000 Hz will probably radiate discernible acoustic noise.

There are two important quantities which describe sound:

1. The amplitude of the pressure fluctuation.
2. The frequency of the pressure fluctuation.

The pressure amplitude and resulting sound pressure level (L_p) are measured in terms of **decibels**, which is a logarithmic scale with a reference of 2×10^{-5} N/m².

$$L_p = 20 \log_{10} \frac{p}{2 \times 10^{-5}} \text{ dB} . \tag{18.1}$$

This reference level is the threshold of normal hearing at 2 kHz.

A sound pressure level equivalent to 0.1 bar is 174 dB (about the noise level emitted from a large magnum rifle). Thus a change from 0 to 174 dB represents a range of 500 million times.

The human ear does not perceive sound at all frequencies to be of equal L_p values, and from Fig. 18.2 it is seen that it is most sensitive in the 1–5 kHz region. It is for this reason that the 'A' scale is introduced in sound measurement. This effectively filters the sound received by a measuring instrument so that it reacts in the same way as a normal human ear.

The preferred reference quantities used to assess the decibel level when dealing with noise and vibration assessments are listed in Table 18.1.

Table 18.1.

Name	Definition	Preferred reference (SI units)
Sound pressure level (gas)	$L_p = 20 \log p/p_o$	$P_o = 2 \times 10^{-5}$ N/m²
Sound pressure level (liquid)	$L_p = 20 \log p/p_o$	$P_o = 10^{-6}$ N/m²
Power level	$L_w = 10 \log W/W_o$	$W_o = 10^{-12}$ W
Intensity level	$I = 10 \log I/I_o$	$I_o = 10^{-12}$ W/m²
Acceleration level	$L_a = 20 \log a/a_o$	$a_o = 10^{-5}$ m/s²
Velocity level	$L_v = 20 \log v/v_o$	$v_o = 10^{-8}$ m/s
Displacement level	$L_d = 20 \log d/d_o$	$d_o = 10^{-11}$ m
Force level	$L_f = 20 \log F/F_o$	$F_o = 10^{-6}$ N

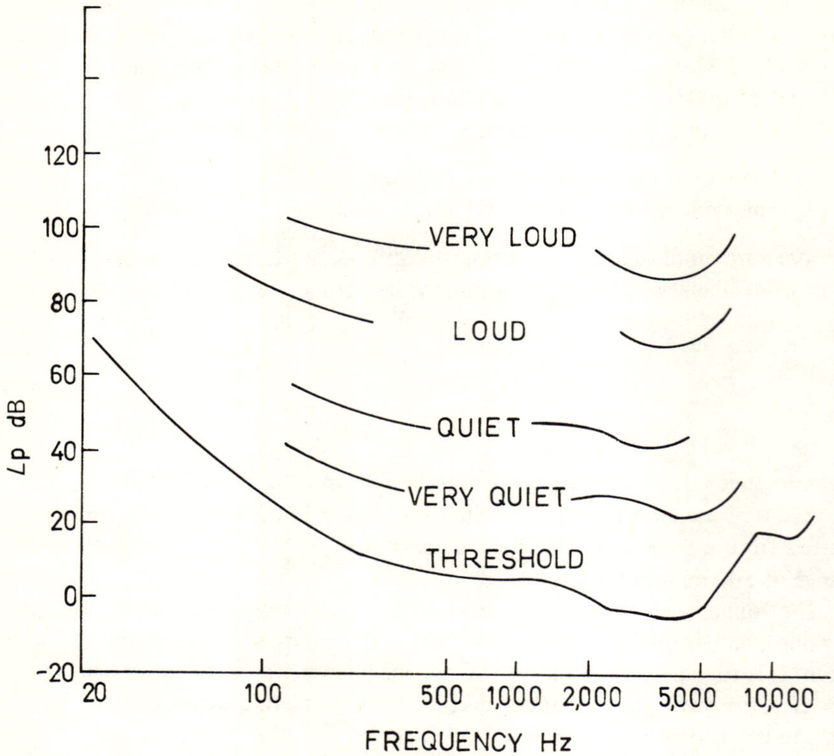

Fig. 18.2 – Frequency sensitivity of the human ear.

Under some conditions acoustic energy loss from a system can be related to system inefficiency. However, it would be incorrect to assume that a quiet system, as a result of adding sound absorbent materials or barrier enclosures, is an efficient one. The key to the problem is in the initial design concept. Noise from equipment can be classified into three general types.

1. Discrete frequency noise characterised by pure tones and usually associated with rotating equipment.
2. Broadband noise resulting from the discharge of high velocity jets, cavitation, etc.
3. Impact noise.

Hence noise is categorised by its magnitude and its spectral (frequency) distribution. It can be shown (1) that the acoustical intensity (I) emitted from a component is related to pressure fluctuations by:

$$I = \frac{p^2}{\rho c} \tag{18.2}$$

where p = rms value $p_{\text{peak}}/\sqrt{2}$
ρ = density of air at 22°C and atmospheric pressure (1.18 kg/m^3)
c = speed of sound at 22°C in air (344.4 m/s)
ρc = characteristic impendance (Ns/m^3) (ρc = 406 Ns/m^3 or mks rayl).

This is the average rate of flow of energy through a unit area of the surface of the component and normal to the direction of wave propagation. The effective excess fluctuation p about the steady atmospheric pressure level, has to be assessed on the basis of its root mean square value. The product ρc is defined as the characteristic impedance and is the ratio between the actual particle velocity u in the medium and the dynamic pressure fluctuation p.

For an omnidirectional monopole source, the energy diverges spherically from the centre, and the acoustical intensity or power per unit area will diminish. Hence the intensity at distance r from the source will be:

$$I = \frac{W}{4\pi r^2} . \tag{18.3}$$

Combining equations (18.2) and (18.3):

$$p^2/\rho c = W/4\pi r^2 . \tag{18.4}$$

There is no instrument available which can directly read sound power or sound intensity. Sound pressure is the parameter measured, and that is a pressure fluctuation above the static atmospheric pressure, 101.34×10^3 N/m^2.

Sound power level is measured in decibels, which is a measure related to some reference value; in this case 10^{-12} W is generally accepted (see Table 18.1).

Sound power level is a useful way of labelling a sound source since it remains constant while the sound pressure level is dependant on distance from the source.

$$\text{Sound power level } (L_W) = 10 \log \frac{W}{10^{-12}} \text{ dB} . \tag{18.5}$$

Sound pressure level can be related to sound power level. From equations (18.4) and (18.2):

$$p^2 = \frac{W}{r^2} \times \frac{\rho c}{4\pi} \tag{18.6}$$

or for two sound source at the same r value:

$$\frac{W_2}{W_1} = \left\{\frac{p_2}{p_1}\right\}^2 . \tag{18.7}$$

The corresponding logarithmic form of equation (18.6) is:

$$L_p = (L_W - 20 \log r - 11)\,\text{dB} . \tag{18.8}$$

This represents free-field measurements performed in spherical space. Suppose the sound pressure level L_{p1} is measured at distance r_1 from a source, and let L_{p2} be the level measure at a distance r_2 from the same source, then:

$$L_{p1} - L_{p2} = 20 \log r_2/r_1 . \tag{18.9}$$

If $r_2 = 2r_1$:

$$L_{p1} - L_{p2} = 20 \log 2 = 6\ \text{dB} .$$

Hence we have the important result that the doubling of distance from a spherical point source changes the sound pressure level by 6 dB. The use of sound power level for rating an acoustic source can be likened to rating an electric light bulb in watts. This rating does not depend on where the observer is situated in relation to the source. Unfortunately, equation (18.8) does not represent the real situation. It represents a machine suspended in mid-air, far away from reflecting surfaces including the ground. A more realistic model of the situation is shown in Fig. 18.3. The only part where the inverse square law, equation (18.9),

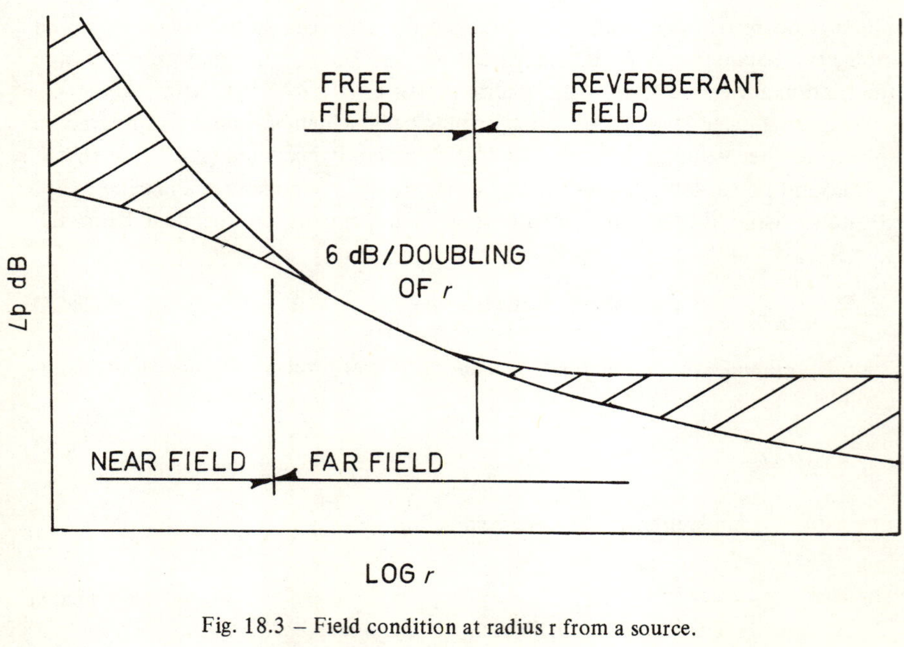

Fig. 18.3 – Field condition at radius r from a source.

actually applies is in a small region between the near field and the reverberant field. Noise is produced in the near field by vibration of various parts of the machine. Not all the parts vibrate in phase with each other. Higher frequencies can be produced by very complex vibration mode shapes, so the situation is not at all clear. In the far field the picture is complicated by reflections from other objects.

In complex situations such as described above, noise sources are not just described in terms of monopole sources of sound energy.

Another source classification is the dipole, which can be thought of as two monopoles radiating with equal strength, but opposite phase summarised in Fig. 18.4. Most industrial equipment is comprised of many mono and dipole radiators.

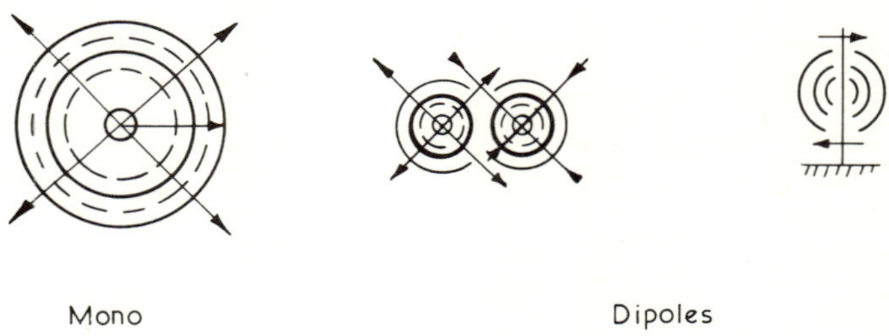

Fig. 18.4 – Types of acoustic sources.

18.3 SIGNAL ANALYSIS

Signal analysis [3, 4] is an essential part of industrial noise analysis and control, and is concerned with the extraction of useful information about the system under investigation. This information may be derived from the total signal or from some filtered version.

In sound and vibration work, one useful measure of a steady-state signal is the mean square value since the logarithm of this quantity is related to a level, such as sound pressure level. Where signals are harmonic in nature the repetition rate is useful, while for random signals, probalistic parameters are used. For short transient signals the peak value is a useful measure, while sound levels that do not commence or cease abruptly but rise and decay exponentially are best described by time rates of rise and decay.

Most signals for analysis will be obtained from a transducer of some sort and will therefore be in the form of voltage. It may even have been transported from the source to the analyser by means of a magnetic tape recorder.

A measure basic to nearly all signal analysis in sound and vibration is the spectral analysis. This is the distribution of energy in the signal as a function of frequency.

A signal is a variation of a significant physical quantity with time and can fall into many classifications as can be seen from Fig. 18.5. The simplest signal is a sine wave which can be physically realised as a pure tone sound, Fig. 18.1. Signals of this type are termed **deterministic** because it is possible to derive an equation which will predict their future behaviour. On the other hand a **random** signal has to be assessed on a statistical basis and is much more complicated to handle.

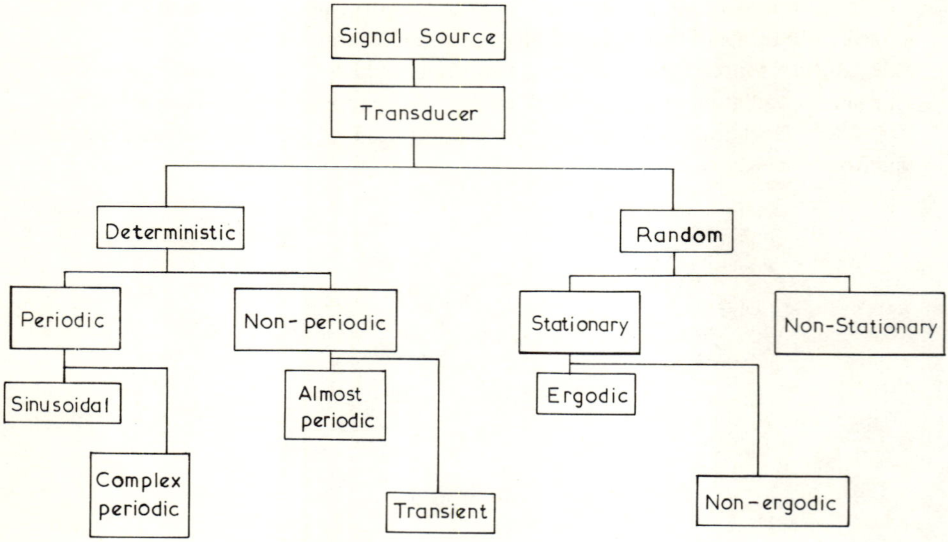

Fig. 18.5 – Signal classifications.

Even simple signals of the sinusoidal type are affected when transmitted through a system. These resultant distortions and attenuations are often used to derive information about the system to which the signal is applied. For example, consider a hydraulic pump end plate modelled as a spring-mass-damper arrangement as shown in Fig. 18.6(a). The lumped mass represents the distributed material of the panel and the lumped spring represents the distributed elasticity of the panel; the damping would be fairly small in this case.

The dynamic behaviour of this system can be found by equating the forces acting on it. Hence if $F(t)$ is the input forcing function and $y(t)$ is the resultant panel displacement, then:

$$F(t) = m\ddot{y}(t) + f\dot{y}(t) + ky(t)\,. \tag{18.10}$$

The first point to note is that the end plate has a particular frequency 'label' of its own, the natural frequency given by:

$$f_n = \frac{1}{2\pi}\sqrt{\frac{k}{m}}$$

(in practice a section of casing will have many such modes of vibration due to its distributed nature). With small damping present, a forcing function of this frequency would use very little energy to cause the end plate to vibrate violently. A forcing function of a higher frequency would require more energy to make it move. Hence the section of casing is acting as a filter to some frequencies and a window to one special frequency. This situation is presented in the form of a spectral distribution, frequency response, or simply how the panel reacts to different excitation frequencies, Fig. 18.6(b).

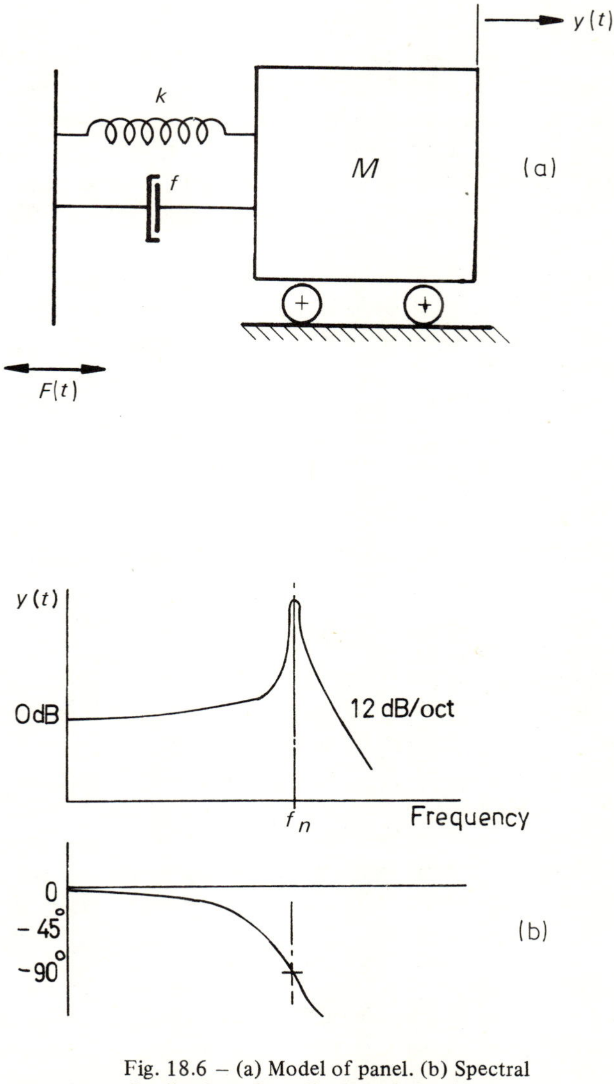

Fig. 18.6 – (a) Model of panel. (b) Spectral distribution of amplitude and phase.

In practice few systems are subjected to the simple harmonic pure tone type of signal. However, without getting involved in random signal analysis, there is a large family of complex deterministic signals that have strong harmonic content. These complex signals can be broken down into frequency components using Fourier analysis [5]. By using this approach, the total response of a system to a complex excitation can be considered as the sum of the responses to separate component responses. This is useful when looking at the way noise is radiated from a structure, and shows how a complex situation can be examined in terms of a parallel sum of simpler response terms.

In practice signals are not 'clean' like their theoretical descriptions but will often contain unwanted noise. The type of signal processing used will therefore relate to the required ratio of wanted to unwanted signal. In certain cases there may also be a need to process rapidly on site and to transmit the processed data for further analysis at some other central control point. These factors together with the nature of the signals to be processed, whether periodic, narrow- or wide-band random, impulsive or transient, will determine the choice of instrumentation to be used.

Visual observation of the signal with an oscilloscope as well as listening to it with earphones are useful first steps in assessing the quality of the signal. The presence of mains frequency contamination can quickly be detected by this means, for example.

It is not proposed to discuss the commercial equipment available for detailed noise and vibration analysis as there are many excellent handbooks available from the various manufacturers [6, 7].

18.4 ENERGY PROPAGATION THROUGH SYSTEMS AND ITS DETECTION

Noise energy propagation mechanisms can be broadly classified as:

(a) **Structure-borne**: noise which is the result of vibration due to the structural components of a system. Structures can radiate acoustic energy by transverse vibration or propagate mechanical energy by longitudinal vibration. Energy sources derive from eccentricity or imbalance of rotating parts or reciprocating motion.

(b) **Liquid-borne**: noise which is the result of pressure fluctuations causing fluid column oscillations, pulsating flow, flow noise, and cavitation.

(c) **Air-borne**: noise which is the propagation of pressure fluctuations through the air to the operator.

Consider a generalised model of a machine, Fig. 18.7, where the energy originates at a fluid-borne source. The terms ΔP_F, ΔP_s, ΔP_A are associated with pressure fluctuation transformed at each boundary interface into the appropriate energy form. Fluid-borne noise, ΔP_F, is associated with the pressure fluctuations in the working fluid as it passes through the system. Because of the distributed mass-stiffness characteristics of component housings and pipelines, the outer surfaces can be excited into complex vibratory modes which are interpreted as structure-borne noise ΔP_S. These structural vibrations in turn interact with the surrounding air causing pressure fluctuations of air-borne noise ΔP_A.

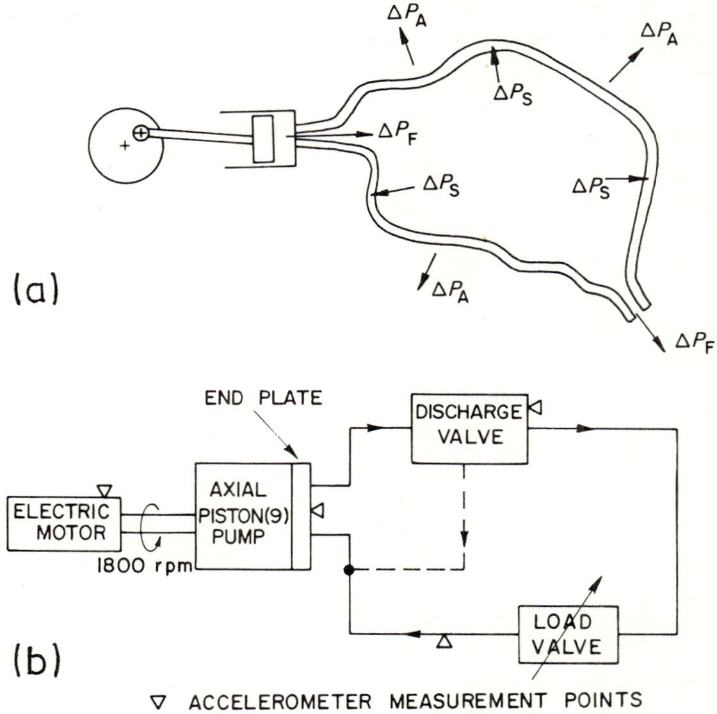

Fig. 18.7 – Hydraulic circuit layout.

Considering ΔP_F as the system input, there will be a correlation with ΔP_s and ΔP_A, through the dynamics of the system. Since ΔP_A is the result of some input energy passing through the system, it should contain information about the system. It should therefore be possible to derive some type of input/output relationship which is capable of defining spectral and transient response behaviour.

For example, in the case of a fluid power component such as a solenoid operated discharge valve, it is reasonable to assume that information about that component will be carried in the energy emitted from a pipe leading from it. Consider Fig. 18.8 which shows the difference in vibration level measured at the output of such a valve when the oil is dumped to tank and when the oil is passed through the valve. In the latter case, valve on, the vibrational level is seen to be much higher, above 600 Hz. The spectral signature below this frequency is dominated by the pump characteristics, with a main pumping frequency of 250 Hz. The energy around 6–10 kHz is due to a casing mode being excited by a harmonic of the pumping frequency. Note also the increase in pumping frequency level when the system becomes loaded by the discharge valve. Subtraction of these two cases shows that the contribution of the discharge valve is essentially broad-band energy as would be expected from noise resulting from turbulent flow through a passive element. Information above 25 kHz cannot be relied upon because the accelerometer used for this experiment has a limited bandwidth.

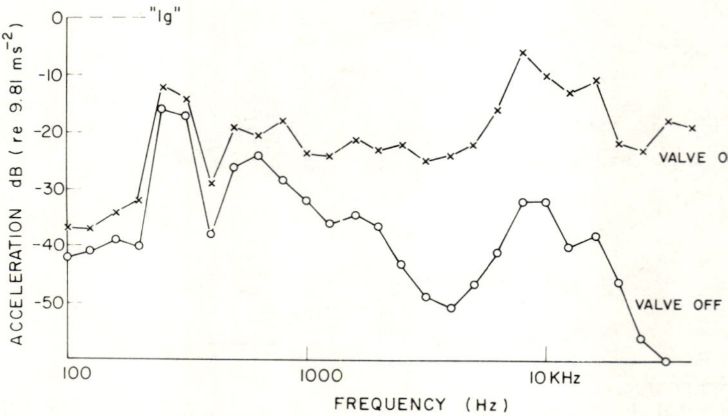

Fig. 18.8 – Vibration levels due to discharge valve operation.

Fig. 18.9(a) shows an idealised diagram of the monitoring arrangement for the results shown in Fig. 18.8. The fluid energy ΔP_F leaving the component, carries within its energy spectrum information relating to the performance of the components upstream. Hence in order to obtain information about the component described by $KG(s)$, a 'with' and 'without' type comparison test must be carried out. Once the norm has been established further comparison tests will monitor any deterioration in the component. It then becomes necessary to decide how to extract the information. From Fig. 18.9(b) it can be seen that information can be lost or distorted at the interfaces. A hydrophone will receive information at an early stage, but will reject the information carried structurally in the pipe wall. The accelerometer will have a measure of both fluid-borne and structure-borne energy but is subjected to one stage of filtering $KG_1(s)$. The accelerometer is very robust and readily available commercially. The microphone suffers from two stages of filtering and interference from external noise sources.

It is not possible to obtain simple mathematical relationships between wall vibrations and sound pressure level at a given distance from a machine.

However, if a machine could be treated in a simplified way, the wave propagation theory discussed in Chapter 10 can be used to derive an expression for the incident pressure wave at a receiver:

$$p = \hat{p} \sin \omega(t - x/c) \tag{18.11}$$

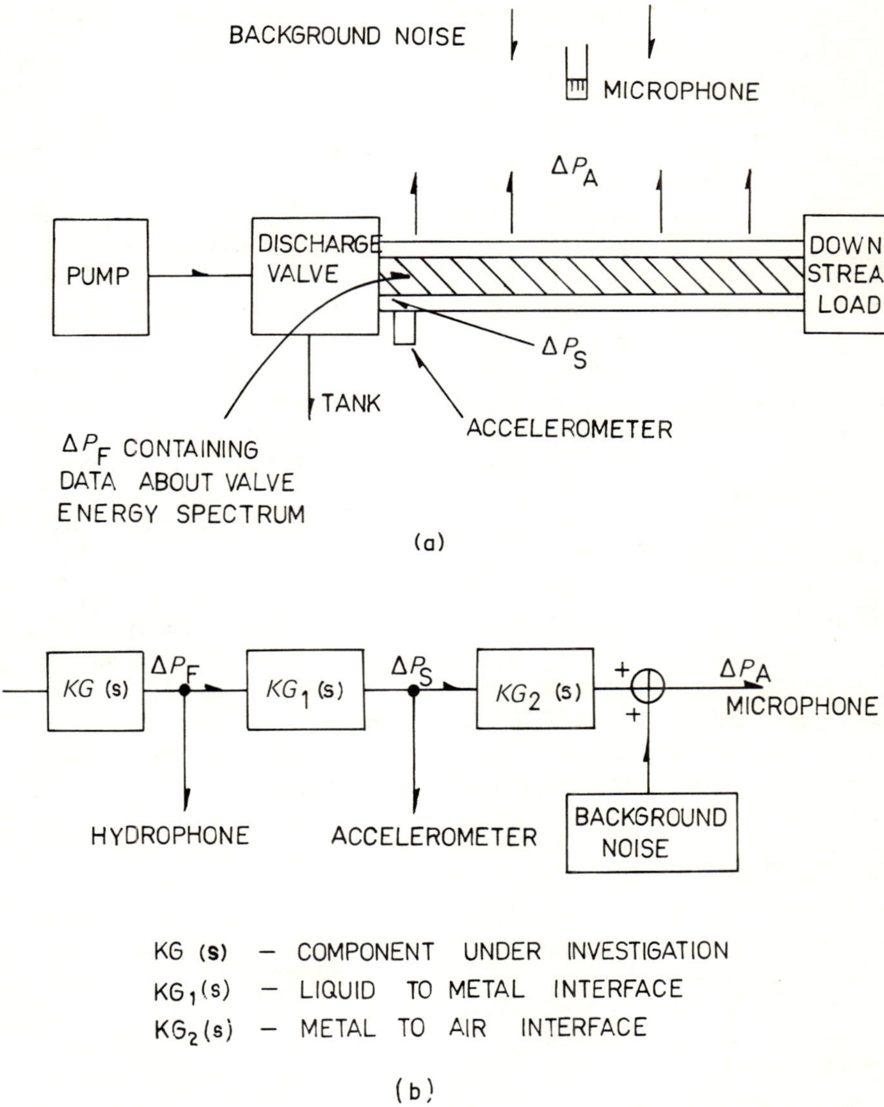

Fig. 18.9 – Propagation of energy through a pump set.

where ω is the frequency of the vibration and c is the speed of sound. Using Newton's second law, the relationship between the wave pressure and its velocity will be:

$$\rho \frac{\partial v}{\partial t} = -\frac{\partial p}{\partial x} . \tag{18.12}$$

Substituting a space differentiated form of (18.11) and (18.12) and integrating with respect to time gives:

$$v = \frac{p}{\rho c} \sin \omega (t - x/c) . \tag{18.13}$$

This shows that sound pressure is closely correlated to machine structure surface velocity, through a term ρc, known as the **characteristic acoustic impedance**. This is true for the case in which the machine radiates plane waves, but under spherical wave conditions phase relationships can greatly alter the situation; the complex modal patterns of panels further complicates the analysis.

The accelerometer is a good choice as a monitoring transducer because, since the signal is proportional to (frequency)^2, it is more sensitive to the high frequency end of the spectral distribution than either velocity, displacement, or sound pressure level. A comparison between Fig. 18.8 and Fig. 18.10 will demonstrate this point.

A detailed examination of these relationships is shown in Fig. 18.11, where measurements were obtained from the end plate of an axial piston pump. The general trend of L_p and velocity is very similar, while the acceleration data rise with frequency at 6 dB/octave. The velocity data were obtained by electronically integrating acceleration, and in both these results the main pumping frequency with the first two harmonics, together with the main end plate resonant frequency and the reasonant peak of the accelerometer are clearly identified. The microphone measurements smooth most of the high frequency detail away, since Fig. 18.9(b) shows several stages of inherent filtering.

18.5 PUMP NOISE

All positive displacement pumps function by moving discrete blocks of fluid from the inlet to the outlet. However, some of the energy is converted into noise in the form of:

(a) fluid pressure pulsations,
(b) audible sound,
(c) casing vibrations.

The audible sound is generally the most annoying. However, fluid pulsations and housing vibrations have much greater power levels if transmitted to a good radiating surface.

It has been shown earlier that piston pumps generate fluid pulsations and the delivery from each cylinder varies as a half sine wave. The combined deliveries

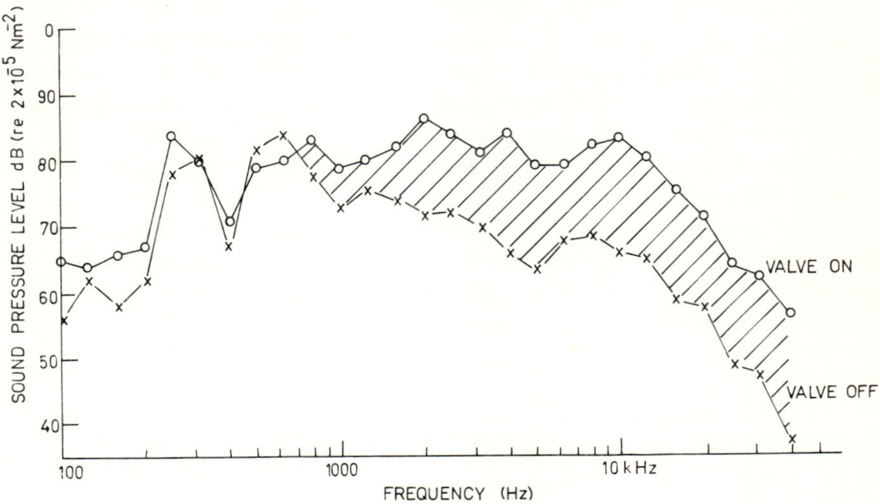

Fig. 18.10 – Sound pressure level at 0.3 m for Fig. 18.8.

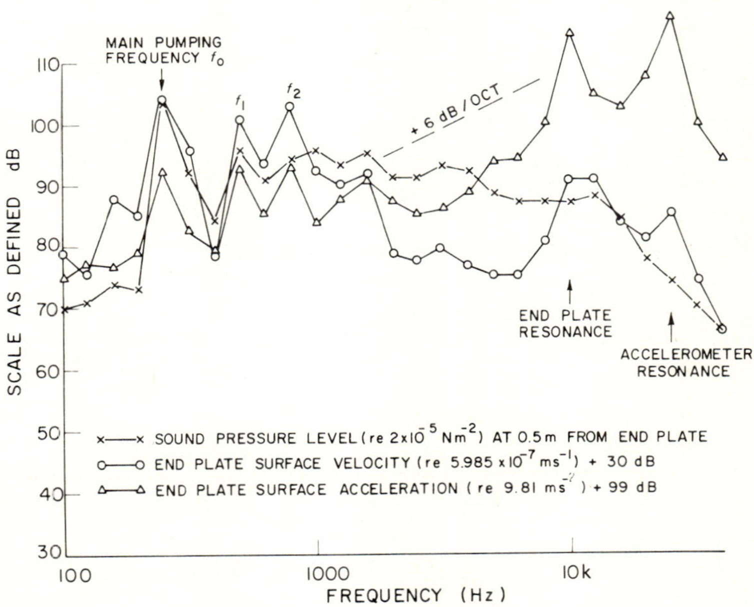

Fig. 18.11 – Comparison of L_p, L_v, and L_a for a piston pump end plate.

from several cylinders results in a steady flow plus a superimposed ripple. An odd number of cylinders gives much less ripple than an even number. For example, a nine piston pump generates one fifth as much ripple as an eight piston pump of equal displacement.

Audible sound from a piston is generated by local vibrations of the casing in response to internal forces. One of the major forces results from the cyclic variation in the number of cylinders pressurised, alternating from 4 to 5 in a nine piston pump. Little can be done to alter the magnitude of the force fluctuations, but by arranging the port tuning for gradual cylinder pressure changes, the shape of the force fluctuations can be modified [8]. Use of a frequency analyser shows that this approach reduces the energy levels in the more undesirable higher harmonics, see Fig. 18.12.

It has been shown in a paper by Helgestead [8] that as the body of an axial pump rotates there are alternatively 4 and 5 of nine cylinders pressurised with respectively 5 and 4 chambers exhausting. The net effect is that a square wave pressure pulse train is acting to excite the pump casing.

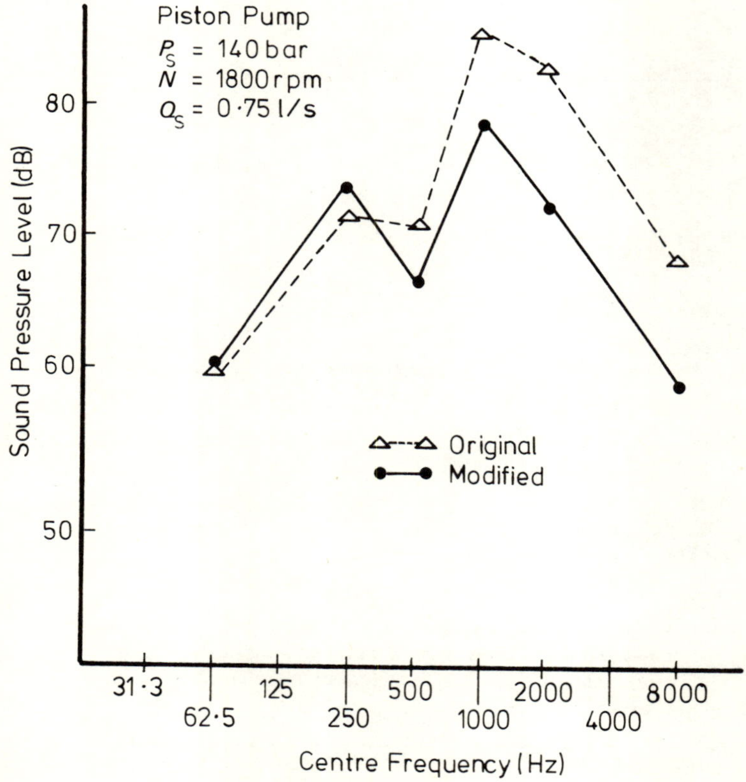

Fig. 18.12 – Noise level reduction by port shaping.

A spectrum of this effect, obtained from a real time analyer, is shown in Fig. 18.13. The axial pump under test is operating under a full load condition. The fundamental pumping occurs at 263.39 Hz and the harmonics of this fundamental up to the eighteenth are shown in brackets. The experimental plot shows that most of these are present in the end plate vibrations.

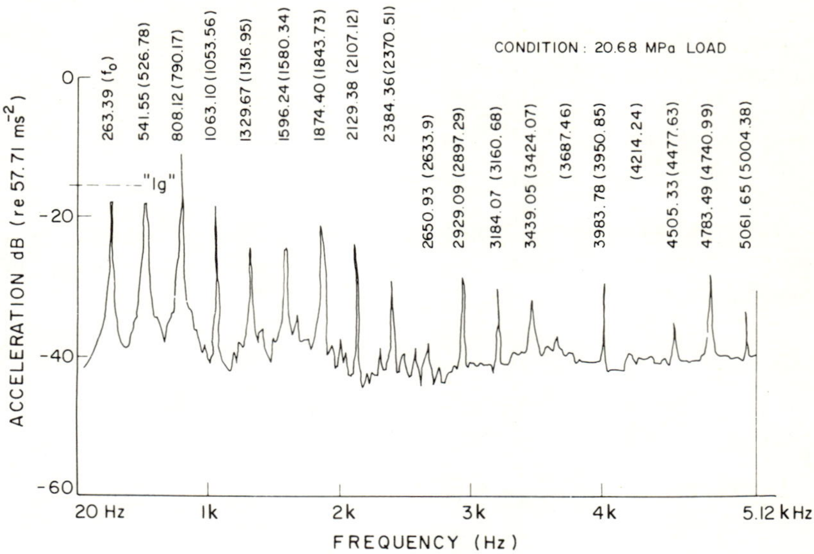

Fig. 18.13 – Narrow band analysis of axial pump end plate.

Increasing the hydraulic loading has a significant effect on the vibration level and has a resultant effect on the radiated sound level. Fig. 18.14 shows the overall increase in average vibration level as the load pressure was increased. The resultant increase in sound pressure level is shown in Fig. 18.15. The major effect is seen to occur at the pumping frequency where there is a dramatic rise in L_p from 84 dB to 104 dB.

It has been suggested that the approximate sound power level of a pump can be estimated from:

$$L_w = 97 + 10 \log HP(1 - 0.5E) \qquad (18.14)$$

where HP = hydraulic horsepower ($1 HP = 0.746$ kW)
E = pump efficiency.

An efficiency of 90 per cent ($E = 0.9$) can be expected from axial piston pumps and therefore over the loading range in Fig. 18.14, the noise power levels will be:

Table 18.2.

Condition	Load pressure	E	Watts	HP
No load	3.1 MN/m^2	0.9	5.875×10^3	7.879
Full load	20.68 MN/m^2	0.9	3.917×10^4	52.5

Hence from equation (18.14):

$$L_{w1} = 103.36\,\text{dB (re } 10^{-12}\text{) for no load}$$

$$L_{w2} = 116.61\,\text{dB (re } 10^{-12}\text{) for full load.}$$

Considering the measurements taken at 1.12 metres from the end plate of the pump, and hemisperical free field conditions above a reflecting surface exit, then the sound pressure level L_p is:

$$L_p = L_w - 20 \log r - 8.0 \tag{18.15}$$

where r = distance from the source in metres.

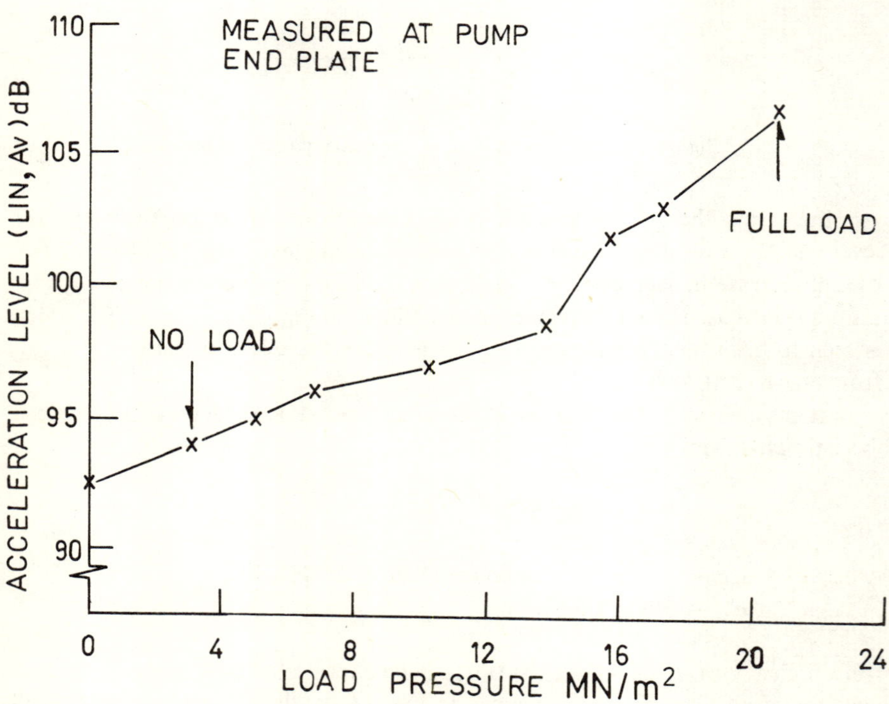

Fig. 18.14 – Effect of hydraulic loading.

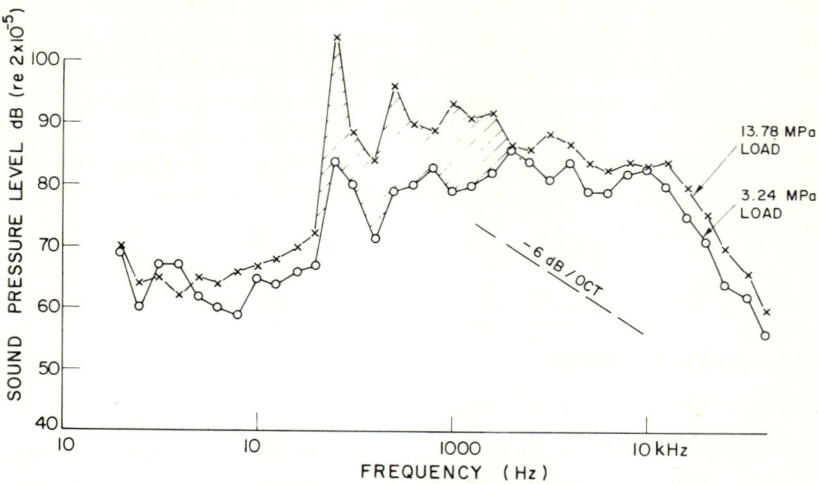

Fig. 18.15 – Effect of hydraulic loading on radiated sound.

Hence:

$$L_{p1} = 94.37\,\text{dB}$$

$$L_{p2} = 107.62\,\text{dB}\,.$$

These correspond reasonably well with the average values measured under test conditions.

The general pattern of frequency distribution of radiated noises from four other axial pumps of a similar power output is shown in Fig. 18.16. A trend can

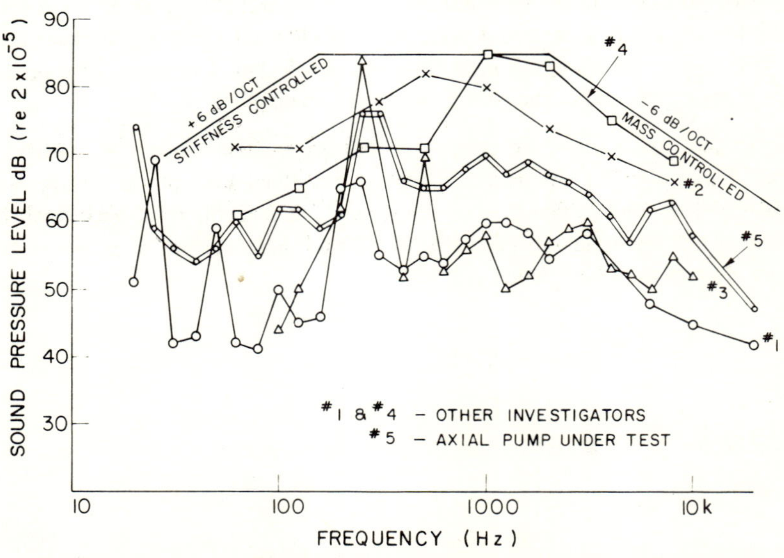

Fig. 18.16 – Comparison of axial pumps of similar power output level.

be seen in that below the pumping frequency the noise levels increase at about 6 dB/oct, then there is a plateau and then a decrease at 6 dB/oct. This has similar characteristics to a vibrating panel where below resonance the structure is stiffness controlled, and above resonance the structure is mass controlled.

By analogy therefore, the plateau where most of the energy is developed can only be effectively reduced in level by the addition of damping. The other alternative is to alter the pressure-time characteristics in the pump valving as described in [8].

18.6 GEAR PUMP NOISE

In gear pumps the major sources of noise are:

1. Engagement of the teeth, when deflection occurs in a tangential direction. When the load comes off, as the gear rotates, the teeth return to their original position. The resultant noise is periodic at the gear meshing frequency.
2. Pitch circle noise is due to load deflection during the engagement cycle where perfect rolling contact is rarely achieved. Fluctuating forces due to the sliding action induce radial teeth deflections. This noise is also periodic at the gear meshing frequency.
3. Gear wheel radiation arises when the gear wheel itself acts as an acoustical radiator. The magnitude of this radiated noise depends on the relationship between the excitation frequencies and the gear wheel natural frequencies, characterised by the natural bell tones of the wheel.
4. Oil and air jets created during the engagement cycle of the gear teeth can result in velocities near the sonic level. The formation of such jets is again periodic at the gear meshing frequency.

Since the dominant noise from gear pumps is deterministic in nature, its spectral charactertistics are useful. Consider a typical gear pump arrangement as shown in Fig. 18.17. The driver gear is rotating at 1800 rpm, with both the idler and driver gears having 16 teeth. Since each gear meshes once per revolution, the gear meshing frequency will be:

$$f_1 = \frac{1800}{60} \times 16 = 480\,\text{Hz}.$$

This will be the fundamental frequency, with harmonics at:

$$f_n = \frac{nkN}{60}, \quad n = 1, 2, 3, \ldots \qquad (18.16)$$

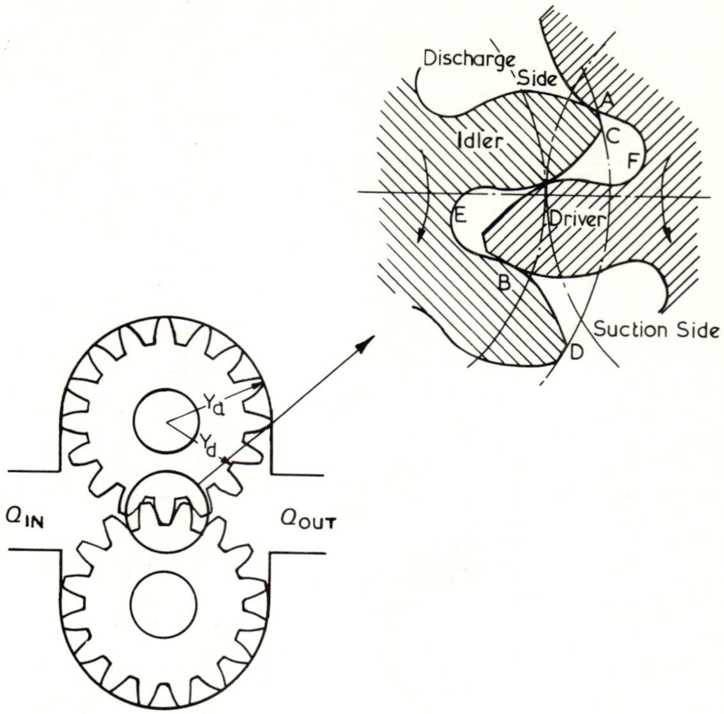

Fig. 18.17 – Simple gear pump.

where n = harmonic number
k = number of gear teeth
N = shaft rotational speed.

A typical spectral distribution is shown in Fig. 18.18, showing that the magnitude of the harmonics generally decreases as n increases. It should be noted however that higher order tones in the 1–4 kHz range may be much more trouble than a low fundamental as far as the noise regulations are concerned, because of the A scale weighting.

In addition to the discrete tones generated by gear meshing, sideband tones can also develop. These frequencies are generally due to periodic fluctuating loads associated with lack of geometric symmetry, and are found on either side of the gear meshing tones.

In Fig. 18.17, the fundamental gear mesh tones can be represented by:

$$a(t) = \hat{a} \sin \omega t \tag{18.17}$$

and from equation (18.16) with $n = 1$:

$$a(t) = \hat{a} \sin 2\pi \frac{N}{60} kt \tag{18.18}$$

where $\hat{a}$ = amplitude of the noise signal.

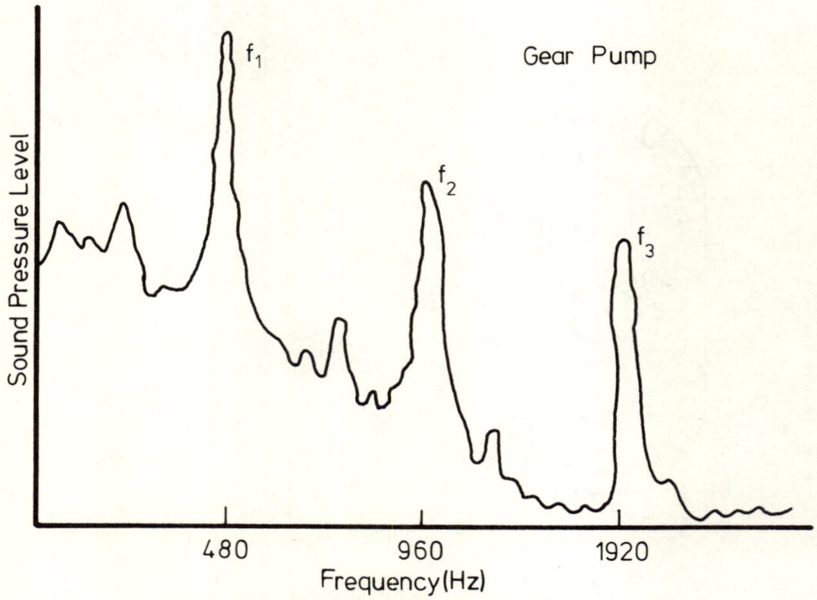

Fig. 18.18 – Narrow band spectral analysis of a gear pump.

If either gear in the pump is eccentric the amplitude of the gear meshing noise will not be constant but could, in the simple case, also be harmonic. Hence equation (18.18) becomes:

$$a(t) = \hat{a}_1 \sin \frac{2\pi Nt}{60} \sin \frac{2\pi Nkt}{60} . \tag{18.19}$$

Using an identity expansion, equation (18.19) becomes:

$$a(t) = \frac{1}{2}\left\{\hat{a}_1 \cos \frac{2\pi}{60}(kN+N)t + \hat{a}_1 \cos \frac{2\pi}{60}(kN-N)t\right\} . \tag{18.20}$$

This represents a pair of discrete frequencies either side of the gear mesh frequency, displaced at ± drive shaft rotational speed.

18.7 PIPE NETWORKS

The connecting network between components is the piping system, which also acts as a pathway for acoustical energy. The character of pressure pulsations in the pipe can either be random or pure tone. Random noise is associated with cavitation and with turbulence noise fluid flow through the pipes. Pure tones of any sizeable amplitude are usually due to hydraulic resonance. Apart from the potential damage of large amplitude pure tones, these are also very disturbing to personnel in the vicinity.

The condition of hydraulic resonance, a result of pressure pulse reinforcement, is caused by the system reflecting pulses repeatedly within the pipe. An example of this was discussed in Chapter 10. When resonance occurs the first reflected pulse arrives at the pump and adds to the pump pulse, yielding a new pulse of increased amplitude. This pulse, when reflected, causes another increase in amplitude at the pump, and so on. Since the pump pulsations are produced at the rate of several hundred per second, this reinforcement process very rapidly builds up large pressure pulsation amplitudes. Fortunately, this process is self-limiting because of friction losses and the containing pipe wall flexure. However, it is possible that the pressure levels can be increased up to fifty times or more by resonance.

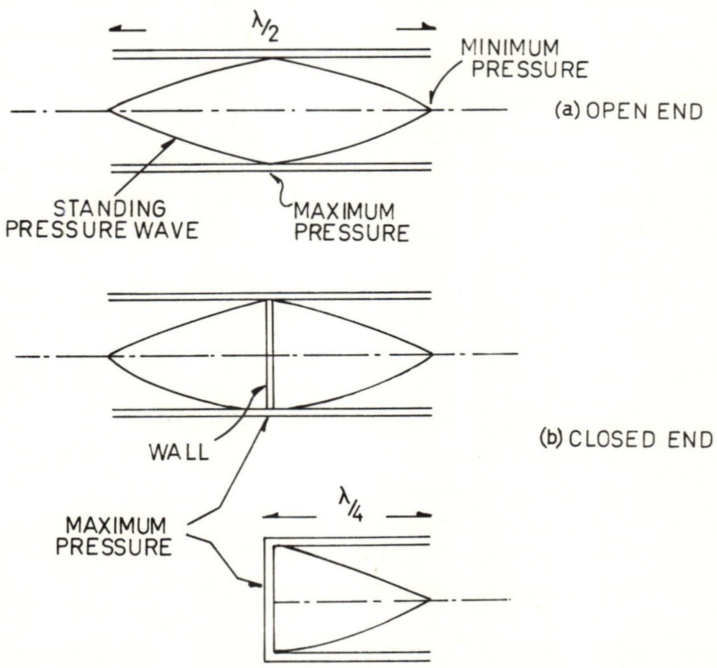

Fig. 18.19 – Pipe resonance.

Fig. 18.19 shows the pressure profile of a standing wave in an open-ended tube. Note that the standing wave amplitude is maximum at the centre of the pipe and minimum at each end, since the open end cannot support a pressure difference. Fig. 18.19(b) shows how closed end pipe resonance is developed; the detailed mathematics for this case were discussed in Chapter 10. By putting a wall at the position of maximum pressure, the standing wave is unaltered even though the wall is present. Since maximum pressure already exists at the centre of the pipe the wall has no effect.

The open pipe contains a standing wave $\lambda/2$ long, which is the shortest fraction of wavelengths possible in a tube with both ends open. In the closed end tube $\lambda/4$ is the shortest possible resonance condition. As a general rule:

(a) Pipes with both ends open or closed resonant with an even number of quarter waves contained within the pipe.
(b) Pipes with one end open and the other end closed resonate with an odd number of quarter wavelengths within the pipe.

Vibrational response of piping walls will also be capable of radiating an acoustic field. This may also be complicated by the mechanical vibrational modes of the pipework. The excitation is caused by fluid/structure interaction. The fluid energy, flow noise, is generated by pressure fluctuations associated with unsteady or oscillating flow fields, especially when generated through an upstream valve or orifice. Actually, only a small fraction of the energy associated with these pressure fluctuations in the turbulent flow field are directly radiated away from the boundary layer as true sound [9, 10].

It was shown in equation (18.13) that, ideally, pressure in an acoustic wave is proportional to the particle velcoity. Following the arguments given by Fagerlund [9] the acoustic pressure at a point in space, r distance from the centre line of the pipe, is:

$$p_{\text{rms}}^2 = (\rho c v)^2 \frac{d}{2r} \xi \tag{18.21}$$

for free field conditions, where

p_{rms} = root mean square of acoustic pressure
ρ = density of air
c = velocity of sound in air
v = surface velocity of pipe wall
d = mean diameter if pipe
ξ = acoustic radiation efficiency.

The **radiation efficiency** is defined as the ratio of acoustic power at the source, to the actual acoustic power radiated. It has been shown that the efficiency with which a surface will radiate sound can be determined as a function of frequency up to the coincidence where it becomes unity. A **coincident frequency** f_c is defined as the frequency at which the propagation velocity of a flexural wave in the pipe wall is equal to the velocity of sound in the acoustic medium. These frequencies are shown in Table 18.3 for common hydraulic piping determined from:

$$f = \frac{3c^2}{\pi t c_s} = \frac{500}{t}\,\text{Hz} \tag{18.22}$$

where t = pipe wall thickness
c_f = velocity of sound in the fluid (1368 m/s)
c_s = velocity of sound in steel (4712 m/s)
c = velocity of sound in air (343 m/s).

There are two other characteristic frequencies. Firstly the **acoustic cut-off**, f_c, which is the lower limiting frequency corresponding to the first shell mode. Below the acoustic cut-off frequency, energy propagates as a plane wave. At low frequencies the shell modes are not present and the pipe can vibrate as a beam with quite high amplitudes:

$$f_c = \frac{0.586}{d} c_f\,\text{Hz}\,. \tag{18.23}$$

Secondly, the **ring frequency** f_r, is the frequency at which the longitudinal wavelength in the pipe material equals the pipe circumference:

$$f_r = \frac{c_s}{\pi d}\,. \tag{18.24}$$

Note that the ratio:

$$\left[\frac{f_r}{f_c}\right]_{\text{air}} = 0.54\frac{c_s}{c_f} = 7.48$$

$$\left[\frac{f_r}{f_c}\right]_{\text{oil}} = 1.86\,.$$

Some results for standard hydraulic pipe are tabulated in Table 18.3. As a result of this behaviour, equation (18.21) can be rewritten as:

$$p^2 = (\rho c v)^2 \frac{d}{2r}\frac{f}{f_c} \quad \text{for } f < f_c$$

and

$$p^2 = (\rho c v)^2 \frac{d}{2r} \qquad \text{for } f > f_c\,. \tag{18.25}$$

Since velocity amplitude is acceleration amplitude divided by frequency it can be shown from equation (18.25) that:

$$f < f_c \qquad L_p = 10\log\frac{a^2}{10^{-6}} + 10\log\frac{d}{2r} - 10\log f f_c + 30.4 \tag{18.26(a)}$$

$$f > f_c \qquad L_p = 10\log\frac{a^2}{10^{-6}} + 10\log\frac{d}{2r} - 10\log f + 30.4\,. \tag{18.26(b)}$$

The propagation of energy through the circuit pipework can be appreciated from Fig. 18.20. The lower curve is a spectral distribution of acceleration levels at the pump end plate. The upper curve is the spectral distribution of energy at the load valve some distance along the pipework from the end plate, (see Fig. 18.7(b) for measurement points). This plot contains the vibrational level from the pump, lifted by the flow noise generated by the downstream valve. The flow is also exciting one of the modes of the pump end plate at 9–10 kHz. The analysis of a shock test on the end plate, using a real time analyser, shows the existence of the mode in Fig. 18.21.

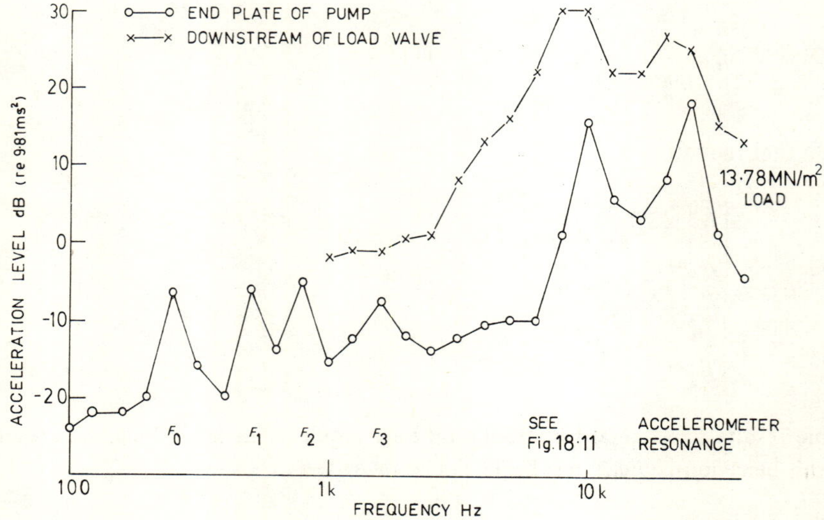

Fig. 18.20 – Noise propagation in circuit pipework.

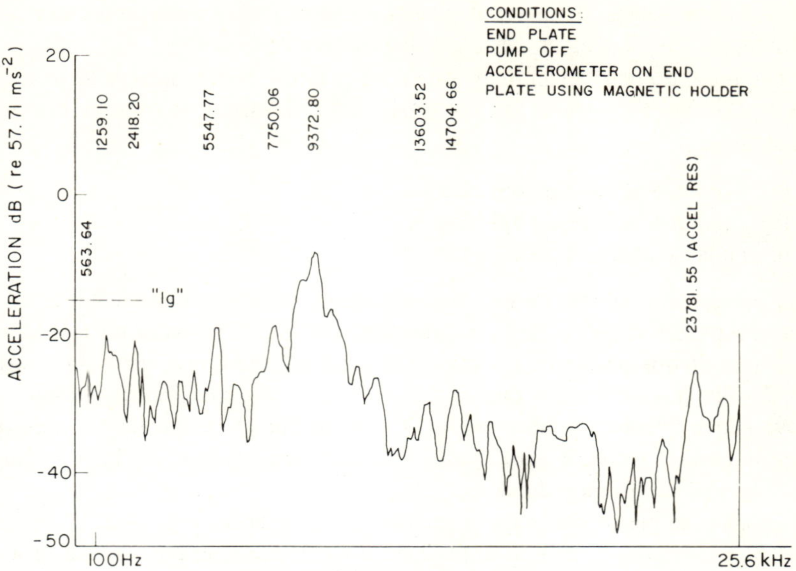

Fig. 18.21 – Shock test on pump end plate.

18.8 IDENTIFICATION OF CAVITATION NOISE

Cavitation will occur in any component with liquid flowing through it, a pipe, orifice, relief valve, hydraulic motor, propeller, if the local pressure field falls below the vapour pressure of the liquid. It is well established (see Chapter 2)

Table 18.3 – Hydraulic piping, Schedule 40.

Pipe NB		O/D (mm)	t(mm)	f_c	f_r
in	mm				
$\frac{1}{8}$	3	10.405	3.4	7352	148 261
$\frac{1}{4}$	6	13.7	4.5	5682	109 695
$\frac{3}{8}$	10	17.1	4.6	5494	87 756
$\frac{1}{2}$	13	21.3	5.5	4587	70 518
$\frac{3}{4}$	20	26.7	5.8	4425	56 414
1	25	33.4	6.8	3759	45 046
$1\frac{1}{4}$	32	42.2	7.1	3571	35 684
$1\frac{1}{2}$	38	48.3	7.4	3448	31 176
2	50	60.3	7.8	3246	24 941

From ASTM A53-68.

that cavitation will erode material, generate a high level of noise, and create very high temperatures locally in the liquid due to cavity collapse.

The rapid collapse of bubbles generates pressure waves of very large amplitude, but of short duration. Studies show that noise from this source has three possible modes:

(a) dissolved air coming out of solution,
(b) resonance of the gas bubbles,
(c) collapse of the vapour cavities.

These correspond to the various stages a liquid experiences when subjected to tension. As the bubble or cavity moves downstream from the local low pressure site, it is subsequently under the influence of an increased pressure field, causing it to shrink. The work done during this process is concentrated on a volume of vapour which becomes smaller so that at the instant of final collapse, relatively large energies are stored in very small volumes. This generates very high pressures in the form of a hydraulic shock (see Chapter 2).

As an example of the marked increase in noise level, Fig. 18.22 shows the effect of a cavitating orifice in water in a 25.4 mm diameter pipe. These data are based on the results presented by Rogers [11].

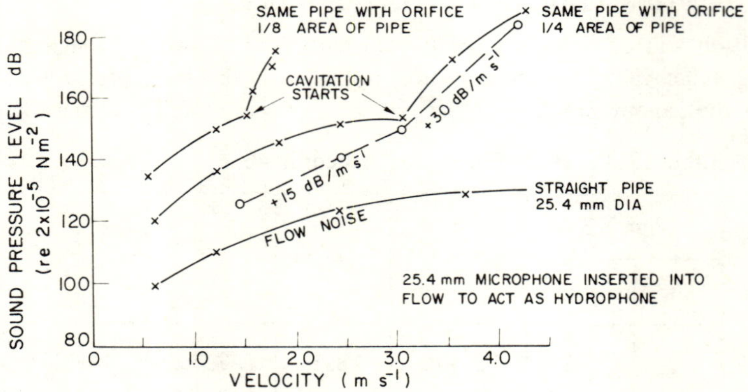

Fig. 18.22 – Sound level increase due to cavitation (ref. [11]).

In the straight pipe there is an increase of flow noise with water velocity at a rate of around 15 dBs/m. Orifices of area 126 and 63 mm^2 were introduced into the pipe and it was found that up to 3.0 m/s and 1.5 m/s respectively the noise level increased with flow, but when cavitation occurred at the orifice there was a significant doubling of the noise level rate of increase to 30 dBs/m.

The noise level recorded seemed, on first glance, to be very high, but it should be noted that these measurements were taken with the 25.4 mm microphone in contact with the liquid, so acting as a hydrophone. Airborne levels would be greatly reduced because of attenuation through the pipe wall and the atmosphere.

The spectral distributions for the same experiment are shown in Fig. 18.23. The straight pipe carrying flow at velocity of 3.0 m/s without the orifice shows a fall off at 9 dB/oct above 400 Hz. Referring to Fig. 18.22, for 3.0 m/s the L_p is about 126 dB, indicating that the higher energy levels in Fig. 18.23 are at the lower frequencies. This characteristic is quite different when a cavitating orifice is in the pipe circuit. Here it is seen that a significant rate of increase of energy is associated with the higher frequencies. Fig. 18.22 shows that at 3.35 m/s the average L_p is 165 dB which is consistent with the peak of 161 dB at 1000 to 1500 Hz in Fig. 18.23. Hence cavitation occurring in machinery has a distinct effect. Experimental evidence shows that the potential noise increase is 10-15 dB over the frequency range 1-20 kHz.

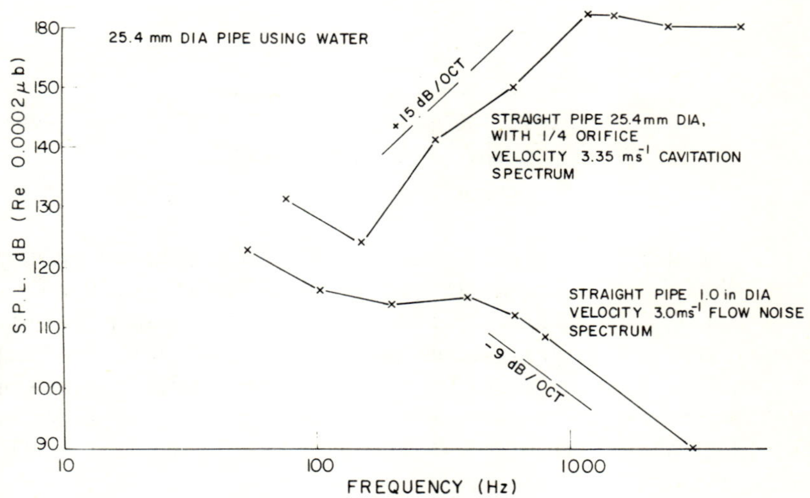

Fig. 18.23 – Spectral distribution change due to the onset of cavitation.

18.9 PNEUMATIC CIRCUITS

By far the most significant source of noise radiation is the discharge of air from pneumatic exhaust ports. The sound pressure level near a typical 6.2 mm diameter shop air nozzle can be as high as 107 dB.

Fig. 18.24 shows the structure of the typical high velocity air jet. It has been shown earlier (Chapter 4) that the volumetric flow rate is:

$$Q = \frac{C_q A}{\rho} \frac{Pu}{\sqrt{T_u}} C_m \; . \tag{18.27}$$

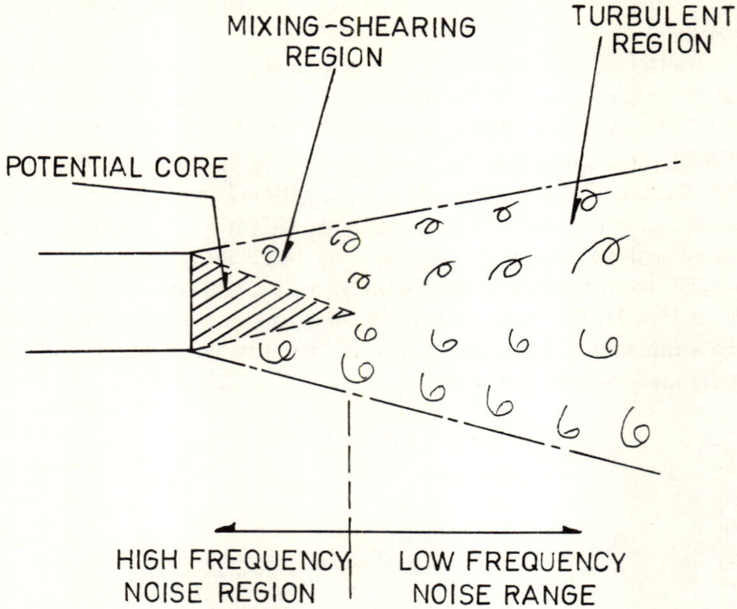

Fig. 18.24 – Jet structure.

It was seen that for valves of *pd*/*pu* less than 0.528, the flow through the nozzle becomes choked, that is, reaches the speed of sound. Further reduction of this pressure ratio does not significantly increase the velocity. Since shop air pressure is typically around 7 bar (g) and the ambient pressure is 1.013 bar (a) the choked flow condition is very common. The noise from these sources is broadband, biased towards the high frequency range. A typical octave band analysis is shown in Fig. 18.25, where it can be seen that the noise level peaks around 100 dB in the range of 2 kHz.

The creation of the noise is due to fluctuations of pressure resulting from turbulence and shearing as the high velocity air interacts with the still surrounding atmosphere. High frequencies are generated in the mixing shearing region and lower frequencies downstream in the turbulent region. The work of Lighthill [12] indicates that the overall sound power (W) from a jet is given by:

$$W = \frac{K\rho a v^8}{c^5} \text{ watts} \qquad (18.28)$$

where a = area of jet nozzle (m^2)
ρ = density of air (kg/m^3)
v = average jet velocity (m/s)
c = speed of sound (m/s)
K = constant (3×10^{-5}).

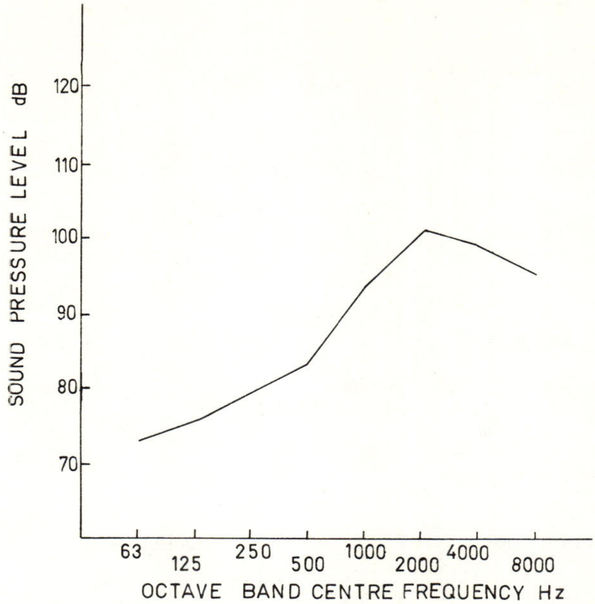

Fig. 18.25 – Octave band analysis of noise from 7 mm air nozzle.

It can also be shown that the total radiated acoustical power is proportional to the kinetic energy of the jet stream. Hence the total radiated power is a fraction of the total power discharged by the nozzle. This expression is a good first-order approximation when the average velocity is in the range $0.15c < v < c$.

The peak frequency of the broadband noise can be estimated from the equation:

$$f_{\text{peak}} = \frac{Sv}{D} \tag{18.29}$$

where S = Strouhal number = 0.2
v = exit velocity (m/s)
D = nozzle diameter (m).

Example

The exit port from a pneumatic drill is 9.2 mm, with an exit air velocity of 343 m/s. Calculate:

(a) frequency of the peak noise level,
(b) the acoustical power level.

From equation (18.29):

$$f_{\text{peak}} = \frac{0.2 \times 343}{9.2 \times 10^{-3}} = 7456 \text{ Hz} .$$

From equation (18.28):

$$W = \frac{K\rho a v^8}{c^5}$$

Let

$$v = 343 \text{ m/s}$$

$$a = \frac{\pi \times (9.2 \times 10^{-3})^2}{4} = 6.64 \times 10^{-5} \text{ m}^2$$

$$\rho = 1.187 \text{ kg/m}^3$$

$$c = 343 \text{ m/s}$$

$$W = 3 \times 10^{-5} \times 1.187 \times 6.64 \times 10^{-5} \times (343)^3 = 0.071 \text{ W}.$$

The sound power level (re 10^{-12}) L_W is:

$$L_W = 10 \log \frac{0.071}{10^{-12}} = 108.5 \text{ dB} .$$

Using equation (18.15), the sound pressure level at 2.0 m from the exhaust nozzle will be:

$$L_p = 108.5 - 20 \log 2.0 - 8.0$$

$$= 24.5 \text{ dB}.$$

18.10 CONTROLLING NOISE AT THE DESIGN STAGE

Eliminating all sound is unnecessary as well as impossible, but controlling that portion which becomes noise is important. Since sound is a form of mechanical energy, originating at a source, travelling via a path to a receiver, there are three places at which noise may be controlled:

1. At the source – by initial design awareness.
2. Along the path – by modifying or eliminating the acoustic transmission path.
3. At the receiver – by shielding.

It has been suggested that noise sources can be classified into five types, namely airflow, friction, impact, magnetic and mechanical, forced vibrations.

The ways in which energy can be propagated through products are shown in Fig. 18.26.

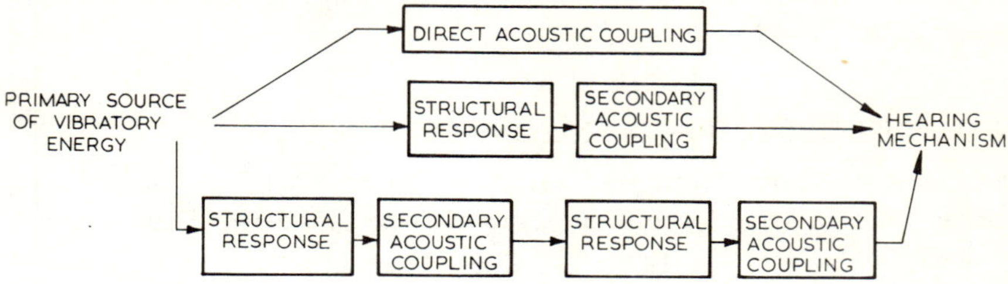

Fig. 18.26 – Flow chart of the basic nature of noise generation.

While it is dangerous to generalise, experience has shown some useful guidelines relating the vibration of sound power level L_p with a particular operating parameter. These are listed below:

Fans:	$10 \log_{10}$ (increase in power)
	$40 \log_{10}$ (increase in power)
Air flow:	α(velocity)^8 at > Mach 1
	α(velocity)^6 at < Mach 1
	α(velocity)^3 at < Mach 1
Liquid flow:	$60 \log_{10}$ (velocity increase without cavitation)
	$120 \log_{10}$ (velocity increase with cavitation).

Hence low pressures and low velocities are desirable from noise elimination considerations.

Lower speed fans, moving air at lower velocity, may require the use of a little more sheet metal and a slightly larger fan, but large gains in simplification of noise and vibration problems result.

Impact processes are particularly serious noise sources, in which their noise output varies directly with the rate with which the energy is used in the impacts. Squeeze processes are usually much less noisy.

Noise in fluid flow results principally from vortex shedding during turbulent flow, which is often indicative of inefficient flow, and power wastage. Smooth hydrodynamic design of all elements of an hydraulic system, for example, will significantly reduce noise and vibration as well as power losses. Where practicable, using lower tip speeds for fans and pumps, lower flow velocities and pressures for fluid flow, and avoiding cavitation and starved inlet conditions will significantly reduce noise and vibration.

Excessive throttling of fluid flow, quick closing valves, and other poor design features of systems account for the 'scream' of valves and the often dangerous 'water hammer' in pipelines.

It is important to note that proper lubrication, adjustment, dynamic balance, and maintenance of equipment are effective noise control measures.

The surfaces and panels attached to a source, the structure on which it is supported, and the earth on which the entire assembly rests, the pipes, ducts, cables, and belts attached to it, and even the fluids surrounding it act as transmission paths.

18.11 NOISE REDUCTION IN EXISTING EQUIPMENT

In cases where further noise level reduction is necessary in order to protect a human operator, some form of enclosure must be considered. There are three basic ways of tackling this problem, by using:

1. sound absorption materials,
2. barrier materials,
3. damping materials.

In fact, modern enclosure construction may use combinations of the above methods [14].

Sound absorption materials effectively provide a resistance to the fluctuating air movement and so convert part of the acoustic energy into heat. Fibres or foam in bulk are generally used and may be grouped either as unbounded fibres, blankets, or as boards. In the blanket form, the binder partially cements lightly packed fibres together to produce a soft pliable material ideal for wrapping pipes and ducts. In the board form, the binder rigidly cements larger and more densely packed fibres together to produce a material which is quite stiff.

A barrier works on the inertia principle, that if a force is applied to a solid material (that is, acoustic noise pressure) then the greater the mass of the material the less likely it is to be accelerated. Clearly therefore, lead makes an ideal but expensive, barrier material.

Damping materials are usually viscoelastic compound applied to a board, and their behaviour is not fully understood; however, an excellent treatment is given in [15]. All these basic techniques are frequency dependent, and it is necessary to select commercially available materials to suit the situation.

The design of suitable enclosures is a topic in its own right, since the complex dynamic behaviour of the structure can make it effective at controlling noise in one frequency range while acting as an acoustic window at other frequencies. Another important practical point is that unless the enclosure is completely airtight, the effectiveness is controlling noise will be greatly reduced.

18.12 REFERENCES

[1] Diehl, G. M. (1973), *Machinery Acoustics*, Wiley.
[2] Morse, P. M. and Ingard, K. V. (1966), *Theoretical Acoustics*, McGraw-Hill.
[3] Beauchamp, K. G. (1973), *Signal Processing*, Allen and Unwin.

[4] Bendat, J. S. and Piersal, A. G. (1971 ed.), *Measurement and Analysis of Random Data*, Wiley.

[5] Beranek, L. L. (1971), *Noise and Vibration Control*, McGraw-Hill.

[6] Brüel and Kjaer (1973), *Acoustic Noise Measurements Application Manual*, Brüel and Kjaer.

[7] General Radio (1972), *Handbook of Noise Measurement*, General Radio.

[8] Helgestead, B., Foster, K., and Bannister, F. (1974), Pressure Transients in an Axial Piston Hydraulic Pump, *Proc. Inst. Mech. Engrs.*, **188**.

[9] Fagerlund, A. C. (1974), Use of Pipe Wall Vibrations to Measure Valve Noise. *Proc. ISA Conf. New York*, Paper 74-829;

[10] Roth, B. and Martin, H. R. (1975), Some Spectral Characteristics of Flow Noise in Pipes, *B.H.R.A. Fluid Power Conf.*, Sheffield.

[11] Rogers, W. L. (1954), Sound Pressure Levels and Frequencies Produced by Flow of Water Through Pipes and Fittings, *Trans. A.S.H.V.E.*, **60**.

[12] Lighthill, M. J. (1954), On Sound Generated Aerodynamically, *Proc. Royal Soc.*, **222A**.

[13] Taylor, R. (1970), *Noise*, Penguin Books.

[14] Beranek, L. L. (1960), *Noise Reduction*, McGraw-Hill.

[15] Faulkner, L. (1976), *Handbook of Industrial Noise Control*, Industrail Press, New York.

Appendix
Graphical symbols for fluid power components

The European Oil Hydraulic and Pneumatic Committee (CETOP) published a proposed set of symbols in 1964. These were accepted by the International Organisation for Standardisation (ISO) and, after some revision, they have been published as ISO 1219:1976 – *Fluid power systems and components – graphical symbols.* The British Standards Institution have recognised these symbols and have published them in BS 2917:1977 – *Specification for graphical symbols used on diagrams for fluid power systems and components.* The following pages list a selection of approved symbols which have been employed in the text. All these symbols have also been introduced at the appropriate place in the book.

In the United States the various manufacturing groups comprising the Joint Industry Conference (JIC) published a set of symbols for fluid components in 1948. These have been developed and amended by the American Standards Association (ASA). The American National Standards Institute continues to develop the standards for the fluid power industry (see *Standard graphic symbols for fluid power diagrams* USA SY32-10-1967, published by the American Society of Mechanical Engineers).

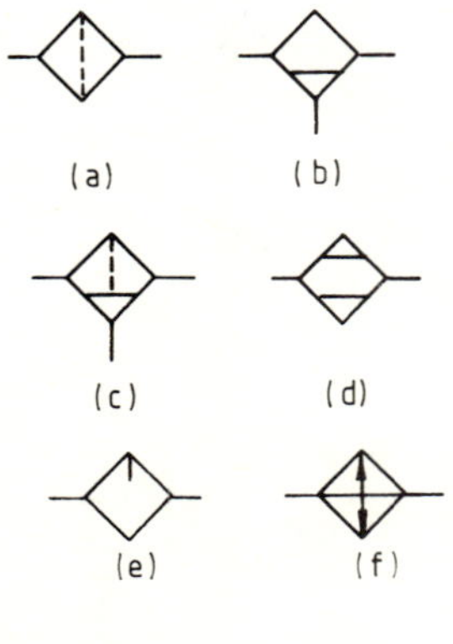

(a) Filter or strainer.
(b) Manually controlled water trap.

(c) Manually controlled filter with water trap.
(d) Air dryer.

(e) Lubricator.
(f) Cooler.

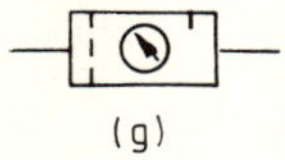

(g) Simplified symbol for conditioning unit consisting of filter, pressure regulator, pressure gauge and lubricator.

Directional Control Valves

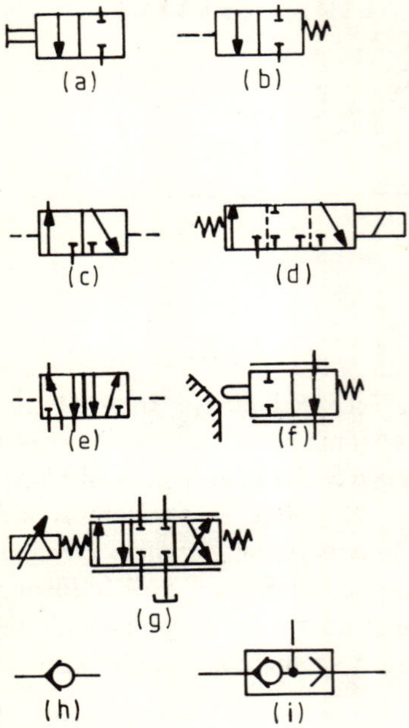

(a) Manually controlled 2/2 directional control valve.
(b) Pilot pressure operated, spring return, normally closed, 2/2 directional control valve.
(c) Double pressure operated 3/2 directional control valve.
(d) Solenoid operated, spring return 3/2 valve, with intermediate condition.
(e) Double pilot controlled 5/2 directional control valve.
(f) Throttling directional control valve.
(g) Single stage electro-hydraulic servo valve.
(h) Non-return valve.
(i) Shuttle valve.

Control Methods

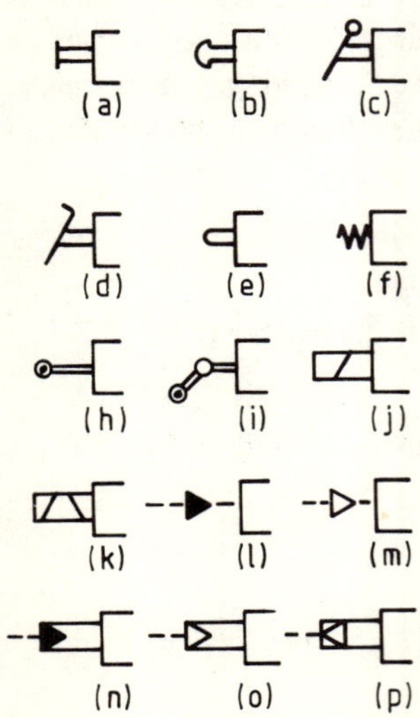

(a) General symbol idicating muscular control.
(b) Push button.
(c) Lever.
(d) Pedal.
(e) Plunger.
(f) Spring.
(h) Roller.
(i) One way roller.
(j) Solenoid (one winding).
(k) Solenoid (two windings).
(l) Oil pressure.
(m) Air pressure.
(n) Pilot oil pressure.
(o) Pilot air pressure.
(p) Release of pilot pressure.

Pressure Control

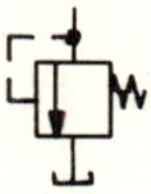
(a)

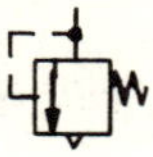
(b)

(a) Hydraulic relief valve.
(b) Pneumatic relief valve (safety valve).

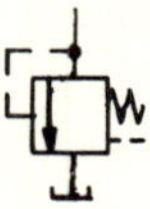
(c)

(c) Relief valve with remote pilot control.

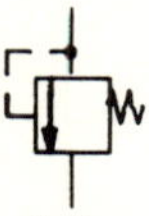
(d)

(d) Sequence valve.

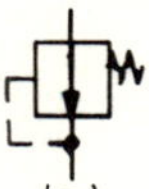
(e)

(e) Pressure regulator or reducing valve.

Flow Control

(a)

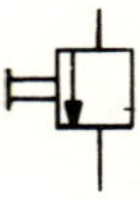
(b)

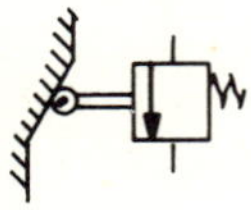
(c)

(a) Simplified symbol for throttle valve.
(b) Manually controlled throttle valve.
(c) Mechanically controlled throttle valve with return spring.

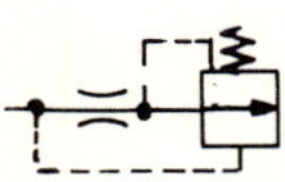
(d)

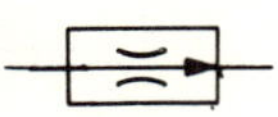
(e)

(d) Flow control valve with fixed output.
(e) Simplified symbol for (d).

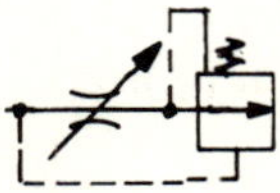
(f)

(g)

(f) Flow control valve with variable output.
(g) Simplified symbol for (f).

(h)

(h) Shut-off valve.

Cylinders

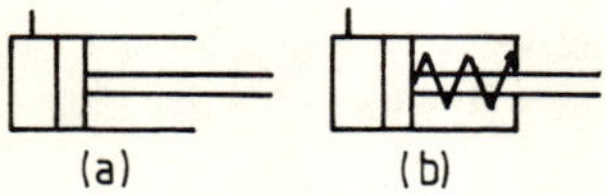

(a) Single acting cylinder.
(b) Single acting cylinder with spring return.

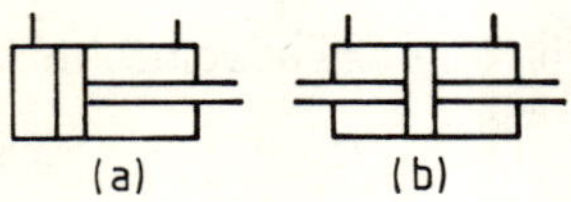

(a) Double acting cylinder with single piston rod.
(b) Double acting cylinder with double ended piston rod.

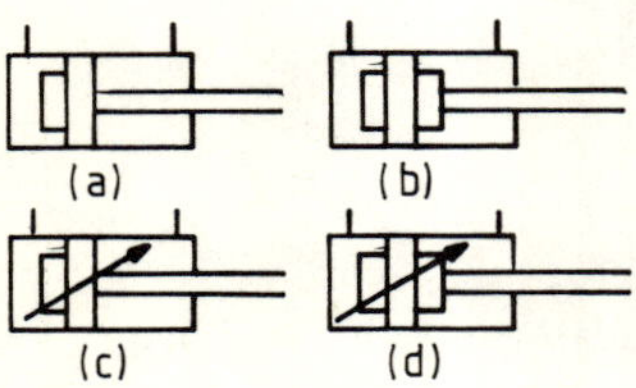

Cylinder with:
(a) Single fixed cushion.
(b) Double fixed cushion.
(c) Single adjustable cushion.
(d) Double adjustable cushion.

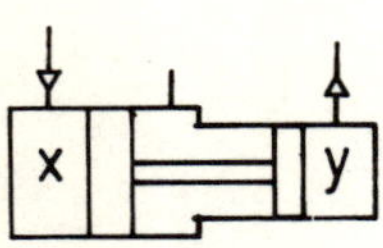

Pressure intensifier.

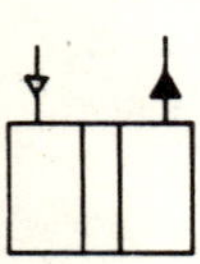

Air-oil actuator.

Energy Converters

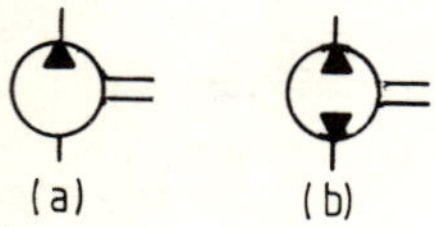

(a) Fixed capacity hydraulic pump.
(b) Fixed capacity hydraulic pump with two directions of flow.

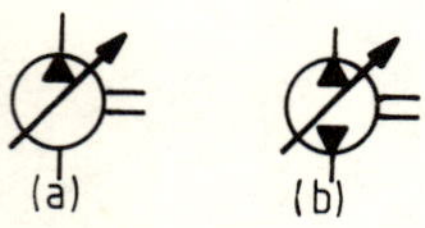

(a) Variable capacity hydraulic pump.
(b) Variable capacity hydraulic pump with two directions of flow.

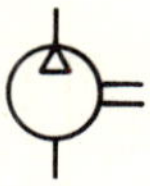

Fixed capacity compressor.

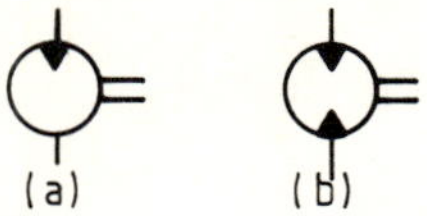

(a) Fixed capacity hydraulic motor.
(b) Fixed capacity hydraulic motor with two directions of flow.

(a) Variable capacity hydraulic motor.
(b) Variable capacity hydraulic motor with two directions of flow.

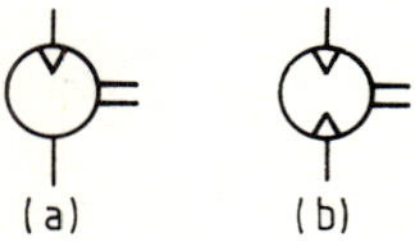

(a) Fixed capacity pneumatic motor.
(b) Fixed capacity pneumatic motor with two directions of flow.

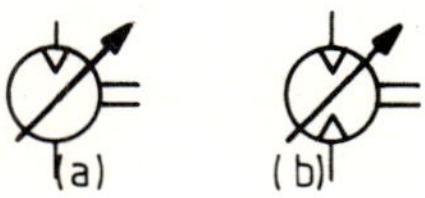

(a) Variable capacity pneumatic motor.
(b) Variable capacity pneumatic motor with two directions of flow.

Index

A

acceleration feedback 334, 348
accumulator 57, 223
acoustics 460
actuator 68
addition, pneumatic 129
additives 27
advantages of
  fluidics 17
  hydraulics 12
  pneumatics 14
aeration 36
air
  consumption 53, 58, 80, 370
  entrainment 39, 40
  in solution 36
amplifier
  bistable 16
  proportional 15
  power 124
  servo 299, 302, 334
  turbulence 385, 389
analogies 229, 234, 240
AND logic 382, 393
angles ports 185
area-displacement gain 81
automatic control 19, 23
axial forces in valves 111, 181, 299, 315

B

back pressure sensor 110, 118, 123
balanced valve 183, 189, 192
ball valve 77, 83, 385
bandwidth 277
beat frequency detector 130
Bernoulli force 182
binary arithmetic 381
binary code 116
bistable amplifier 16, 386
bleed-off speed control 221, 367
Boolean algebra 381
boost pump 149
bubbles, entrained 39, 57
bubbly oil 37
bulk modulus 31
button vortex amplifier 142

C

cam shaft programmer 408
capacitance, fluid 236
capacitive transducer 117
cascade method 422
cavitation 29, 41
  boundaries 291
  in orifices 79, 96
  noise 485
  number 98
C.E.T.O.P. symbols 495
chamber geometry 96
characteristic impedance 250
characteristics, valve 159, 267
check valve 144
choked flow 81, 161
closed centre valve 136, 138
closed loop control 19, 261, 299
Coanda effect 16
coded circuits 400
compensation of flow forces 181
compressibility 27, 32
compressible flow 80
compressors 58
cone sensor 115

conical poppet valve 77, 82, 90, 100
contamination 43
contraction coefficient 78
controllers 126
control
  of position 21, 261, 299, 337
  of pressure 147, 359
  of speed 144, 365
  on-off 442, 449
  optimal 449
  valves 133
coolers 55, 61
Coulomb friction 125, 162, 323
counter, binary 116, 400, 411
cracking pressure 147
critical pressure ratio 80
cushioning 71
cylinders
  double acting 70
  single acting 69

D

damping length 190
dead zone 323
deadband 323
decoding 402, 411
delivery of compressors 58
  of pumps 66
De Morgan's theorem 388
density 29
diameter, hydraulic 92
digital
  actuator 402, 440
  control, advantages 435
  control systems 435
  to analogue conversion 435
directional control valves 133
dirt 43
discharge coefficients 78, 86, 92
discontinuous control systems 435
distributed parameters 237, 245
dither 326
double-acting cylinders 70
dynamic
  flow in orifices 100
  stiffness 276
  viscosity 30

E

ear, its response 462
efficiency
  circuit 217, 220, 223
  mechanical 68
  overall 68
  volumetric 67
efflux angle of jet 87, 183, 189
electric field valve 46, 152
electrohydraulic
  servo 299
  valve 11, 124, 303
electrosensitive fluids 46
encoder
  binary coded 116
  Gray coded 119
  incremental 116
entrained air 39
entrainment 39, 40
erosion 42
equivalent circuits 238, 244
exclusive OR gate 397

F

feedback 19
  acceleration 334, 348
  position 262, 303
  pressure 333
  velocity 334
filters 43, 53, 57
five-port valves 140, 399
fixed displacement pumps 62
flapper-nozzle valves 77, 91, 101, 113, 116, 139, 173, 194
flip-flop 386
flow
  coefficient 94
  control 144, 365
  control valves 133
  forces 181, 315
  gain 169
  gain, fluidic 16, 141
  number 95
  orifice 79
  pressure coefficient 169
fluerics 15, 140, 381
flueric valves 140
fluid
  capacitance 236
  compressibility 32, 265
  impedance 236
  inductance 235
  power 12, 217, 219

properties 26
transmission line 203, 229
fluidics 15, 140, 381
foaming 28, 40
forces, axial in valves 11, 181, 299, 315
four-port valve 140
free air delivery 58
frequency response 281, 286, 302, 315

G

gain
flow 169
margin 272
gas
constant 80
flow 80, 162
gaseous cavitation 40
gear pump 62

H

Heat dissipation 56, 61, 217
heating, due to aeration 38
high temperature systems 45, 353
hot gas servos 353
humidity 53
hydraulic
diameter 92
fluids 27
motors 68
natural frequency 269
power 12, 217
pumps 62
servo 21, 261

I

impedance 236
inductance, fluid 235
inductive transducer 116
inertia, effective 269
interlocking 414
interruptible jet sensor 114
invertors 381

J

jet
deflection amplifier 15
geometry 84
pipe valve 139, 174
pump 175
tube devices 84, 126

K

Karnaugh maps 412
Kearfott ball element 386
Kinematic viscosity 30

L

laminar flow 31
in orifices 92, 102
in pipes 206
lap, valve 138, 169, 285, 291, 294, 325
leakage 44, 59, 313, 333
limited period signals 425
linearisation
of servo system 266, 342
of valve characteristics 164, 267
load disturbances 275, 346
locked on signals 416
logic
functions 381
rules 387
loss coefficients 205
losses 205, 207
lubricators 55
lubricity 28
lumped parameters 236, 244
L.V.D.T. 116

M

Martonair cascade method 422
mass flow
parameter 80
rate 80, 162
mechanical efficiency 68
meter
-in speed control 365, 369
-out speed control 146, 365, 371, 376
mineral oils 27
modulation
differential-pulse-width 445
pulse-width 443
modulo *n* logic 397
multipliers 126

N

natural frequency 269, 345
calculations 279
needle valve 18, 77, 145
negative force port 187
noise 459
cavitation 485
in pumps 472

in valves 469
reduction 490
non-return valve 144, 392, 394
NOT logic 381
nucleation centre 42

O

oil
column stiffness 268, 279
properties 27
on-off control 22, 442, 449
open
centre valve 284, 294
loop system 19, 308
optimal control 449
optimum response 271
OR-NOR gates 384, 385
orifice
cavitation 79, 96
flow 78, 80
types 75
oscillators 131, 446
overlap 138, 284

P

pertubation, small 164, 266, 342
phosphate esters 27
pneumatic servo, analysis 335
poppet valve 19, 77, 82, 90, 98, 105
ports, circular 81, 94, 284
posicast control 448
positive displacement
compressors 60
pumps 62
position transducers 115
positions, valve 137
potentiometers 115
power, hydraulic 12, 217
pressure
compensated flow control valves 146, 357
compensated pump 65
control 147, 359
control valves 147
feedback 333
peaks 204, 295, 320
recovery 174
reducing valve 149
regulator 151
relief valve 147
propagation constant 250
properties of fluid 27
proportional
amplifier 15
control valves 21, 134
proximity detection 112
pulse
-controls 438
-width modulation 443
pumps, hydraulic 62

Q

quick exhaust valve 147

R

radial hole orifices 184
rate measurement 120
reattachment 16, 96, 105
reducing valve 150
regulators 151
relief valves 147, 359
reservoirs 40, 55
resistance, fluid 235
response
time optimal 271, 449
hydraulic speed control 376
pneumatic speed control 367
pressure control valve 361
relief valve 361
servovalve 303
torque motor 299
reverse flow 308, 455
Reynolds number 92
roots blower 60
ring counter 403

S

safety 14, 396, 414
selector valve 416
sensitivity to load disturbances 275, 346
sensors
force 122
position 115
proximity 112
velocity 120
sequence circuits
counter controlled 410
electric 416
step completion based 414
time based 408
variable programmed 429
sequence valve 150, 414

servovalves 21, 111, 124, 139, 143, 152
short tube orifices 103
shuttle valves 144, 384, 394
signal analysis 465
single-acting cylinder 69
single-stage valve 111
small signal response 322
solubility 36
speed
control 144, 365, 369, 376
measurement 120
spool
shaping 187
valves 75, 133, 160
spring logic element 382
square root calculator 128
stabilisation 332, 347
stanford ball element 385
static pressure balance 183, 192
stepping motor 439, 440

T

tandem centre 139
Taylor's expansion 165
temperature, effect on viscosity 30
three-port valves 135, 139
threshold 322
time optimal control 448
torque motor 111, 299
transducers 115
transient pressure feedback 333
transmission line 245, 253
trip-operated valve 426
truth table 383
turbulence amplifier 382
turn-down ratio 141
two-stage
servovalve, analysis 303
valves 124, 148

U

underlap 138, 169, 284, 294
unloading valve 149

V

valve
axial forces 181
characteristics 159
check 144, 392, 394
electric field 152
five-port 136, 399
flapper-nozzle 77, 91, 101, 113, 139, 173, 194
flow control 133
flueric 140
four-port 136, 399
jet pipe 139, 141, 174
needle 18, 77, 145
non-return 144, 392, 394
pressure-flow characteristics 159
relief 147
sequence 150
servo 139
three-port 135, 395
trip operated 426
two-port 133, 393
unloading 149
vane pump 63
vapour pressure 41
vaporous cavitation 40
variable displacement pump 64, 220
velocity
coefficient 78
transducers 120
vena contracta 78
vent 142
viscosity 30
index 30
volumetric efficiency 67
vortex
amplifier 141
rate sensor 120
servo valve 142

W

water hydraulics 27
wave equations 249
Winslow effect 46